THE SHILOAH CENTER FOR MIDDLE EASTERN AND AFRICAN STUDIES
THE MONOGRAPH SERIES

THE MIDDLE EAST, OIL AND THE GREAT POWERS

The Shiloah Center for Middle Eastern and African Studies
Tel Aviv University

The Shiloah Center is, with the Department of Middle Eastern and African History, a part of the School of History at Tel Aviv University. Its main purpose is to contribute, by research and documentation, to the dissemination of knowledge and understanding of the modern history and current affairs of the Middle East and Africa. Emphasis is laid on fields where Israeli scholarship is in a position to make a special contribution and on subjects relevant to the needs of society and the teaching requirements of the University.

The Monograph Series

The studies published in this series are the work of the Research Associates and Visiting Research Associates at the Shiloah Center. The views expressed in these publications are entirely those of the authors.

Uriel Dann / IRAQ UNDER QASSEM

David Kimche / THE AFRO-ASIAN MOVEMENT

Itamar Rabinovich / SYRIA UNDER THE BA'TH 1963–66

Aryeh Yodfat / ARAB POLITICS IN THE SOVIET MIRROR

Benjamin Shwadran / THE MIDDLE EAST, OIL

AND THE GREAT POWERS

Nissim Rejwan / NASSERIST IDEOLOGY

THE MIDDLE EAST, OIL AND THE GREAT POWERS

Benjamin Shwadran

A HALSTED PRESS BOOK

JOHN WILEY & SONS
New York · Toronto

ISRAEL UNIVERSITIES PRESS
Jerusalem

Copyright © 1973 by Benjamin Shwadran

First Published 1955
Second Printing 1956
British Edition 1956
Second Edition 1959
Third Edition, revised and enlarged 1973

ISRAEL UNIVERSITIES PRESS
is a publishing division of
KETER PUBLISHING HOUSE JERUSALEM LTD.
P. O. Box 7145, Jerusalem, Israel
IUP Cat. No. 25087

Published in the Western Hemisphere and Japan by
HALSTED PRESS, a division of
JOHN WILEY & SONS, INC., NEW YORK

Library of Congress Cataloging in Publication Data

Shwadran, Benjamin.
    The Middle East, oil, and the great powers.

    "A Halsted Press book."
    Bibliography: p.
    1. Petroleum industry and trade – Near East.
I. Title.
HD9576.N36S54   1974   338.2'7'2820956   73–10181
ISBN 0–470–79000–8

PRINTED IN ISRAEL

To my daughter Avivah
and her husband Ari Hillel
who realized my life's dream

CONTENTS

Section Seven — CONCLUSIONS

TABLES

MAPS

PREFACE TO THIRD EDITION

The exhaustion of the supply of the second edition was very gratifying and an encouraging temptation to revise the book and bring it up to date. However, the tremendous accumulation of new and varied research materials — documentary and secondary, institutional and partisan, the steady and ever increasing flow of periodic general as well as specialized literature and new books, the output of old and new company-reports, and the additional, constantly expanding reports of the producing countries — acted as somewhat of a deterrent. Nevertheless, the urgent request of colleagues and, most decisively, of his students — in the United States and in Israel — finally persuaded the author to prepare this third revised and enlarged edition.

It seemed that the special historical-political approach of the book with its narrative organization, which was favorably received by readers and critics of the first two editions, should also characterize the updated presentation. The author felt that the historical-evolvement framework should be preserved as much as possible in order to give the issues and problems their proper perspective. Many chapters, therefore, except for factual updating and for more precise historical timing, remained intact. The developments since the second edition and the evaluation and assessment of the earlier events were also cast in a historical-political structure. The concluding chapters were completely rewritten, enlarged and reorganized, not only in view of the new developments but in the perspective of the intra-regional, inter-regional and global realignments.

In the presentation of collective statistical data the author limited himself to the end of 1971, but he tried, as much as the publishing date permitted, to include developments of 1972 and early 1973.

The author hopes that, like the previous editions, this one will be of interest to Middle East academic specialists, governmental and other professionals dealing with the oil of the region and readers concerned with contemporary international affairs in general and the Middle East in particular. The last chapters should be of interest to every oil consumer anywhere in the world.

The author is more than deeply grateful to his students, undergraduate and graduate, for stimulating and fructifying, through searching questions and wide-based discussions, his thinking, organization and presentation of the issues and aspects of the complex problem of Middle East oil. He tried out the material on them, and their reactions served as helpful and highly appreciated guides.

In gathering materials for this edition the author was graciously supplied with a great variety of primary and secondary sources by the companies, by most of the producing countries and by the Secretariat of the League of Arab States (they appear in the footnotes and bibliography), for which he expresses his sincere gratitude.

It would have been desirable to have included Libya in a study of Middle East oil, and the temptation was indeed very strong. But consideration of space and compositional structure persuaded the author to forego specific treatment. Libya, however, was included in the discussion of issues between the producing countries and the concessionaire companies, and her role in OPEC. Indeed, the conclusions of the last section of the book are applicable equally to Libya as to the other countries of the region.

More than any one person, the author's wife, Helen, made the completion of this edition possible. She typed the draft and final typescript of the entire book, and most decisively encouraged him throughout the undertaking to bring it to a success-ful completion, for which he is profoundly thankful.

B.S.

Hofstra University, March 20, 1973

PREFACE TO SECOND EDITION

It was with a sense of gratitude and satisfaction that the author undertook to revise and bring up to date the Middle East oil story. The warm reception the book received from reviewers, both in the general as well as in the scholarly literature, and from teachers and scholars in many parts of the world, confirmed his belief that there was a basic need for such a book. It is sincerely hoped that the revised edition will continue to serve the needs of students of the Middle East in general and of the Middle East oil problem in particular, and lead to a deeper appreciation of the region's role in the international world scene.

Some reviewers of the first edition regretted the fact that the book did not deal more extensively with the economic problems and patterns of the region, and that it did not go into the problem of sterling-versus-dollar oil, with all the ramifications of these problems. Others felt that the social problems of the region, as necessary background for the impact of the oil industry on the populations, should have been dealt with more fully; they indicated that the author was more concerned with the Great Power policies than with the peoples of the area and that not enough emphasis was placed on inter-Arab relations as part of the oil problem.

Granting that all of these aspects, and others of which the author is aware and which were not mentioned by the reviewers, would help to a better understanding of the complex problems of the region, including that of oil, it must be obvious that one book of limited length could not possibly deal successfully and comprehensively with the subject at hand, as well as with all the other subjects that touch on it and affect it. After having determined the scope of the subject, namely, the historical diplomatic story, the author had to restrict himself to the treatment of oil within the set framework, and limit all other aspects to proportions which would not distort or disturb the smooth organization and presentation of the major topic. The book would otherwise not only have had to be expanded to two or three times normal size, but the material would have become unmanageable, the major theme diffused, and the structural unity disrupted. As great as was the temptation to go more fully into other phases, it had to be firmly resisted.

In view of financial limitations, the revision operation presented a number of difficulties. A complete rewriting of the entire book was ruled out from the outset, and a procedure that retained the maximum of the existing text had to be followed with the hope that the organizational pattern and structural unity would not be affected. As the reader will notice, sections under the general heading "1955–1958" have been added at the end of each major division, and at the end of the chapters dealing with specific countries. The last chapter of the first edition has been revised, and a new chapter analyzing some of the basic issues, as well as developments since 1955, added. For similar reasons a supplemental bibliography has been appended; it consists of some entries published since the first edition appeared, and others which should have been included in the original list.

The author fully agrees with some of the reviewers of the original edition that large detailed colored maps would have enhanced the usefulness of the book; unfortunately, financial considerations made such an addition impossible.

Last, but not least, the author wished to express his deep appreciation to all who helped make the writing and revising of the book possible — and, above all, to the many readers for whom the book was primarily written.

B.S.

March 1, 1959

Section One

INTRODUCTION

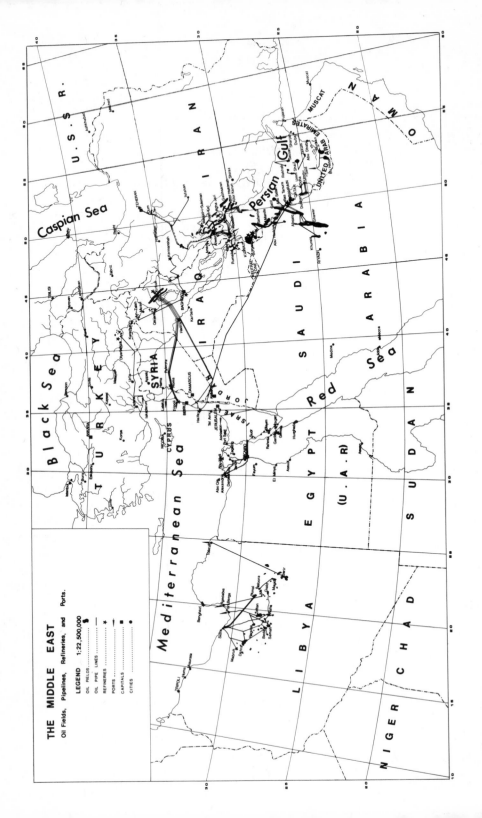

THE MIDDLE EAST

Oil Fields, Pipelines, Refineries, and Ports.

LEGEND   1:22,500,000

OIL FIELDS ....
OIL PIPE LINES ....
REFINERIES .... +
PORTS .... ■
CAPITALS .... ■
CITIES .... ●

Chapter I
INTRODUCTORY

The Middle East embraces about 100 million people and covers an area of some six and one-half million square kilometers, located at the crossroads of the world between East and West and at the juncture of three continents. Except for Turkey, and to a lesser extent Iran, the region has only recently emerged, as a result of the two World Wars, into independent states, but it is still politically, economically and socially underdeveloped.[1] Most of these states, lacking in experience as well as appreciation and even understanding of the importance and basic significance as well as responsibilities of independence, and in the understanding of economic determinants and social forces, suffer from instability. There can be no doubt that part of this instability is the result of the rigors and strains that our age imposes on independent states, even mighty world powers, but the primary reason for the difficulty in the Middle East must be sought elsewhere.

The most outstanding aspect of the history of the Middle East in the last hundred and twenty years has been the impact of the West on the society of the region, and the intensity of the impact has increased in geometric progression with the passage of time. There have been two manifestations, one positive and voluntary, and one negative and involuntary.

The Western education afforded to some of the Middle Easterners by the Christian missionaries in the latter part of the last century opened up to them the gates to modern Western civilization and created a deep discontent with their Eastern patterns of life which slowly matured into a desire to "enlighten" their peoples and lead them from their backward and static position into the light of European civilization. The notions of nationalism on the one hand, freedom of the individual on the other, and the other liberal concepts of late nineteenth century Europe captivated the newly educated and "liberated" Middle Easterners and they began to dream — not unlike some of their more romantic Western colleagues — of a harmonious synthesis of the best in the East and the West. They planned for a grafting, on a basic Eastern civilization, of the attractive and inspiring ideas of the West. But they failed to understand the underlying forces which had created the Western-type civilization and the forces that had brought about the Eastern type, that the values evolved by each were the results of processes and forces that could not be manipulated and changed by a mere decision of an elite group. They laid emphasis on education, either along Western lines in the East or, whenever possible, in the West proper. The result was the emergence of a Western-educated element whose strength was proportionate to the general standard of education in their respective countries, but not rooted in the realities of their immediate environment.

This voluntary and positive impact of the West was soon to encounter a negative and involuntary one. Egypt came under British rule at the end of the century, and France and Britain were to follow as mandatories over the territories detached from

3

the Ottoman Empire after World War I. The very Western-educated element which advocated Western civilization became the most vigorous and determined opponent of Western political penetration. The attacks on the Western mandatory-type colonialism by Western-trained Middle Easterners inevitably carried with them attacks on basic Western concepts which could but arouse the non-Western-influenced elements in the various countries of the area. The reactions against the foreign powers were not restricted to the governmental controls which those powers exercised in the area but became a rejection of practically everything Western in concept and practice and set in motion a movement towards strict orientalism.[2] There was only one manifestation of Westernism that all elements were willing, nay, anxious, to adopt.

The impact of the West and the experiences of two world wars introduced into the area many innovations of a technological character which in many respects changed the daily mode of life even of the most humble elements of Middle East society. In many minds confusion prevailed. Westernism was taken to mean technological advances, through which alone the standard of living could be improved, and poverty, hunger and disease which afflicted the overwhelming majority of the populations of the various Middle Eastern countries eliminated. Failing to realize the forces underlying the social and cultural values of the concepts of individual development that were behind these technological advances, Middle Easterners accused the foreign political rulers of denying the peoples of the area technological innovations in order to maintain their own supremacy and control, and the removal of foreign political control was demanded as the means for the rapid transformation of the backward societies into modern advanced communities. Operating, however, basically on the patterns of their established societies, the newly independent regimes — all the foreign powers having been eliminated — adopted the outer manifestations and practices of Westernism; since the inner contents were lacking, failure was inevitable.

It is in this misunderstanding that we should seek the reason for the prevailing instability in the Middle East. The Middle Easterners' attempt to organize governments outwardly based on Western democratic principles, on the assumption that by incorporating the forms of Western governmental organization and by officially adopting the technological advances of our age, progressive, advanced societies could be developed in short order out of the very deeply rooted Eastern communities, could not but have failed.[3]

In the last 25 years a further and more dramatic breakdown of the social, economic and political-institutional structure has taken place. It has polarized, to some extent, the Arab area into a socialist-revolutionary orientation on the one hand, and the old autocratic-conservative establishment on the other. This phenomenon, which no doubt had some indigenous regional roots, has been abnormally exaggerated and affected by the impact of the global struggle between the United States and the Soviet Union and the rather rapid unexpected penetration of the latter into the area. Yet it would seem that the almost externally unrestricted revolutionary undertakings have failed to achieve the stated objectives.

In order to understand the difficulties and failures of many of the experiments undertaken by the Middle Eastern governments, we must carefully examine and study the old established institutions, practices and traditions as they clashed with the new – the Eastern versus the Western. Any study of the problems of the modern Middle East of necessity prescribes a basic knowledge of the past, not only for the more comprehensive understanding of the present as it evolved from that past, but for discovering the clashes between past practices and the modern usages adopted by the new Middle East society.

One of the factors which has seriously affected the Middle East has been the discovery and production of petroleum. In attempting to evaluate the contribution of this factor to the modern development of the area, we have nothing to fall back on in the past to contrast with the present. Oil production is purely a product of modern times and must be studied as such in its effects on the East as well as on the West.

Petroleum has been a factor in revolutionizing modern civilization; it is the energy and motive power in the home, in industry and in transportation, and it has been the most decisive force in war. The specific aspect of the petroleum impact on the Middle East has been the huge quantities of petroleum discovered in the area which have become a major source of supply to the world, and as such the mainspring of fabulous wealth to the region.

The petroleum of the area also presented problems to the Great Powers who in one form or another had been involved in the area for the last hundred and forty years. For the Middle East, because of its location and strategic significance, has played an important role in international affairs; it was here that great power interests clashed, and it was here that decisive battles were fought. British policy since the development of the empire in India regarded the Middle East as the lifeline to the empire, and attempted to control all possible approaches to the area and prevent the penetration by other Great Powers. England clashed frequently with Russia over Persia; in 1907 a *modus vivendi* was worked out which divided Persia between Britain and Russia; this division gave Britain control over southern Persia. By maintaining the integrity of the Ottoman Empire, England kept both France and Russia from breaking through into the Middle East. In the meantime oil had been discovered and developed by the British in Persia, and the area assumed new importance. Germany became a threat in her efforts to penetrate into the Middle East by way of the Berlin-to-Baghdad railway, and during the first fourteen years of this century British policy veered between completely opposing Germany and cooperating with her; this culminated in the agreement of 1914. With the collapse of the Ottoman Empire, the defeat of Germany, and the elimination of Russia, for a while, from the region, England occupied the area directly. Reluctantly, if not resentfully, she had to permit France to gain control over part of the region and admit her to a share of Iraqi oil.

The experience of World War I clearly demonstrated the strategic importance of the Middle East in terms of world conflict, and the decisiveness of oil in the prosecution of a war; a clash ensued between the British and the Americans for the

fruits of victory in the form of oil. During World War II the Middle East assumed a
new role in the relations between the Great Powers. It was imperative to hold the
area for the Allies, but it was even more important not to let it fall into the hands
of the enemy, for Germany would not only have gained a strategic position, but,
above all, would have acquired a supply of the lifeblood necessary for the prosecution of global war. Thus, while peacetime objectives might dictate control of the
sources of the area's wealth, war objectives would dictate even sacrifices to guarantee a solid front against possible enemies. Great Britain, the United States and the
Soviet Union cooperated in the Middle East against the common enemy into whose
hands the oil fields might have fallen. When the hostilities were over, all realized
that the oil resources of the region would have to be shared by the victors on the
basis of new arrangements. The Soviet Union tried to gain advantages for itself in
northern Iran, and the erstwhile Allies fought a battle for Iranian oil before the
Security Council of the newly organized United Nations.

As the world was rapidly consolidating into two great power blocs, Western
policy on the Middle East and its oil became of supreme importance. The region had
to be held by the West as a vital strategic base and as an invaluable source of
petroleum with which to oil a possible future war; conversely, it was not to be
allowed to fall into the hands of the enemy.

However, since the end of World War II great and far-reaching changes have
taken place in the Middle East. Britain has withdrawn from the entire area, including her traditionally exclusive sphere – the Persian Gulf. The only interest
Britain still has are the oil resources, which she needs to meet her own needs and as
a source of income from selling the surplus. France, though hopefully dreaming of
some miraculous return to the area, has for all intents and purposes been removed
from the Middle East and North Africa, and her interests are practically limited to
the oil resources. The United States, on the other hand, has assumed by design as
well as by circumstances the role of the major Western power in the area while the
Soviet Union has become the major Eastern power in the region.

For the student of modern history, economics and politics, therefore, a study of
oil developments in the Middle East is important, significant and instructive; for it
involves a complex multiplicity of issues affecting international, economic, financial
and technological aspects of modern society. Such a study must deal with intra-
regional relations between the various states: between those states which have oil in
great abundance and those states through whose territories the oil is only transited
on its way to world markets; it must deal with boundary disputes between various
states both onshore and especially offshore; between those who are Arab countries
and between those who are non-Arab; between those who consider themselves
socialistic and revolutionary and between those who consider themselves traditional
conservative; it must deal with the efforts of all the major crude producers, joined
by some outsiders, to form a united exporting group: Organization of Petroleum
Exporting Countries (OPEC); and with the attempt of some Arab countries of the
area to protect themselves against Arab non-producing as well as non-Arab coun-

tries in the area by uniting as a separate Arab group: Organization of Arab Petroleum Exporting Countries (OAPEC).

Such a study must also deal with the relations between the rulers and the peoples of the countries in the utilization of the income from the oil; the governmental and institutional procedures and methods of economic, educational and social programs of the oil-producing countries, and the resultant transformation of the societies of the entire area.

It must deal with inter-regional relations. Middle East and North African oil has been, is and will become even more in the future the vital source of supply for the ever-growing energy needs of Europe, the Far East, and ultimately the United States.

It must deal with global superpower relations. The strategic importance of the region and its abundant oil resources have assumed great if not primary importance in the struggle for world influence and control.

It must deal with the long, almost never-ending negotiations between the concessionaire foreign companies and the producing countries, negotiations undertaken singly or collectively between diverse groups. The struggle between the two is of epic proportions and testifies to the extent of human endurance, for both the profits of the companies and the payments to the governments reach astronomical figures. The most outstanding aspect of those almost constantly on-going negotiations is the long series of achievements of the governments. During the last 60 years the producing countries have advanced from a 16% share of the profits of production, when the concessionaire had practically a country-wide exclusive concession free of all taxes with exemption from all customs duties on imports and employed very few natives in the staff categories, to a stage where the producing counties formally receive 55% of the profits, and in some cases as much as 79%; they forced the concessionaires to relinquish major portions of their concession areas; the companies are subject to taxes and customs duties; the royalty payments – running from 12½% to 15% – are expensed; the percentage of native employees, including supervisory personnel, is constantly rising and by far outstripping the foreign staff. Indeed the producing countries have a say in the determination of the posted prices on the basis of which the profits are calculated. Moreover, the tendency in the Middle East seems to be in the direction of creeping nationalization of the oil industry, while the concessionaire companies are helpless.

While within the different governments of the area there seemed to be a basic difference of opinion between a gradual 51% participation in the companies' ownership advocated by the moderates and outright nationalization advocated by the extremists, the producing countries have succeded in persuading the companies to grant an immediate 25% participation with a definite schedule for attaining 51% interest.

Another aspect that must be dealt with is technological development. The technology which the oil companies brought to the area resulted in tremendous changes in the industry itself, and as the years progressed had far-reaching and irreversible technological repercussions outside the industry. With the tremendous growth in production it was inevitable that newer, greater and more imaginative facilities would be developed. As the oil flowed from newer and more distant wells in ever more enormous quantities, gathering centers, pipelines of ever-expanding diameters, looplines, separation stations (oil from gas), tank-farms, terminals, underwater pipelines, greater harbors, huge artificial island terminals, and gigantic bottomless storage tanks far out in the sea had to be constructed; with offshore development, tremendous exploration and production stations were built and with the advance of the new, huge 500,000-ton oil tankers, great berthing and pumping facilities were installed.

Practically every Middle Eastern oil-producing country has organized a national oil company. These companies, in smaller or larger degrees, began to take over and run some phases of the industry, in some cases exclusively and in others partially, from exploration through development, production, refining and transportation to marketing abroad.

The natural gas aspect of the Middle East oil issue is another victory for the producing countries. At first the natural gas, after it was separated from the oil, was flared and therefore completely wasted. After many protests and prolonged negotiations, a part of the natural gas became integrated in the great wealth-producing pattern of the oil story. It was first utilized for generating local power for the companies — in the great variety of needs of the industry — and for the governments. Later it was expanded to supply the needs of local developing industries and home use. Finally it became a regular export item, both raw through pipelines or processed and shipped abroad by special tankers.

From the utilization of the natural gas and other oil byproducts emerged a flourishing petrochemical industry in practically every Middle Eastern oil-producing country.

The producing countries have succeeded in breaking up the monopoly of the major international companies over the Middle East, and have brought into the region a great number of smaller independent companies which were granted limited areas for exploration and development on highly favorable terms. Indeed, as the majors relinquished great chunks of their concessions, they were broken up into small units and offered as new concessions to the new companies.

The great significance of Middle East oil is its role in the supply of world energy in the next 20 years. In 1971 the Middle East's share was about 33% of world production, and its proven published reserves were conservatively estimated as 58% of world reserves. The indications are, from the general information available, that the Middle East might become, in the next decade or two, the major petroleum supplier not only of Europe but also to a very large extent of the United States. The economic, financial and political implications of such a situation are clear; however, it is not at all clear what the Western companies and governments would do about it.

Neither is it clear what the aim and objectives of the Soviet Union are. Conflicting estimates are given of Communist Bloc oil and natural gas reserves, nor is there agreement on the Communist Bloc production capacity. Various and different assessments are made of the present and future oil and natural gas needs of the Communist countries. The real purpose of Soviet efforts in developing the oil resources of the Middle East producing countries has not been manifestly established, nor has there been agreement as to Russia's possible control of the European oil market.

This study of Middle East oil and the issues outlined in this chapter are delimited by the historical and political-international framework of the development of the Middle East and of its peoples and their repercussions on the world community.

NOTES

1. Included are Turkey, Syria, Lebanon, Israel, Jordan, Iraq, Saudi Arabia, Iran, the Persian Gulf Shaikhdoms, Yemen and Egypt.
2. For an analysis of this phenomenon see Nadav Safran, *Egypt in Search of Political Community* (Cambridge, 1961).
3. For a very interesting and brilliant attempt (unfortunately belied by subsequent events) to explain the differences between the Turks, who apparently succeeded with democratic government, and the Arab-Islamic states, which failed, see E.A. Speiser, "Cultural Factors in Social Dynamics in the Near East," *The Middle East Journal*, VII, 133–152, Spring, 1953.

Section Two

IRAN

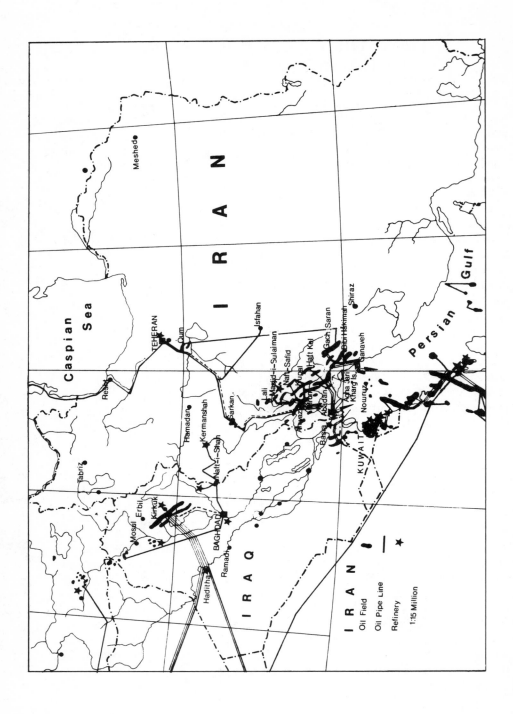

Caspian Sea

Meshed●

I R A N

TEHERAN
Qum

Isfahan

Resht

Tabriz

Hamadan●
Kermanshah●
Naft-i-Shah

Persian Gulf

Shiraz
Gach Saran
Masjid-i-Sulaiman
Naft-Safid
Haft Kel
Lali
Dezful

Ahwaz
Agha Jari
Bandar Hakimat
Kharg Is.
Ganaveh
Nowruz

Basra
Abadan

KUWAIT

Mosul
Erbil
Kirkuk
BAGHDAD
Ramadi

Haditha

I R A Q

I R A N

Oil Field

Oil Pipe Line

★ Refinery

1:15 Million

Chapter II
FROM THE D'ARCY CONCESSION
TO THE RISE OF RIZA KHAN

In dealing with the issues and problems involved in the development of the oil resources of Iran and the subsequent effects on the political, social and economic patterns of the country, as well as on the relations between the great Powers who were vying among themselves for concessions and their respective relations with Iran, one must keep in mind the difference in civilizations and standards of values between those who granted concessions and those who obtained them. It is only with such a basic approach that it is possible to understand the nature of the concessions originally granted, and the subsequent cancellations, expropriation and nationalization. Only these differences — in standard of personal and public behavior and general knowledge, and above all in scientific knowledge — between the leaders of the Iranian Government on the one side and the representatives of the concessionaires on the other can expain the fact that the concessions were granted at all; the nature and character of each concession were determined by the stage of the general development of the Iranian leadership. As the country advanced and it was realized what had been signed away, the old concession was cancelled and a new one was elaborated which served until a still higher stage of progress was attained. The later manifestation of this process were the agreements with the Anglo-Iranian Oil Company and the international Iranian oil Consortium. This lasted until 1973 when the relationship between Iran and the Consortium changed to total ownership and operation by the National Iranian Oil Company, and to oil purchaser by the Consortium. The full impact of the new arrangements on Iran, the Consortium, the Middle East oil industry and the world energy supply is yet to be unfolded.

EARLY CONCESSIONS

The first major concession to exploit the mineral resources of Persia was granted by Shah Nasr ed-Din on July 25, 1872 to a British national, Baron Julius de Reuter. One of the most sweeping concessions of all times, it covered all of Persia and granted de Reuter a seventy-year monopoly to construct railroads and streetcar lines, and exclusive rights to exploit all mineral resources except gold, silver and precious stones. As a token of good faith, de Reuter deposited £40,000 with the Shah.[1] The following year, when the Shah visited St. Petersburg, the displeasure of the Russian court with the de Reuter concession was made effectively clear to him, and in 1873 the concession was cancelled and the deposit confiscated.[2]

Although Lord Curzon claims that England's lack of support for the concession was as much a reason for the Shah's cancelling it as was Russian opposition, it was British refusal to recognize the cancellation that caused the Shah, in 1889, to grant a new sixty-year concession to de Reuter to open a bank and to exploit many minerals throughout Persia, including petroleum. The Persian Government was to receive 16 per cent of the net profits. Consequently the Imperial Bank of Persia was formed,[3] and in 1890 the Persian Bank Mining Rights Corporation was organized to work the mineral deposits. During 1891–1893 the Corporation conducted searches for oil at Daliki on the road between Shiraz and Bushire, but without success. In 1899 the Persian Government declared the minerals concession no longer valid, and in 1901 the Corporation was liquidated.[4]

In the meantime, in 1891, the Governor of Kermanshah, who was convinced that there was oil in his province, asked Jacques de Morgan, head of a French archeological expedition in Persia, to search for oil. De Morgan published his findings in *Les Annales de Mines* (1892); he claimed that oil existed in the Qasr-i-Shirin region near the Persian-Mesopotamian border.[5]

THE D'ARCY CONCESSION

In 1900, at the Paris Exposition, Edouard Cotte, one-time secretary to Baron de Reuter, called the attention of General Kitabji Khan, Persia's Commissioner General at the Exposition, to de Morgan's findings. Kitabji's efforts to interest French fianciers in the exploitation of Persian oil proving unsuccessful, he turned to the former British Minister in Teheran, Sir Henry Drummond Wolff, and asked him to find someone in London who would be willing to invest in such exploitation. Some time later General Kitabji was summoned to London, where he proposed his plan to William Knox D'Arcy. After D'Arcy had had the possibilities of the Persian fields investigated by a geologist, H.T. Burls, he sent to Teheran his representative, Alfred M. Marriot, accompanied by General Kitabji and Mr. Cotte, to obtain a concession in his name.[6]

Sir Henry Drummond Wolff asked his successor as Minister to Persia, Sir Arthur Hardinge. to intervene with the Persian Government on D'Arcy's behalf. Hardinge was advised to try to obtain the concession by assigning shares in the proposed company to some of the most influential Persian ministers, particularly the Grand Vizier himself. Hardinge accordingly strongly urged the Grand Vizier to grant the concession to D'Arcy. The Grand Vizier, however, saw Russian objections as a stumbling block and offered Hardinge a scheme for overcoming this difficulty. Hardinge reports: "The Grand Vizier declared himself prepared to fall in with the project, but he suggested that a letter – to be written by me, in the Persian language, embodying its main features – should be immediately drawn up for submission to the Russian Legation. He was aware that M. Argyropula could not read Persian, more especially in the written or 'shikaste' character, which is illegible, owing to its peculiar abbreviations, even to scholars familiar with the printed language. He also knew from his own spies that the Russian Oriental Secretary, M. Stritter, who alone could read it, was about to leave Zergendeh, the summer residence of the Russian Legation, for a short sporting excursion in the neighboring hills. He therefore sent the letter to Zergendeh, where it lay several days untranslated, awaiting M. Stritter's return, and as no objection to the proposal contained in it was made by the Russian Minister, who could not read it, and never suspected the importance of its contents, all the Persian members of the Government supported the Grand Vizier's decision to assign the concession to Mr. D'Arcy. M. Argyropula was far from pleased when he learnt what had actually happened; but the Grand Vizier could not be blamed for the accidental and temporary absence of the Legation's Persian translator, and the Russian Minister accordingly adopted the sensible course of accepting the accomplished fact."[7]

Even though Russian opposition was thus circumvented, Russia's position in Persia had nevertheless to be taken into consideration, and the five major northern provinces — Azerbaijan, Gilan, Mazanderan, Khorasan and Astrabad — were excluded from the concession. The concession was signed on May 28, 1901 for D'Arcy by his attorney, Alfred Marriot, and for Persia by Shah Mozaffar ed-Din. The pertinent provisions were:

1. D'Arcy was to have a special and exclusive privilege to search for, obtain, exploit, render suitable for trade, carry away and sell natural gas, petroleum, asphalt and ozocerite throughout the whole extent of the Persian Empire for a term of sixty years from the date of signature (Article 1).

2. D'Arcy was to have the exclusive right to lay the necessary pipelines from the place where any of the above products were discovered to the Persian Gulf, as well as distributing branches (Article 2).

3. The Government granted gratuitously to the concessionaire all uncultivable lands belonging to the State which the concessionaire's engineers might deem necessary for the construction of the whole or any part of the works of the exploitation of the concession. The concessionaire was to purchase cultivable lands belonging to the State "at the fair and current price of the Province." The Government also granted to the concessionaire the right to acquire all other lands or buildings necessary for his purpose, with the consent of the proprietors, on conditions mutually agreed to, without, however, the latter having been "allowed to make demands of a nature to surcharge the prices ordinarily current for lands situate in their respective localities" (Article 3).

4. Notwithstanding the provision of Article 1 as to the extent of the concession, the provinces of Azerbaijan, Gilan, Mazanderan, Astrabad and Khorasan were excluded, but on the express condition that the Imperial Persian Government would not grant to any other person the right of constructing a pipeline to the southern rivers or to the south coast of Persia (Article 6).

5. All lands acquired through the concession were to be free of all imposts and taxes during the life of the concession; and all material imported for the operation, maintenance and development of the concession was to be free of all taxes and custom duties (Article 7).

6. The concessionaire was granted the right to found one or several companies for the working of the concession, and these companies were to enjoy all the rights and privileges granted to the concessionaire and to assume all his responsibilities and engagements (Article 9).

7. The first exploitation company to be organized by the concessionaire was

to pay the Persian Government, within one month of its formation, £20,000 in cash and an additional £20,000 in paid-up shares. It was also to "pay the said Government annually a sum equal to 16 percent of the annual net profits of any company or companies" that were formed (Article 10).

8. The Persian Government was to appoint a commissioner to advise, and to be consulted by, the concessionaire and the directors of the companies. He was to establish by agreement with the concessionaire such supervision as he might deem "expedient to safeguared the interests of the Imperial Government." The concessionaire was to pay the commissioner an annual sum of £ 1,000 for his services, beginning with the date of the formation of the first company (Article 11).

9. The workmen employed in the services of the company were to be "subjects of his Imperial Majesty the Shah, except the technical staff such as the managers, engineers, borers and foremen" (Article 12).

10. The Persian Government bound itself "to take all and any necessary measures to secure the safety and the carrying out of the object of this concession, of the plant and of the apparatuses of which mention" was made for the purposes of the undertaking of the company and to protect the representatives, agents, and servants of the company. The Imperial Government having thus fulfilled its engagement, the concessionaire and the companies created by him were not to have "power under any pretext whatever to claim damages from the Persian Government" (Article 14).

11. With the expiration of the concession all the materials, buildings, apparatuses then used by the company were to revert to the Persian Government, and the company was to have no right to indemnity (Article 15).

12. Any dispute which might arise as to the interpretation of the concession was to be given to a board of arbitration in Teheran, composed of one representative each of the company and the Government, and a third member chosen by the other two, and the board's decision to be final (Article 17).[8]

In accordance with Articles 9 and 16, D'Arcy organized the First Exploitation Company in May 1903, with a capital of £600,000 in shares of £1 each, largely subscribed by himself. The Persian Government received 20,000 shares. The Company began operations about 100 miles north of Baghdad, but production was not of a magnitude to justify commercial operations. The actual holding of the First Exploitation Company was limited to one square mile in the Maidan-i-Naftun field, in the territory of the Bakhtiari khans. In order to obtain the cooperation of the khans, an agreement was signed with them giving them 3 per cent of the shares in any company which exploited the oil in their territory. To effectuate the agreement, the Bakhtiari Oil Company Ltd., with 400,000 shares of £1 each, was formed to work the oil-bearing lands in the Bakhtiari territory other than those operated by the First Exploration Company.[9]

Before the oil fields began producing in sufficient quantities, D'Arcy's original investment was almost exhausted and he called for financial assistance. The British Admiralty became interested, apparently fearing that the concession might fall into the hands of the American or Dutch oil trusts, and it asked D'Arcy to defer negotiations with foreigners until British interests could be found to invest in his undertaking. The Admiralty advised Lord Strathcona of the Burmah Oil Company to cooperate with D'Arcy, and in May 1905 the Concessions Syndicate Ltd. was formed, with D'Arcy as a director. It took over the assets of the First Exploitation Company and provided financial resources for continuing operations.[10]

The additional resources were soon exhausted, but in the middle of 1908 oil in commercial quantities was discovered at Masjid-i-Sulaiman, and on April 14, 1909 the Anglo-Persian Oil Company was formed, with an initial capital of £2,000,000, with Lord Strathcona as chairman and D'Arcy as a director.[11]

During the early years of the concession, two important events took place which deeply affected the future history of Persia as well as the development of oil exploitation. On August 5, 1906, Shah Mozaffar ed-Din issued a proclamation granting constitutional government,[12] and on December 30 of the same year the fundamental laws (constitution) were enacted.[13] On August 31, 1907, an agreement between England and Russia was signed dividing Persia into three zones: the northern, under Russia influence and reserved for Russian nationals to seek concessions, etc.; the southern, under British influence and reserved for British nationals; and the center, a neutral zone as regards these two great rival Powers.

The granting of constitutional government, while it introduced restraints on the Shah and established the Majlis with powers and controls formerly exercised by the ruler, tended to antagonize the tribal areas. The local chieftains became more than ever rebellious against the Teheran government and the control of the central government over the outlying provinces was considerably weakened.

The Anglo-Russian agreement of 1907 did bring peace between the two great-power rivals, but, although its stated objective was "to respect the integrity and independence of Persia," it neither admitted Persia to partnership nor even to consultation. The net result for Persia was further entrenchment of foreign powers on her soil and a further weakening of the control of the Teheran government over the zones under the influence of Russia and Great Britain.[14] Thus, when prospectors working the D'Arcy concession penetrated into the tribal areas in southern Iran, they soon realized that they would have to come to some direct understanding not only with the tribal chiefs (who were to receive 3 per cent interest as promised in the agreement with the Bakhtiari khans), but also with the local people. At the end of 1907 the Indian government sent Lt. A.T. Wilson with a detachment of soldiers to the neighborhood of Shustar, ostensibly to guard the British consulate in Ahwaz, but actually to protect the drillers from the local inhabitants.[15]

Next to the problem of the Bakhtiari was that of the Shaikh of Muhammarah. This was, from the point of view of the British, more complex, and in subsequent years more troublesome. Muhammarah is the territory in southern Khuzistan bordering on Iraq and inhabited mostly by Arabs. Its ruler at that time was Shaikh Khazaal, who exercised legal control over the entire area and was the recognized head of a number of Arab tribes located along the Karun River on the eastern bank of the Shatt-al-Arab. Khazaal's relations with the Teheran government were those of a semi-independent emir. During the constitutional revolution, the situation in the southern part of the country was chaotic, with Khazaal and the Bakhtiari openly opposing the Teheran government.

In May 1909, a month after the formation of the Anglo-Persian Oil Company, Sir Percy Cox, British resident at Bushire, went to Muhammarah to negotiate an agreement with Shaikh Khazaal on behalf of the Company. By this time the oil wells of southern Persia were producing in commercial quantities and the oil had to be brought to a port. The Company selected a site on Abadan island, which was in Shaikh Khazaal's territory. Cox, after prolonged and skillful negotiations, obtained an agreement from the Shaikh granting the Company a right of way for the pipeline to carry the oil, and the right to buy navigable frontage on the Shatt-al-Arab to build a refinery, depot and storehouses. The Shaikh was also to provide, at the Company's expense, guards for the pipeline and buildings. He was to receive an annual rent of £650 for ten years, in advance. Should the Persian Government renew the concession after ten years, his annual rent thereafter would be £1,500. The British guaranteed the rights of Khazaal and his successors to Muhammarah. On July 16, 1909, after signing the agreement, the Shaikh received £6,500, the advance rent, and a loan of £10,000, nominally from the British Government but actually from the Anglo-Persian Oil Company.[16]

By 1912 a refinery had been erected on Abadan island, to which a pipeline brought the oil from the fields. At this time the British Admiralty was actively discussing the possibility of converting at least part of the British Navy from coal to oil, and by 1913 it had accomplished the conversion. Negotiations between the Admiralty and the Anglo-Persian Oil Company for buying into the Company were thereupon initiated. The Admiralty sent out a commission of experts, headed by Admiral Sir Edmond S. W. Slade, to investigate the oil fields covered by the Company's concession. After the commission reported favorably and enthusiastically on the Persian fields, an agreement between the Company and the Admiralty and the Treasury was signed on May 20, 1914, making the British Government the major and controlling partner in the Company.[17] The agreement provided:

That the Government was to buy:

Ordinary shares	£2,000,000
Preferred	1,000
First debentures	199,000
	£2,200,000

The Company was to increase its capital by £2,000,000, giving the Government a majority of 2,000 shares.

That the Government was to appoint two *ex officio* directors "to enable Government interests in the Company to be duly protected, but with the minimum of interference with the conduct of its ordinary business." A power of veto over all acts of the Board and Committees of the Company and its subsidiaries was conferred upon the two *ex officio* directors. At the same time, the

Government emphasized that it did not surrender its right to be represented on the Board by a larger number of directors, proportionate to its share in the Company, should such a step become necessary. [18] The Government gave assurances to the Company, however, that its veto would be exercised with due regard to the financial and commercial interests of the Company and "only in respect of matters of general policy such as the supervision of the activities of the Company" as they might "affect questions of foreign and military policy, any proposed sale of the undertaking or change of status of the Company and new exploitation and other matters directly bearing on its fulfillment of current contracts for the Admiralty." [19]

That the Company must always remain an independent British concern, registered in England and with its head office in England, and every director must be a British subject.

An important and decisive provision of the agreement was that the Government was to enter into a contract with the Company for supplying fuel oil to the Admiralty. [20]

On June 17, 1914 the agreement came up for a full-dress debate in the House of Commons. Its staunchest advocate was, of course, Winston Churchill, First Lord of the Admiralty. His major arguments were that since a good part of the Navy was oil-burning, the Admiralty must have direct access to at least part of its oil. Private companies, to be sure, were ready to supply oil, but "at a price." Churchill claimed that the private companies were squeezing the Government, hence if the Government could obtain oil at a good price from the Persian fields, it would also reduce the price of oil charged by private companies operating in other fields. He maintained that the price the Government paid for the shares in the Anglo-Persian Oil Company was reasonable and that the Government stood to gain a great deal from the agreement.

The arguments against the agreement were that the Government should not enter into business; that it was questionable whether there really was as much oil as was hoped for; that in time of war, because of the disturbed conditions of the area, the oil could not be brought out, and as far as peace time was concerned, there was no need for the Government to take any risk when oil could be obtained from private companies; that it would antagonize Russia and increase the tension between that country and Great Britain; that it was desirable to develop the oil resources in the British Empire rather than those in a foreign territory. All these arguments, however, did not affect the final vote; the agreement was approved 254–18. [21] Moreover, all attempts — and they were as devious as they were varied — to learn from the Government the price paid for the oil for the Navy in the contract with the Company failed. The Government persistently refused to divulge the secret; it would not even give a hint of the approximate price. The closest hint as to the basis for establishing the price was given by Austen Chamberlain in 1919 when he declared: "The price ruling under the Admiralty contract for fuel oil is subject to reduction in proportion to the profits realized above a certain limit." [22]

TRANSFERRED TERRITORIES

Between Turkey and Persia there existed a long-standing border dispute; although Persia was in *de facto* possession of some of the frontier territories, for a long time Turkey claimed *de jure* rights over them. On December 17, 1913, a Persian-Turkish frontier protocol was signed in Constantinople by the British, Russian and Persian Ambassadors and the Turkish Grand Vizier. While the Qasr-i-Shirin area covered by the D'Arcy concession was awarded to Persia, the Khanaqin area, where the Anglo-Persian Oil Company had oil wells, was transferred to Turkey. This protocol became known as the Transferred Territories Agreement. However, the British Government obtained a formal undertaking from Turkey that the D'Arcy concession would be recognized in the transferred territories. In fact, so far as the Khanaqin area was concerned, not only did Turkey recognize the concession, but also granted the Anglo-Persian Oil Company the additional privilege of constructing pipelines through Turkish territory to the sea.[23] When, therefore, the concession of the Turkish Petroleum Company (which later became the Iraq Petroleum Company) was under consideration, the "transferred territories" were consistently exempted.

WORLD WAR I

With the outbreak of the war, the Persian Government proclaimed its neutrality. However, since the Government could not protect its territory against the belligerents, Persia soon became a battleground between the Central Powers — Turkey, assisted by Germany and Austria, and the Entente Powers — Russia, with considerable participation of British forces.[24] Both Russia and Britain raised local levies under respective Russian and British commands and officers. The Russians used the existing Persian Cossack Brigade, while the British, under Sir Percy Sykes, organized the South Persia Rifles. Between them they were in actual military control of the country; the Russians agreed to the extension of British control in the neutral zone, in return for a free hand in their zone, which was interpreted as practical annexation.[25] There was a considerable pro-Central Powers' attitude among the younger elements of the Persian population who resented the 1907 Anglo-Russian agreement. The Swedish officers of the gendarmerie were also inclined towards the Central Powers and so influenced the men under them.

At the end of 1915 most of the cabinet members left Teheran, which was in danger of occupation by Russian and British forces, and they hoped that the Shah would soon follow and establish the new seat of government either in Qum or some other locality. The Shah was prevented at the last moment, however, by the persuasion of the Russian and British Ministers, from leaving Teheran. A rival government was therefore set up at Kermanshah and the diplomatic agents of the Central Powers followed that government and established themselves at Qum. The Shah and his government were thereafter under Entente influence.

In March 1917 the South Persia Rifles were officially recognized by the Persian

Government, but the cabinet of Ala es-Sultana Dowleh, which came to power in June, refused to recognize the force. The British pressed for its recognition and after considerable maneuvering on the part of the British, the Shah changed his cabinet. On August 9, 1919, an Anglo-Persian agreement was signed.[26]

As for the oil fields and refineries, on the whole they escaped serious damage, except that during the first week of February 1915 the pipeline was cut. It was completely repaired by the middle of June.[27]

THE 1919 ANGLO-PERSIAN AGREEMENT

With the collapse of the Russian Government and the defeat of the Turks, the British found themselves the only force in Persia and they were determined to establish themselves militarily. "To secure a strong, stable and friendly state on the flanks of India and Mesopotamia, Great Britain aimed at controlling the armed forces of Persia after the armistice."[28]

It is in the light of these considerations, as well as the desire to guarantee their oil, that we must understand the methods and techniques which the British employed in obtaining the Anglo-Persian agreement, and the strenuous objections to the agreement not only of France, but also the United States.

The father of the idea of the agreement was Lord Curzon, considered at the time the great expert on Persia,[29] and he was motivated by the security of India and Mesopotamia, flanking Persia. Curzon was aware of the corruptibility of Persia's political leaders and he felt that a stable and peaceful regime could be brought about by Britain's controlling the country, thus guaranteeing the British position in the Persian Gulf. A memorandum, dated August 9, 1919, which was circulated among British cabinet members, concluded thus: "Lastly, we possess in the southwestern corner of Persia great assets in the shape of the oil fields, which are worked for the British Navy, and which give us a commanding interest in that part of the world."[30]

In order to achieve their objective, the British agents persuaded the Shah to dismiss Premier Ala es-Sultana Dowleh and to appoint Vossugh ed-Dowleh. This persuasion was achieved by a monthly subsidy of 15,000 *tomans* to the Shah. After the 1919 agreement was signed, Lord Curzon was determined to stop the subsidy and only after considerable pleading by Sir Percy Cox was he induced to continue it so that Vossugh would stay in office.[31]

The Persian triumvirate responsible for the agreement were the Premier, Foreign Minister Prince Firooz Mirza Nusret ed-Dowleh, and the Finance Minister, Akbar Mirza Sarem ed-Dowleh. These three were persuaded to work for the agreement by the payment of subsidies to Premier Vossugh and to Farman Farma, the father of Nusret.[32] They must have had a feeling that their signing the agreement might endanger them with the Persian people, for, besides arresting and exiling their more vocal opponents, the three obtained a promise, addressed by Sir Percy Cox to each one, and made a secret enclosure to the agreement, which read: "It gives me much pleasure to inform Your Highness that His Majesty's Government authorize me to intimate that, in view of the agreement concluded this day, the 9th August, 1919, between His Majesty's Government and the Persian Government, they are prepared to extend to Your Highness their good offices and support in case of need, and further to afford Your Highness asylum in the British Empire should necessity arise." [33]

After reiterating "in the most categorical manner" Great Britain's undertaking "to respect absolutely the independence and integrity of Persia," the agreement provided that the British were to supply, at the Persian Government's expense, all advisers; that the British were to supply, at the Persian Government's expense, officers, munitions and equipment, to be agreed to by a joint commission, for the formation of a uniformed force to be created by the Persian Government; that the British were to offer a substantial loan, to be backed by adequate security (customs receipts and other revenue) for the purpose of paying the advisers and reorganizing the uniformed force; that Anglo-Persian enterprise for developing and improving communications in Persia, such as railways and other means of communication, was to be encouraged; and that a joint committee of experts to examine and revise the existing tariffs was to be appointed. A separate contract provided for a £2,000,000 loan at 7 per cent, redeemable in twenty years. [34]

There was considerable opposition to the agreement not only in Persia, where some elements saw in it an attempt by the British to establish a protectorate over their country, especially in view of the actual British military occupation, but also on the part of the French and the Americans. The French opposition was vocal both in the press and in parliament, and the French Legation in Teheran was actively engaged in working against the agreement. [35]

The French also saw in the agreement a British effort to establish a protectorate over Persia, and already embittered by their difficulties with the British in Syria, they were determined to thwart the realization of the agreement. [36] The Americans were disturbed not only by the secrecy employed by the British in obtaining the agreement, but by the purpose behind it. As early as August 13, 1919, the American Minister in Teheran informed the American Secretary of State that public sentiment against the cabinet was rising because of the Anglo-Persian agreement; and in a subsequent dispatch (August 16) he explained the lack of vocal opposition to the agreement because of martial law, a controlled press and occupation of the country by the British Army. [37]

The British tried to counteract the efforts of the French Legation and asked for the assistance of the American Minister in Teheran. In reply to this request, Secretary of State Lansing wrote to the American Ambassador in London, on August 20, 1919: "The new Anglo-Persian agreement has caused a very unfavorable impression upon both the President and me and we are not disposed to ask our Minister at Teheran to assist the British Government or to ask him to preserve a friendly attitude towards this agreement." While the British were negotiating the agreement, Arthur James Balfour, the British Foreign Minister and chief of the British delegation to the Paris Peace Conference, three times turned down Lansing's suggestion that the Persian delegation be heard; after the publication of the agreement Lansing understood the motive behind the British refusal and he advised the American Ambassador: "We cannot and will not do anything to encourage such secret negotiations or to assist in allaying the suspicion and dissatisfaction which we share as to an agreement negotiated in this manner."[38]

The Persian opponents to the agreement apparently taunted their Government for not asking the American Government for assistance, instead of signing a protectorate with the British, for a Government spokesman published an article in *Raad* of Teheran which declared that the United States had refused to help Persia and that the Anglo-Persian agreement was therefore the best available. After consulting with Washington and receiving specific instructions from Secretary of State Lansing, the American Minister in Teheran issued a communique — which was sent to the Government as well as to the local press — in which he denied that the United States had refused help to Persia, and he vigorously disassociated the United States Government from the agreement.

The French Minister also issued a communiqué expressing readiness to assist the Persian peace delegation in Paris, to whom the British consistently refused a hearing. Lord Curzon was greatly exercised over the American communiqué. About the two communiqués, Sir Percy Cox wrote to Curzon on September 11, 1919: "This is all part of a combined French and American intrigue against Vossugh and agreement."[39]

After the American Ambassador in London had repeatedly asked the Secretary of State what the American position was on the substance of the agreement, Lansing telegraphed him on October 4, 1919 that the United States made its approval of the agreement conditional "until and unless it is clear that the Government and the people of Persia are united in their approval and support of this undertaking."[40] Besides insisting upon Persian approval, the American opposition had been based upon apparent exclusion of Americans from Persian employ. The British Ambassador in Washington, Lord Grey, complained to Lord Curzon that he had to remain silent "when Americans say that even if Persian Government wanted to employ individual Americans they would not be permitted under Anglo-Persia agreement though Belgians or others may be allowed." To this Lord Curzon replied that Belgians and Frenchmen were survivors of continuous pre-war employment, while American appointments would be a fresh start, "but if Persian Government desired to make it, with our approval, there is no obstacle." To which Lord Grey, in a dispatch dated October 27, 1919, remarked that this qualification — that Persian appointments must have British approval — was the very objection Americans had raised against the Anglo-Persian agreement. "Indeed if I repeat it here, it would confirm impression that we treat Persia as a protectorate."[41]

A further American motivation for objecting to the agreement was brought out clearly in a telegram, dated March 17, 1920, from Secretary of State Colby to the American Minister in Teheran: "Department is concerned by the possibility that confirmation Anglo-Persian Agreement by Medjlis may make more difficult the obtaining of petroleum concessions by American companies."[42]

Since the position of the French and the Americans encouraged the local opposition, the Vossugh government did not rush to convene the Majlis to ratify the agreement. To be sure, the Shah had gone to London ostensibly to discuss the agreement with the British; but he did not arrive there until October 31, 1919 and since he left on November 17, his stay was more of a social visit than a diplomatic mission to complete negotiations on the agreement. After the Shah returned to Persia, the Government of Vossugh fell and Mushir ed-Dowleh of the opposition formed a new government in July 1920.[43] The new government decided not to act on the agreement until it had been ratified by the Majlis, and hence it refused to accept any British advances on the £2,000,000 proposed loan.[44] In November the Shah clashed with the Premier because of the latter's objections to the dismissal of the Russian officers in command of the Persian Cossack Brigade and their substitution by British officers, and he nominated a new cabinet under Sipahdar Azem.[45]

Meanwhile, on August 19, 1920 the American Minister in Teheran reported that as a result of the Government's action on the agreement, some of the British advisers, who had been working in Persia for some time under the provisions of the treaty, had left the country "apparently convinced that said treaty will never be accepted by the Persians."[46]

It has already been mentioned above that Lord Curzon himself gave British oil interests in Persia as a reason for the agreement. In a memorandum prepared for Clemenceau and Berthelot late in 1919, Henry Berenger, the French oil expert, declared that British policy in Asia Minor, and from the Caucasus to India, in Persia and Mesopotamia, had not been formed or pursued as a purely territorial policy but essentially as a petroleum policy. Great Britain had displayed great activity in securing control over oil fields in Asiatic countries; she was determined to free herself of dependence on the United States for all the oil supplies required by the British Navy and by military and commercial enterprises. He thus explained "the recent Anglo-Persian agreement, which was negotiated and concluded without the knowledge of the United States or of the European Powers."[47]

The external forces, combined with internal difficulties, prevented the agreement from being confirmed. In the meantime, a new force was emerging in Persia. Detachments of the Persian Cossack Brigade under Riza Khan marched from Kazvin and entered Teheran on February 21, 1921. They met with little resistence and were joined by the local police and gendarmerie. They took over all the government departments, and caused the collapse of the Sipahdar Azem government. On March 8, the Undersecretary of State for Foreign Affairs announced in the House of Commons that the new Persian Government had issued a proclamation "containing a statement of policy in the course of which the Anglo-Persian agreement concluded between their predecessors and His Majesty's Government of 9 August 1919 is denounced by them."[48]

The Anglo-Persian agreement had, however, one important consequence which affected the petroleum development of Persia — the appointment of Sidney Armitage-Smith as financial adviser to the Persian Government.[49]

THE ARMITAGE-SMITH AGREEMENT

During the war years two major issues had developed between the Persian Government and the Anglo-Persian Oil Company, both arising from interpretation of the original D'Arcy concession. Article 10 provided that the Company was to pay the Government 16 per cent of the annual net profits of any company or companies that might be formed. The Persian Government maintained that this applied to all companies, even those not operating in Persia, while the Company maintained that it covered only those operating in Persia under the direct organization of the D'Arcy concession. The other issue concerned Article 14, in which the Persian Government undertook to protect the properties of the Company and its personnel. The Company argued that the Government had failed to protect the pipeline and was therefore responsible for the cutting of the line by the Bakhtiari tribes, through incitation by Turkey, in February 1915, and it subsequently withheld the royalties due the Government against payment for the damage. [50]

The Persian Government, on the other hand, argued that under Article 14 of the concession it was not liable for loss or damage caused by acts beyond its control, and, in accordance with Article 17, it asked for arbitration of the dispute. The British Minister in Teheran, Charles M. Marling, informed the Persian Government that if it was the intention to confine the appeal to arbitration to the examination of the correctness of the amount claimed as reparation by the Company, the latter was prepared to accept the proposal. If, however, the Government wished to refer to arbitration the question of the Company's right to withhold the Government's dues pending the settlement of its claim, or the Company's right to reparation for the losses it had sustained, the Company could not accept the demand, "since the arbitration clause does not operate in respect of these questions." [51]

The Persian Minister argued, in September 1917, that Article 17 provided for arbitration of disputes of every kind in respect of the interpretation of the concession. He protested the withholding by the Company of the Government's royalty payments of £19,600 for 1915 and £24,747 for 1916, and declared the action illegal inasmuch as the Company had put itself up as a judge in a case in which it was itself interested. He ended his note with the statement that the Persian Government was ready "to pay any sum that may be found due by it in consequence of the arbitral award." [52]

The amount of the damage was also an issue. The Company claimed £600,000. In a statement in the House of Commons on December 11, 1919, the amount was estimated as £300–£400,000, while accountants appointed by the Persian Government estimated the damage at not more than £20,000. [53]

Not being able to arrive at any solution as to either the bases for calculating the 16 per cent net profit due the Government, or the issue of damages for the cutting of the pipeline in 1915, the Undersecretary of the Persian Ministry of Finance announced, in a letter dated August 29, 1920, the appointment of Sidney Armitage-Smith as the representative of the Persian Government to "finally adjust all questions in dispute between the Anglo-Persian Oil Company and the Imperial Government of Persia."[54]

The result was an agreement signed on December 22, 1920 between the Persian Government, represented by Armitage-Smith, and the Anglo-Persian Oil Company.

The agreement defined "subsidiary company" as any company of which APOC owned, whether directly or through some other subsidiary company, shares sufficient to give it control, or any company more than half of whose directors were appointed or nominated by the APOC, or any company managed by the APOC.

It provided that the Persian Government was entitled to receive a 16 per cent royalty of all annual "net profits arising from the mining, refining and marketing of Persian oil whether all the stages of the above processes" were handled by the Company itself or through subsidiary companies, whether in Persia or outside, with one exception: the Government was "not to receive royalty on the profits arising from the transporting of oil by means of ships." However, in ascertaining the net profits from Persian oil, freight costs were not to be the actual prices charged by the Company tankers or any subsidiary transportation company, but the regular market rates charged by tankers similar to those which were employed in carrying the oil.

Articles 3 and 4 provided for specified deductions for various oil and oil products before calculation of net profits for the purpose of allotting the 16 per cent net profits to the Persian Government, and were applicable to "subsidiary companies refining, distributing or dealing with Persian oil outside Persia, and to any other company refining, distributing, or dealing with Persian oil outside Persia where 'the Company' " was able to procure the necessary accounts prepared by such company and the necessary facilities for inspection to be given by such company to the Government. These two articles also stipulated the methods for ascertaining the profits of the refining and distributing companies handling other oil and oil products in addition to Persian oil.

Article 7 provided that net profits were to be taken as adjusted for income tax purposes; no deductions were to be made from net profits for interest or dividends, and interest and dividends received were to be excluded from the profits on which royalty was payable; they would not apply to the period prior to March 31, 1919.

Article 9 provided that the Company was to prepare each year a statement of royalty payable, such statement to be submitted to a person to be designated for that purpose by the Persian Government fourteen days before the Company's annual meeting. If within six months that person had not challenged the statement, it was to be considered correct. If the statement of royalty was, in the opinion of the Government's nominee, not sufficient to enable him to judge whether the terms of the concession and the agreement had been fulfilled, then the Company undertook to give said nominee access to all information which he might reasonably require. Should a dispute arise, it was to be submitted for arbitration to a chartered accountant nominated by the Institute of Chartered Accountants in England, whose decision was to be final.

Article 10 read: "The Government undertakes to use its best endeavor to facilitate the work of 'the Company' and its subsidiary companies, and the Company agrees that it will not enter into any fictitious or artificial transaction which would have the effect of reducing the amount of royalty payable." [55]

On the same day the agreement was signed, a collateral agreement was also concluded between Armitage-Smith and the Company under which the latter agreed to pay £1,000,000 in settlement of all outstanding questions between the Company and the Government, including such royalties as were due up to March 31, 1919.

With the signing of the agreements it seemed as if the two issues which had disturbed the relations between the Persian Government and the Anglo-Persian Oil Company had been satisfactorily solved. [56] However, the rise to power of Riza Khan in 1921 and the Russo-Persian agreement of that year (which will be discussed in Chap. IV), by which the Soviet Government renounced all its concessions and privileges in Persia, cast a shadow over the Armitage-Smith agreement. Nevertheless, for a number of years, perhaps until Riza Khan had entrenched himself and become Shah Pahlevi, the agreement was the only basis of relations between the Persian Government and the Anglo-Persian Oil Company.

THE RISE OF RIZA KHAN

Events moved rapidly after the fall of Sipahdar Azem. Zia ed-Din formed a new government in August 1921, with Riza Khan as Minister of War. Soon, however, Zia was forced to resign and in 1923 Riza Khan became Premier; Shah Ahmed left for an extended stay in Europe. [57] The following year a republican sentiment began to develop in Persia and Riza took to modeling himself after Mustafa Kemal in Turkey. However, it soon became apparent that neither Persia, nor Riza himself, was ready for the secularization of the country at Kemal's pace. Realizing the influence of the Mujtahides, Riza declared that a republic was against the Shia religion, and in April 1924 he issued a proclamation enjoining the Persian people to cease discussing the proposed introduction of a republican regime. [58] In 1925 Riza was made dictator, and in October the Shah was deposed by the Majlis. A new Majlis elected Riza Shah in December 1925 and he assumed the name of Pahlevi. Before long the oil issue occupied his attention.

SUMMARY

It has been suggested that the initial support given by the British to the D'Arcy concession was a countermeasure to the German Baghdad railway concession, [59] hence the Government's desire to control the concession. Be that as it may, it is clear that by the time the British were ready to sign the secret Anglo-German convention on the railway, they were already majority owners in the Anglo-Persian Oil Company, and the Anglo-Persian Oil Company was the majority owner of the Turkish Petroleum Company. [60] It is also clear that the Persians who signed the

D'Arcy concession did not know, perhaps were not capable of knowing, and very likely did not care, what they were signing, and had no notion of the extent of their country's potential wealth. [61]

D'Arcy in his first venture took a considerable risk, and the Burmah Oil Company under Lord Strathcona had also, at the behest of the British Admiralty, risked additional capital and formed the Concessions Syndicate; in fact, the Burmah Oil Company had paid the dividends on the Anglo-Persian's preferred shares until the Company yielded profits. As First Lord of the Admiralty, Winston Churchill reported in the House of Commons that Burmah Oil had advanced £44,848 for that purpose by 1914. [62] However, by the time the Anglo-Persian Oil Company had begun to operate, the risk had been more than cleared. When the British Government began to negotiate for buying into the Company, it was not taking any risk whatsoever. It had the report of the experts of the Slade commission which assured the Government of the potentialities of the Persian oil fields. After a careful scientific examination of the fields, the commission concluded: "We are satisfied that the Company's Concession is a most valuable one and, providing no unforeseen factor intervenes, the existing field is capable, with proper development, of supplying a large proportion of the requirements of the Admiralty for a considerable period, while the whole Concession, judiciously worked, would probably safeguard the fuel supply of His Majesty's Navy." [63]

So valuable did the British consider the Persian oil that at the outbreak of the First World War they sent an expeditionary force from India into Mesopotamia to protect — at least it was a major consideration — the oil fields and pipelines and the refinery in Abadan. During the war the British began to appreciate fully the vital aspects of oil to the success of the war, as illustrated by the now famous statement of Lord Curzon in a speech in Lancaster House on November 21, 1918: "Truly posterity will say that the Allies floated to victory on a wave of oil." [64] They also realized their dependence on American oil. They were therefore determined to guarantee the oil necessary for peacetime as well as wartime needs. The area for this purpose was the Middle East and especially Persia, where they were already operating and reaping such extraordinary profits. [65] The British Government decided, therefore, to eliminate all other nationals from exploiting the oil resources of Persia. At least one of the major considerations in forcing upon the Persian Government the 1919 agreement was the protection of the British oil interests in south Persia and even its extension into north Persia. [66]

The lengths to which the British Government went to obtain the 1919 agreement caused suspicion both in France and in the United States of the objective they aimed to achieve. The American suspicions, perhaps more than the French, were justified. The British hoped to control entirely the Persian state, to the exclusion of all others, primarily — after consideration of imperial strategic objectives — for the exploitation of the oil fields, south and north.

The Armitage-Smith agreement must be admitted to have diverged basically from some of the provisions of the D'Arcy concession. The letter from the Persian Minister of Finance authorizing Armitage-Smith to negotiate was of such a broad character that he could not be accused of exceeding his authority when he concluded the agreement, but the agreement as such certainly granted the Company privileges that were not included in the original concession. It was therefore inevitable that as soon as the Persians felt they were in a position to challenge the validity of the Armitage-Smith agreement, they would do so.[67]

NOTES

1. Said Lord Curzon about the concession: "When published to the world, it was found to contain the most complete and extraordinary surrender of the entire industrial resources of a kingdom into foreign hands that has probably ever been dreamt of, much less accomplished in history." George N. Curzon, *Persia and the Persian Question* (London, 1892), I, 480. See also a discussion, in the light of the times, of the concession in Sir Henry Rawlinson, *England and Russia in the East* (London, 1875), 124–130; an abstract of the concession, 391–394.

2. International Court of Justice, *Pleadings, Oral Arguments, Documents. Anglo-Iranian Oil Co. Case (United Kingdom v. Iran),* 1952, 185. (Henceforth cited *Pleadings.*) A sensational incident of how the Shah snubbed and embarrassed de Reuter in London is told in Edmund Jaraljmek, *Das andere Iran* (Munich, 1951), 244–245.

3. Sir Percy Sykes reports, on the basis of information obtained from the manager of the Bank, that on the constitution of the bank de Reuter's "good faith" money of £40,000 from the first concession was returned to him, and that he also received a premium of £2 per share when the capital was issued. Percy Sykes, *A History of Persia* (London, 1930), II, 372.

4. M. Nakhai, *Le Pétrole en Iran* (Brussels, 1938), 29–31; Curzon, *Persia and the Persian Question,* I, 481; Azami Zangneneh (Abdul-Hamid), *Le Pétrole en Perse* (Paris, 1933), 80–83; *Pleadings,* 185–186.

5. Henry Filmer, pseudonym for James Rives Childs, reports that the Anglo-Persian Oil Co., in appreciation of de Morgan's part in the discovery of oil, granted him a handsome pension, *The Pageant of Persia* (New York, 1936), 56.

6. Nakhai, *Le Pétrole en Iran,* 31–32. William Knox D'Arcy was born at Newton Abbot in Devonshire in 1849, and educated at Westminster School. In 1866, when D'Arcy was seventeen, his father, a lawyer, emigrated with his whole family to Queensland. There he established himself in his profession in the town of Rockhampton. A client by the name of Sandy Morgan showed young D'Arcy some pieces of gold quartz and asked him what it was; he told him there was a whole mountain of the stuff back of his place. This was the beginning of the Mount Morgan gold rush in Australian history. D'Arcy, who bought shares in the mine, became a millionaire and returned to England. William Henry Beable, *Romance of Great Business* (London, 1926), 219–220; Azami, *Le Pétrole en Perse,* 84.

7. Sir Arthur H. Hardinge, *A Diplomatist in the East* (London, 1928), 278–279. Referring to this detailed story, Sir Arnold T. Wilson, who later became a high official in the Anglo-Persian Oil Co., says in a footnote: "It must, however, be stated here that his detailed recollections are not borne out by such documen-

tary evidence as is available." Arnold T. Wilson, *Persia* (London, 1932), 90. This seems to be a rather weak attempt to question Hardinge's veracity.

8. The full text of the concession is to be found in: League of Nations, *Official Journal,* 13th Year, 1932, 2305–2308; United States, Department of State, *Papers Relating to the Foreign Relations of the United States 1920* (Washington, 1936), III, 347–351 (henceforth cited *Foreign Relations,* and year); *Pleadings,* 224–227.

9. *Agreement with the Anglo-Persian Oil Company Ltd.,* Cmd. 7419, 1914, 24; Wallace E. Pratt and Dorothy Good (eds.), *World Geography of Petroleum* (Princeton, 1950), 172.

10. *Parliamentary Debates, House of Commons* (henceforth cited *PDC* – Commons, and *PDL* – Lords), 63, cols. 1190–1191, June 17, 1914; *The Anglo-Iranian Oil Company and Iran,* 1; Azami, *Le Pétrole en Perse,* 96; United States Senate, *American Petroleum Interests in Foreign Countries.* Hearings before a Special Committee Investigating Petroleum Reserves, Seventy-Ninth Congress, First Session (Washington, 1946), 342–343.

11. At the time of its formation, the Anglo-Persian Oil Company came into possession of all the shares of the two companies not held in Persia. The shareholding was thus as follows:

	Shares	Per Cent
First Exploitation Co.		
Anglo-Persion	478,400	87.95
Persian shareholders	65,540	12.05
Bakhtiari Oil Co.		
Anglo-Persian	388,000	97.00
Persian shareholders	12,000	3.00

Cmd. 7419, 1914, 24. Winston Churchill, then First Lord of the Admiralty, reported in 1914 that the Burmah Oil Co. held 970,000 shares of £1 each of the ordinary shares in the Anglo-Persian Oil Co. and Lord Strathcona held the balance of 1,000,000 shares. The Burmah Co. received half its holdings in payment of its share in the original concession and the balance it obtained from D'Arcy in return for Burmah Oil Co. shares. *PDC,* 63, col. 723, June 15, 1914.

12. Edward G. Browne, *The Persian Revolution of 1905–1909* (Cambridge, 1919), 353–354.

13. *Ibid.,* 352, 371; *Correspondence Respecting the Affairs of Persia December 1906 to December 1908,* Cmd. 4581, 1909, 10–14; W. Morgan Shuster, *The Strangling of Persia* (New York, 1912), 337–345; Helen Miller Davis, *Constitutions, Electoral Laws, Treaties of States in the Near and Middle East* (Durham, 1947), 69–78.

14. Shuster, *The Strangling of Persia,* xxvii. Full text of the agreement, xxv–xxix.

15. Lt. Wilson was later to play important and varied roles, including the managership of the Anglo-Persian Co. in the Persian Gulf area. Long before retirement age he applied for a pension from Government service to accept the manager-

ship of the Company, *PDC*, 139, cols. 651–653, Mar. 10, 1921; *PDC*, 191, col. 2223, Feb. 18, 1926.

16. Sir Arnold T. Wilson, *SW. Persia* (London, 1941), 17–18, 92–93; Philip Graves, *The Life of Sir Percy Cox* (London, 1941), 123 *et passim;* Elgin Groseclose, *Introduction to Iran* (New York, 1947), 123. That these activities of the Company were the beginning of a plan aimed at undermining the authority of the central government at Teheran so that the Company could better exploit Persia on its own terms became an argument advanced by the Iranian Government for the nationalization of the oil industry. "Iran Presents Its Case for Nationalization," *The Oil Forum*, VI, 79–94, Mar., 1952. This is categorically denied in "Anglo-Iranian Answers Iran with Facts," *The Oil Forum*, VI, iii–v, Apr., 1952. That these agreements with the local chiefs tended to weaken the central government in Teheran was expressed by Ramsay MacDonald as far back as early 1914, and his concern was shared by other members of Parliament. *PDC*, 63, col. 1165 *et passim*, June 17, 1914.

   While denying that it had ever attempted to undermine the loyalty of the local tribes to the Teheran government, the Company admitted that since the Teheran government had no authority in Khuzistan, the Company had no option but to deal directly with the local tribal chiefs and the "Company felt justified in making these deductions [payments to the local tribal chiefs] from the royalty." "Anglo-Iranian Answers Iran with Facts," *The Oil Forum*, VI, iv, Apr., 1952.

17. Cmd. 7419, 1914.

18. This method apparently proved very successful. For years the Government was constantly asked to augment its representation on the Board but refused, and in November 1952, when it was suggested in Parliament that Government representation be increased in accordance with its financial holdings, the reply was in the negative. The Government declared that its representatives "have powers additional to those possessed by ordinary directors and that they have the right of veto over any decision of the Board." *PDC*, 508, col. 611, Nov. 27, 1952.

19. It should be stated that all the various governments – Liberal, Conservative, Labor and National – consistently refused to interfere with the commercial operations of the Company.

20. Cmd. 7419, 1914, 5.

21. *PDC*, 63, cols. 1131–1250, June 17, 1914. This vote was in the House sitting as a committee; the final vote in the House sitting as the House was 228–48. *PDC*, 64, cols. 1033–1054, July 7, 1914.

   Throughout the entire period of the existence of the Company only once, in 1923, did the Government consider selling its holding. *PDC*, 168, col. 350, Nov. 15, 1923. The offer came from the Director of the Burmah Oil Co. *PDC*, 169, col. 523, Jan. 21, 1924; the price offered for each ordinary share was between £3 15 shillings and £4. However, in the middle of February the Government announced that after a full examination of the proposal by the departments concerned, it was satisfied that it would not be in the national interest to sell the shares. *PDC*, 169, cols. 1038–1039, Feb. 14, 1924. To appreciate this change, one should keep in mind that the consideration to dispose of the Government holding was made by the Conservative government, and the decision not to sell was made by the Labor government which had come to power in the meantime.

22. *PDC*, 122, col. 915, Dec. 8, 1919. For Churchill's evaluation of the Government's purchase, see his *The World Crisis* (Charles Scribner's Sons, N. Y., 1923), I, 134–140 *et passim*. Not only did he calculate that the dividends, savings on price paid for the oil, appreciation of the Government's shares, etc. would be about £40 million, but he made the following statement: "On this basis it may be said that the aggregate profits, realised and potential, of this investment may be estimated at a sum not merely sufficient to pay all the programme of ships, great and small of that year and for the whole pre-war oil fuel installation; but are such that we may not unreasonably expect that one day we shall be entitled also to claim that the mighty fleets laid down in 1912, 1913 and 1914, the greatest ever built by any power in an equal period, were added to the British Navy without costing a single penny to the taxpayer."

  Said Churchill as Chancellor of the Exchequer: "The original cost of the holding of His Majesty's Government in the Anglo-Persian Oil Company was £5,200,000 and the value today at approximate market prices is £23,600,000." *PDC*, 211, col. 1574, Dec. 8, 1927. Two years later he reported that the market value of the Government's holding was approximately £33,242,000. *PDC*, 224, cols. 1795–1796, Feb. 6, 1929. In 1932, Anthony Eden stated that the total amount which the Government had received from the Company in dividends and interest up to then was £9,997,344. *PDC*, 273, col. 8, Dec. 12, 1932. In 1937, Sir John Simon stated that the total amount of interest and dividends received up to then was £15,989,354. *PDC*, 330, col. 1337, Dec. 16, 1937.

23. Wilson, *SW. Persia*, 215 *et passim; PDC*, 63, col. 1810, June 24, 1914; *PDC*, 153, col. 1987, May 9, 1922.

24. In spite of pious expressions from apologists, there can be no doubt that one of the major, if not *the* major, considerations for the British invasion of Mesopotamia was to protect the oil fields, pipeline and refinery at Abadan. Sir Percy Sykes, *Persia* (Oxford, 1922), 155; *–A History of Persia*, II, 440; G.E. Hubbard, *From the Gulf of Ararat* (Edinburgh, 1916), 150; H.W.V. Temperley, *A History of the Peace Conference of Paris* (London, 1920–24), VI, 209. "The importance of retention of the source of oil supply was among the motives which induced the British Government to occupy Lower Mesopotamia directly the Turks joined the war on Germany's side." Olaf Caroe, *Wells of Power* (London, 1951), 84.

25. Temperley, *History of Peace Conference*, VI, 208.

26. Sykes, *A History*, II, 444 *et passim;* Groseclose, *Introduction*, 71–76; Nasrollah Saifpour Fatemi, *A Diplomatic History of Persia 1917–1923* (New York, 1952), 6–22 *et passim.*

27. *PDC*, 71, col. 576, Apr. 27, 1915; *PDC*, 73, col. 2121, July 27, 1915; *PDC*, 122, col. 1763, Dec. 11, 1919; Wilson, *Persia*, 91.

28. Temperley, *History of Peace Conference*, VI, 210.

29. Graves, *Life of Sir Percy*, 248–260.

30. Great Britain, Foreign Office, *Documents on British Foreign Policy 1919–1939*, E. L. Woodward and Rohan Butler (eds.), (henceforth cited *Documents*), 1st Series, IV (London, 1952), 1119–1122.

31. *Ibid.*, 1125–1126.

32. *Ibid.*, 1190–1191.

33. *Ibid.*, 1142. For a discussion of the personalities involved and the general atmosphere in Teheran at the time the agreement was signed, see J.M. Balfour, *Recent Happenings in Persia* (London, 1922), 122 *et passim.* Balfour was chief assistant to the financial adviser to the Persian Government under the 1919 agreement.

34. *Agreement between His Britannic Majesty's Government and the Persian Government signed at Teheran August 9, 1919,* Cmd. 300, 1919.

35. *Documents,* 1133–1135 *et passim.*

36. A partisan French presentation of the issue is given by Emile Lesueur, *Les Anglais en Perse* (Paris, 1921). Lesueur was one of the law professors in Teheran engaged by the Persian Government, to all of whom Lord Curzon took exception in principle on the basis of the agreement. *Documents,* 660–661, 1128 *et passim.*

37. *Foreign Relations 1919,* II (Washington, 1934), 699.

38. *Ibid.,* 700.

39. *Documents,* 1163.

40. *Foreign Relations 1919,* II, 717. Third Assistant Secretary of State Long reported on December 22, 1919 a conversation he had with the Persian Minister in Washington, who was on his way to Switzerland to see the Shah. The Minister asked Long, since the agreement had to be ratified by the Majlis and the Shah was shortly returning to Persia, what attitude the United States would take should an arrangement be made by the Shah to prevent ratification. "In response I told him that it was a matter which was purely Persian and that we could not take any attitude but that we would feel very sorry to see an arrangement made by the terms of which Persia would lose part or the whole of its sovereignty." *Ibid.,* 719.

41. *Documents,* 1205, 1212, 1214.

42. *Foreign Relations 1920,* III, 355. The American Petroleum Institute submitted a memorandum to the State Department which described the agreement as placing under British supervision and direction Persian civil, military and financial administration. The memorandum maintained that the agreement strengthened the exclusion of American citizens from petroleum development in the country where the Anglo-Persian Oil Company had exclusive rights. *Foreign Relations 1920,* I, 365.

43. *Documents,* 1235; *PDL,* 42, col. 285, Nov. 16, 1920.

44. *PDC,* 132, col. 2109, August 2, 1920; *PDL,* 42, col. 285, Nov. 16, 1920; Balfour, *Recent Happenings,* 132–134.

45. *PDC,* 134, col. 1520, Nov. 15, 1920; *PDL,* 42, col. 288, Nov. 16, 1920.

46. *Foreign Relations 1920,* III, 346.

47. *Documents,* 1111–1113.

48. *PDC,* 138, cols. 1413–1414, Feb. 28, 1921; *PDC,* 139, col. 274, Mar. 8, 1921. It was not until May 24, 1921, however, that the Undersecretary of State for Foreign Affairs reluctantly admitted: "I think it [the agreement] can be regarded as having lapsed." *PDC,* 142, col. 34, May 24, 1921. For a Persian point of view of the agreement, see Fatemi, *Diplomatic History.*

49. Sir Percy Sykes declares that Armitage-Smith, a Treasury official without

experience of the East, "was engaged by the Persian Government in 1920." Lord Curzon, however, stated in the House of Lords: "We sent out to Persia an able Treasury official, Mr. Armitage-Smith, to act as head of the financial commission for the reorganization of Persian finances," and he took up his duties at Teheran on May 6. Sykes, *A History,* II, 523; *PDC,* 134, cols. 969–970, Nov. 9, 1920; *PDL,* 42, col. 284, Nov. 16, 1920.

50. The Company alleges that the Persian Minister for Foreign Affairs at first made a verbal admission of his Government's responsibility in the matter, but that it was later repudiated in writing. During the period of the dispute, February 1915 to the end of 1919, the Company paid the Government £325,552 in royalties. "Anglo-Iranian Answers Iran with Facts" *The Oil Forum,* VI, iv, Apr., 1952.

51. League of Nations, *Official Journal,* 14th Year, 1933, 295.

52. *Ibid.,* 295–296.

53. *PDC,* 122, col. 1673, Dec. 11, 1919; League of Nations, *Official Journal,* 14th Year, 1933, 290; in Anglo-Persian Oil Company, *Report of the Directors and Balance Sheet to 31st March, 1916,* the loss sustained as a result of damage to the pipeline is given as £160,000; in the *Report* for the year ending March 31, 1917 the loss was estimated as £402,887. Arnold Toynbee explained the basis of the Company's claims: "£26,000 appears to have been the Anglo-Persian Oil Company's estimate of the net cost of oil lost. The £600,000 included consequential losses arising from the interruption of all oil supplies to the refinery, breaking of contracts, and so on." *Survey of International Affairs 1934,* 230.

54. *Pleadings,* 189–190. The letter appears, *ibid.,* 228. In 1952 the Iranian Government maintained that Armitage-Smith had gone beyond the powers granted in the letter in concluding the agreement. In a communication from the Iranian Minister of Foreign Affairs to the President of the International Court of Justice, it is stated: "The Iranian Government empowered the said Mr. Armitage-Smith to audit the accounts of the oil Company, and to settle the existing differences." *Ibid.,* 673. A reading of the August 29, 1920 letter will, however, not give that impression. The Iranian Government did not challenge the authenticity of the letter.

55. *Pleadings,* 229–235.

56. Since the basis of Armitage-Smith's appointment was the Anglo-Persian agreement of 1919, and since that agreement had "lapsed," Armitage-Smith had to leave the employ of the Persian Government.

57. *PDC,* 140, col. 82, Apr. 5, 1921; *PDC,* 142, col. 1022, June 1, 1921.

58. *PDC,* 176, col. 1293, July 23, 1924.

59. Nakhai, *Le Pétrole en Iran,* 32–33.

60. George E. Kirk, *A Short History of the Middle East* (London, 1948), 95; E. M. Earle, "The Secret Anglo-German Convention of 1914 regarding Asiatic Turkey," *Political Science Quarterly,* XXXVIII, 24–44, Jan., 1923.

61. The table in Footnote 11 indicates that the advice given to Hardinge, to bribe high Persian officials in order to obtain the D'Arcy concession, was followed, for the number of shares held by Persians in D'Arcy's First Exploitation Co. at the time the Anglo-Persian Oil Co. was organized was 65,540, 12.05 per cent of the total. The Persian Government, in accordance with the concession, was given 20,000 shares, and the rest must be considered as bribes given to high

officials in order to obtain the concession.

62. *PDC*, 63, col. 1087–1088, June 17, 1914; *Report of the Directors for the year ending 31st March 1914*, [8].

63. Cmd. 7419, 1914, 28; *PDC*, 63, col. 1141, June 17, 1914. It should be noted that Admiral Slade subsequently was, for a number of years, vice-chairman of the board of directors of the APOC, and that another member of the commission, John Cadman, Professor of Mining at the University of Birmingham and petroleum adviser to the Colonial Office, was later to become chairman of the board of the APOC, Petroleum Executive in the British cabinet during World War I, and chief negotiator between the British and American companies following World War I.

64. Pierre L'Espagnol de la Tramerye, *The World Struggle for Oil* (New York, 1924), 109.

65. The net profit of the Anglo-Persian Oil Co. for the year 1918, for instance, after providing for administration, expenses, debenture interest, royalty and depreciation, was £1,308,558. *PDC*, 143, col. 4, June 13, 1921.

66. E. Weekly, the petroleum expert of the British Foreign Office, in a minute of Dec. 12, 1919 on the Berenger memorandum regarding the request for a French share in the oil exploitation of the Middle East, said that there could be no question of admitting French participation in oil areas covered by the Anglo-Persian Oil Co.'s concession situated in Persia proper, or in territories transferred from Persia to Turkey. "North Persia is, as far as we know, an open field, but no doubt the Anglo-Persian Oil Company have not lost sight of the possibilities of that field." *Documents*, 1113.

67. What was said about those who signed the D'Arcy concession could equally well be said about those who agreed to the Armitage-Smith agreement. The main issue here was the damage caused by the cutting of the pipeline rather than the basis of calculating the 16 per cent net profits. The British refused to arbitrate the issue and withheld the royalty, at least the greater part of it, due the Persian Government. The inducement held out to the Persians was the dropping of the claims for damages, and for this concessions were asked. When the Persian Foreign Minister was in England in the latter part of 1919 to discuss the new Persian agreement, a concession was squeezed out of him. The files of the British Foreign Office merely indicate that a provisional agreement concerning the basis of calculating the 16 per cent net profits due the Persian Government was signed between Prince Firooz Mirza Nusret ed-Dowleh and representatives of the Anglo-Persian Oil Co.; no details are given. *Documents*, 1182. However, in the debate of 1952, the Company representative revealed that this provisional agreement provided for the consent of the Persian Government to the exclusion of the profits of the tanker company from the 16 per cent net profit due the Persian Government, in return for the withdrawal by the Company of its claim for compensation for the losses arising out of the cutting of its pipeline in 1915. "Anglo-Iranian Answers Iran with Facts," *The Oil Forum*, VI, v, Apr., 1952. Both of these — the dropping of the claims on the part of the APOC, and the consent of the Persian Government not to claim net profit from the tanker company — formed part of the Armitage-Smith agreement. The fact that Persia received £1,000,000 for past royalties made the Armitage-Smith agreement palatable and acceptable at the time.

Chapter III
FROM THE RISE OF RIZA SHAH
TO THE END OF WORLD WAR II

In 1921, after his march on Teheran, Riza Khan achieved a double objective: the 1919 Anglo-Persian agreement was renounced, and the agreement with Soviet Russia was ultimately ratified. That same year, Riza Khan engaged William McClintock, a London chartered accountant, to study the financial relations between the Anglo-Persian Oil Company and the Persian Government. On the basis of McClintock's report, which listed a number of serious grievances against the Company, and on direct complaints from Persia's Oil Commissioner, Riza Khan, who had in the meantime gained full control over Persia, turned to the oil question. The Armitage-Smith agreement became a major issue in his relations with the British Government.

As early as December 22, 1926, the Persian Oil Commissioner, Mirza Eissa Khan, wrote to Sir John Cadman, Chairman of the Anglo-Persian Oil Company, and asked his intervention in bringing about an early settlement of all points under dispute pertaining to the accounts for the years ending March 1924, 1925, 1926, and 1927.[1] On May 9, 1928, he wrote to the Company: "I have frequently stated, and now repeat, that Mr. Armitage-Smith was not instructed or authorized to draw up an agreement effecting changes in the terms of the D'Arcy concession."[2] Soon, however, the Persian Government was demanding not only the invalidation of the Armitage-Smith agreement, but the D'Arcy concession itself.

Early in 1928 Sir John Cadman had suggested to the Persian Minister of Court, Prince Teymourtache, that an extension of the concession period would be necessary if the requisite capital was to be obtained.[3] On August 12, 1928, Teymourtache declared in a letter to Sir John Cadman that the D'Arcy concession had been obtained at a time when the government of the Kajars did not realize what was being demanded from it and what it was giving away, and that the Persian Government was prepared to negotiate with a view to revising the concession. The following year Sir John Cadman went to Teheran to discuss the draft of a new concession. After negotiations which lasted over a period of two years, he informed the Minister of Court, on August 7, 1931, that the revision of the concession could no longer be contemplated. The Company later reported that it had refused to consider the revision of the concession because "the demands of the Persian Government were greatly in excess of anything which the Company could accept."[4]

Discussions therefore took place on a new formula for calculating the 16 per cent net profit to which the Persian Government was entitled. The Minister of Court and Sir John Cadman took up the matter first in Switzerland and later in London, and a draft agreement was arrived at. After this had been studied in Teheran, the Government asked the Company to send a representative to Teheran to clarify and elucidate further some points, but the Company felt that there was no justification in incurring the expense of sending its experts to Teheran and suggested that a Persian representative be sent to London; this the Government refused.

When the world depression set in after the 1929 crash, the Company's profits declined sharply, and the annual statement for 1931, completed on June 3, 1932, showed that the royalty payments for that year were £306,872, compared with £1,288,312 the previous year. The Persian Government protested to the Company's local representative, and on June 29 refused to accept the royalty. Later the Government indicated that new proposals might be submitted to the Company. On November 27, however, the Minister of Finance, Hasan Taqizadeh, notified the Resident Director of the Company of the cancellation of the D'Arcy concession.[5] In his letter he declared that the concession did not protect the interests of the Persian Government; that the Government could not legally and logically be bound by the provisions of a concession grated prior to the establishment of the constitutional regime; and that since the Company had persistently refused to recognize the Government's rightful complaints and its efforts to protect Persian interests, there was no alternative but to cancel the concession. However, the Government was willing to enter into negotiations for granting the Company a new concession.[6]

T. L. Jacks, Resident Director of the Company, replied on the following day, November 28, 1932, in the name of the Board of Directors of the Anglo-Persian Oil Company. He denied that the terms of the concession did not protect the interests of the Persian Government and said that the Persian Government had no right to cancel the concession. He argued that successive governments, both before and after the establishment of the constitutional regime, had recognized the D'Arcy concession; that the contention of the Persian Government that it could cancel the concession had no foundation either in law or in equity. Relying on the good faith of the Persian Government, the Company had expended many millions of pounds sterling in Persia, and the benefits which the Persian Government had derived therefrom could not be denied. Moreover, the Company had always indicated its willingness to negotiate with the Government, thus recognizing the needs of Persia. Jacks' letter concluded with the warning that publication in the press of the cancellation of the concession might be damaging to the Company, and that the responsibility for such an eventuality rested with the Persian Government.[7]

On December 1, the Finance Minister issued a denial of the contentions contained in the Company's letter and asserted that the Persian Government had full right to cancel the concession, that such a step remained the final decision of the Government.[8] The next day the British Minister in Teheran handed a note to the Foreign Minister in which he declared that cancellation of the concession was an inadmissible breach of the terms of the concession, and demanded immediate withdrawal of the notification issued to the Company. Noting that while the British Government hoped the Persian Government would be at pains to reach an amicable settlement of the issue in direct negotiations with the Company, he warned in strong terms that if the necessity arose the British Government would not hesitate "to take all legitimate measures to protect their just and indisputable interests," and that the Persian Government would be held responsible for any damage the Company might sustain as a result of the cancellation.[9]

On the 3rd, the Persian Foreign Minister addressed a note to the British Minister at Teheran repeating the previous contentions of the Persian Government[10] and on the 8th the British Minister presented a second note. He again denied the right of the Persian Government to cancel the concession unilaterally on the ground that such action was confiscation and a breach of international law directed against a British company which the British Government was bound to protect. The British Government was repeating its request that the notification to the Company be withdrawn. If, however, within a week the Persian Government had not complied with the request, the British Government would have no alternative but to refer the dispute to the Permanent Court of International Justice at The Hague, as a matter of urgency under the optional clause, and to ask the Court to indicate provisional measures to preserve the Company's rights.[11]

Four days later, on December 12, the Persian Foreign Minister, Ali Khan Furughi, in a rather lengthy note to the British Minister in Teheran, outlined the reasons for his Government's action. He maintained that the circumstances under which the D'Arcy concession had been granted were inimical to the interests of Persia; not only had those who had granted the concession been unmindful of the welfare of the country, but those who had obtained the concession had taken advantage of the ignorance of the authorities and had used all sorts of threats and pressure to gain their end. He implied that the concession had been obtained under duress and therefore had no validity. Moreover, the Company had not even lived up to the provisions of the concession: it had never allowed the Government to supervise its expenditures in order to safeguard Persia's 16 per cent net profit; it had never submitted to the Persian Government or its representative any detailed accounts or other evidence of its own expenditures and the expenditures of all its subsidiaries, which would have enabled the Government to check the calculation of its royalties; it had refused to pay the Government its share of the profits of the subsidiary companies. The Foreign Minister declared that by granting large subsidies to its subsidiaries, the Company had reduced the net profit share of the Persian Government. Furthermore, during the war prices of petroleum and petroleum products had risen very sharply and production had increased, yet the Company did not pay the Government its share; indeed, contrary to the provisions of the concession, it refused to pay income taxes (introduced in 1930). The Government was particularly disturbed by the Company's policy not to develop Persia's great petroleum potentialities, but to restrict operations to a very small area. Nevertheless, the Government had been willing to discuss a new formula for the basis of calculating the 16 per cent net profit, but the Company had refused to send representatives to Teheran, and no alternative was left but to cancel the concession.

As regards the British threat to refer the dispute to the Permanent Court, the Foreign Minister pointed out that the Court was not competent to deal with the case since it did not come within the terms of the optional clause. Turning to the possibility of direct negotiations between the Persian Government and the

Company, the Minister declared that it seemed, unfortunately, that the British Government was not in favor of an agreement between the Persian Government and the Company, because the threats and intimidations it was bringing to bear on the Persian Government and the unacceptable demands that were being put forward prevented either the Persian Government or the Company from taking a single step towards reaching a mutual understanding. The Foreign Minister's note concluded with the statement that such an attitude was incompatible with the spirit and the desire for peace which should prevail among friendly powers and members of the League of Nations, and that the Persian Government therefore felt that it would be within its rights in bringing to the notice of the Council of the League of Nations the threats and pressures which had been directed against it. [12]

BEFORE THE COUNCIL OF THE LEAGUE

The British reply to this letter was a request, on December 14, to the Secretary General of the League of Nations to insert in the agenda of the Council, under Article 15 of the Covenant of the League: "Dispute which has arisen between His Majesty's Government in the United Kingdom and the Imperial Government of Persia in consequence of the Persian Government's action in purporting to cancel the concession held by the Anglo-Persian Oil Company, a British company." [13] On the 19th, the Secretary General informed the Persian Government that the Council had decided to take up the case on January 23, 1933. [14] Both governments submitted written memoranda and later argued the case orally before the Council.

THE ISSUES

The major issues of the dispute were of a substantive as well as a technical nature. One of the first issues was the Armitage-Smith agreement: was it merely an interpretation of the D'Arcy concession needing no ratification by the Majlis and hence valid, or did it introduce changes in the original concession of such a magnitude as to require ratification, without which it was invalid? Another was cancellation of the D'Arcy concession itself, and this brought into focus the methods by which the concession had been obtained. A legal point at issue was the competency of the Permanent Court under the optional clause: was the dispute one between two governments and hence one which could properly come before the Court, or was it only between the Persian Government and a private foreign company subject to the jurisdiction of the Persian courts? Finally, there was the question of whether the dispute came under Article 15 of the Covenant: was the claim of denial of justice to the Company of such a nature that the British had no alternative but to bring the case to the Council of the League as a threat of a rupture of peace; or should the Company have first tried the local Persian courts, and only after failing there appealed for diplomatic action under Article 15?

In a memorandum submitted to the Council of the League, the British maintained that the Company had cooperated wholeheartedly with the Persian Government in an effort to arrive at a mutual understanding, but that the Persian Government, without cause, had unilaterally and suddenly cancelled the concession. In 1929 the Company representatives had gone to Teheran, at the insistence of the Persian Government, to negotiate for the revision of the concession, but no agreement had been arrived at. In 1932 a new royalty payments agreement had been initialed in London by Mirza Eissa Khan, Persian Oil Commissioner, appointed under Article 11 of the concession, on behalf of the Persian Government, but the Government had refused to ratify it. The memorandum asserted that the exploitation of the oil by the Company had brought nothing but good to Persia, as evidenced by the more than £11,000,000 royalties paid to the Persian Government by the Company.[15]

In the oral presentation, Sir John Simon, British Foreign Minister, argued that the British Government had no alternative but to bring the case before the Council, under Article 15, for, since the Majlis had approved the Government's action in cancelling the concession, the cancellation had become the law of the land and the Company could not have had recourse to the Persian courts. It was the British Government's duty to take up the wrong done to a British company, which was within international law, and since there was danger of a rupture, Article 15 was invoked.

As to the nature of the concession and the interpretation of the Armitage-Smith agreement, Sir John Simon contended that the Persian Government had been quite satisfied with both and had accepted the royalties over the years without questioning either the original concession or the Armitage-Smith agreement. The real reason, according to Sir John, why the Persian Government had repudiated the agreement, arrived at in London for a new formula for calculating the net profits, was the low amount to which it was entitled as the result of the decreased Company profits during the year 1931. The only legal basis for resolving differences between the two parties — as provided in the concession — was arbitration: no other courts had jurisdiction: the Government, however — the Majlis approving — had cancelled the concession and the Company was left without any local legal recourse.[16]

The Persian Government argued that the D'Arcy concession was invalid because it had been granted by a non-constitutional government, acting under foreign influence — British and Russian: that the Company had not complied with the provisions of the concession: and had refused arbitration and had limited the area of exploitation, thus depriving the Persian Government of its proper revenue. Since the Company had consistently refused to respond to the legitimate requests of the Persian Government, the latter felt justified, under Persian law, in cancelling the concession. Since the Armitage-Smith agreement modified the D'Arcy concession it required parliamentary approval, and lacking this was invalid. As to the £11,000,000 which the Persian Government had received in revenue, the memorandum pointed out that if the concession had been granted gratis, the custom duty which the Company would have had to pay for the period 1901–1932 would have amounted to £19,998,509 and 16 shillings.

In his oral arguments the Persian representative proposed to ask the Permanent

Court for an advisory opinion whether the action of the Persian Government could be considered a violation of international law before the remedies of local law had been exhausted, and whether there was ground for intervention by way of diplomatic protection. [17]

Two days before the Council opened the oral arguments, it appointed as Rapporteur Edvard Beneš of Czechoslovakia. On February 3 Beneš reported to the Council that both parties had agreed to suspend proceedings before the Council until the May 1933 session, with an option of extending this time limit still further if necessary; that both parties had agreed that the Company would enter immediately into negotiations with the Persian Government; and that while negotiations were under way the work and operations of the Company in Persia were to be carried on as they had been prior to November 22, 1932. The Council approved this report. [18]

The negotiations were begun in Geneva on February 4 and continued in Paris on the 10th and 11th, between M. Davar and Hussein Ala, representing the Persian Government, and representatives of the Company. However, no tangible results were obtained, and later in the month the British Government and the Company agreed to continue discussions in Teheran on condition that the Rapporteur continue to play his role. [19] These began on April 4, 1933, with the Government's delegation headed by Hasan Taqizadeh, and that of the Company by William Fraser, Deputy Chairman of the Board of Directors; at later stages both the Shah and Sir John Cadman participated. On April 29, 1933, a new concession was signed.

On May 26, Beneš reported to the Council that on May 1 the Persian Foreign Minister, M. Furughi, had informed the Secretary General that negotiations between his Government and the Company had resulted in the conclusion of a new agreement. It was at this juncture that the Persian representative, M. Sepahbodi, stated: "It is with great pleasure that I have the honor to inform the Council that the differences between my Government and the Anglo-Persian Oil Company have been definitely settled and that a new concession has been signed between the Persian Government and the authorized representative of the Company." [20] The Persian Government informed the Secretary General on June 5 that the Majlis had ratified the new concession. In his final report as Rapporteur, Dr. Beneš stated: "I am happy to say that the Council may take it that the dispute between His Majesty's Government in the United Kingdom and the Imperial Government of Persia which was brought before it last December is now finally settled." Both the representative of Persia and the representative of Great Britain expressed satisfaction in the new concession and praised the Rapporteur for his contribution in bringing the dispute to a successful solution. [21]

No decision, however, was reached on the issues raised — legal and otherwise — as to the Persian Government's right to cancel the D'Arcy concession, the validity of the Armitage-Smith agreement, the competence of the Permanent Court to deal with the case, and whether or not the dispute came under Article 15 of the Covenant.

THE NEW CONCESSION

In order to evaluate the new concession, the practical issues between the Company and the Persian Government must be considered. One can easily dismiss the arguments of the Persian representative that Armitage-Smith had overstepped his authority in signing the agreement, since (as was mentioned in the previous chapter) the letter from the Persian Finance Minister which authorized him to negotiate with the representative of the Company put practically no limitations on him. On the other hand, the British argument that the Armitage-Smith agreement was merely an interpretation and did not modify the D'Arcy concession does not seem to hold water. Nor can the argument that the Persians themselves, by accepting the royalty payments, never questioned the agreement, be upheld. To be sure, they accepted the royalties, for they had no alternative, particularly after the havoc wrought in their country during the First World War, although Persia was not a belligerent, and the determination of the British not to allow them to present their case before the Peace Conference. However, as early as 1926, perhaps even earlier, the Persians began to question the validity of the Armitage-Smith agreement.

The practical issue was the basis of calculating the 16 per cent net profit due the Government, or perhaps finding a different method than profits as a basis for royalty. There are two bases for calculating royalties: per ton production, and profit-sharing. In time of prosperity and high prices, the profit basis is preferable, for regardless of production policies, the Government would share in the profits; furthermore, in the case of the Anglo-Persian Oil Company, the Government would also share in the profits of some of the subsidiary companies. In time of depression and low prices, the per ton production basis is preferable, for even if production should be reduced, there would still be a solid guaranteed base of income, and for this the Government would be willing to forego the uncertain profits from the subsidiary companies. As indicated above (footnote 4), one of the stumbling blocks, if not *the* stumbling block, in the negotiations for a new concession in 1931 was the Persian Government's demand for an annual guarantee of £2,700,000. There can be no doubt that the immediate cause of the cancellation of the D'Arcy concession, the last of a long series of grievances, was the unusual shrinkage in the Government's share of the profits from petroleum for the year 1931. A method had therefore to be found which would guarantee the Government a minimum income on the basis of per ton production, and in addition, some share (considerably smaller, to be sure) in the profits.

The duration of the concession was an issue of primary importance for the Company. In view of the cancellation of the old concession and the elaboration of a new one, the Company wanted to extend the life of the concession, and this was indicated by Sir John Cadman as early as 1928. The Government, on its part, sought to reduce the area of the concession; it would then be able to grant concessions to others, and the exclusive right of transportation could no longer be granted to the Anglo-Persian Oil Company. The two remaining issues were the question of the price of oil and oil products consumed locally, which the Government wanted to be as low as possible, and the employment of Persians in the industry, which the Government wanted to be as high as possible.

It is in the light of these issues that the 1933 agreement must be viewed. It should also be remembered that while there was a considerable risk involved when D'Arcy signed his concession, in 1933 when the new agreement was signed the Anglo-Persian Oil Company entered into one of the surest and most profitable commercial ventures of modern times. [22]

OUTLINE OF THE CONCESSION

The agreement, written in French and dated April 29, 1933, consisted of thirty-seven articles. It was signed by S. H. Taqizadeh, for Persia; and John Cadman, Chairman, and W. Fraser, Deputy Chairman, for the Anglo-Persian Oil Company. The major provisions were:

1.   The Company was granted exclusive right, within the area of the concession, to search for and extract petroleum as well as refine or treat it in any other manner and render suitable for commerce the petroleum obtained by it; [23] and non-exclusive rights throughout Persia to transport petroleum, refine or treat it in any other manner and to render it suitable for commerce as well as to sell it in Persia and export it (Article 1). The Company was also to have the non-exclusive right to construct and own pipelines (Article 3).

2.   The area of the concession was to be reduced to 100,000 square miles, which the Company was to select within the area of the former D'Arcy concession not later than December 31, 1938 (Article 2).

3.   The right of and the conditions under which the Company could obtain land owned by the Government but not utilized, Government utilized, and private land necessary for the operation of the Company, were about the same as those in the D'Arcy concession, except that the procedure of obtaining the land was more precisely defined (Article 4).

4.   The Company was granted the right to effect without special license all imports necessary for the exclusive needs of its employees, on payment of the custom duties and other duties and taxes in force at the time of importation; the right to import, without special license, the equipment, material, medicinal and surgical instruments and pharmaceutical products necessary for its dispensaries and hospitals in Persia, exempt from any custom duties and any other duties and taxes in force at the time of importation, or payments of any nature whatever to the Persian state or to the local authorities. The Company was to have the right to import, without any license and exempt from any custom duties and from any taxes or payments of any nature whatever to the Persian state or to the local authorities,"anything necessary exclusively for the operations of the Company in Persia." The export of petroleum was to enjoy customs immunity and exemption from all central or local taxes (Article 6).

5.   The Company was to proceed forthwith with its operations in the

province of Kermanshah to produce and refine petroleum (Article 9).

    6.   The bases of payment to the Persian Government were:

        a.   Four shillings per ton petroleum either consumed in Persia or exported, beginning as of January 1, 1933.

        b.   A sum equal to 20 percent of the distribution to ordinary stockholders of the Anglo-Persian Oil Company in excess of £671,250 – whether the distribution was as dividends for any one year or whether it related to the reserves of the Company, exceeding the reserves of December 31, 1932.

           The annual payment of the above two items was guaranteed at a minimum of £750,000.

        c.   On the expiration of the concession or its surrender previous to expiry, the Company was to pay 20 per cent of the surplus difference between the amount of the general reserve of the Anglo-Persian Oil Company at the date of the expiration or surrender and on December 31, 1932, and 20 per cent of the surplus difference of the balance of the Company between those two dates.

        d.   The Government was to have the right to check the returns on the production of oil. Provision was made for adjustment in payment due to the fluctuation of the gold value of the pound sterling (Article 10).

    7.   The Company was to be completely exempt, for the first thirty years of its operations in Persia, from any taxation "present or future of the State and of local authorities," in consideration of the following payments:

        a.   During the first fifteen years of the concession, 9 pence per ton of the first six million tons of petroleum, and 6 pence per ton in excess of six million tons. The Company was to guarantee under this heading a minimum of £225,000.

        b.   During the fifteen years following, one shilling for each of the first six million tons, and 9 pence per ton in excess of six million tons. The Company was to guarantee under this heading a minimum annual income of £300,000 (Article 11).

    8.   The Government was to have a right to appoint a representative, to be designated as "Delegate of the Imperial Government," with the rights: to obtain from the Company all the information to which stockholders of the Company were entitled; to be present at all meetings of the Board of Directors, its committees, and all meetings of stockholders "which have been convened to consider any question arising out of the relations between" the Government and the Company; to request that special meetings of the Board of Directors be convened at any time to consider any proposal that the Government should submit to it. He was to be paid by the Company £2,000 a year (Article 15).

    9.   The Company was to recruit its artisans, as well as its technical and commercial staff, from among Persian nationals to the extent that it should "find any Persian persons who" possess the requisite competence and experience. The unskilled staff was to be composed exclusively of Persian nationals. Both parties were to study and prepare a general plan for yearly

progressive reduction of the non-Persian employees and their replacement "in the shortest possible time and progressively by Persian nationals." The Company was to make an annual grant of £10,000 to provide Persian students with professional education necessary for the oil industry in institutions in Great Britain (Article 16).

10. The price of petroleum and petroleum products for internal consumption was to be based on the price f.o.b. Rumania or the Gulf of Mexico, whichever was the lower; the Government "for its own needs and not for resale" was to pay 25 per cent less than the above price, other Persian customers 10 per cent less (Article 19).

11. Article 21 provided: "This concession shall not be annulled by the Government and the terms therein contained shall not be altered either by general or special legislation in the future, or by administrative measures or any other acts whatever of the executive authorities."

12. Differences between the parties of any nature whatever and especially differences in interpretation of the agreement were to be settled by arbitration. If the arbitrators — each nominated by the parties — were not able to name an umpire, the President of the Permanent Court of International Justice was to nominate him at the request of either of the parties. If one of the parties failed to name an arbitrator, the other was to apply to the President of the Permanent Court to name a sole arbitrator. There was to be no appeal against the award (Article 22).

13. The Company was to pay the Government £1,000,000 in settlement of all its claims, of any nature, except in regard to Persian taxation, in respect of the past until the date the new agreement entered into force. The payment to the Government for the years 1931 and 1932 was to be on the new basis and not on the basis of the D'Arcy concession (Article 23).

14. The period of the concession was to be from the time the agreement entered into force to December 31, 1993.

15. Article 26 stated: "Before the date of December 31, 1993, this concession can only come to an end in the case that the Company should surrender the concession or in the case that the Arbitration Court should declare the concession annulled as a consequence of default of the Company in the performance of the present agreement." [24]

EVALUATION OF THE CONCESSION

Some have suggested that the new agreement was a victory for Riza Shah,[25] others maintain that it was a victory for the Company.[26] A careful analysis of the terms of the old and new concessions reveals that neither could claim a victory. There can be no doubt that the area of the concession was considerably reduced; that the royalty payments calculated on the basis of per ton production, with minimum guarantees, would have been a considerable advantage to the Persian Government in times of depression. However, the reduction of the area of the concession cannot be considered a too serious disadvantage, for the Company had free choice of its 100,000 square miles, and it knew where the best areas of potential oil reserves were located. While the fixed per ton royalty payment may have guaranteed the Government a fixed amount in times of low prices, in times of prosperity the Government was entitled to only 20 per cent of the dividends distributed to ordinary stockholders over the amount of £671,250. The Company was also exempted from all taxation against an annual payment with minimum guarantees, and it was not obliged to surrender its foreign exchange. But the great advantage that it obtained was the extension of the concession for an additional thirty-two years. Thus it would seem that neither the Persians nor the Company emerged from the agreement with decided advantages one over the other, as compared with the D'Arcy concession. [27]

INTER-WAR PERIOD [28]

The years that followed the new agreement witnessed steady progress in oil production, refining capacity and royalty payments. In accordance with Article 9 of the new concession, the oil field of Naft-i-Shah in Kermanshah province in northwestern Iran was developed by the subsidiary, the Kermanshah Petroleum Company, Ltd. A pipeline was constructed to Kermanshah and a refinery erected which supplied the northern and northwestern parts of the country with petroleum and petroleum products.

According to the Company, the one major difference between itself and the Government during the period, until the outbreak of the war, was the definition of the word "ton" in the concession. The Company maintained that the English long ton of 2,240 pounds was meant, while the Government maintained that the ton of 2,000 pounds was meant. On July 30, 1936 the Company agreed to pay royalty on the basis of the ordinary ton, though it continued to maintain that it was the English ton that had been understood. In 1938 production dropped; the Government's revenue decreased, and discontent was expressed. However, next year the Second World War broke out and greater and more serious difficulties arose.

WORLD WAR II

In consolidating his position in Iran, Riza Khan had two immediate objectives: strengthening the power of the central government and increasing its control over the outlying provinces and tribes, [29] one of the major factors in Persian instability; and weakening the influence of outside powers, especially Russia and Great Britain, rivals but oftentimes collaborators. At the beginning Russia was no problem, as its puppet regimes in Gilan and Khorasan soon collapsed and the 1921 Perso-Russian agreement was ratified only after the complete withdrawal of the rebel troops. As regards Britain, the 1919 Anglo-Persian agreement was renounced, British military and financial advisers left Persia, and the Persian Government began to employ American financial advisers, though only in a private capacity. [30] The attempts of the Persian Government to attract American oil companies to northern Persia as a counterbalance to the British in southern Persia failed (as will be discussed in the next chapter).

Economically, therefore, Persia turned to Soviet Russia. However, with the emergence of Nazism in the early 'thirties, a doctrine which held a great attraction for the Persian ruler, himself a dictator, Riza Shah became more and more anti-Soviet. He turned for his economic needs to Germany. Not only did he see in German technical know-how, scientific knowledge and ideology advantages for his scheme of modernizing and secularizing his new Iran along the lines of a nationalist state, but also an excellent neutralizer against British influence and economic power in the country. Thus, when World War II broke out, Iran proclaimed her neutrality (September 4, 1939), as she had done during the First World War. The great number of Germans in Iran, employed by the Government and others, became a serious menace to the Allies, especially to the British oil installations in the south. However, so long as Germany did not attack Russia — in which case Iran would have become a base of operations — the Allies were willing to let Iran keep her neutrality.

In the meantime, relations between Iran and Great Britain began to deteriorate. In 1938, Iran's commercial treaty with Soviet Russia had lapsed and her trade with Germany was by sea rather than overland via Russia; this was seriously interfered with by the British naval blockade of Germany. Moreover, the rate of oil production had begun to decline, even before the war, and with the outbreak of hostilities, because of tanker shortages, it was further curtailed. Both of these points of friction, however, were soon removed: a new commercial treaty signed with Russia on March 25, 1940 greatly increased trade with Germany; and the Anglo-Iranian Oil Company agreed, in the summer of that year, to guarantee to the Iranian Government an annual income of £4,000,000 for the years 1940 and 1941, an indemnity of £1,500,000 covering the years 1938–1939. [31]

OCCUPATION OF IRAN

The German Army invaded Russia on June 22, 1941, and four days later the Soviet Government warned Teheran of the danger of the Germans to Iranian security. After Great Britain and Russia signed their Mutual Assistance Agreement on July 12, 1941, the importance of Iran to the Allies assumed both negative and positive aspects. The security of the Allied position in Iran required the ejection of the Germans — many of whom were no doubt spies — from the country. [32] Because of the uncertainty of the Arctic route of supply to Russia, Iran was to become the artery through which military supplies were to be funneled, and hence the country would have to come under Allied control. [33] On July 19, therefore, both Great Britain and Soviet Russia demanded of the Iranian Government that it deport all Germans who could not give satisfactory reasons for their presence in Iran. Riza Shah was not willing to comply with this request, and after strong notes had been delivered to the Iranian Government, British and Russian troops marched into Iran on August 25. [34] At first Riza Shah's army offered resistance; but after two days the cabinet resigned and the new Premier, Muhammad Ali Furughi, ordered the army to lay down its arms. British forces occupied the southern part of Iran, while the Russians occupied the northern provinces; on September 3 they made contact south of Kazvin. When the Iranian Government did not comply with the Allied request to close the legations of the Axis powers and their satellites, both armies marched on Teheran. [35] On September 17, Riza Shah notified the Majlis that he had abdicated, and that his son, Muhammad Riza Pahlevi, had succeeded him. [36] The realtions between the Allies and Iran were formalized in a tripartite treaty drafted by the Allies and submitted to the Majlis on December 21, 1941. [37] It was approved on January 26, 1942 and signed on the 29th.

The treaty was based on the principles of the Atlantic Charter, which had been announced by the British Prime Minister and the President of the United States on August 14, 1941, and to which the Shah announced his agreement, and the Soviet Government its adherence, on September 24, 1941. The treaty declared that Great Britain and Soviet Russia jointly and severally undertook to respect the territorial integrity, sovereignty and the political independence of Iran; the Allied powers jointly and severally undertook to defend Iran by all means at their command from all aggression on the part of Germany or any other power. The Shah, on his part, promised to cooperate in every way possible with the Allied powers with all the means at his command, so that the Allies could fulfill their undertaking. The assistance of Iranian forces, however, was limited to the maintenance of internal security. He also promised to secure the passage of troops or supplies from one Allied power to the other; he was to grant the unrestricted right to raise, maintain, guard and in case of military necessity control in any way they might require, all means of communications throughout Iran, including railways, roads, rivers, aerodromes, ports, pipelines and telegraph and radio installations; and to help in obtaining material and recruiting labor for maintaining and repairing the means of communication. In addition, he was to establish and maintain such measures of censorship control as the Allies might require.

The treaty further provided that the Allied powers might maintain in Iranian territory such land, sea and air forces as they considered necessary; it underlined that it was "understood that the presence of these forces on Iranian territory" did not constitute a military occupation and would "disturb as little as possible the administration and the security forces of Iran, the economic life of the country, normal movements of the population, and the application of Iranian laws and regulations."

A very important provision, which subsequently caused serious difficulties, was enunciated in Article 5: "The forces of the Allied powers shall be withdrawn from Iranian territory not later than six months after all hostilities between the Allied powers and Germany and her associates have been suspended by the conclusion of an armistice or armistices, or on the conclusion of peace between them, whichever is the earlier." The treaty declared that the Allied powers would use their best endeavors to safeguard the economic existence of the Iranian people against deprivation and difficulties arising as a result of the war. [38]

In the meantime, the collapse of Riza Shah's army and administration created not only administrative chaos in the outlying areas, which necessitated the taking

over of the actual administrative control by the British and the Russians in their
respective zones, but also economic hardships which aroused bitter resentment
against the occupying powers, with serious consequences both to the Allies and to
Iran.

One of the most outstanding achievements of Riza Shah's modernization of his
country was the completion, in 1938, of the Trans-Iranian Railway from Bandar
Shahpur on the Persian Gulf to Bandar Shah on the Caspian Sea. This became the
main line of supply for war material to Russia. The Allies not only increased the
locomotives and rolling stock, and improved the technical and managerial personnel
of the railway, but also built branch lines and improved the motor roads and port
facilities. In August 1942 it was decided that the United States Army should take
over the operation of the ports and railways and the bulk of the road haulage in the
country. The United States Persian Gulf Service Command was organized in
November of that year; it was composed of between 20,000 and 30,000 men.

From the very beginning the Russians practically sealed off the northern part
of Iran. While Russian officers could circulate freely in the British southern zone,
the northern provinces were not easily accessible to British officers or American
personnel. Moreover, despite the provisions of the tripartite treaty, the movement
of Iranian officials and trade between the northern provinces and the other parts of
the country were seriously slowed down and interfered with.

With the British and the Russians, both old hands at occupation, in control of
considerable portions of the country, and especially disturbed over the separatist
activities of the Russians in the northern provinces, the Iranian Government wanted
the United States to become a signatory to the tripartite treaty and thus guarantee
the execution of its provisions as regards Iran's territorial integrity and independence. The United States, however, refused, and in a telegram dated February 6,
1942 from President Roosevelt to the Shah, the United States merely took note of
the Anglo-Soviet assurances contained in the treaty. [39] When the United States
Persian Gulf Service Command was established in November 1942, the Iranian
Government again proposed that the United States become a party to the tripartite
treaty to formalize the status of American troops in Iran. Instead, the United States
proposed a separate agreement, and negotiations to that effect were initiated. By
December, however, they had broken down and the possibility of a separate treaty
was abandoned.

After sending General Patrick J. Hurley to the Near East and studying his
report, the President presented to the Big Three meeting in Teheran a special
declaration on Iran. [40] The Teheran Declaration, signed by Churchill, Roosevelt and
Stalin, stated that after consulting among themselves and with the Premier of Iran:
"The Governments of the United States, USSR and the United Kingdom recognize
the assistance which Iran has given in the prosecution of the war against the
common enemy, particularly by facilitating the transportation of supplies overseas
to the Soviet Union.

"The three Governments realize that the war has caused special economic
difficulties for Iran, and they are agreed that they will continue to make available

to the Government of Iran such economic assistance as may be possible, having regard to the heavy demands made upon them by their world-wide military operations and to the world-wide shortage of transport, raw materials and supply for civilian consumption.

"With respect to the post-war period, the Governments of the United States, USSR and the United Kingdom are in accord with the Government of Iran that any economic problems confronting Iran at the close of hostilities should receive full consideration, along with those of other members of the United Nations, by conferences or international agencies held or created to deal with international economic matters.

"The Governments of the United States, USSR and the United Kingdom are at one with the Government of Iran in their desire for the maintenance of the independence, sovereignty and territorial integrity of Iran. They count upon the participation of Iran, together with all other peace-loving nations, in the establishment of international peace, security and prosperity after the war, in accordance with the principles of the Atlantic Charter, to which all four Governments have continued to subscribe." [41]

In the meantime, the Tudeh party (at first under Communist influence and then domination) emerged as an important force in the life of the country; it was particularly active in Azerbaijan and among the Kurds, fostering and encouraging separatist tendencies and activities. Disturbances in those areas, combined with the general economic deprivation, brought about the fall of Furughi. He was succeeded, in July 1942, by Ahmed Ghavan es-Sultaneh. In response to the new Premier's request for American advisers, Dr. Arthur C. Millspaugh went to Iran as Director General of Finance. He arrived in Teheran on January 29, 1943 at the head of a sixty-man financial mission. [42] On March 19, 1944, Muhammad Saed Maraghai became Premier.

OIL CONCESSION HUNTING

The importance of oil to the conduct of the war and its decisive value to state power in time of peace had become more than ever evident to the Iranians. Moreover, the gigantic consumption of American oil and the consequent fear of depletion of American oil resources soon emerged as seriously disturbing elements in the relations between the British and the Americans, on the one hand, and the British, Americans and Russians on the other hand. The oil of the Middle East, as the great yet unexploited source, inevitably became the subject of discussion among the Big Three, and it was reported that at the Teheran conference of 1943 this subject, and especially the oil of Iran, had been an item on the agenda. [43]

In spite of the principles of the Atlantic Charter, and in spite of the spirit of cooperation between the Allies so highly advocated by all concerned, when it came to oil the dominating principle seemed to have been: everyone for himself. Thus, in the middle of 1943 a representative of the Anglo-Shell Oil Company arrived in Teheran to negotiate oil concessions in southeast Persia. [44] The American Socony-

Vacuum Oil Company thereupon made inquiries at the State Department whether the American Government would object to the Company's seeking concessions in Iran. On November 15, Secretary of State Cordell Hull replied that the United States favored the development of all possible outside sources of petroleum, because of the importance of the product "both for war purposes and for the long-range point of view," and that the Department therefore had no objections to negotiations between Socony-Vacuum and the Iranian Government. Hull added that the British already had extensive oil concessions in Iran, and since the United States had no agreement with Britain not to seek oil concessions there, he saw no reason why the Company should not go ahead with its plans, and he notified the American Legation in Iran accordingly. [45]

Early in 1944 the Socony-Vacuum Oil Company and the Sinclair Consolidated Oil Company sent representatives to Iran to obtain oil concessions. This apparently disturbed the Russians and at the end of February the Soviet Embassy in Teheran stated that the USSR had prior rights to the exploitation of oil in northern Iran. This was the first shot in the tense battle for oil concessions in Iran during 1944. According to Kirk, the negotiations between the British and the American companies and the Iranian Government got nowhere, for in April the Iranian Government engaged a firm of American oil consultants to draft a standard concession for all applicants, and at the end of July Herbert Hoover, Jr. and A. A. Curtice arrived in Teheran to advise the Government on the development of oil resources. [46] Hull, however, states that the negotiations of the American and British oil interests with the Iranian Government continued satisfactorily throughout the summer of 1944, until the arrival of the Soviet delegation in September. [47]

Early in September, the Soviet Ambassador in Teheran informed the Iranian Government that the USSR was sending a mission to Iran to discuss oil concessions, and in the middle of the month a delegation, headed by Deputy Foreign Commissar Sergei I. Kavtaradze, arrived at the capital. On the 22nd, Kavtaradze demanded from the Iranian Premier an oil concession in the five northern provinces. About a month later it was publicly announced that the Iranian Government had decided — on September 2, 1944 — not to grant new concessions to foreigners until after the war. A violent press campaign undertaken by the Soviet mission and abetted by the Tudeh party and its "sympathizers" put almost unbearable pressure on the Premier. In an interview on October 24, after attacking the integrity of the Premier, Kavtaradze argued that "if the Iranian Government chose to grant a concession to the Americans in southeast Iran, they should give the Soviets a concession in the north."

The American Government's reaction to the Iranian Government's decision to postpone all oil concession negotiations until after the war was that, although the American companies were disappointed, the Iranian Government was acting in good faith, and the American Government naturally expected that when and if the Iranian Government were ready to consider applications for oil concessions, those received from Americans would be given no less favorable consideration that those from the "government or nationals of any other country." Moreover, the American

Ambassador in Moscow, W. Averell Harriman, called the attention of the Russian Government to the Teheran Declaration on Iran and declared that the United States could not agree to any action constituting undue interference in internal Iranian affairs. The Soviets, however, were not to be deterred.

The press campaign, both in Iran and in Russia, led to mass demonstrations in Teheran and in other cities against Premier Saed's policy. Saed resigned on November 9, and on the 20th, Mortaza Ghali Bayat formed a new government. On December 2, 1944, the Majlis passed a bill that no premier or other minister of the Government might enter into negotiations for oil concessions with any foreign government or with any foreign oil company, or sign any concession or agreement relating to oil. The penalty for violation of this law was solitary confinement for periods ranging from three to eight years, with permanent dismissal from Government service. A week later the Kavtaradze mission returned to Moscow. [48]

Thus came to an end the first encounter between the Allies for oil exploitation in Iran during the war. The motive of the Americans in seeking concessions was obvious — to increase oil resources outside the United States. Moreover, as the main supplier of petroleum for the war machine, compared with the relatively small supply from the British Middle East oil, the United States considered that a concession in Iran would be merely a just compensation for the American contribution. The British, on their part, also sought to increase their oil supply, depending as they do for their oil on sources mostly outside the Empire. But what was the motivation of the Russians in seeking concessions in north Persia, especially after they had given up the concession in 1921?

Different explanations have been given. One was that they were as disturbed as the Americans by the depletion of their oil resources — which are limited — as a result of the war efforts, and were seeking additional sources to replenish them. Another was that they were not so much interested in increasing their oil supply as in stopping the Americans and the British from obtaining concessions in Iran and thus establishing themselves firmly along the Russian-Iranian border, which they always considered their sphere of influence. If the latter be the case, then the law against granting oil concessions to foreigners until after the war was in reality the primary objective of Soviet Russia, and as soon as this was accomplished, the Kavtaradze mission could withdraw and return to Moscow as victors. [49]

Whatever the motives of the Soviets, the British and the Americans, by encouraging their nationals to seek concessions in Iran without consulting either between themselves or with Russia, precipitated the first phase of a crisis which created tension, with serious consequences in the relations between Russia and Iran, as well as in the relations between Russia and its Western allies, and was to reappear in a much sharper clash immediately after the termination of the war. [50]

WITHDRAWAL OF ALLIED TROOPS

The question of the withdrawal of foreign troops from Iran became an issue between the Allies even before the date specified in the tripartite treaty. At the

Yalta conference (February 1945) British Foreign Secretary Anthony Eden and United States Secretary of State Cordell Hull proposed to Foreign Commissar V. Molotov that the Teheran declaration at least be reaffirmed. Molotov asked for time to consider the proposal; he refused to discuss it at the conference or to make any reference to Iran in the communiqué issued at the close of the conference. [51] On May 3, 1945, Ibrahim Hakimi succeeded Bayat as Premier of Iran, and began to press for the withdrawal of all foreign troops. The Americans and British were disposed to comply with the request, and consultations between the three Allied powers took place during the month of June. [52]

At the Potsdam conference (July—August 1945) it was decided that Allied troops be immediately withdrawn from Teheran, while the withdrawal from the rest of the country would be discussed at the Foreign Ministers' conference scheduled to meet in London in September. [53] According to the treaty, the date of evacuation — six months after the signing of the armistice with Japan — was March 2, 1946. In a letter to Commissar Molotov, Foreign Secretary Ernest Bevin proposed that the date be advanced to the middle of December, with the exception of the southern oil area (British), and Azerbaijan (Russian). Molotov, however, declined the proposal and refused to discuss the issue in the Foreign Ministers' Council. The Council therefore merely took note that both had agreed to withdraw completely on March 2, 1946. [54]

Iranian authority in the northern provinces was by this time considerably reduced, and the Tudeh party began to agitate for the establishment of provincial councils, especially in Azerbaijan, as provided in the constitution. The Democrat party — incorporating the Tudeh — was organized under the leadership of Jafar Pishevari, and on November 16, 1945 openly rebelled against the Teheran government. The latter sent troops against the rebels but they were stopped by the Russians at Kazvin. The Iranian Government appealed to the United States and Great Britain. The American Ambassador in Moscow handed a note to the Soviet Government on November 24 in which it was proposed that all foreign forces be completely withdrawn from Iran by January 1, 1946. [55] Two days later the British asked the Russians to withdraw their troops by March 2, 1946, as provided in the treaty. [56] In reply, the Soviet Government declared that it saw no reason for advancing the date of withdrawal. Moreover, it informed the Iranian Government that the dispatch of additional Iranian troops to northern Iran would increase the disorder and bloodshed there, and would necessitate sending more Russian troops to preserve order and ensure the security of the Soviet garrison. At the same time, the Russian Government officially informed Iran that it supported the demands of the Azerbaijan Assembly for autonomy. [57]

On December 15 the Iranian Foreign Minister addressed notes to the British, American and Russian Ambassadors in Teheran requesting that the question of the withdrawal of all foreign forces from his country be taken up at the forthcoming conference in Moscow of the Council of Ministers of Foreign Affairs. He declared that no military or non-military reason any longer existed which would justify the further stay in Iran of the three Allied powers, even for one day. [58] As if in reply, on the following day the establishment of the National Government of Persian Azerbaijan was officially announced.

When the Foreign Ministers' Council met on December 19, Premier Stalin intimated to Secretary of State Byrnes that the withdrawal of Russian troops from Iran by March 2 would depend on the conduct of the Iranian Government. He stated that the Baku oil fields were near the Iranian border and must be safeguarded against possible acts by Iran against the Soviet Union, and he recalled that the 1921 Irano-Russian treaty authorized the Soviet Government to send troops to Iran if there were a threat to Russian security. Byrnes, anxious to avoid having the Iranian case brought before the United Nations Security Council, as it would not only reflect on Allied disunity but perhaps put undue strain on the new and inexperienced international organization, pleaded with Stalin not to force the issue upon the United Nations, for the United States would have to support Iran against the USSR. [59] Bevin, on his part, proposed that an international commission of inquiry be sent to Azerbaijan to make recommendations on the autonomy issue. This was fully supported by Byrnes, but Molotov, although at first indicating willingness to consider it, ultimately rejected the proposal and declared that the Iranian issue was not properly on the agenda of the Council of Foreign Ministers and therefore could not be considered. [60]

BEFORE THE SECURITY COUNCIL – I

Having been rebuffed in Moscow, the Iranian Government decided to appeal to the United Nations. On January 19, 1946, at the first session of the UN General Assembly in London, H. Taqizadeh, head of the Iranian delegation, lodged a complaint with the Security Council that the Soviet Government was interfering in the internal affairs of Iran. [61] The Soviet reply, on the 24th, argued that the issue did not belong to the Security Council, for it could and should be settled by bilateral negotiations. [62] On the 26th, in a letter to the President of the Security Council, the Iranian delegate repeated the original complaint and asserted that the Soviet denials notwithstanding, the USSR had publicly admitted interfering with internal Iranian affairs when it had said – on November 26, 1945 – that it would not permit the passage of Iranian reinforcements to suppress the revolt in Azerbaijan. The letter declared that the Soviet action was a clear breach of the 1942 Tripartite Treaty of Alliance and the Declaration of Teheran. [63]

When the Council opened hearings on the issue, the Iranian representative asked that the Soviet authorities cease interfering in the internal affairs of Iran; that they cease preventing the free movement of Iranian forces and officials in and out of the territory in which Soviet forces were stationed; and that the Soviet Government give instructions for the complete evacuation of its troops from Iran on March 2, 1946, as provided in the treaty. [64]

The Russian representative, Andrei Vyshinsky, asked the Council to dismiss the case since the governments of the Soviet Union and Iran were negotiating a settlement and the Iranian Government had indicated satisfaction with the negotiations, hence the Security Council had no legal basis to take up the case. [65] The Iranian delegate pleaded with the Council not to dismiss the case. He declared that his country was willing to enter into direct negotiations, if it were the recommendation of the Council, and under Council supervision. This the Soviet delegate rejected.

The issue soon became clear: both parties were willing to negotiate directly, but the Soviet Union wanted to be free from any outside influence, while Iran, as a small power treating with its powerful northern neighbor, wanted to be fortified by the Security Council of the new international world organization.

The immediate problem was whether the issue should remain on the agenda of the Council or be removed. After a lengthy debate, the Council adopted a resolution taking note of the readiness of both parties to negotiate for a solution of the dispute, and requesting them to inform the Council of the results of their negotiations. At the same time the Council retained the right to request information at any time on the progress of the negotiations. [66] In this first tussle, neither Iran nor the Soviet Union came out victorious; the Council merely considered the readiness of both states to negotiate; it made no recommendations nor offers to supervise the negotiations: only in a limited sense did it keep the item on the agenda.

Meanwhile, the Hakimi cabinet fell and on January 26, 1946 the Majlis again elected Ahmed Ghavam es-Sultaneh as Premier. He was considered at the time friendly to Russia, and worked with the Tudeh party in spite of the fact that he was a rich landowner. In the middle of February, he left at the head of a delegation for Moscow to negotiate a settlement, for the question of withdrawal was becoming more and more tense. By January 1 all American troops were out of Iran; the British announced that their troops would leave by March 2, as provided in the treaty. Radio Moscow, however, stated on March 1 that Soviet troops would be withdrawn as of the 2nd from the northeastern province "where the situation is relatively quiet," and that "the Soviet forces in other parts of Iran will remain there pending clarification of the situation." [67]

BEFORE THE SECURITY COUNCIL – II

The United States informed the Soviet Government on March 6 that it could no longer remain indifferent to the decision of the Soviet Government not to comply with the provisions of the tripartite treaty, and about a week later it warned Russia that if the dispute were not settled by March 25 (the date the Security Council was to convene in New York) and if Iran would not herself do so, the United States would raise the issue with the Council. At the same time, Ghavam returned to Teheran and reported that his mission had been a failure. Five days later the Iranian Ambassador to Washington, Hussein Ala, asked the Security Council to put the Iranian issue on its agenda, on the bases of violation of the Tripartite Treaty of Alliance, the Teheran Declaration, and the Charter of the United Nations. [68]

The Soviet delegate, Andrei Gromyko, asked the Secretariat on the following day to postpone until April 10 the meeting of the Security Council which was to deal with the Iranian issue, on the ground that the Iranian request had not been expected by the Soviet Union since negotiations between itself and the Iranian Government were actually going on, and that the USSR needed additional time to prepare for discussion of the issue. [69]

On the 20th, the United States notified the Secretary General that when the Security Council met, its representative, Edward R. Stettinius, would move that Iran's request be put at the head of the agenda, and that Iran and the USSR be requested to report on the negotiations which had taken place between them "in accordance with the resolution of the Council adopted 30th January 1946." The same day Iran asked the Security Council not to grant the Soviet Union's request for postponement. It argued that the obligation of the Soviet Union to withdraw its forces from Iran was not a proper subject for negotiations under the Charter of the United Nations or the constitution of Iran. [70]

On March 26, 1946, the Security Council, meeting in New York, took up the issue of Iran for the second time. Unfortunately, during all the proceedings before the Council only procedural matters were debated; the substance of the case never reached the stage of open discussion, although of course it was always in the background of the procedural wrangling. The representative of the Soviet Union, Andrei Gromyko, argued that the case should not have been included on the agenda of the Council because an understanding between the Soviet Union and Iran had already been reached and because the Security Council resolution of January 30 had already produced positive results. The American delegate, James F. Byrnes, contended that the two parties had not reported agreement, as requested by the January 30 resolution; on the contrary, complaints had been lodged about the non-fulfillment of the 1942 treaty of alliance, and the Council must consider the issue. [71] The Council voted, 9–2, to place the Iranian case on its agenda. The Soviet delegate then attempted to obtain a postponement until April 10, as he had previously suggested; he warned that if the request were denied, his Government would not be able to participate in the discussions. The Council, however, rejected the Soviet proposal for postponement, and Gromyko left the Council chamber. The Iranian representative was subsequently asked to participate in the discussion as to whether or not to postpone the debate. [72]

There were some discrepancies, if not actual differences, between the statements made by the Iranian representative at the United Nations, Hussein Ala, as to the progress of the negotiations and the relations between the Iranian Government and the Soviet authorities, and the news reports from Teheran which supposedly originated with Premier Ghavam. These differences are perhaps made understandable by the fact that Premier Ghavam, under immediate pressure of the impact of the Soviet forces in Iran and the weight of Soviet Russia on Iran's borders, uttered statements which were less decisively critical of the Soviets, and in a sense even critical of the actions of his own representative in New York, a fact which the Soviet delegate utilized very effectively in his statements before the Council. Hussein Ala, on the other hand, not under immediate pressure of Soviet power and definitely encouraged by the Western powers to act freely, was not only critical of the Soviets in his complaint to the Security Council, but clearly implied that the Soviets had put pressure on the Premier in Iran. Thus, when he was called to the Council table, he categorically denied Soviet claims that there was already an agreement between his Government and the USSR on the withdrawal of Soviet troops. He declared that the basic Iranian requests were the withdrawal of Soviet troops, and non-interference by the USSR in internal Iranian affairs. The Soviets' demand of Premier Ghavam while he was in Moscow in February were that USSR troops remain indefinitely in some parts of Iran; that Iran recognize the internal autonomy of Azerbaijan; and that Iran either grant the USSR an oil concession, or that the Soviet Government have control of a joint Iranian-Russian company for the exploitation of Iranian oil. The negotiations in Moscow had broken down, with no positive results achieved. He concluded with the statement that until definite assurances had been given to the Council that the complete evacuation of Russian troops from Iran would take place within a brief and fixed period of time, and that such evacuation would not be conditional on any foreseen or unforeseen circumstances or agreements, Iran believed that postponement would not be in accord with the principles of the UN Charter. [73]

It would seem that by rejecting the Soviet request to postpone consideration of the issue, and then listening to the arguments of the Iranian representative against postponement while the Soviet representative was absent, the Council was rapidly reaching an impasse, for no one was prepared to take action against the Soviet Union. Secretary Byrnes saved the situation by proposing that the Secretary General obtain reports from both Governments as to the existing status of the negotiations between them, ascertain from their representatives whether or not the reported withdrawal of troops was conditional upon the conclusion of agreements between the two Governments or other subjects, and submit a report to the Council on April 3. For this escape avenue various members of the Council expressed gratitude to the American delegate and gladly voted for the proposal. [74]

On April 3, Andrei Gromyko reported that the negotiations had already led to an understanding regarding the withdrawal of Soviet troops from Iran and that this would be completed within one-and-a-half months. As for the other issues, they were in no way connected with the withdrawal of troops. Gromyko stated that it was well known that the question of an oil concession or of a mixed joint stock company had been raised in 1944 independently of the question of the evacuation of USSR troops.

In contrast, Hussein Ala's letter to the Secretary General asserted that negotiations pursuant to the resolution of January 30, 1946 had achieved no positive results; that Russian agents, officials and armed forces were continuing to interfere in the internal affairs of his country; and as regards the conditions for withdrawal, Ala declared that on March 24 the Soviet Ambassador in Teheran had called on the Premier and handed him three memoranda: 1) on evacuation of Soviet troops, which was beginning that day (March 24) and would continue for six weeks; 2) on the formation of a joint Iranian-Russian corporation for the extraction of oil; 3) on the organization of the local government of Azerbaijan. The Soviet Ambassador later informed the Iranian Premier that his instructions from Moscow as regards the withdrawal of the troops were that it would proceed only "on the condition that no unforeseen circumstances should occur." Three days later the Soviet Ambassador told the Premier that if agreement could be reached on the other two subjects, there would be no further cause for anxiety and no unforeseen circumstances would take place. [75]

The Council was again coming to an impasse, and again Byrnes came to the rescue. He asked the Iranian representative what he would suggest the Council should do. Ala replied that if the condition of "unforeseen circumstances" were withdrawn, and if the Soviet Government gave the Council assurances that the unconditional withdrawal of all its forces would be completed by May 6, Iran would be willing not to press further at this time for consideration of the issue by the Council, provided of course that it remain on the agenda for consideration at any future time. Subsequently, Byrnes introduced a resolution along those lines, deferring until May 6 consideration by the Council of the Iranian issue, at which time both Governments were to report whether the withdrawal of USSR troops from the whole of Iran had been completed. [76] The Council adopted the resolution by nine votes. [77]

The Soviet Union, however, did not wait till May 6; on April 6 it asked the Council to remove the issue from its agenda. [78] Three days later, the Iranian delegate asked that the issue be retained. [79] On April 15, the Iranian representative told the Council that his Government had the previous day advised him that as a result of an agreement between Iran and the USSR the Red Army would evacuate all of Iran by May 6, 1946, and had instructed him to inform the Council that while it had no doubt this agreement would be carried out, it did not have the right to fix the course the Security Council should take. However, the following day he received a telegram which stated: "It is necessary that you immediately inform the Security Council that the Iranian Government has complete confidence in the word and pledge of the USSR Government and for this reason withdraws its complaint from the Security Council." [80]

A new procedural wrangle developed. The Soviet delegate maintained that the withdrawal of the complaint by Iran automatically removed the item from the agenda of the Council, while the United States maintained that it remained on the agenda at least until May 6. The Council soon plunged into a legal debate on this point; a memorandum from the Secretary General and a report by a committee of experts of the Council could not bring about any agreement. [81] The Soviet proposal to have the issue removed failed, and Gromyko notified the Council that the Soviet delegation would not participate in the Council sessions which would deal with the Iranian issue. [82] Finally, on May 8, the Council took up the Iranian issue again. It had before it a letter of May 6 from the Iranian delegate that USSR troops had been completely evacuated from four of the northern provinces; as regards Azerbaijan, the Iranian Government had been informed that the evacuation there would be completed by May 14. [83] These reports had not been verified by direct observation or otherwise of the Iranian Government. The USSR did not submit a report, and the Council decided to put off consideration of the matter until May 20. [84]

On May 22 the Council had before it two letters, dated May 20 and 21. [85] The first declared that the conditions laid down by the Security Council had not been fulfilled by the Soviet Government. The one of May 21 declared that the Russian troops had been fully evacuated. The note left so much to be desired as to clarity and decisiveness that the United States delegate, Edward R. Stettinius, was not willing to accept it as final and proposed to postpone consideration of the Iranian issue for a while. [86]

With this came to an end consideration of the issue by the Security Council. With its potentialities for raising basic principles involved in the realtions between nations, and the functions and responsibilities of the Security Council, this case could have disrupted the new, yet untried, international organization. But as it turned out, the issue was not allowed to reach the action stage. [87]

In the meantime, on April 4, 1946, a joint communiqué by Premier Ghavam and the Soviet Ambassador in Teheran, Ivan V. Sadchikov, declared that agreement had been reached by the two Governments on the following bases: Iranian territory to be evacuated completely within six weeks of March 24; Azerbaijan to be recognized as an internal problem with which the Iranian Government was to deal benevolently and with consideration for the need for reforms under the existing laws. A joint Irano-Russian oil company, for a period of fifty years, was to be organized. For the first twenty-five years Russia was to own 51 per cent of the shares and Iran 49 per cent; for the second twenty-five years each was to own 50 per cent; profits were to be divided according to the shares held. The Government was to submit a bill on the organization of this company to the Iranian Parliament within seven months from March 24. [88]

Negotiations between the Teheran government and the Azerbaijan Democrat Committee broke down and the Democrat leaders prepared for war against Teheran, but on June 13 an agreement with the Azerbaijanis was concluded. Meanwhile, the evacuation of the Soviet troops had been completed. It was not, however, until December 12, 1947, when the Democrat regime resisted the Teheran forces which had come to supervise the provincial elections that the Democrat forces completely collapsed; the Teheran government gained full control of Azerbaijan, and Jafar Pishevari fled to Soviet Russia. Thus, after a long and arduous struggle in which the United States played a not unimportant supporting role, Iran succeeded in clearing her territory of all foreign troops, but not without paying a price – an oil concession to the Soviet Union.

In the second phase of the oil battle, Soviet Russia emerged the victor, at least at the time, although it had to back down in the issue of evacuation of Azerbaijan. Had the oil project materialized, the USSR would have held 51 per cent of the shares in the company, and at the same time would have prevented its Western allies from obtaining oil concessions which would have established them firmly in northern Iran.

EVALUATION

In the years following the 1933 concession, relations between the Iranian Government and the Anglo-Iranian Oil Company were tolerably good. The new bases for royalty payments, plus increased production, supplied Iran with a considerable source of revenue which was badly needed by Riza Shah for his scheme of modernizing the country, as well as for expanding and modernizing the army. However, the importance of oil to war efforts could not have escaped the Iranians, who had experienced a second, if not a third, occupation of their country in forty years by allied yet rival powers because of oil, primarily, in one form or another. They also could not have failed to realize that rivalry for oil was far greater than the force for unity between the allies, even in the midst of war, and that at least one of the reasons behind the Soviet Union's tactics in Azerbaijan was a desire for oil concessions. At the same time, seeing the enormous quantities of oil produced during the latter part of the war, at least some Iranians began fully to comprehend the great potentialities of their oil resources for the welfare of their country. Fully aware of the antagonism generated by the Russians on the one side, and the United States and Great Britain on the other side, just because of oil concessions, the Iranians probably had no intention of ratifying the agreement signed between the Ghavam government and the Soviet Government on April 4. For in spite of the provision in the agreement that the proposal be submitted to the Majlis within seven months from March 24, the Government indicated no intention whatsoever of doing so, and on September 15, 1947, the Soviet Ambassador handed Ghavam a very strong note on the matter. [89] At the same time the Iranians, stirred up by the Tudeh party, began to realize the inadequacy of the royalties from the Anglo-Iranian Oil Company. While the profits of the Company were becoming fantastically high, Iran was experiencing, in addition to her traditional poverty, many hardships and difficulties arising from the war. [90] Serious labor troubles broke out throughout the Anglo-Iranian Oil Company's fields and installations. [91] To meet both situations — the demand of the Russian Government for the oil concession, and the desire for increased royalties from the Anglo-Iranian Oil Company — the Majlis adopted the following one-article law on October 22, 1947:

(a) In view of the fact that the Prime Minister, acting in good faith and upon his inference from the provisions of Article 2 of the law of 2nd December 1944, entered into negotiations and drew up an agreement under the date of 4th April 1946, concerning the creation of a mixed Irano-Soviet Oil Company, and whereas the Iranian Majlis does not deem the said inference to be consistent with the true purport and intent of the above-mentioned law, it therefore considers the said negotiations and agreement as null and void.

(b) The Government is required to make arrangements for a technical and scientific research to be made for the exploitation of petroleum mines and to draw up and prepare within a period of five years full technical and scientific plans of the oil-bearing zones of the country, whereafter the Majlis may, with full knowledge that oil exists in sufficient quantities, arrange for the commercial exploitation of these national resources through the enactment of the necessary laws.

(c) The grant of any concession for the exploitation of oil and its derivatives in the country to foreigners and the creation of any kind of company for this purpose in which foreigners may have a share in any way whatsoever is absolutely forbidden.

(d) If, after the technical investigations mentioned in paragraph (b) above, the existence of oil in commercial quantities in the northern areas of Iran is proved, the Government is hereby authorized to enter into negotiations with the USSR for the sale of oil products, informing the Majlis of the result.

(e) In all cases where the rights of the Iranian nation, in respect of the country's natural resources, whether underground or otherwise, have been impaired, particularly in regard to the southern oil, the Government is required to enter into such negotiations and take such measures as are necessary to regain the national rights and inform the Majlis of the result. [92]

This law was not only the death blow to the Soviet claim, but also the background to the negotiations with the Anglo-Iranian Oil Comapny which led to the supplemental agreement.

It would seem the Iranians had advanced to a new stage in their development. The country's basic economic needs and the seven-year plan prepared by American engineering firms for the development of Iran with the oil royalties as an important item of revenue, [93] together with the difficulties that the Government was encountering in obtaining the financial and economic aid it had expected from the United States, all combined to set the stage for the next phase: nationalization.

NOTES

1. League of Nations, *Official Journal,* 14th year, 1933, 298—299. From this letter it would not appear that the Persian Government questioned the validity of the Armitage-Smith agreement; the extent of the differences to which the Persian Oil Commissioner took exception would seem to have been the amounts calculated by the Company as the 16 per cent net profit.
2. *Ibid.*
3. *Pleadings,* 191. As a matter of fact, in reply to a question in the House of Commons whether any revision of the agreement between the Persian Government and the Company was contemplated, Winston Churchill, then Chancellor of the Exchequer, said that he understood that no such question had arisen. *PDC,* 211, col. 1574, Dec. 8, 1927.
4. *Pleadings,* 191. The excessive demand, according to Sir John Simon, was a minimum annual guaranteed income of £2,700,000. League of Nations, *Official Journal,* 14th Year, 1933, 202.
5. League of Nations, *Official Journal,* 14th Year, 1933, 289—295; *Pleadings,* 190—192.

6. League of Nations, *Official Journal*, 13th Year, 1932, 2301; *Pleadings*, 235–236.

7. League of Nations, *Official Journal*, 13th Year, 1932, 2301; *Pleadings*, 236–237.

8. League of Nations, *Official Journal*, 13th Year, 1932, 2302; *Pleadings*, 237.

9. League of Nations, *Official Journal*, 1932, 2302; *PDC*, 272, cols. 1790–1791, Dec. 8, 1932; *Pleadings*, 237.

10. League of Nations, *Official Journal*, 1932, 2302; *PDC*, 272, col. 1791, Dec. 8, 1932; *Pleadings*, 238.

11. League of Nations, *Official Journal*, 1932, 2303; *Pleadings*, 238–239; *PDC*, 272, cols. 1791–1792, Dec. 8, 1932. In reply to a question whether this note meant armed measures against Persia, Anthony Eden, Undersecretary of State for Foreign Affairs, replied: "I should have thought that the position is quite clear. We hold the Persian Government responsible for protecting the rights of the British Company." *PDC*, 272, col. 1793, Dec. 9, 1932. However, eleven days later the First Lord of the Admiralty categorically stated, in reply to an inquiry: "No war vessels have been despatched to Persian waters for this purpose." *PDC*, 273, col. 740, Dec. 19, 1932. But British warships did appear in the Persian Gulf, and the Persians could not miss the implications. See Alan W. Ford, *The Anglo-Iranian Oil Dispute of 1951–1952* (Berkeley, 1954), 17.

12. League of Nations, *Official Journal*, 1933, 300–303; *Pleadings*, 239–244.

13. League of Nations, *Official Journal*, 1932, 2296–2297; *Pleadings*, 244.

14. League of Nations, *Official Journal*, 1933, 288.

15. League of Nations, *Official Journal*, 1932, 2298–2305.

16. League of Nations, *Official Journal*, 14th Year, 1933, 197–204, 208–210.

17. League of Nations, *Official Journal*, 14th Year, 1933, 204–208, 210–211, 289, 295.

While the argument advanced in 1952 by the representative of the Iranian Government — that the cancellation of the D'Arcy concession had been contrived by the Company in order to obtain a better concession, and only after the cancellation was announced did the British Government raise an alarm, as if it had come suddenly and unexpectedly, and brought the case before the Council of the League, only to be persuaded later to negotiate directly with the Persian Government — cannot be substantiated. There was, however, a decided difference in attitude between the Persian and British representatives arguing the case before the Council. The Persian representative, M. Davar, in dead earnest, labored the issue of the granting of the D'Arcy concession and the nature of the Armitage-Smith agreement. The British representative displayed a certain levity in his attitude. He ridiculed the preciseness of the amount, calculated down to shillings, that would have been due to the Persian Government for custom duty. He dismissed many issues as ancient history and emphasized that it was necessary to look towards a future solution. One could discern a certain confidence in Sir John Simon's manner that a new concession would be worked out. Though at the time Persia was more anxious for a new concession than was Britain, her representative displayed no such feeling of assurance.

That the Company was desirous of revising the concession is evidenced from the statement of the Chairman at the twenty-first annual meeting held on July 11, 1933: "It should not be imagined for one moment that the desire for a new and more modern form of agreement was felt only by the Persian Government." Anglo-Persian Oil Co., *Twenty-Fourth Ordinary General Meeting 1933*, 2.

18. League of Nations, *Official Journal*, 14th Year, 1933, 252–253.
19. See Dr. Beneš' letters to the Persian representatives at Geneva, *Pleadings*, 245–246.
20. League of Nations, *Official Journal*, 1933, 827–828.
21. *Ibid.*, 996, 1606.
22. For a discussion of the problems and difficulties during the period 1919–1933, see Nakhai, *Le Pétrole en Iran*, 49–71; Azami, *Le Pétrole en Perse*, 111–128; "Iran Presents Its Case for Nationalization," *The Oil Forum*, VI, Mar., 1952 (Persian point of view); and "Anglo-Iranian Answers Iran with Facts," *The Oil Forum*, VI, vi–vii, Apr., 1952; *Pleadings*, 190–195 (Company point of view).
23. Petroleum was defined as "crude oil, natural gases, asphalt, ozokerite, as well as all products obtained either from these substances or by mixing these substances with other substances."
24. League of Nations, *Official Journal*, 1933, 1153–1160; *Pleadings*, 247–270. The Majlis ratified the agreement on May 28, 1933 and the Royal assent was given on the following day. In identical letters dated Aug. 17, 1933, the Persian Government and the British Government registered the agreement with the Permanent Court for the purpose of the action of the President of the Court, as provided in the agreement.
25. Ford, *The Anglo-Iranian Oil Company Dispute*, 19; Arthur C. Millspaugh, *Americans in Persia* (Washington, 1946), 28.
26. "Iran Presents Its Case for Nationalization," *The Oil Forum*, VI, 83–85, Mar., 1952.
27. See the evaluation in *Survey of International Affairs 1934*, 243–245.
28. On Dec. 25, 1934, the Persian Ministry for Foreign Affairs addressed a circular memorandum to the foreign diplomatic missions in Teheran requesting that beginning Mar. 1, 1935 the terms Iran and Iranian be used in official correspondence, instead of Persia and Persian. *PDC*, 298, col. 351, Feb. 20, 1935. There can be no doubt that the desire of the Persian Government to change the name to Iran was motivated not only by the wish to have a name more inclusive of the country than a single province (Fars, from which the name Persia was derived), but also to identify the Iranians as the racial Aryans of Nazi racial ideology. This of course caused some opposition to the new name in some liberal circles in England, but eventually the name Iran was officially used, and the name of the Anglo-Persian Oil Co. was changed to Anglo-Iranian Oil Co. When Riza abdicated and his son succeeded him and restored constitutional government, the use of Persia and Persian was restored. *PDC*, 377, col. 1818, Feb. 18, 1942. After nationalization the British reverted to Persia, which irritated the Iranians. See Ford, *The Anglo-Iranian Oil Company Dispute*, 54–55.
29. The specific case of the Shaikh of Muhammarah, Khazaal, not only affected the relations between the semi-independent Shaikh and Riza Shah, but also the

relations between the Shaikh and Great Britain, and the latter's efforts with the Teheran authorities on behalf of the Shaikh. The undertaking given to the Shaikh by the British Government in October 1909, as mentioned above (Chap. II), to protect him and his heirs against encroachment by the Persian Government on his jurisdiction and recognized rights, was confirmed in 1914. Until 1924 Khazaal was the *de facto* ruler of the left bank of the Shatt-al-Arab and the navigable section of the Karun River from Muhammarah to Ahwaz. That year Khazaal, fearing that Riza Khan was about to cancel all the rights which he held from former Shahs, publicly denounced Riza and his policy, although he had previously reached an agreement with him. This gave Riza the opportunity he had been looking for, and he opened military operations against the Shaikh. Khazaal capitulated and in April 1925 he was removed to Teheran, while Riza's forces pacified the area and brought it under strict control of the central government. Though the Shaikh was treated well, he was never permitted to return to Muhammarah and regain even part of his former rights; he died in Teheran a few years later. Arnold J. Toynbee, *Survey of International Affairs 1925,* I. *The Islamic World* (London, 1927), 539–543; Sykes, *A History,* II, 547; Groseclose, *Introduction,* 123; *PDC,* 183, col. 581, May 4, 1925. Though the British were uncomfortable about this obligation to the Shaikh, they were determined to make terms with the new ruler of Persia. In reply to a question in the House of Commons by Anthony Eden whether any undertaking as to the Shaikh's personal security had been given to him by the British Government, Austen Chamberlain replied that no such undertaking had been given "nor was any required, since the settlement of differences between the Shaikh and the Persian Prime Minister was agreed upon in December 1924." *PDC,* 183, col. 581, May 4, 1925. Two years later, however, Chamberlain stated that the British Government had repeatedly urged the Persian Government to put into practice the assurances that the Shaikh's affairs would be settled on a just and equitable basis. *PDC,* 209, cols. 381–382, July 20, 1927. He subsequently declared that Khazaal's nationality was Persian and that the interest of the British Government in him was merely "based on long-standing friendship." *PDC,* 213, col. 1572, Feb. 22, 1928. No obligations, no assurances, no promises. A diametrically opposite position was taken by Lord Lamington, *PDL,* 61, cols. 353–354, May 19, 1925. Writing about the year 1925, Toynbee remarked: "Evidently the British Government did not consider that its assurance to Shaykh Khaz'al had become operative in the circumstances." *International Survey 1925,* I, 542. Yet in 1927 Toynbee reported that it was rumored that the British Government had presented a note to the Persian Government asking for the Shaikh's rehabilitation. *Survey of International Affairs 1928* (London, 1929), 353.

30. The experiences of the American Administrator General of the Finances of Persia (1922 to 1927) are described in A. C. Millspaugh, *The American Task in Persia* (New York, 1925); – *Americans in Persia,* 22–26 et passim. Millspaugh's mission was the second American group to go to Persia; the first was headed by Morgan Shuster, whose experiences are reported in his book, *The Strangling of Persia,* mentioned above.

31. George Kirk, *Survey of International Affairs, 1939–1946. The Middle East in*

The War (London, 1952), 130—131 (henceforth cited *Survey 1939—1946. The Middle East*); Anglo-Iranian Oil Co., *Report and Balance Sheet at 31st December 1942*, [3].

32. Some of the German agents worked with prominent Iranians against the Allies even after the fall of Riza Shah. During 1942 a German spy, Mayr, went to Isfahan and worked out plans with Gen. Fazlollah Zehedi (who was to become Premier in 1953) for the cooperation of the southern tribes and for revolt when German troops arrived at Iran's borders. Kirk, *Survey 1939—1946. The Middle East*, 157.

33. Lord Moyne stated in the House of Lords that the advantage of the route of supply to Russia by way of Iran was difficult to overestimate. *PDL*, 120, col. 9, Sept. 9, 1941.

34. The full texts of the British and Soviet notes of Aug. 25 are in Leland M. Goodrich (ed.), *Documents on American Foreign Relations*, IV, *July 1941— June 1942* (Boston, World Peace Foundation, 1942), 674—681.

35. For the role the US played in the hectic events of the three weeks of Aug. 25—Sept. 17, and the efforts of Riza Shah with President Roosevelt, see Cordell Hull, *Memoirs* (New York, 1948), II, 1501—1502. The British explanation for the drastic measures taken by the Allies is given by Prime Minister Churchill, *PDC*, 374, cols. 79—80, Sept. 9, 1941; and by Lord Moyne in *PDL*, 120, col. 9, Sept. 9, 1941.

36. Riza Shah was taken to Mauritius and from there he went to South Africa — he was not permitted to go to Karachi — where he died on July 26, 1944. Sir Reader Bullard, who was British Minister at the time and who subsequently signed the tripartite treaty of alliance, declared emphatically: "The common belief that the Allies called upon the Shah to abdicate is unfounded. The Shah was probably aware of his unpopularity among his own people," and therefore resigned. *Britain and the Middle East from the Earliest Times to 1950* (London, 1951), 134. Sir Reader also criticized J. C. Hurewitz for stating in his book, *Middle East Dilemmas*, that Riza Shah was forced to abdicate, *International Affairs*, XXX, 252, Apr., 1954. Against this Arthur C. Millspaugh, who was the head of a financial mission to Iran, stated: "The Soviet Ambassador and the British Minister demanded the Shah's abdication. Riza necessarily agreed." *Americans in Persia* (Washington, 1946), 39. And Prime Minister Winston Churchill declared in the House of Commons: "We have chased a dictator into exile, and installed a constitutional Sovereign pledged to a whole catalogue of long-delayed, seriously-needed reforms and reparations." *PDC*, cols. 518—519, Sept. 30, 1941.

37. *PDC*, 376, col. 1503, Dec. 10, 1941; *PDC*, 377, cols. 1155—1156, Feb. 4, 1942.

38. *Treaty of Alliance between the United Kingdom and the Soviet Union and Iran, Teheran, January 29, 1942*. Cmd. 6335, 1942; *Department of State Bulletin*, VI, 249—252, Mar. 21, 1942; United Nations, Security Council, *Official Records*, First Year: First Series, Supplement No. 1, 43—48 (henceforth cited UNSC, *Official Records*, Supplement 1).

39. Hull, *Memoirs*, II, 1501—1502.

40. *Ibid.*, 1498—1510.

41. *Department of State Bulletin,* IX, 409–410, Dec. 11, 1943; UNSC, *Official Records,* Supplement 1, 49–50.

42. For the complete story of the mission, its functions, activities, difficulties and final dismissal (from the point of view of its head), see Arthur C. Millspaugh, *Americans in Persia.*

43. Kirk, *Survey 1939–1946. The Middle East,* 474.

44. By the end of 1938 the Anglo-Iranian Oil Co. had selected its area, limited by the 1933 concession, and the other companies were vying for concessions in the rest of the area originally covered by the AIOC concession.

45. Hull, *Memoirs,* II, 1509; Kirk, *Survey 1939–1946. The Middle East,* 474. The prime motivation, according to Hull, was the alarming statistics on oil depletion in the US, which had come to the attention of the Army and the Navy.

46. Kirk, *Survey 1939–1946. The Middle East,* 474; George Lenczowski, *Russia and the West in Iran 1918–1948. A Study in Big-Power Rivalry* (Ithaca, 1949), 216.

47. Hull, *Memoirs,* II, 1510. The impression one gathers from Hull's statement is that American and British interests were working as a team. This is not borne out in any other source. See the pertinent remark of Millspaugh, *Americans in Persia,* 233.

48. Kirk, *Survey 1939–1946. The Middle East,* 475–480; Hull, *Memoirs,* II, 1508–1510; Millspaugh, *Americans in Persia,* 187–190; Lenczowski, *Russia and the West,* 218–223; the full text of the law in *Report on Seven Year Development Plan for the Plan Organization of the Imperial Government of Iran* (New York, 1949), IV, 242.

49. Lenczowski, *Russia and the West,* 217–218; Kirk, *Survey 1939–1946. The Middle East,* 480–481; Millspaugh believes that the USSR had both objectives: to increase its oil resources, and to prevent the Western powers from obtaining concessions.

50. Kirk presents a different motivation for Allied action, *Survey 1939–1946. The Middle East,* 480–481. But his explanation is *ex post facto.*

51. Robert E. Sherwood, *Roosevelt and Hopkins. An Intimate History* (New York, 1948), 865.

52. *PDC,* 411, cols. 858–859, June 6, 1945.

53. *Protocol of the Proceedings of the Berlin Conference. Berlin 2nd August, 1945,* Cmd. 7087, 1945; *PDC,* 413, col. 297, Aug. 20, 1945.

54. *PDC,* 414, cols. 245–247, Oct. 10, 1945. The full text of the correspondence is given there.

55. Full text of the note in UNSC, *Official Records,* Supplement 1, 53–55.

56. George Kirk, *Survey of International Affairs. The Middle East 1945–1950* (London, 1954), 57–60 (henceforth cited *Survey. The Middle East 1945–1950*); Millspaugh, *Americans in Persia,* 194; James F. Byrnes, *Speaking Frankly* (New York, 1947), 118.

57. UNSC, *Official Records,* Supplement 1, 57–58, full text of the note.

58. UNSC, *Official Records,* Supplement 1, 60, full text of the note.

59. Byrnes, *Speaking Frankly,* 118–120.

60. *PDC,* 419, cols. 1357–1358, Feb. 21, 1946; Byrnes, *Speaking Frankly,* 121 *et*

passim; Kirk, *Survey. The Middle East 1945–1950,* 63.
61. UNSC, *Official Records,* Supplement 1, 16–17. Byrnes reported that the Iranian representative had come to see him and asked whether he should file the complaint with the Security Council on two counts: the presence of Soviet troops; and Soviet interference in the internal affairs of his country. Byrnes hesitated but promised to listen to his case and advise him the following day. However, the Iranian delegate did not wait and filed the complaint. Byrnes, *Speaking Frankly,* 123.
62. UNSC, *Official Records,* Supplement 1, 17–19.
63. *Ibid.,* 19–24.
64. United Nations Security Council, *Official Records,* First Year: First Series, No. 1, 32–38 (henceforth cited UNSC, *Official Records,* No. 1).
65. *Ibid.,* 46–49.
66. *Ibid.,* 54–71.
67. Kirk, *Survey. The Middle East 1945–1950,* 66.
68. United Nations Security Council, *Official Records,* First Year: First Series, Supplement 2, 43–45 (henceforth cited UNSC, *Official Records,* Supplement 2).
69. *Ibid.,* 44.
70. *Ibid.,* 44–45.
71. United Nations Security Council, *Official Records,* First Year: First Series, No. 2, 11–14 (henceforth cited UNSC, *Official Records,* 2). Sec. Byrnes declared that the battle was fought over a subsidiary issue — Iran's right to present her case to the Security Council. He felt that if a precedent were established that denied any country the right of speedy access to the Security Council, the United Nations would be crippled from birth. And he added: "I felt so strongly, in fact, about both the issue of Iranian sovereignty and the issue of ready access to the Security Council that I personally argued the American case before the Security Council." Byrnes, *Speaking Frankly,* 304.
72. UNSC, *Official Records,* 2, 21–62.
73. *Ibid.,* 62–72.
74. *Ibid.,* 75–82.
75. *Ibid.,* 85–86.
76. See a very interesting editorial, "United Face Saving," *The New Statesman and Nation,* 257, Apr. 13, 1946.
77. UNSC, *Official Records,* No. 2, 87–97.
78. UNSC, *Official Records,* Supplement 2, 46–47.
79. *Ibid.,* 47.
80. UNSC, *Official Records,* No. 2, 123.
81. For the full text of the committee report, see UNSC, *Official Records,* Supplement 2, 47–50.
82. UNSC, *Official Records,* No. 2, 145–152, 200–214.
83. UNSC, *Official Records,* Supplement 2, 50–51.
84. UNSC, *Official Records,* No. 2, 246–252.
85. UNSC, *Official Records,* Supplement 2, 52–54.
86. UNSC, *Official Records,* No. 2, 285–305.
87. Remarked Sec. Byrnes: "The Security Council never took action in a formal sense but it was the forum which made it possible for Iran to appeal to the

conscience of the world." Byrnes, *Speaking Frankly,* 304.

88. Kirk, *Survey. The Middle East 1945–1950;* 70–77; Millspaugh, *Americans in Persia,* 197.

89. For the attitude of the British Government to the refusal of the Iranian Government to confirm the agreement as reflecting a fear for the British concession in the south, and the advice it offered to the Iranian Government see footnote 18 of the following chapter.

90. The statement of the Chairman of the Company to the stockholders, published on July 5, 1947, covering the year 1946, declared that the trading profit of £12,199,247 was the highest figure yet attained, and this was after payment of interest on preferred stock and payment of all accounts to the Iranian Government of £7,131,669. The dividend on ordinary stock was 30 per cent. Anglo-Iranian Oil Co., *Report of Balance Sheet at 31st December 1946,* 6–7; *Pleadings,* 210.

91. *PDC,* 425, cols. 1224–1225, July 17, 1946; cols. 1683–1684, July 22, 1946.

92. *Pleadings,* 273; Kirk, *Survey. The Middle East 1945–1950,* 88. On November 5, the Shah ratified the action of the Majlis. For the role played by the United States, especially the statement of Ambassador George V. Allen before the Iran-American Relations Society on September 11, in the determination of the Majlis to adopt the above law see "U.S. Bids Iran Resist Threats as Debate on Soviet Oil Nears," *The New York Times,* Sept. 12, 1947.

93. *Report on Seven Year Development Plan for the Plan Organization of the Imperial Government of Iran* (New York, 1949), 5 vols.

Chapter IV
THE NORTHERN PROVINCES

In approaching the issue of oil development in northern Iran, one must keep in mind the physical conditions involved. Should oil be found in the north, there are only two ways of bringing it to international markets. One is over the Zagros Mountains to a port on the Persian Gulf, which would be so expensive as to make the undertaking almost prohibitive; the other is to Baku and across the Caucasus to the Black Sea, and thence by tanker through the Straits to the Mediterranean — this would of course depend on Russian consent and goodwill. [1] Whenever, therefore, the issue of the development of petroleum came into consideration, physical as well as political factors had to be reckoned with.

As will be recalled, the original D'Arcy concession exempted the five northern provinces from the concession because they were considered, even as early as 1901, under Russian influence. The Russians on their part, it would seem, made no attempt to develop the oil resources of these provinces and consequently sought no concessions, at least until 1916.

THE KHOSHTARIA CONCESSION

On March 9 of that year Premier Sepahsalar granted A. M. Khoshtaria, at that time a Russian subject, a seventy-year concession to exploit petroleum and natural gases in the provinces of Gilan, Mazanderan and Astrabad. [2] Vossugh ed-Dowleh, who became Premier the following year, confirmed the concession. [3]

In the meantime, the Russian Revolution had broken out, and in 1918, when Vossugh ed-Dowleh again became Premier, both the Russian and Persian Governments considered the concession invalid. Vossugh declared that in a letter to him the British Minister in Teheran agreed with the Persian Government that the concession was invalid. [4] However, on May 8, 1920 the Anglo-Persian Oil Company bought from Khoshtaria, in consideration of £100,000, his concession in northern Persia, and organized a subsidiary, the North Persia Oil Company, Ltd., with a £3,000,000 authorized capital. [5]

Some nine months later, on February 26, 1921, Russia signed a treaty of alliance with Persia. It reaffirmed previous renunciations by the Russian Republic of the whole body of treaties and conventions concluded with Persia by the Czarist government aimed against the rights of the Persian people and declared them null and void; it similarly declared null and void the whole body of treaties and conventions concluded by the former Russian Government with third parties respecting Persia; it banned intervention by one in the internal affairs of the other, or commission of hostile military acts, or permission to a third party to operate against the other. It provided that should Russia or the Russian frontier be menaced by a third power, Russia was to have the right to advance into the Persian interior for military operations. Russia renounced all debts due the Czarist government from Persia, as well as other economic rights and concessions in Persia granted Czarist Russia, and ceded to the Persian Government some of Russia's communication installations. Article XII of the treaty stated: "The other concessions [in addition to those listed in previous Articles] obtained by force by the Czarist Government and its subjects shall also be regarded as null and void. In conformity with which the Russian Federal Government restores, as from the date of the signing of the present treaty, to the Persian Government, as representing the Persian people, all the concessions in question, whether already worked or not, together with all land taken over in virtue of those concessions." Article XIII stated that the Persian Government on its part promised not to cede to a third power, or to its subjects, the concessions and property restored to Persia by virtue of the treaty, and to maintain those rights for the Persian people. [6]

As a countermeasure to British and Russian influence in his country, the Persian Minister in Washington tried to interest American companies in exploiting the oil resources of the northern provinces. [7] In a note dated August 12, 1920, to the Persian Minister in Washington, the Third Secretary of the Department of State wrote that the United States Legation in Teheran had been notified that the Department believed that American companies would seek concessions in the northern Persian provinces and that the Department hoped that American companies might obtain such concessions. Moreover, he informed the Minister that the Standard Oil Company of New Jersey had indicated that it would consider a proposal to operate in northern Persia should a satisfactory agreement be reached. [8]

Two days later the Secretary of State informed the American Minister in Teheran that the British had formed the North Persia Oil Company, Ltd. "to work in conjunction with the Anglo-Persian Oil Company," and to develop further concessions obtained from the Persian Government, and that "the Department has taken the position that the monopolization of the production of an essential raw material, such as petroleum, by means of exclusive concessions or other arrangements " was in effect contrary to the principle of equal treatment of the nationals of all foreign countries. He was of the opinion that from the standpoint of international economic relations, the Persian Government should postpone granting oil concessions to the British until an opportunity had been given to American companies to enter into negotiations for such grants. On September 11, 1920, the American Minister in Teheran informed the Secretary of State that the Persian Government had not received any proposals for oil concessions in the northern provinces. As for the Khoshtaria concession, the Government considered it invalid, but its confirmation or rejection could not be definitely determined until the Majlis had acted on it.

Secretary of State Bainbridge Colby was apprehensive that the 1919 Anglo-Persian agreement would make it difficult for American companies to obtain concessions in north Persia, [9] but since the British disclaimed any such implication in the agreement, he wrote to the American Minister in Teheran late in November that the Persian Foreign Office might give consideration to American interests in the northern provinces "in case Russian concessions should legally and definitely be abrogated." He advised strong representation should the British claim prior rights. When Mr. Colby was informed that H. E. Nichols of the Anglo-Persian Oil Company was proceeding to Teheran to conclude the negotiations for the Russian concession in northern Persia, he instructed the American Minister: "Make sure that the new Foreign Minister clearly understands the position of this Government and of the American companies." [10]

The financial and economic pressure which the British, as the owners of the Khoshtaria concession, exerted on the Persian Government was apparently very heavy. In order to be able to withstand it, the Persian Minister in Washington suggested to the State Department at the end of 1921 that a loan from the United States would make it possible for his Government to grant the concessions to Americans. This information was passed on to the representative of the Standard Oil Company of New Jersey. From then on the question of granting a concession in northern Persia to an American company was always linked with the issue of a loan.

During the first half of 1921, the Anglo-Persian Oil Company's representative pressed for recognition by the Persian Government of its ex-Russian concession, but with very little success. When it seemed probable that the concession might be granted to Americans, the British protested vigorously, and the battle for the Khoshtaria concession was drawn. On September 21, 1921, the Persian Government replied to the British protest that it had never recognized the Russian concession and that the British claim was therefore invalid. About two weeks later, the British Ambassador in Washington officially informed the Secretary of State that the Russian rights had been taken over by British interests "in proper form"; and contrary to the Persian Government's contention that the Russian rights had lapsed and the concession had therefore been offered to American concerns, the British Government "had left the Persian Government in no doubt that the British right to the concession" was valid and would receive official support. This strong warning was rejected by Secretary of State Charles Evans Hughes on October 15, 1921; he challenged the British to state that all legal requirements necessary for the granting or transfer of the concession – especially the approval of the Majlis – had been met. Moreover, since the Anglo-Persian Oil Company already had a monopoly in southern Persia, the new concession would grant it a monopoly in all of Persia, to the exclusion of the Americans from the petroleum field. [11]

On November 22, 1921, the Majlis declared the Khoshtaria concession invalid and voted unanimously to grant the Standard Oil Company of New Jersey a fifty-year concession for petroleum exploitation in the five provinces of northern Persia. The Persian Government was to receive fifteen per cent of gross earnings. Other details were to be worked out with the Company and submitted to the Majlis for approval. Nevertheless, Article 5 of the Majlis bill stated: "The Standard Oil Company of New Jersey cannot under any circumstances assign or transfer this concession to any foreign government or company or individual, and likewise partnership with other firms or capitalists is subject to the approval of the Madjless.

Non-observance of this article will entail the invalidity of the concession." [12]

The reaction came immediately from two directions – the Soviet Union and Great Britain. The Russian Minister to Persia, T. Rothstein, in a note of December 23, 1921, objected to the granting of the concession to Standard on two grounds: 1) since the Russo-Persian Treaty of February 1921 had not been ratified, all Russian rights in Persia were in full force; [13] 2) even had the treaty been ratified, the concession violated Article 13, which forbade the granting of a concession formerly held by a Russian subject to a foreign national. [14] The British Minister protested that the concession was an unfriendly act towards Britain. [15]

The Persian Government's reply to the Russian objection was that the Khoshtaria concession had nothing to do with the Russo-Persian Treaty, for the concession, not having been ratified by the Majlis, "was legally null and void," and was as if it had never existed. A similar reply was made to the British. [16]

Meanwhile, the Anglo-Persian Oil Company not only protested, through the British Government, to the Department of State against the concession the Persians had granted to Standard Oil of New Jersey, but put pressure on Standard itself. [17] The APOC no doubt employed two arguments: 1) the Persian Government must under no circumstances be allowed to invalidate any concession it had issued, regardless of constitutional formalities, for otherwise all concessions would be endangered if they were subject to the whims of succeeding Persian governments, since legal and constitutional complications could always be found; [18] 2) the only way for Standard to bring the oil out of Persia was by pipeline to a port on the Persian Gulf; the Anglo-Persian Oil Company had the exclusive right to oil transportation throughout Persia, with the exception of the five northern provinces, and without its participation Standard's oil could not reach world commercial markets.

Because of the struggle for control of oil resources which was going on at the time between American and British companies, a struggle aggravated by the San Remo agreement which divided the oil of Mesopotamia between England and France to the exclusion of the Americans (this will be treated in the next section), the State Department was inclined to support the Standard Oil Company against the Anglo-Persian Oil Company. The British constantly and vigorously argued that the Persian grant of concessions to American companies was motivated by the desire to play up the British against the Americans to the advantage of the Persians alone and to the discomfiture of both British and American interests. The only way to safeguard and guarantee their investments was by Americans and Britons uniting against the machinations of the Persians to protect the existing concessions.

Since Standard's enthusiasm for the concession was not very great, particularly because of the condition that it obtain a loan for the Persian Government, British pressure was able to bring about a change in the Company's policy. On February 28, 1922, an agreement between the two companies, elaborated by Morgan Shuster, the fiscal agent of the Persian Government, was submitted by the American chargé d'affaires to the Persian Premier, Mushir ed-Dowleh. It provided for joint participation of Standard and Anglo-Persian in a new Perso-American Petroleum Company, but it stated specifically – to meet Persian objections – that Standard was to have voting control of the Board of Directors and of the management of the company. A note accompanying the agreement declared that the Department of State had not participated in the negotiations concerning this concession, but had been kept fully informed of their progress, that while the Department could not predict whether the Standard Oil Company would exercise its option of withdrawal, it felt that the proposed arrangement was a feasible plan for bona fide American participation. The Department believed that the arrangement made it possible to begin prospecting and development in the northern Persian provinces on a practicable basis without delay, and that it would avoid protracted controversy which would be injurious to Persia as well as to Anglo-American relations. [19]

At first the Persian Minister in Washington seemed to be in favor of the proposal, and in a sense he was even instrumental in its formulation, but the general political atmosphere in Persia was not conducive to such an arrangement. There can be no doubt that the Persians feared that this new proposal was a subterfuge for the Anglo-Persian Oil Company to obtain the concession. It has also been suggested that the Russians warned the Persians that they would not object to American capital being invested in the northern provinces, but that they would strenuously oppose British capital in the area. Be that as it may, on March 4, 1922 the Premier informed the American chargé d'affaires that the Standard–APOC agreement was causing "political difficulties in Persia." [20]

It is definitely clear that the attitude of the State Department toward the whole issue of the northern Persian concession had changed basically; what is not quite clear is the cause for that change. To be sure, Secretary of State Hughes never categorically denied the legal basis of the Khoshtaria concession; even in his note of October 15, 1921, mentioned above, he indicated that he was not in full command of the pertinent legal facts, but he raised grave legal doubts as to the validity of the concession. However, after the Standard–Anglo-Persian Oil Company agreement, when the Persian Minister in Washington requested Secretary Hughes to deny a report in *The New York Times* that there had been American and British governmental pressure on Persia not to deny the Khoshtaria concession, Hughes stated, on March 14, 1922: "The Department's position with regard to claims to concessions in northern Persia has been and is that it is not sufficiently informed at this time to express an opinion on the legal status of any of the contracts to which Khoshtaria may have been a party or which may have been transferred [21] to him." [22]

At the same time Sinclair, which had developed commercial oil relations with the Soviet Union, began to seek a concession in north Persia. [23] After the American chargé informed the Secretary of State of the arrival in Teheran of a representative of Sinclair to negotiate for a concession, apparently aware of the concession to Standard, Secretary of State Hughes instructed him, on December 20, 1921, to "observe strict impartiality as regards the two companies." [24]

The Persian Government rejected the proposed Standard–Anglo-Persian Oil Company arrangement, and on June 11, 1922 the Majlis voted an amendment to its previous resolution, empowering the Government to negotiate a concession in northern Persia with any independent and responsible American company. Subsequently, the Government approached not only Standard of New Jersey, but also the Sinclair Consolidated Oil Corporation. Standard indicated willingness to conform to the Majlis' resolution and operate the concession entirely on its own account. In August 1922 Standard and Sinclair submitted detailed draft concessions, both of which were rejected.[25]

On June 14, 1923, the Majlis passed a bill authorizing the Government to offer the north Persian oil concession to an American company, conditional upon the latter's arranging for a $10,000,000 loan to the Persian Government. Article 14 of the bill provided that the concessionaire could, under no circumstances, assign or transfer his rights to any foreign government, or to one or more foreign citizens.[26]

In September, Standard submitted counterproposals in Washington, but they were unsatisfactory. Sinclair's revised draft concession met with the requirements of the law and a concession was finally signed on December 20, 1923.[27] It covered four of the five northern provinces; the Persian Government reserved the fifth for exploration either by itself or by a Persian company.[28]

The Standard Oil Company, however, was not ready to give up its claims under its agreement with the Anglo-Persian Oil Company, and in a rather novel way it made its claim known to the Persian Government. On January 2, 1924, less than two weeks from the date of the signing of the Sinclair concession, a Mr. A. G. Berger from Gildford, Montana, addressed a letter to Standard asking the Company to enlighten him about reports that Standard had associated with a British syndicate for the purpose of excluding an American company from procuring concessions in Persia. "In view of an apparent necessity of concerted action in the United States to preserve and assure the future supply of petroleum products," he asked the Company to supply him with the correct information. On the 18th the Company replied that it was common knowledge that three and one-quarter provinces of the four granted to Sinclair were covered by a prior concession granted by the Persian Government — in 1896 to Sepahsalar. These rights had been passed on to the Anglo-Persian Oil Company in 1920, through a purchase from Khoshtaria and his associates by its subsidiary, the North Persia Oil, Ltd. While the Persian Government claimed these grants to be invalid, the British Foreign Office asserted their validity. When Standard was negotiating in 1921 with the Persian Government for the concession, the British company, supported by the British Government, protested on the ground that the offer of the Persian Government was a violation of its previous grants, which were valid. The State Department was consulted, and while it did not pass on the validity of the grants, it favored a policy of cooperation instead of controversy, and the Company decided not to accept the proffered concession except on the basis of recognition of the prior grants. After negotiations with representatives of the Anglo-Persian Oil Company and the Persian Government, a plan was worked out by which Standard would accept the concession and manage and control the operations under it, provided the British company had a half interest. The new concession would take the place of the Khoshtaria concession "in which, in connection with the plan, the Standard Oil Company (N. J.) had acquired under option a one-half interest by purchase from the Anglo-Persian

Oil Company, Limited." The Persian Government did not oppose this plan. Standard Oil therefore claimed that since its plan, agreed to by the Persian Minister in Washington and the special representative of the Persian Government empowered to negotiate the concession, had been completed in February 1922, previous to the time when Sinclair representatives began to negotiate with the Persian Government, it was disproven that Standard had associated with a British syndicate to exclude an American company from obtaining concessions in Persia. The reply concluded with this warning: "The Standard Oil Company (N. J.) meantime holds jointly with the Anglo-Persian Oil Company, Limited, a one-half interest in these earlier Persian grants covering approximately three and one-quarter provinces in North Persia, and will take the proper steps to protect its rights and to develop a petroleum production."

This correspondence was to appear in the current issue of Standard's *The Lamp*, but a digest of it appeared earlier in *The New York Times*, on February 24, 1924, with headlines considered very damaging by the Persian Minister, Hussein Alai. Three days later *The New York Times* published a statement by Hussein Alai rejecting the British claim on the ground that the Khoshtaria grants were invalid. He outlined the development of events from the Persian point of view. [29] Suspecting that this newspaper correspondence might only be a prelude to United States Government action, the Persian Minister pleaded with the State Department to take into consideration the facts as well as the motivations of the Persian Government and to make it possible for a purely American concession to operate in Persia. [30]

The Persian Government was nevertheless ready to proceed with the development of the northern fields and obtained the Majlis' approval of the Sinclair concession. Sinclair, however, was confronted with two difficulties: securing the $10,000,000 loan for Persia, and obtaining from the Soviet Union permission to transport the oil from the Persian fields through Russian territory to world markets. As regards the latter, Sinclair was in a rather favorable position for it had received an oil concession on Sakhalin Island from the Soviet Union and the right to sell Soviet petroleum products in world markets, and in Teheran it was believed that the company had an understanding with the USSR on the north Persian exploitation. [31] The question of the loan, however, was much more difficult. Even under the Standard concession, American financial houses had shown no enthusiasm for granting Persia loans of any considerable size. J. P. Morgan and Co. made two conditions for such loans: that British financial houses share equally with the Americans; and that the royalties from the Anglo-Persian Oil Company serve as security. [32]

On April 1, 1924, James G. Forbes, representing Blair and Company, arrived in Teheran to study Persia's securities for the loan, and on the 19th a bill to ratify the Sinclair concession was introduced in the Majlis. [33] However, in the middle of May the British made it clear that they could not consent to hypothecation of Anglo-Persian's royalties and (southern) customs duties for a loan in the United States so long as no settlement had been reached concerning Persia's debts to Great Britain. [34]

The Persians were subsequently willing to drop the loan provision and it looked as if the concession might be approved. However, Sinclair's position vis-à-vis the Soviet Union changed; in the summer of 1924 it lost its concession on Sakhalin Island, as well as the right to market Soviet petroleum products. Accordingly, Sinclair notified the Persian Government early in 1925 that the attitude of Soviet Russia prevented it from pursuing further its concession in northern Persia. [35]

SEMNAN

A concession in a very small area (not now part of the northern provinces) had serious repercussions involving the Soviet Union. It goes back to 1878 when Shah Nasr ed-Din granted a concession to Ali Akbar Aminé Maaden covering the district of Semnan, east of Teheran. Forty-six years later, in 1924, this concession was confirmed by the Persian Minister of Public Works in the names of Mirza Ali Akbar Khan Sotudeh and Mirza Abdol Hosein Khan Aminé Maaden. The Anglo-Persian Oil Company at the time protested the confirmation on the ground that Semnan was included in the D'Arcy concession, since it was not included in the five northern provinces exempted in the concession. The Persian Government, however, countered that while in 1924 the Semnan district was not part of the northern provinces, in 1901, when the D'Arcy concession was granted, it was part of Khorasan province.

In 1925 an Iranian corporation, the Kevir Kurian company, was organized, with a capital of fifty million rials, for the purpose of exploiting the Semnan district, and it acquired twenty-five per cent of the rights of the concessionaires. Through Khoshtaria as intermediary, the Soviet Government obtained 65 per cent of the company's stock. [36] In spite of some attempts at drilling, no positive results were ever reported. However, in 1944, when the Soviet Government decided to seek oil concessions in Iran and sent Kavtaradze to Teheran, the Semnan concession re-emerged. Kavtaradze intimated that the subject he wished to discuss with the Iranian Premier was the Kevir Kurian company; the Premier willingly consented, to discover soon afterwards that the Semnan concession was only a come-on and that the real issue was the northern provinces. In the controversy over the northern provinces which ensued, the Semnan concession was again submerged.

AMIRANIAN

In 1936 a Maryland state trooper, ignorant of international law, arrested the Iranian Minister to Washington in a traffic incident. Though the Minister was released as soon as the State Department instructed the Maryland police of his diplomatic immunity, Riza Shah was personally offended and ordered the Iranian Legation in Washington to be closed and the Minister to return to Teheran. In addition, criticism of Riza Shah in the American press brought about an almost complete embargo of American magazines and books. This caused relations between Iran and the United States to become strained. [37]

In spite of these diplomatic difficulties, in spite of the failures of the Standard–Anglo-Iranian Oil Company project based on the Khoshtaria concession, the failure of Sinclair and the failure of the Kevir Kurian Company, an American concern — the Amiranian Oil Company — a subsidiary of the Seaboard Oil Company of Delaware, was seeking a concession in northern Iran in 1936. After obtaining a concession in Afghanistan, Charles C. Hart, representing Amiranian, went to Teheran; by the end of the year he expected to sign an agreement with the Iranian Government which would have covered the province of Khorasan and one other province — either Astrabad, Mazanderan, or Semnan. [38]

On January 3, 1937, Hart signed two concessions with the Iranian Government — one for oil exploration and development; the other for the construction and operation of a pipeline. After subsequent modification, the first concession was granted for a period of sixty years; it covered an area of 200,000 square miles, including Gorgan (old Astrabad), most of Khorasan, northeast Kerman, all of Sistan, and northern Mokram. On March 9, 1937 Hart was informed that the Shah had signed both concessions. [39]

Amiranian faced the same problem that Sinclair had faced in the 'twenties: transportation, an outlet for the pipeline which it would build and operate. Hart made many efforts, through the intervention of the American diplomatic agent in Teheran, to obtain Russian cooperation in the operation of the concession. On March 17 he discussed the concession with the Soviet Embassy Counsellor in the American Legation in Teheran. M. Kartachov, the Counsellor, was disturbed by the old fear of British infiltration; he wanted to know to what extent AIOC was participating or would participate in the new company. Although Hart assured him that Amiranian was purely American, that it had no contact either with the AIOC or with the Germans and gave him the names of the interested American groups, and although he pointed out that the concession provided that the majority of the shares must be held by either Americans or Iranians, Kartachov was not convinced. He pointed out that it would have been necessary for Amiranian to reach some kind of an arrangement with the AIOC whereby the oil would be allowed to go out of Iran by way of the Persian Gulf. However, Hart's explanation, that the Iranian Government felt that no such arrangement was necessary, that it was free to grant such an outlet without permission from Anglo-Iranian, and that the Shah would indeed have been outraged at the implication that he depended on the consent of Anglo-Iranian, did not convince the Soviet Embassy's Counsellor; neither did the statement that Amiranian intended to have its outlet not on the Persian Gulf, but at the port of Chahbar on the Indian Ocean. [40]

The relations between the United States and the Soviet Government at that time were friendly and there was less fear in Moscow of American capital than of British penetration on the borders of the Soviet Union. The Russians suspected that the Amiranian Oil Company was actually a cover-up for the Anglo-Iranian Oil Company. They reasoned that the only company that could successfully operate the northern Iranian fields would be the AIOC, for it had the equipment, pipelines to the Persian Gulf, and the marketing facilities of the area; without the complete cooperation of and use of the facilities available to the British no new company could successfully operate in northern Iran.

In September the Soviets made a friendly gesture to the American company by granting it permission to ship through the Soviet Union supplies and machinery destined for Iran, although no Irano-Russian treaty covering such privileges existed. However, the American chargé in Moscow, Loy W. Henderson, advised: "It would be definitely unwise for the Amiranian company to place itself in such a position as to make it dependent upon the good-will of the Soviet Union for the successful exploitation of its concession."[41]

The attitude of the Anglo-Iranian Oil Company was voiced by its Chairman, Sir John Cadman, during his visit to the United States in the early part of 1937, when he said that the fact that the Amiranian concession was closely modeled on the AIOC concession indicated that the Iranian Government had adopted a stable policy with regard to oil concessions which was reassuring to his company. Moreover, with another concern entering on the scene, AIOC would cease to be the single object of critical attacks, and it would not have to defend its every move. In general, the British Government felt, according to the American Minister in Teheran, C. Van H. Engert, that if the British could not obtain the concession themselves, they would want to see none other than Americans in Iran, and they wanted close cooperation between the AIOC and Amiranian, or the Iranians would constantly try to play up one against the other. He was convinced that the British Government had not made any official protest. The Anglo-Iranian Oil Company, on its part, did make formal representation to the Iranian Government against granting the concession to Amiranian, raising again the Khoshtaria concession. This move, according to Engert, was with an eye to the possibility of arguing the claim of North Persia Oil, Ltd. should the Americans abandon their concession.[42]

Amiranian entered into extensive geologic survey work and by the end of 1938 gave up the concession. The reasons given differed. Some maintained that it was purely for commercial causes; though the production potentialities were excellent, the expense involved in carrying the oil across the Zagros Mountains to the Persian Gulf would have been prohibitive, and northern Iranian oil could not compete with other oil produced in the area; in Iraq, Saudi Arabia, southern Iran.[43] Alan W. Ford emphatically stated that the concession was abandoned for commercial rather than political reasons. In contrast, Herbert Feis, who was economic adviser to the State Department, declared: "The supporters of the venture faltered over the economic and political hazards."[44] According to Feis, the death of Ogden Mills, former Secretary of the Treasury, and the withdrawal of the inheritance by his heirs, caused the concession to be abandoned.[45] The Company, in its note of renunciation to the Iranian Government, declared that its decision was due to commercial reasons as well as to the general political uncertainty.[46]

It would seem that the commercial considerations could not have been the decisive determinant, for the difficulty of transportation was well known to the Company even when it began to negotiate for the concession, and it was for this purpose that it had obtained a special pipeline concession. Moreover, if the British were as eager to see Amiranian established in northern Persia as Sir John Cadman said and as the American Minister Engert reported, they would have cooperated with Amiranian in the transportation of oil. To be sure, a disturbing factor in the Middle East oil market was Saudi Arabia, but production had not reached such proportions in 1938 as to disturb the market seriously and eliminate, through the fear of competition, the possibility of marketing oil from northern Persia. It would appear that the decisive element — except perhaps the death of Ogden Mills and the unwillingness of his heirs to enter the venture, or Caltex's refusal to go further — in the renunciation of the concession was the unwillingness of the Soviet Union to permit the transportation of the oil and necessary equipment through its territory; and above all, the uncertainty and precariousness of a Soviet agreement or arrangement, as well as the general practice and policies of the Soviet Government along Russia's border. Moreover, as the international political situation began to deteriorate rapidly, Amiranian's risk became greater. It would seem that from the very beginning Amiranian had based its undertaking on Soviet cooperation, especially after relations between the United States and Russia improved. However, as the difficulties with the Soviet Government mounted and the Second World War approached, the American company decided to withdraw from Iran.

There were, however, others who were willing to venture into the field, in spite of the difficulties. In May 1939 a Dutch company was granted an exploration license covering the five northern provinces. It was to be given a sixty-year concession; should oil be discovered, only Dutch and Iranian nationals were to be shareholders; if public subscriptions were invited, Iranians were to be permitted to subscribe up to 40 per cent; the Iranian Government was to receive 50 per cent of the net profits. The war, however, intervened, and the license was abandoned in 1944.[47] The attempts to obtain oil concessions during the war were outlined in the previous chapter.

1955–1971

After the oil industry was nationalized, and the relations between the Iranian Government and the Western oil companies assumed the form of the international consortium which was to run the oil operations of exploration, development, production and refining on behalf of the National Iranian Oil Company, there were no foreign efforts to obtain any type of oil agreement in the north.[48] It would also appear that the Soviet Union made no attempt to exploit the northern fields.

The July 1957 new petroleum law enacted by the Majlis gave NIOC new broad powers to expand petroleum production as rapidly as possible throughout Iran, its territorial waters and offshore to the continental shelf, outside the consortium area. It was obvious to all, both the West and the Soviet Union, that the national concern was not ready to grant long-term large-area concessions. The Soviet Union, for its own purposes, would have preferred that the Iranian northern provinces remain unexploited: at worst they should be developed by Iranians rather than by foreign interests. The limited technological and financial resources would not hasten rapid oil development along its borders with Iran.[49]

In the summer of 1959, the Iranian government opened areas in the northern part of the country, especially in the northern desert, Dasht-e-Kavir, for competitive bidding for oil exploration, but no tenders were offered.[50]

In 1961, NIOC reported that it was drilling for oil in the Gorgan province (old Astrabad excluded from the D'Arcy concession).[51] However, no subsequent discovery or development was reported.

NIOC entered into an agreement in 1966 with the Soviet Union to supply the latter with huge quantities of natural gas in 1971, and to be expanded into larger quantities in 1977. To meet this commitment a trunk pipeline was built. The noteworthy aspect of this extraordinary deal was the report that the pipeline from Saveh in Central Iran to Astra on the Russian border was built by the Soviet Union.[52] The same report stated that the greatest gas discovery in Iran had been made in Sarakhs in the northeast: in Khorasan province; the hope was expressed that it could supply natural gas to Khorasan, and the nearby provinces of Gorgan and Mazanderan.[53]

SUMMARY

There are indications that northern Iran may be potentially rich in petroleum resources, yet throughout the last sixty-odd years all attempts to explore for and produce oil there have failed.

The basic consideration was political-physical: transporting the oil to the world markets. This was inevitably dependent on the good-will of Russia, a very difficult and unstable foundation. Moreover, Soviet Russia looked upon northern Iran not only as her sphere of influence, but also as a strategic area in which no foreign power must be allowed to establish itself. Prior to the outbreak of World War II, the foreign power especially feared was Great Britain, an old partner as well as a rival in Iranian spheres of influence; later this fear was extended to Germany. Russia was less afraid of an American company, but suspecting that an American company might be a subterfuge for British penetration, it held back cooperation with American concerns – in the 'twenties with Sinclair and in the 'thirties with Amiranian. Beginning with the 'forties, Soviet Russia became suspicious also of the United States, and in 1944 blocked both British and American efforts to obtain concessions in northern Persia by demanding the concessions for herself. After the war, Russia attempted to fortify her position in the north – through puppet regimes in Azerbaijan and by means of an oil concession in the northern provinces through a joint Iranian-Russian corporation. Under pressure from the United States, through the United Nations, Russia had to withdraw from Azerbaijan, but she obtained the oil concession she was after and prevented the Western powers from obtaining concessions. Because of changed international conditions, and encouragement – no doubt from the United States – the concession was not effectuated, but

Soviet Russia had succeeded in preventing the establishment of other foreign companies in northern Iran.

The Iranians, on their part, tried very hard to induce American companies to exploit their northern fields. This would not only have increased their income, but would have established a strong, far-away power along the border of her ambitious northern neighbor, as well as a counterbalance to British influence. However, neither Soviet Russia nor Great Britain was willing to permit the Americans a dominant position in northern Iran. In fact, the American State Department, though at first determined to oppose monopolistic British domination of Iranian oil, later gave in to British demands by not opposing the Khoshtaria concession and by not extending full support to Sinclair in its competition with AIOC — Standard of New Jersey.

The British were first willing to let the Americans come into north Persia, as they expected the Anglo-Persian agreement to be confirmed. However, when it became evident that the agreement was a dead letter, and after the AIOC had bought the Khoshtaria concession, the British were determined to keep the Americans out of Persia except on their own terms: the Khoshtaria concession. They saw in the concession — all the legal arguments of the Iranian Government against its validity notwithstanding — a safeguard to their own interests in the south. The Iranian Government was under no circumstances, no matter how questionable the legal foundations, to be allowed to cancel unilaterally a concession which it had issued. After a visit by Sir John Cadman to the United States and a meeting of representatives of Standard of New Jersey and of the Anglo-Persian Oil Company in London, the Americans were convinced of the soundness of the British argument, and Standard withdrew. The State Department was no longer fighting for the open door and it began to see greater merit in the Khoshtaria concession and cooperation between the British and American companies.

The desperate efforts of the Iranian Government to grant the concession to Sinclair failed because of the difficulties Sinclair encountered at home as well as from Soviet Russia. The 1937 American interlude was as much a failure as was that of 1924; though their experience with the cancellation of the D'Arcy concession made the British eager to have an American company to bolster their position against the Iranians, they nevertheless talked, even though only formally, of their rights under the Khoshtaria concession. However, the major reason for the failure of Amiranian was no doubt Soviet Russia.

During the war, in 1943–1944, the rivalry between Great Britain and the United States for concessions in northern Persia reappeared, but both were eliminated by the Soviets' tactics. In 1946 Russia outmaneuvered her Western rivals by obtaining an oil concession as the price for withdrawing her troops from Iranian soil. But, encouraged by the United States, Iran refused to consider the concession and subsequently cancelled it.

So far the efforts, whatever they have been, of NIOC to develop the oil and even natural gas potential of the northern provinces have not borne fruit. The difficulties, it would seem, remain the same: physical and political. Whether the great Iranian trunkline carrying natural gas from the southern oil fields to the Soviet border will bring about the reverse development of oil from the north to the south, and whether the Soviet Union will help develop the northern Iranian potential oil fields, if not directly control them, only the future will tell.[54]

NOTES

1. Charles C. Hart, representative of the Amiranian Co. in Iran, assured the Russian Counsellor in Teheran in 1937 that the pipeline his company intended to build would have its terminus at Chahbar on the Indian Ocean.
2. The districts which Shah Mozaffar ed-Din granted to Muhammad Vali Khan Sepahsalar in 1896 were specifically excluded. *Foreign Relations, 1920*, III (1936), 351.
3. The full text of the concession is to be found in *Foreign Relations, 1920*, III, 351–352; letter to the editor of *The Times* (London), reproduced in United States Senate, *Oil Concessions in Foreign Countries*, Sixty-Eighth Congress, First Session, Document No. 97 (Washington, 1924), 106–107. (Henceforth cited US Senate, Document No. 97.)
4. US Senate, Document No. 97, 106–107. The letter, dated Dec. 13, 1919 and signed by Sir Percy Cox, read: "With reference to our recent conversation regarding the Khoshtaria concession, I have the honor to inform your Highness that the British Government have again taken this matter into consideration and maintain the view that they in no wise favour the assignment of the said concession to any company. The British Government prefer to support the standpoint of the Persian Government in that the Khoshtaria concession is invalid. At the same time, I beg to inform your Highness that should the Persian Government desire to grant a new concession in this connection, the British Government hope, in the interests of Persia, that an English Company will be preferred."
5. *Foreign Relations, 1921*, II (1936), 641. The American consul general in London reported: "The vendors of the rights in Persia seem to have been given half of the shares in the new company, and the other half have been taken over principally by the Anglo-Persian Oil Company, in which true control is vested." *Ibid.*, 642.
6. League of Nations, *Treaty Series, 1922*, IX, 383–413.
7. A very interesting explanation by the Persian Foreign Minister for involving American companies in northern Persia for the joint benefit of Persia and Great Britain is recorded by Sir Percy Cox, British Minister in Teheran, in a letter of Dec. 20, 1919 to Lord Curzon, Foreign Secretary. *Documents*, First Series, IV, 1266.
8. *Foreign Relations, 1920*, III, 353.
9. See "The 1919 Anglo-Persian Agreement" in Chapter II above.
10. *Foreign Relations, 1920*, III, 353–356.
11. *Foreign Relations, 1921*, II, 643–649.
12. *Ibid.*, 648–649; US Senate, Document No. 97, 96, 120.
13. The specific item invloved comes under Article XII (quoted above) and the

voiding of the concession was therefore to have been from the date of the signing of the treaty.

14. US Senate, Document No. 97, 94.

15. *Foreign Relations, 1921*, II, 649–650.

16. US Senate, Document No. 97, 94. A full presentation of the Persian Government's argument against the validity of the Khoshtaria concession is given in 95–98.

17. The Federal Trade Commission declared that when Sir John Cadman, representative of the Anglo-Persian Oil Co., visited the US in 1921, an agreement was reached with Standard Oil of N. J. whereby the two companies submitted a proposal to the Persian Government that they be granted a joint concession for the undeveloped north Persian oil fields. United States Federal Trade Commission, *Report of the Federal Trade Commission on Foreign Ownership in the Petroleum Industry* (Washington, 1923), 62–63.

18. In 1947, after having agreed to an oil concession with the Soviet Union in April 1946, the Iranian Government refused to recommend to the Majlis a draft oil agreement proposed by the Soviet Ambassador. The British, though they had cooperated with the Americans in the Security Council in supporting Iran against the Soviet Union in the question of the evacuation of Soviet troops from Iranian territory, were seriously disturbed over the refusal of the Iranian Government to honor its oil agreement with the USSR. George Kirk stated: "The British Government, according to 'well informed quarters' in London, were apprehensive that, if the Persian Government were encouraged to reject outright the Soviet demand for the joint development of the oil resources of northern Persia, their latent nationalism might be tempted to challenge the Anglo-Iranian concession in the south." Early in September 1947 the British Ambassador in Teheran handed Premier Ghavam a note which recommended that the Soviet draft should not be rejected outright, but that the door be kept open for further discussions and an opportunity for revised and better terms to be presented. *Survey. The Middle East 1945–1950*, 86–87.

19. US Senate, Document No. 97, 100.

20. *Ibid.*, 101.

21. This refers to a claim by Khoshtaria that he had obtained the rights of Sepahsalar who was granted a concession by Shah Mozaffar ed-Din in 1896; see footnote 2.

22. US Senate, Document No. 97, 104. It has been suggested that the about face both of the State Dept. and of the Standard Oil Co. was the result of an understanding worked out between the Anglo-Persian Oil Co. through Sir John Cadman, during his visits to the US, and Standard for the latter's sharing in the Turkish Petroleum Co. Louis Fischer, *Oil Imperialism* (New York, 1919), 221–222; Azami, *Le Pétrole en Perse*, 194–208.

However, a Company spokesman declared in 1952: "The granting to the American group of a 23.75 per cent interest in the IPC had no connection with this matter." "Anglo-Iranian Answers Iran with Facts," *The Oil Forum*, VI, Apr., 1952. In reply to the challenge by Secretary of State Hughes, on Dec. 15, 1921, to the British to present the legality of their claims under the Khoshtaria concession, the American Ambassador in London informed him that the British were disinclined to discuss the legality of their claims. He added: "I was told privately by official of Foreign Office that [Charles] Greenway, [Chairman] of Anglo-Persian and [A.C.] Bedford of Standard Oil had met in London recently, and that it was understood they had arrived at informal agreement to operate jointly. This official also told me that in anticipation of such agreement, the British reply to the Secretary of State for their legal grounds, was held up." *Foreign Relations, 1921*, II, 652–653. See also a memorandum of meeting between a representative of the British Embassy in Washington and State Dept. officials, the subject of discussion of which was equality of opportunity for Americans and Britishers in north Persia, Mesopotamia and Palestine. *Ibid.*, 654–655.

23. According to A. G. Veatch, Vice Pres. of Sinclair Consolidated Oil Corp., Sinclair began to negotiate with the Persian authorities "following up a suggestion of the Secretary of Commerce," who at that time was Herbert H. Hoover, USFTC, *Report of the Federal Trade Commission on Foreign Ownership in the Petroleum Industry*, 127.

24. *Foreign Relations, 1921*, II, 652. The attitude of the State Dept. to the Sinclair Co.'s effort to obtain the concession, and the Dept.'s role in Sinclair's failure to operate the concession, are still subjects of serious controversy.

25. US Senate, Document No. 97, 112–113.

26. *Foreign Relations, 1923*, II, 713. The American Consul in Teheran considered this not only an anti-British gesture, but also a recognition of the anti-British attitude of Russia, which was opposed to British capital entering the Russian sphere of influence directly or indirectly. *Ibid.*, 714. See also Lenczowski, *Russia and the West in Iran, 1918–1948*, 81–84; Azami, *Le Pétrole en Perse*, 200.

27. Sinclair complained that the State Dept. favored Standard and was interfering with Sinclair's efforts. This complaint apparently reached public notice and Sec. of State Hughes found it necessary to advise Pres. Coolidge, in a letter dated Nov. 8, 1923, that the Department had been scrupulously impartial between the two companies. *Foreign Relations, 1923*, II, 717–718. This letter, incidentally, outlined the policy and philosophy of the State Dept. supporting American business and commercial enterprises abroad.

28. The full text of the concession is in *Foreign Relations, 1923*, II, 721–736.

29. US Senate, Document No. 97, 115–120. Later on he signed his name Ala.

30. *Ibid.*, 115.

31. The American Consul in Teheran reported on June 27, 1923: "It is already definitely known that the Sinclair Co. has come to an agreement in this regard [transportation] with the Bolsheviks." *Foreign Relations, 1923*, II, 714.

32. *Ibid.*, 713.

33. *Foreign Relations, 1924*, II, 545; Millspaugh, *The American Task in Persia*, 207.

34. *Foreign Relations, 1924*, II, 546.

35. Azami, *Le Pétrole en Perse*, 194–208; Millspaugh, *The American Task in Persia*, 293.

36. Nakhai, *Le Pétrole en Iran*, 104–106; Azami, *Le Pétrole en Perse*, 209; Kirk,

Survey 1939–1946. The Middle East, 475; for French efforts in this area see *Report on Seven Year Development Plan,* IV, 240.

37. Hull, *Memoirs,* II, 1501.
38. *Foreign Relations, 1937,* II (Washington, 1954), 734–735.
39. *Ibid.,* 376–377. It was ratified by the Majlis on Feb. 4. Summary of the concession, 744–747. The full text of the concession, with elaborate interpretation, is given in Nakhai, *Le Pétrole en Iran,* 93–98, 149–171; the text of the pipeline agreement, and elaboration, 99–104, 173–190.
40. *Foreign Relations, 1937,* II, 738–740. As regards the Iranian Government, the Soviet Government maintained that the Khoshtaria concession was valid, at least to the extent that Iran could not grant concessions to foreigners until it had consulted with the USSR. *Ibid.,* 740, 741–742. Incident to the Amiranian concession, the State Dept. prepared a detailed memorandum on the Khoshtaria concession: regrettably, however, it was not made public. *Ibid.,* 747.
41. *Ibid.,* 750–757.
42. *Ibid.,* 758–759.
43. Stephen Hemsley Longrigg, *Oil in the Middle East* (London, 1954), 62.
44. Herbert Feis, *Seen from E.A.* (New York, 1944), 174.
45. G. M. Lees states that the concession was relinquished because the Company, in which Caltex had in the meantime acquired an interest, failed to find prospects of fresh promise to justify development in such an unfavorable geographical situation. *World Geography of Petroleum,* 175. Against this should be mentioned the report of the Federal Trade Commission that the US Bureau of Mines had stated that the northern part of the country contained the most promising oil-bearing lands in Persia. *Report of the Federal Trade Commission on Foreign Ownership in the Petroleum Industry,* 62.
46. The full text of the letter written in June 1938 and signed by Charles Calmer Hart and Frederick Gordon Clapp is in Nakhai, *Le Pétrole en Iran,* 191–193; Lenczowski, *Russia and the West in Iran 1918–1948,* 84–85.
47. Longrigg, *Oil in the Middle East,* 62; *Report on Seven Year Development Plan,* IV, 241.
48. SOFIRAN, contractor to the NIOC-ERAP agreement, conducted seismic operations in the Dasht-e-Kavir area from May 1968 to the end of the year. National Iranian Oil Company, *Annual Report 1968,* 6; *NIOC in 1969* (henceforth cited as NIOC and *Annual Report*). But no subsequent development was reported to the end of 1971.
49. *The Christian Science Monitor* of November 13, 1959, reported that the Soviet Union offered Iran an 85/15 profit-sharing proposal for a Soviet oil concession in northern Iran on condition that no foreign military bases be permitted there. For political reasons the offer was not accepted. Jane Perry Clark Carey and Andrew Galbraith Carey, "Oil and Economic Development in Iran," *Political Science Quarterly,* LXXV, 72, Mar., 1960.
50. Careys, *loc. cit.,* 70; *Middle Eastern Affairs,* XI, 228, Aug.–Sept., 1960.
51. NIOC, *National Iranian Oil Company,* 8.
52. National Oil Company, Affiliated Companies Affairs and International Re-

lations Group, *The Economic Impact of Petroleum Industries on Iran* (Teheran, 1972), 13. (Henceforth cited as NIOC, *Economic Impact.*) For details see Chapter VII, below.
53. *Ibid.;* NIOC, *Annual Report 1968,* 45. *The Petroleum Press Service* described the gas deposit as one of the largest in the world, XXXVI, 453, Dec., 1969.
54. See below, Chapter XXVII.

Chapter V
THE 1948–1953 DISPUTE AND NATIONALIZATION

The Anglo-Iranian oil dispute, which began early in 1950, differed basically in many respects from previous oil disputes. This time Great Britain was confronted not with a mere demand for increased royalties, nor with a request for a greater share in the profits and other concessions — the bones of contention in former disputes — but with nationalization of the oil industry by the Iranian Government, an act which came upon the British unexpectedly and completely upset their methods of negotiation. Nor was the role of the United States that which the British had previously met with in Middle East oil disputes. The Americans did not ask for any share in Iran's oil exploitation, but were offering themselves — to the discomfiture of British diplomatists, negotiators and Company representatives — as a disinterested third party vitally concerned with finding a speedy solution, satisfactory to both parties, to the dispute. At first the Americans displayed a decisive disposition, accompanied by a sense of acute urgency, to force concessions from the British for the Iranians.

The rock on which the negotiations finally foundered and which resulted in a united front of British and Americans against the Iranians was the question of compensation. Not only was there no difference in the British position between the Labor government under Prime Minister Attlee when the dispute began, and the Conservative government under Prime Minister Churchill when the Mossadegh regime fell, but on the demand for compensation — as the British understood it — there was no difference of opinion between the Democratic administration of President Truman and the Republican administration of President Eisenhower. Both advised Premier Mossadegh to recognize the contractual rights of the Anglo-Iranian Oil Company and pay compensation accordingly.

The Iranians, on their part, because of their insistence on their demands, which were heavily charged with emotional nationalism, maneuvered themselves into an impossible position. After nationalization they were neither able to produce oil nor even to market the supply on hand which had been produced by the British company. As time went on and internal conditions in Iran grew rapidly worse because of the stoppage of oil revenues, an already bedeviled political system was further confounded, and this led to the violent overthrow of the hero of nationalization.

THE SUPPLEMENTAL AGREEMENT

During the adjustment following World War II, it became obvious both to the Iranian Government[1] and the Anglo-Iranian Oil Company[2] that some modification would have to be made in the terms of the 1933 oil agreement. This was not only because of the change in general political and economic world conditions, and of the terms of oil exploitation in the neighboring Middle Eastern countries, but because of the Company's policy of limiting the distribution of dividends and building up its general reserve. This affected the Iranian Government adversely, since its royalties were based on the distribution of dividends as well as on tonnage production.[3] The Company therefore offered, early in 1948, to discuss with the Government methods of adjusting its royalty income. The Government on its part raised three issues: increase in tonnage royalties; progressive reduction of the number of non-Iranian employees of the Company in Iran; and basic prices of oil products sold for consumption in Iran. Discussions were carried on during the latter part of 1948 and the early part of 1949.

On May 14, 1949, it was officially announced in Teheran that a basis of agreement on all matters of mutual interest had been fully discussed.[4] Two months later a supplemental agreement to the 1933 concession was signed in Teheran by representatives of the Company and the Iranian Government. Its main provisions were: 1) a per ton royalty increase from 4 to 6 shillings, retroactive to 1948; 2) an annual payment in respect of Iranian taxation of a flat rate of one shilling per ton, retroactive to 1948; 3) easier terms for the share of the Iranian Government in the general reserve of the Company,[5] and a guarantee of a minimum payment of £4 million per annum in respect of dividends and allocation to general reserve; 4) a different basis for fixing prices for oil consumption in Iran.[6] After the supplemental agreement was ratified, the Company was to pay the Government £5,090,909 in respect of the amount standing on the general reserve as of December 31, 1947, and additional sums under the various headings: £18,667,786 for 1948, and £22,890,261 for 1949.[7]

In accordance with constitutional requirements, the Iranian Government submitted the agreement to the Majlis for ratification. On July 28, 1949, however, the Majlis was dissolved; when the new Majlis convened, the agreement met with considerable opposition, and in January 1950 the Government proposed that before the agreement was considered by the Majlis as a whole, it be submitted to a special committee for careful examination. A committee of eighteen members was appointed, under the chairmanship of Dr. Muhammad Mossadegh. In December, the committee reported against the agreement on the ground that it did not safeguard Iranian rights and interests, and at the end of the month the Government withdrew the agreement. On January 11, 1951, the Majlis approved the report of the committee and charged it with preparing a report on the course which the Government should take in the question of the Anglo-Iranian Oil Company.

These developments must be viewed in the light of events in other Middle Eastern countries. It was known then that negotiations were being conducted between the Saudi Arabian Government and Aramco, and between the Iraqi Government and the Iraq Petroleum Company, for revisions in their contracts. Indeed, reports appeared in the press that Aramco and Saudi Arabia had agreed on a 50-50 profit-sharing plan, as of December 31, 1950. To be sure, the Iraq Petroleum Company, in which the Anglo-Iranian was a one-quarter shareholder, had granted in November 1950 the same two shillings per ton royalty increase as was offered to Iran. Nevertheless, the Iraqi Government was dissatisfied with the increase and was agitating for better terms.

When the news of the Aramco-Saudi Arabian agreement was made public, the attitude of the Iranian Majlis stiffened. The Company lost no time in communicating to Premier General Ali Razmara its willingness to examine with the Iranian Government suggestions for a new agreement on lines similar to those of the Saudi Arabian agreement. Moreover, the Company was ready to assist the Iranian Government in its financial difficulties by making an immediate £5 million advance against future royalty payments, and a monthly advance of £2 million for the remainder of 1951. However, these moves had come too late, for on February 19, 1951, Dr. Mossadegh proposed to the oil committee that the industry be nationalized. The committee asked Premier Razmara whether nationalization was practicable; he referred the question to a panel of Iranian advisers who submitted a report rejecting nationalization as impracticable. On March 3, Premier Razmara reported back to the oil committee. Four days later he was assassinated by followers of the Fidaiyin-i-Islam.

The oil committee was not deterred by these events. It adopted the nationalization proposal on March 8, and asked the Majlis for two months to study ways and means for implementing it. On the 15th, the Majlis approved nationalization in a single-article bill, and the Senate approved it five days later.

> For the happiness and prosperity of the Iranian nation and for the purpose of securing world peace, it is hereby resolved that the oil industry throughout all parts of the country, without exception, be nationalized, that is to say, all operations of exploration, extraction and exploitation shall be carried out by the Government.[8]

On April 30, the Majlis passed a nine-article Enabling Law providing for the implementation of nationalization.[9] With the royal assent to both laws, they became effective on May 1, 1951.

With the death of Premier Razmara, the cabinet resigned and the Shah named Hussein Ala, one-time Premier and former Ambassador to Washington, as Razmara's successor.

The British Government, meanwhile, realizing the implications of the impending development, sent a note to the Iranian Government on March 14 expressing concern over the intention of the Majlis' oil commission to nationalize the oil industry before the expiry of the Company's concession. The note declared that the Company's operations could not be legally terminated by an act of nationalization, and emphasized the Iranian Government's obligation to submit to arbitration, as provided in Article XXII of the 1933 agreement.[10]

Tension in Iran was growing, and general strikes broke out in the southern oil-producing districts. On March 26, Premier Ala proclaimed martial law. On April 8, he replied to the British note of March 14. Rejecting the British protests, he declared that the Company had not heeded changed world conditions and public opinion in Iran in presenting the supplemental agreement, that the Iranian parliament had adopted nationalization and a special oil commission was engaged in formulating proposals for implementation. "At present the Government's only obligation is to await the result of the Commission's deliberations."[11] The note also questioned the right of the British Government to intervene in a matter which was only the concern of the Iranian Government and the Anglo-Iranian Oil Company.[12]

The refinery in Abadan was shut down on April 15, and on the 27th Premier Ala presented his resignation to the Shah. On the 28th, the Majlis voted unanimously the immediate seizure of the Company's properties in Iran; it also voted to ask the Shah to name Dr. Muhammad Mossadegh, of the National Front, Premier. On May 8, 1951, the Company formally notified the Iranian Government that it requested arbitration, in accordance with the provisions of Article XXII and XXVI of the 1933 agreement and that on its part it appointed Lord Radcliffe as arbitrator.

In the meantime, Foreign Secretary Herbert Morrison sent a personal message to Premier Mossadegh asking the Iranian Government to refrain from unilateral action against the Company and suggesting negotiations. To this Premier Mossadegh replied, on the 8th, that in nationalizing the oil industry the Iranian people were exercising their sovereign rights, but that under the articles of implementation, the Iranian Government was prepared to examine the Company's claim. He rejected any arbitration on the issue of nationalization by declaring that nationalization of industries was the sovereign right of every nation. "Assuming that agreements or concessions have been concluded with persons or private companies in respect of these industries and assuming that from a juridical aspect these agreements and concessions are considered to be valid, the fact remains that they cannot form a barrier against the exercising of national sovereign rights nor is any international office competent to consider such cases."[13]

Ten days later, on May 18, the British Foreign Secretary addressed another note to the Iranian Premier, protesting the intended action of the Iranian Government. Herbert Morrison declared that, without wishing to interfere with the exercise of sovereign rights by the Iranian people, the proposed action against the Company was not a legitimate exercise of such rights. "The 1933 Agreement is a contract between the Iranian Government and a foreign company, concluded under the auspices of the League of Nations." It had been ratified by the Majlis and thus became Iranian law. That agreement provided for arbitration in cases of differences between the parties. The essential point to the British Government was that the "Iranian Government in effect undertook not to exercise those rights and the difference at issue is therefore the wrong done if a sovereign State breaks a contract which it has deliberately made."

As to the right of the British Government to enter the dispute, the Foreign Secretary declared that the Company was a British company registered in the United Kingdom, and that the Government owned a majority of the shares in the Company. It was clear therefore that the British Government had the fullest right to protect its interests in every way it properly could. He intimated that should the Iranian Government not be willing to come to terms by negotiation, the British Government would bring the case to the International Court of Justice. Mr. Morrison offered to send a commission to Teheran to discuss terms of a new agreement. The message concluded with a menacing note: "I should, however, be less than frank if I did not say that a refusal on the part of the Imperial Government to negotiate, or any attempt on their part to proceed by unilateral action to the implementation of recent legislation, could not fail gravely to impair those friendly relations which we both wish to exist, and to have the most serious consequences."[14]

At this stage the British Government still rejected nationalization; it was not willing to discuss or consider compensation, on the ground that the unilateral act of nationalization was confiscation, a breach of contract for which the Iranian Government would be liable.

The American Government, however, was not willing to go along with this position. On the one hand, it urged the British to recognize in principle the Nationalization Law and to work out a plan which would not seriously disturb the production operations of the Company, and on the other hand it urged the Iranian Government to come to some sort of *modus operandi* with the Company on the basis of the Nationalization Law. It seemed, nevertheless, that the American Ambassador, Henry F. Grady, was disposed to give much stronger support to the Iranians than was the State Department.[15]

On May 18, the United States made a statement which brought forth, three days later, a very strong protest from the Iranian Foreign Minister, who declared that while the American authorities in Iran were claiming that they were neutral on the oil issue, the United States Government was now throwing away its neutrality and advising the Iranians to negotiate. This advice was incompatible with the Nationalization Law and therefore must be regarded as interference in the internal affairs of Iran.[16] Five days later the United States replied that it had no intention of interfering in any way in Iran's internal affairs or with Iran's sovereign rights; however, the dispute between Iran and Great Britain had the potentiality of undermining the unity of the free world and seriously weakening it. This was a concern of the United States, and it believed that an issue of this kind could be settled satisfactorily only by negotiation between the parties concerned. While the United States took no position on the details of any arrangement that might be worked out, it "reaffirmed its stand against unilateral cancellation of contractual relationships and actions of a confiscatory nature." The formula which the United States supposedly hoped for was thus stated: "The United States is convinced that through negotiation a settlement can be found which will satisfy the desires of the Iranian people to control their own resources, which will protect legitimate British interests, and which will assure the uninterrupted flow of Iranian oil to its world markets. Such a settlement is, in the opinion of this Government, of the utmost importance not only to the welfare of the two powers concerned, but to that of the entire free world."[17]

NATIONALIZATION

The Iranian Government, nothing daunted, proceeded with nationalization. On May 20, the Minister of Finance stated in a letter to the Company that the Iranian Government had no other duty than of implementing the Nationalization Law, and that the Government was not subject to arbitration. He invited the Company to nominate immediately representatives to attend meetings for the purpose of making arrangements for putting into effect the law relating to nationalization. On the 24th, the Company was warned that unless it appointed representatives within seven days to discuss arrangements for the transfer of its holdings, the State would proceed with its own plans to nationalize the Company's properties. That same day both the Company and the British Government asked the International Court of Justice to rule that Iran must arbitrate or be found guilty of violating international law. [18]

In a letter to the Iranian Minister of Finance, dated May 27, 1951, the Company maintained that the action proposed by the Iranian Government was a breach of its concession; it rejected the contention that nationalization was not subject to arbitration and informed the Government that it had made application to the President of the International Court of Justice at The Hague for the appointment of a sole arbitrator in accordance with Article XXII (D) of the 1933 concession. The letter also stated that in respect to the request of the Iranian Government, the Company was sending a representative to attend meetings. "On the other hand, having regard to the purpose of the discussions, I must state that the representative of the Company will only be in a position to listen to what is said to him and to report the substance to the Company in London." [19]

On the same day, the British Ambassador in Teheran informed the Iranian Foreign Minister that since the Iranian Government had refused arbitration, the British Government had applied to the International Court, in accordance with the provisions of the 1933 agreement. Still hoping for a negotiated solution, the Ambassador informed the Foreign Minister that should the Iranian Government be willing to negotiate, and "should negotiations prove successful, the proceedings in the International Court of Justice could be arrested before judgment was given." [20]

Two days later the Iranian Government gave notice that it did not recognize the competence of the International Court to act in the oil dispute. At the same time, Premier Mossadegh told United States Ambassador Grady that unless the AIOC complied with the Nationalization Law, it would be driven out of Iran by non-military means. Herbert S. Morrison, Foreign Secretary, expressed the hope in the House of Commons that "wiser counsels would prevail in Iran and bring about the speedy settlement of the oil crisis." But at the same time he warned the Iranian Government: "We have every right and indeed a duty to protect British lives" in Iran. [21]

On May 30 the Iranian Finance Minister informed the Company's representative in Teheran that both the act of nationalization of March 15 and the implementation act of April 30 provided fully for the protection of the rights of all concerned. The Iranian Government was concerned *inter alia* with two fundamental matters: rights of previous purchasers and customers; and compensation for the damages caused to the Company through the nationalization of oil. As regards the former, Article 7 of the Enabling Law was quoted.[22] As regards the latter, the Iranian Government expressed its willingness to deposit 25 per cent of the net oil revenues with a bank mutually agreed upon. The Foreign Minister's letter continued: "In view of these premises it will be appreciated that the Iranian Government has absolutely not intended, and does not intend, to requisition the properties of the former oil company, nor does it propose to hinder the sale of oil to former customers."[23] He then outlined the regulations prepared by the Government, under the supervision of the mixed (Senate and Majlis) oil committee, for the execution of the Nationalization Law: a temporary Board of Directors of three nominated by the Government to function under the supervision of the mixed committee; temporary continuation of the rules of the former oil company; "the specialists, employees and workmen of the former Oil Company, Iranian as well as foreign, shall continue in employment as before and shall be regarded *from this date* as employees of the National Oil Company of Iran"; provision for enabling former purchasers to obtain oil in accordance with their previous practices. The letter concluded with an invitation to the former company to make proposals and, if they "do not conflict with the principle of nationalization of oil, the Government will take them into consideration."[24]

At this juncture the President of the United States took a hand and addressed personal letters, on May 31, to both Prime Minister Attlee and Premier Mossadegh. Previous soundings in Teheran and in London had apparently convinced the President that if the British were willing to recognize nationalization, the Iranians would agree to a procedure of operations which would satisfy the British, and he therefore cautiously proposed that negotiations be undertaken on the following basis: "Information received by me makes me believe the Iranian Government is eager, even anxious, to find ways and means with the British Government whereby both British fundamental interests and the Iranian nation's inclinations to nationalize its own oil should be protected. The United States Government has informed the British Government of its view that the Iranian Government has effected an opportunity for negotiations, which should begin as soon as possible."[25] The President had also obtained the consent of the Iranian Government to recognize the representative of the Company as the representative of the British Government, and he stated: "I am sure a solution can be found that can be acceptable both to the British and Iranian Governments. I sincerely hope every effort will be made to this end."[26]

On June 3, 1951, the Company's representative informed the Iranian Foreign Minister, in reply to the latter's letter of May 30, that the Company and the British Government were ready to attempt to solve all difficulties by negotiation. Since, however, the Company felt it could not submit proposals of a complex nature within five days, and since it believed that discussions face to face would be preferable to written communications, while reserving its legal rights, it would send representatives from London to Teheran as soon as possible in order to hold full and frank disscussions with the Iranian Government. [27]

A basic issue in the dispute was the status of the British Government. From the beginning the Iranians took the position that the British Government had no standing, that the dispute was not between two governments but between the Iranian Government and a private company, and therefore the dispute was entirely an internal matter and any interference from the outside would be interference in the internal affairs of a sovereign independent state. However, in a gesture of compromise — to what length it is hard to determine — Premier Mossadegh agreed, at President Truman's urging, to allow the representatives of the Company also to represent the British Government. But how they were to represent the British Government was not clear.

Thus, President Truman cautiously stated in his letter to Premier Mossadegh: "I believe that although Iran's invitation was extended to the Anglo-Iranian Oil Company, Iran has suggested that if AIOC representatives also represent the British Government, Iran would have no objection." [28] However, the May 30 letter of the Iranian Government was addressed to the Company and the invitation extended only to the Company. In his letter of June 3, the Company representative tried to insinuate the British Government into the negotiations, and on June 4 Herbert Morrison emphasized in the House of Commons that the British Government reserved the right "to interfere" in Iran, and that he did not accept the Teheran view that the actual dispute concerned only the Iranian Government and the Anglo-Iranian Oil Company. [29]

The three provisional directors of the National Oil Company of Iran left on June 5 for the oil-producing province of Khuzistan, and after taking over the Anglo-Iranian Oil Company's properties in Ahwaz, they proceeded to Abadan, where they took over the offices of the AIOC and raised the Iranian flag over the Company's main building.

On the 14th, the first meeting between the representatives of the Company, sent specially from London, and the Iranian Government took place in Teheran. The Iranians demanded as a condition for continuing the talks that the Company agree to turn over at once to the Government all revenue derived from the sale of Iranian oil as of March 20, 1951, after deducting expenses and 25 per cent to guarantee the Company's probable claims. After consulting with London, the British representatives submitted counterproposals. The Company would advance the Government £10 million against any sum which might become due to the Government as a result of an eventual agreement between the Government and the Company, "on the understanding that the Government undertakes not to interfere with the Company's operations while discussions are proceeding," [30] and pay the Government £3 million a month from July onwards until an arrangement was reached. For working out a satisfactory arrangement which would maintain the efficiency of the industry and at the same time be consistent with the principle of nationalization, "whilst fully reserving all our rights," the Company presented the following scheme: The Persian assets of the Company would be vested in a Persian national oil company, and "in consideration of such vesting the National Oil Company would grant the use of the assets to a new Company to be established by the Anglo-Iranian Oil Company Limited. The new Company would have a number of Persian directors on its board and would operate on behalf of the Persian National Oil Company. The distribution business in Persia would be transferred to an entirely Persian owned and operated company on favorable terms as regards the transfer of existing assets." [31]

These proposals were rejected by the Iranian Government. The next day Herbert Morrison warned Iran that Britain would send troops to Iran to protect her nationals if their lives were in jeopardy. He also told the House of Commons that the Company representatives had been instructed to return to England, and that the British Government would seek an injunction from the International Court of Justice to preserve British rights in Iranian oil properties pending the Court's ruling on the action brought by the British against Iran. [32] Dean Acheson, United States Secretary of State, appealed to Iran to reconsider her rejection of the Company's offer.

The practical problems involved in the taking-over process created new difficulties and tensions. The oil committee was pressing the British personnel of the AIOC on whether or not they would work for the new Iranian company, a pressure which they resisted. An explosive issue was the question of the loading receipts which the oil committee asked the British tanker captains to sign; these stated that the oil was taken from the National Oil Company of Iran. The captains refused to sign and in some cases the oil had to be unloaded. All efforts to work out a compromise formula failed, [33] and on June 25, Basil R. Jackson, deputy chairman of the AIOC declared that the Company was giving up hope of a negotiated settlement and was resigned to evacuating Iran. The following day the British Government ordered the 8,000-ton cruiser *Mauritius* to the vicinity of Abadan, and the Company withdrew all its tankers from the area. [34]

The attitude of the United States suddenly stiffened against Iran. On the 27th, Secretary of State Acheson denounced Iran's "threat and fear" tactics in trying to force British cooperation with her oil nationalization program, and again urged Iran to come to terms with the British. [35]

The following day Premier Mossadegh addressed a letter to President Truman in which he repeated that his Government was duty bound to implement the nationalization act and pointed to its willingness to discuss terms of compensation with the Company and to employ all the foreign experts, technicians and others presently in the service of the oil industry at the same salaries, allowances and pensions due them, to leave untouched the existing organization and administration of the Company, and to enforce "so far as they may not be contrary to the provisions of the law, the regulations made by that Company." He complained that the Company authorities were "encouraging the employees to leave their services," and were "threatening the Government with their resignation en masse," and "they force the oil tankers to refuse to deliver receipts to the present Board of Directors of the National Oil Company." He declared that any stoppage in the flow of oil would be the result of the Company's tactics, and he warned: "In such an eventuality the responsibility for the grievous and undesirable consequences which might follow will naturally lie upon the shoulders of the former oil company authorities."

Two days later Herbert Morrison informed the Iranian Foreign Minister that since the Iranian Government refused to accept a modified endorsement by the tanker captains on the receipts, reserving the legal rights of the Company over the oil, the Company was withdrawing its tankers, the refinery would have to stop operations and its personnel be withdrawn. He warned the Iranian Government that "the responsibility for withdrawal of tankers and progressive closing down of the Company's installations with consequent loss of revenue to Iran and large-scale unemployment amongst Iranian workers, results solely from the present attitude of the Imperial Government which has not only refused repeated offers to negotiate but has persisted in pursuing, without proper study or previous consultations, a course of action which must have the gravest consequences." [36]

INTERIM INDICATION OF THE INTERNATIONAL COURT

While Iran continually challenged the competence of the International Court, the latter opened hearings on June 30, at the request of Great Britain submitted on June 22, for the indication of interim measures of protection. Great Britain's application stated that should the plans of the Iranian Government materialize, great damage would have been suffered by the AIOC even should the Court decide in favor of the Company. The interim measures were asked to prevent such damages. [37]

The Court had before it a telegram from the Iranian Foreign Minister, addressed to the President of the Court, dated June 29, 1951, acknowledging the notification of the institution of proceedings by Great Britain (May 26), as well as for interim measures (June 22), and denying the Court's competence to deal with the issue. The telegram pointed out: 1) that the 1933 agreement was only between the Iranian Government and the Company; 2) that the 1932 declaration of the Iranian Government on the optional clause of the Court excluded all questions that might have a bearing on the sovereignty of Iran; 3) that the United Nations Charter excluded intervention in matters pertaining to the sovereign rights of any nation; 4) that Article 36 of the Court's Statute limited the jurisdiction of the Court to cases referred to it, or treaties, or conventions, and that none of these were involved in the present issue.

On July 5, 1951 the Court issued an order. Basing itself on Articles 41 and 48 of the Court's Statute and Article 37 of the Rules of the Court, and stressing that the indication of temporary measures would not prejudge the question of the Court's jurisdiction to deal with the case, and that the object of interim measures as provided by the Statute was to preserve the respective rights of the parties, the Court indicated, pending final decision, the following provisional measures which would apply on the basis of reciprocal observance: 1) both governments to ensure that no action be taken which might prejudice the rights of the other party; 2) both governments to ensure that no action be taken which might aggravate the situation; 3) both governments to ensure that "no measure of any kind should be taken designed to hinder the carrying on of the industrial and commercial operations of the Anglo-Iranian Company Ltd., as they were carried on prior to first May 1951"; 4) the operations of the Company to continue under the direction of its management as it was constituted prior to May 1, 1951; 5) in order to ensure the full effect of the above provisions by agreement between the two Governments, a board – to be known as the Board of Supervision – be established, composed of two members appointed by each side, and a fifth member from a third state, chosen by agreement between the two governments or by the President of the Court. The duty of the Board of Supervision was to ensure that the Company's operations were carried out in accordance with the provisions set forth above, to see to it that all revenue in excess of operating expenses were deposited in banks of its own selection, to be disbursed either in accordance with the decisions of the Court or by agreement by the parties. [38]

Meanwhile, on July 1, the Anglo-Iranian Oil Company had ordered a 40 per cent cutback in the Abadan refinery, and on the 3rd had decided to transfer all field operations to the Iranians and to withdraw all British personnel from the oil fields to Abadan, for possible evacuation to Britain. A week later, the British Ambassador informed the Iranian Government that Britain accepted in full the recommendations of the International Court for interim measures. He expressed the hope that the Iranian Government would soon nominate its respective two members to the Board of Supervision, and subsequently make suggestions about the fifth member. [39] Two days later, however, the Iranian Foreign Minister, B. Kazemi, informed the Secretary General of the United Nations that because of the International Court of Justice's order of July 5, the Iranian Government had withdrawn its declaration of October 3, 1930, accepting the compulsory jurisdiction of the Court.

President Truman once again tried his hand. In a personal letter to Premier Mossadegh, dated July 9, 1951, he urged the Iranian Government to re-examine the recommendations of the International Court as a basis for settling the oil dispute. He offered to send W. Averell Harriman as his personal representative to Teheran, "to talk over with you this immediate and pressing situation." [41] Two days later the Premier, replying to President Truman, complained that so far no proposal or suggestion had been made by the "former oil company" denoting acceptance of the principle of nationalization. However, should Iran's right to nationalize the oil industry be accepted in accordance with the enacted laws, she would be ready to enter into immediate discussions with the aim of removing all the disputes so that there might "be no stoppage in the production and exploitation of oil – a situation which the Government of Iran has always been anxious to avoid and which, as you have mentioned, Mr. President is causing losses to all concerned." As to the President's suggestion to send his personal representative to Teheran, the Premier stated that he would welcome Mr. Harriman and hoped to take full advantage of consultations with a man of such high standing. [42]

Meanwhile, the Iranian Government cancelled all the contracts of the Anglo-Iranian Oil Company and offered oil of the nationalized company to all comers on a cash-and-carry basis.

As a result of intensive negotiations between the Iranian authorities and Mr. Harriman, at a joint session of the cabinet and the mixed parliamentary committee,[43] the following formula was submitted to Mr. Harriman on July 23, 1951, as the final view of the Iranian Government:

1. In case the British Government, on behalf of the former Anglo-Iranian Oil Company, recognizes the principle of nationalization of the oil industry of Iran, the Iranian Government would be prepared to enter into negotiations "with representatives of the British Government on behalf of the former Company." 2. Before sending representatives to Teheran, the British Government to make a formal statement on behalf of the former Company of its consent to the principle of nationalization of the oil industry. 3. "By the principle of nationalization of the oil industry is meant the proposal which was approved by the Special Oil Committee of the Majlis and was confirmed by the law of Esfand 29, 1329 (March 20, 1951)." 4. The Iranian Government was prepared to negotiate the manner in which the law would be carried out, in so far as it affected British interests. [44]

Thus, while the Iranians insisted on British acceptance of the Nationalization Law, they made a concession in not insisting on acceptance of the Enabling Law. At the same time, although representatives of the British Government were acceptable to the Iranian Government they were to act only on behalf of the Company. Mr. Harriman volunteered to submit the formula to the British Government and flew for that purpose to London.

The British raised some objections to the proposed statement they were to issue in response to the Iranian proposals. After further discussions, with the intervention of the US Embassy in Teheran, a compromise proposal was worked out and the British officially announced that Richard R. Stokes, Lord Privy Seal, would head the British delegation to Iran and would arrive there on August 4. [45]

Addressing the Majlis on August 5, 1951 on Mr. Harriman's efforts and achievements, Premier Mossadegh stated: "The least tangible result which has been procured through his helpful co-operation is that through negotiations the long standing dispute of 42 years duration with reference to the southern oil concession has been totally and successfully settled." [46]

THE STOKES MISSION

On August 6 discussions between the British and Iranian delegations commenced, and a week later Richard Stokes produced an eight-point proposal. The most pertinent points were: 1) the AIOC to transfer to the National Iranian Oil Company (NIOC) its installations, machinery, plant and stores in Iran, and compensation for the assets in southern Iran to be included in the operating costs of the oil industry in the area; 2) a Purchasing Organization to be formed to provide an assured outlet for Iranian oil, and this Organization to obtain a long-term contract, of about twenty-five years, from the NIOC for the purchase of very large quantities of crude oil and products from southern Iran; 3) the NIOC to make additional sales of oil provided this in no way prejudiced the interests of the Purchasing Organization; 4) in order to assure to the Purchasing Organization the necessary quantities of oil for its committments, "the Purchasing Organization will agree with the National Iranian Oil Company on an Organization which, under the authority of the National Iranian Oil Company, will manage on behalf of the National Iranian Oil Company the operations of searching for, producing, transporting, refining and loading oil within the area. The Purchasing Organization will arange from current proceeds the finances necessary to cover operating expenses"; 5) the Purchasing Organization to buy the oil from the NIOC at commercial prices f.o.b. Iran, less a price discount "equal in the aggregate to the profit remaining to the National Iranian Oil Company after allowing for the discount and for the costs of making the oil available to the Purchasing Organization," 6) Iranians to be represented on the Board of Directors of the operating organization; non-Iranians to be employed to the extent necessary for the efficiency of the operations. [47]

The Iranian cabinet rejected the Stokes plan. On August 17, Mr. Stokes declared in Teheran that his proposals were final and that he would not make any further efforts. Subsequently the Iranians asserted that the Stokes proposals did not conform to the definition of nationalization of the oil industry, as stipulated in Iranian law, and which formed part of the Harriman formula, for they took away from the Iranian Government a substantial part of the powers of management of the oil industry and would only "revive the former Anglo-Iranian Oil Company in a new form."

After criticizing the proposals under the various headings, the Iranian delegation offered the following counterproposals: 1) the Iranian Government to sell oil products to England, on the basis of ordinary commercial contracts, in the quantity which had been supplied in recent years for British consumption, but no monopoly to England for the sale of oil; 2) no discount to the Purchasing Organization; 3) no operating organization; while the Iranian Government was ready to grant proper authority to the foreign experts whom the NIOC would engage for oil production operations, it was not ready to restrict its sovereignty; 4) the Iranian Government was ready to investigate fully and fairly the just claims of the former AIOC, "taking into consideration the claims which the Persian Government has aginst the Company, and in respect of its assets in Persia and outside Persia, at last settling these claims after the rights of both parties have been established." [48]

On August 21, Mr. Stokes addressed a note to the Premier of Iran, informing him that his eight-point proposal was withdrawn. [49] However, he gave the Premier till twelve noon the following day to accept the principles he had outlined, "which would make it possible for the British staff to remain in the refinery and oil fields," and he stated that he would be prepared to resume discussions. [50]

The Premier, in his reply to Stokes of the same date, formulated the three issues involved in the dispute after the principle of nationalization was accepted: sale of oil to former customers; use of foreign technicians; and compensation. The solution to these problems would have to reconcile the interests of the Iranian Government with the interests of the customers and shareholders of the former Company. Iran therefore made the following proposals: as regards the first point, former customers to have priority in the purchase of oil, f.o.b. Iran port. As regards the second, the NIOC to retain in their posts foreign experts, at the same salaries and with the same allowances as they had enjoyed under the former Company, in accordance with annual agreements. The organization and administrative and technical arrangements of the former Company would be maintained. A sufficient number of first-class experts from countries with no special political interests in Iran would be employed as members of the board of management. On the third point, compensation, this was to be based on agreement with the former Oil Company over its claims and just demands. [51]

The negotiations were terminated and Richard R. Stokes returned to England. On the 24th, Averell Harriman left Iran, and on the same day Premier Mossadegh wrote to him in reply to Harriman's exposition of the reasons for the withdrawal of the Stokes plan. [52] From London came the announcement that the British personnel would be withdrawn from the oil fields, and that the British Government would pursue its case against Iran before the International Court of Justice.

ENTER THE UNITED NATIONS

On September 5, 1951, after having obtained a 26–0 vote of confidence in the Senate on his oil policy, Premier Mossadegh threatened to expel from Iran all British oil refinery technicians – the only AIOC employees still remaining – unless the British Government resumed negotiations within fifteen days. The next day the British Foreign Office declared that no further negotiations with the present Iranian Government could produce any results and therefore negotiations were completely broken off. Three days later Premier Mossadegh announced in the Majlis that his warning to Britain, that all British technicians would be expelled from Iran unless negotiations were resumed within fifteen days, would be delivered to the British Government through W. Averell Harriman.

On the 12th the Premier outlined in a letter to Mr. Harriman, the following proposal for Harriman to transmit to the British as a basis for the renewal of negotiations: 1) as to the question of management, the original staff of the AIOC to be retained, in so far as it did not contradict the terms of the Nationalization Law, and managers and responsibles of technical sections in the AIOC to be employed by the NIOC and to have the same authority that they had enjoyed previously. Moreover, the Iranian Government was prepared to take advantage of foreign technicians from neutral countries and to provide in the original law of the NIOC for a mixed executive board composed of such experts and Iranian experts who would jointly manage the administrative and technical affairs of the National Iranian Oil Company; 2) as to compensation he proposed, with due regard to the claims of the Iranian Government, one of the following three methods: a) determination and amount of compensation to be based on quoted value of shares of the former company at prevailing quotations prior to the passage of the oil nationalization law; b) rules and regulations relative to nationalization in general which have been followed in democratic countries to be regarded as the basis for the determination and amount of compensation; or c) any other method which might be adopted by mutual consent by the two parties; 3) as to the sale of oil, the Iranian Government was ready to sell Great Britain 10 million tons of Iranian oil annually (the yearly British internal consumption of Iranian oil) for a period agreed upon by both parties, at prevailing international prices every year on the basis of f.o.b. value in Iranian port.

The suspension of oil operations was causing great hardship to Iran, and the Government was determined to bring to an end the disastrous state of uncertainty. It warned the British that unless they resumed negotiations within fifteen days from the day these proposals were submitted to them, the Iranian Government regretted "to state its compulsion to cancel the residence permits held by the British staff and experts now residing in southern oil fields." [53]

The British on their part began to adopt economic measures to force the Iranians to submit to British demands. On September 10 all special financial and trading facilities for Iran were withdrawn, thus depriving her of practically all dollar exchange. All licenses for export to Iran of scarce material, such as railway track, were revoked, and cargoes of such goods already en route to Iran were requisitioned. [54]

Five days later Mr. Harriman informed Premier Mossadegh that his latest proposals were no different from those the Iranian delegation had proposed to the British delegation and which had been rejected, that transmission of those proposals to Great Britain would aggravate rather than facilitate a solution to the problem, and that he therefore was unable, regretfully, to transmit them to the British Government. [55]

Having failed with Harriman, Premier Mossadegh tried a different method. The Iranian Minister of Court, Hussein Ala, handed to the British Ambassador, on September 19, a document which contained, with some modifications, the proposals Mossadegh had submitted to Harriman. The document was not signed, and it did not contain the ultimatum. Three days later F. M. Shepherd, the British Ambassador in Teheran, handed Hussein Ala a letter in which he pointed out that the document containing the proposals, which Ala had left with him, was not signed, nor dated, nor drawn up on official paper; and as to contents, it constituted a retrogression rather than an advance over former proposals. Moreover, the British Government was in full agreement with the Harriman reply to the Premier's proposals. [56]

Consequently, on September 25, the Iranian Government ordered the last three hundred British oil technicians in Abadan to leave the country before October 4; two days later Iranian soldiers took over the Abadan refinery and locked out the British technicians. Realizing the explosive nature of the situation, President Truman appealed to the Iranian Government to cancel its plan to expel the British technicians, and to the British Government not to use armed force to prevent the Iranians from carrying out their order.

On the 28th, the British announced that they would call on the Security Council to compel Iran to abide by the interim decision of the International Court of Justice and to stop the expulsion of the British technicians.

That same day the British representative at the United Nations addressed a letter to the President of the Security Council and the Secretary General asking them to place on the agenda of the Security Council "Complaint of failure by the Iranian Government to comply with the provisional measures indicated by the International Court of Justice in the Anglo-Iranian Oil Company case.' [57] With the letter, the British submitted a draft resolution calling on the Government of Iran "to act in all respects in conformity with the provisional measures indicated by the Court and in particular to permit the continued residence at Abadan of the staff affected by the recent expulsion orders." [58]

The Council took up the British request on October 1. Immediately a procedural wrangle developed. The Soviet Union maintained that what the United Kingdom was asking the Security Council to do was to interfere in Iran's internal affairs, since the nationalization of the oil industry was entirely within the province of Iran's own affairs, and the complaint was therefore an infringement on the sovereign rights of the Iranian people and inadmissible for the Security Council to discuss. The other members of the Council, except for Yugoslavia, maintained that since there was a danger to international peace the Council must consider the complaint, without, however, in any way passing judgment on the substance of the case. A vote to put the complaint on the agenda was adopted, 9–2, and the issue was taken up by the Council. [59]

Sir Gladwyn Jebb, the permanent British representative, argued that it was intolerable that one party to a matter laid before the International Court should be allowed to flout the Court's findings "and to impose unilaterally its own will in regard to this matter." After presenting at considerable length the substance of the case from its inception to the expulsion order, he urged: "By adopting the draft resolution which we have before us, the Security Council will make it plain that it is determined to uphold the rule of law in international affairs, to say nothing of the prevalence of reason; it will assert its authority not on behalf of the powerful against the weak but on behalf of intelligent progress against blind and unintelligent reaction." [60]

Speaking for Iran, Ali Ghali Ardalan stated that since Iran had contested the competence of the Court from the very beginning, he was surprised to see that the United Kingdom was bringing the complaint against his country before the Council. Moreover, since Britain recognized the principle of nationalization, his Government believed that there was no ground for discussion of this question in the Council. However, since the Council had decided to take up the complaint, Iran would be ready to argue the case, but a least ten days would be required to enable the representative of Iran to come from Teheran to New York. He therefore asked the Council to adjourn the discussion for ten days. This evoked opposition from the British delegate – for in ten days the issue of expulsion of the British staff would be dead – but in view of the opinions of the other members of the Council, Sir Gladwyn agreed to the postponement. [61]

The British draft resolution had apparently met with little enthusiasm or support in the Security Council, as was revealed in the initial debate on the question of putting the complaint on the agenda. [62] When the Council met on October 15, therefore, the British submitted a revised resolution. The question of the expulsion of the British technicians from Abadan no longer existed, and instead of calling upon Iran "to act in all respects in conformity with the provisional measures indicated by the Court," the revised draft merely called for "the resumption of negotiations at the earliest practicable moment in order to make further efforts to resolve the differences between the parties in accordance with the principles of the provisional measures indicated by the International Court of Justice, unless mutually agreeable arrangements are made consistent with the Purposes and Principles of the United Nations Charter." [63] Moreover, the British representative indicated that his government did not simply and purely insist on the return to the status quo which had existed before May 1; he emphasized the extent to which Great Britain was willing to go, provisionally, in order to enable the flow of oil to be resumed. He explained that it was "with the object of upholding the rule of Law that we have retained some reference to the International Court in the preamble." [64] He appealed strongly to the members of the Council, as well as to Premier Mossadegh, to accept the softened-down resolution. However, in spite of the watering down of the resolution, England was bound to meet with failure before the Security Council.

Premier Mossadegh flew to New York to present personally the Iranian case before the Security Council. He too, like the British delegate, presented the case in considerable detail, reviewing the tortuous story from the beginning up to nationalization. But his major legal argument was that the issue was purely an Iranian domestic affair, and that neither the Court nor the Security Council had any competence. "It is a settled principle of international law that in matters of domestic concern, to which this question eminently relates, their exercise [of rights] can neither be abridged nor interfered with by any foreign sovereign or international body." [65] The Premier of Iran rejected, in a strongly worded speech which ran for two days, any possible action by the Security Council on the British plan. While the Iranian representative maintained his intransigence, the British representative displayed considerable forebearance and was willing to water down his resolution even further in order to obtain some action, any action, from the Security Council. Nevertheless, the resistance was strong.

The Indian and Yugoslav delegates introduced, on October 16, several amendments to the revised British resolution, in an effort to safeguard the legitimate position of each party and to offer a possibility for the resumption of negotiations in a favorable atmosphere. The major change was in the operative article, which called for "the resumption of negotiations at the earliest practicable moment in order to make further efforts to resolve the differences between the parties in accordance with the Purposes and Principles of the United Nations Charter." The reference to the International Court was omitted. [66]

Both India and Yugoslavia were aware that the competence of the International Court was a basic issue, and by omitting reference to the Court they hoped to eliminate the whole issue of competence. Premier Mossadegh, however, maintained that the Security Council was also not competent to deal with the question, and consequently he opposed "any draft resolution, however amended." [67]

While the Soviet Union supported the Iranian position — that the question was an internal Iranian affair and therefore could not be interfered with by the Security Council — the British delegate declared that the Indian and Yugoslav amendments diminished very considerably the force and utility of "the already very watered-down draft resolution" which he had submitted. Nevertheless, his delegation was prepared, "though with the greatest reluctance, to accept the amendments put forward jointly in the names of the Indian and Yugoslav delegations." [68] But even with the acceptance of these amendments, the representative of Ecuador stated that the Council was not competent to deal with the issue, and therefore Ecuador would not vote for the resolution. He they introduced a new draft resolution, the operative clause of which read:

> "The Security Council,
> "*Without deciding* on the question of its own competence,
> "*Advises* the parties concerned to reopen negotiations as soon as possible with a view to making a fresh attempt to settle their differences in accordance with the Purposes and Principles of the United Nations Charter." [69]

The United States delegate maintained that since the dispute was between states, as Iran's willingness to negotiate with Great Britain indicated, and consequently a danger to international peace was involved, the Security Council was competent to deal with the issue. He declared that the United States would support the revised British draft resolution. The Iranian Premier, however, continued to argue that the Council had no competence since the dispute was not between Iran and the United Kingdom, but between Iran and the Anglo-Iranian Oil Company.

The death blow to the British re-revised draft resolution was delivered on October 19 when the French delegate stated: "It seems to me that the Security Council had better adjourn its debate on the draft resolution now before it . . . until the International Court of Justice has ruled on its own competence in the matter." [70] Subsequently Yugoslavia withdrew her support from the British draft as amended by herself, and the British saw the handwriting on the wall. Sir Gladwyn declared: "This minority [who doubt the competence of the Council] — there is no disguising of the fact — is sufficient in size to prevent the adoption by the Council of the greatly watered-down draft resolution which we have submitted. . . . Therefore, since we cannot get the consent of the Council even to the greatly diminished draft resolution which we have submitted, so far as we are concerned we can only agree — and I hope that the whole Council agrees — to the intelligent suggestion just made by the representative of France." [71]

But the Soviet Union objected even to the French proposal, and when the Council voted on the French motion, it was adopted 8–1 with 2 abstentions, the

Soviet Union voting against, and the United Kingdom and Yugoslavia abstaining. [72]

The attempt of the British to force the issue through the United Nations must be written off as a failure. The practical immediate issue of stopping Iran from expelling the British workers from Abadan was dead when the Security Council began to debate the matter. That Russia should fully support Iran was to be expected, but that the hesitations on the question of competency should so paralyze the Council as to make it unable to take any action was certainly unforeseen. The British failed to obtain from the Security Council any resolution, regardless of how weak and conciliatory, which would have strengthened the British position. It would be no exaggeration, on the other hand, to say that although the Iranians could not boast of a victory, their experience in the Security Council made them if anything more determined than ever to resist any pressure by the British and to persist in their basic demands.

The financial situation of the Govenrment of Iran resulting from the cessation of oil royalties as well as from other measures of financial pressure applied by the British became more pressing as time went on, and Premier Mossadegh was as anxious — for different motives — as were the British and the Americans to start the oil flowing again. The Americans, on their part were eager to solve the oil problem and pull the British out of their ever deteriorating position. Advantage was therefore taken of the presence of Premier Mossadegh in the United States, to hold conferences between the Premier and President Truman on October 23; on the 24th between the Premier and Secretary of State Dean Acheson; and subsequently with Assistant Secretary of State George C. McGhee. These proved fruitless, and on November 13 the State Department announced that despite intense efforts, no new basis of settlement had been reached. [73] Premier Mossadegh, realizing that all attempts at a solution had failed, appealed to President Truman for a $120 million loan; all he could obtain from the President was a statement, on the 16th, that his plea for a loan would be considered expeditiously. Three days later the Premier left the United States.

THE INTERNATIONAL BANK

While Dr. Mossadegh was in Washington, neutral diplomats suggested that the International Bank for Reconstruction and Development might possibly assist in a settlement of the Anglo-Iranian oil controversy and act as an intermediary between the two member-states. Bank representatives discussed the matter with the United States Government officials since they had been dealing with the subject, and the Americans encouraged the Bank to make its services available to the two parties. Mr. M. A. H. Ispahani, the Pakistan Ambassador to Washington, called on Premier Mossadegh and suggested that the Bank's good offices be enlisted; the Premier expressed interest. Thereupon, the Vice-President of the Bank, Robert L. Garner, called on the Premier (November 10). Mr. Garner made it clear that the Bank would avoid passing judgment on the merits of the controversy and would be prepared to act only upon the invitation of both Iran and Great Britain. He outlined in general terms the possible procedure by which the Bank might assist in restoring the flow of oil. The basic points of his proposal were: 1) the Bank to set up, in agreement with both parties, for a limited time, say two years, a temporary management for the operations of the oil properties; the management to be responsible to the Bank and headed by nationals of countries not party to the dispute; 2) the Bank to arrange a contract for the sale of oil to the Anglo-Iranian Oil Company on terms satisfactory to both parties, and part of the proceeds of sale to be held by the Bank in escrow pending final settlement; 3) the Bank to undertake to see that necessary funds were provided for the resumption of operations, to be reimbursable from oil revenues.

Dr. Mossadegh expressed willingness to have the Bank attempt to work out a solution which would be acceptable to all concerned, and Mr. Garner informed the British of his conversation with the Iranian Premier. While the Premier was in Washington the British did not react, but subsequently Garner held discussions in London with Foreign Secretary Anthony Eden and other members of the cabinet, as well as with officials of AIOC. The Foreign Secretary indicated interest in having the Bank use its good offices. For practical information the Bank obtained as adviser Torkild Rieber, president of the Barber Oil Corporation of New York and former chairman of the Board of the Texas Company.

In January 1952, a Bank mission — Rieber and Hector Prud'homme of the staff of the Bank — visited Iran as guests of the Iranian Government and inspected the Abadan refinery and the oil fields. They found the properties in good physical order and declared that they could be put into operation without delay. Meanwhile, the Bank addressed letters to the Governments of Iran and Great Britain formulating its proposal. The reply from Premier Mossadegh raised a number of questions, the most important of which related to the employment of British nationals. The British Government, on its part, expressed accord with the proposal in general.

A second mission, headed by Robert Garner, went to Iran on February 11, 1952. The mission felt that for practical reasons, as well as because of the neutral character of the Bank, AIOC technicians would have to be employed if the oil were to flow again. Premier Mossadegh, however, stated that while he understood the Bank's position, he must ask the Bank to recognize that under the current political tensions his Government could not agree to the employment of any British nationals. Even more difficult was the Premier's demand that the Bank state publicly that it would be operating the oil properties "for Iran's account." This the Bank felt it could not do in view of its neutral character. Another stumbling block was the price of oil. The difference between what the AIOC offered and what the Iranian Government demanded was so great that the Bank could not arrive at any practicable solution.

On February 20, Garner, Rieber and Prud'homme left Teheran for London where they had further discussions with Government and Company representatives. Subsequently, Garner and Rieber returned to Washington, and Prud'homme returned to Teheran.

However, since no progress had been made in reconciling the opposing views on the major points at issue, it was mutually decided to adjourn the talks. [74]

THE INTERNATIONAL COURT'S DECISION

After a number of delays, the International Court of Justice began hearings on the Iranian case, on June 9, 1952 — not on the substance, but on the competence of the Court to deal with the case. Although the Iranian Government had at first refused to participate in the arguments before the Court, it subsequently changed its position and on February 11 filed with the Court a document: "Preliminary Observations: Refusal of the Imperial Government to recognize the jurisdiction of the Court." This document concluded with the notice that not all the Iranian arguments against the competence of the Court had been included, and that Premier Mossadegh would appear personally before the Court to present additional oral observations. [75]

On June 9, Premier Mossadegh came before the Court and made a short statement, which was followed by oral arguments and rejoinders by Professor Henri Rolin, a Belgian international lawyer, on June 9, 10, 11, 18 and 19. [76] The British side was contained in a document: "Observations and Submissions," dated March 24, 1952; oral pleadings were delivered by Sir Lionel Heald on June 13 and 14, and arguments and rejoinder by Sir Eric Beckett on June 16, 17, 21 and 23. [77]

Although the issue was primarily the competence of the Court, both sides debated the substance of the case and submitted auxiliary documents dealing with the problem. The immediate legal issues were reduced to two specific items: the character and limitation of the Iranian declaration of 1932 on adherence to the Court's jurisdiction; and the character of the 1933 agreement between Iran and the Anglo-Iranian Oil Company.

The Court ruled that the phrase in Iran's declaration of adherence to the Court's Statute, Article 36 (2), was limited to disputes relating to the application of treaties and conventions accepted by Iran after ratification of the declaration in September 1932. As to the 1933 agreement, the Court declared that it could not accept the view that the agreement had a double character — both with the Company and with the Government of the United Kingdom. "It is nothing more than a concessionary contract between a Government and a foreign corporation. The United Kingdom Government is not a party to the contract." In concluding its reasoning, the Court declared: "Accordingly, the Court has arrived at the conclusion that it has no jurisdiction to deal with the case submitted to it by the Application of the Government of the United Kingdom dated May 26, 1951." The Court's order of July 5, 1951 became inoperative. [78]

EVENTS IN IRAN

Events in Iran were progressing as hectically as the international negotiations of the Premier. After Mossadegh returned as a victor from the United States, the Majlis voted, 90–1, on November 25, 1951 (this was a reversal of a previous vote), and the Senate the following day, 36–0, for immediate parliamentary elections. This was no doubt an attempt by Mossadegh to obtain popular approval of his oil policy, for some members both in the Majlis and in the Senate were beginning to question, and even to challenge, his policy because of the rapidly deteriorating financial and economic situation. On December 11, opposition and Government Deputies actually engaged in fisticuffs inside the Majlis; two days later the Government announced that elections would start on December 18.

On January 9, 1952 the Foreign Minister presented a note to the British Embassy in Teheran, accusing British officials of interfering in Iranian affairs and threatening to take the most serious measures unless such interference ceased. Three days later the Foreign Ministry ordered that all British consulates in Iran be closed by January 21. Although the British protested the order as a violation of the Anglo-Persian Treaty of March 4, 1857, they nevertheless complied with the order and closed their nine consulates and vice-consulates on January 20. At the end of January, all foreign cultural, information and educational centers — British, American and Russian — in Iran outside Teheran were ordered closed.

On February 19 the old Majlis was dissolved, and on April 27 the new Majlis was opened by the Shah. Premier Mossadegh handed in his resignation, in accordance with constitutional practice, after the reorganization of Parliament on July 5. The next day the Majlis voted, by a large majority, to request the Shah to reinstate Dr. Mossadegh as Premier. After the Senate fell into line — though at first hesitating — and made a similar request, Mossadegh agreed to form a new government. Now assured of his strength, Premier Mossadegh asked Parliament on July 13 for unprecedented powers to rule Iran for six months, in order to solve the economic crisis of the country. When this was refused, he returned his mandate to the Shah, and the latter asked Ahmed Ghavam es-Sultaneh to form a new government.

The new Premier promised an all-out drive to settle the oil dispute. However, on July 19 violence (apparently instigated by Mossadegh's supporters) broke out in Teheran and other cities against the Ghavam government; the religious leader, Ayatollah Kashani, denounced the new Premier, and troops and police clashed with the rioters. Ghavam was determined to employ strong measures to suppress the demonstrations, but Parliament refused him full powers to quell the strikes and riots, and on the 21st he handed in his resignation. The toll of casualties from the riots in Teheran alone was twenty dead and a hundred wounded.

On the same day that the International Court decided that it had no jurisdiction in the Anglo-Iranian case, the Shah called on Mossadegh to form a new govenrment. After presenting his cabinet, in which he was also Minister of War — a major issue between him and the Shah — Mossadegh received a unanimous vote of confidence in the Majlis (July 29). On August 3 the Majlis granted him dictatorial powers to rule the country for six months; the same was reluctantly granted by the Senate on August 11.

On the 7th, Premier Mossadegh sent a note to the British Government demanding that the Anglo-Iranian Oil Company pay the amounts which were due from the Company when nationalization was enacted; the sum mentioned amounted to some 40 million pounds.

About three weeks later the American and British Ambassadors presented to the Premier a compromise plan worked out by President Truman and Prime Minister Churchill for restoring oil operations. The following main provisions were primarily in reply to the August 7 demand: 1) the question of compensation to be paid for the nationalization of the oil industry to be submitted to the International Court, "having regard to the legal position of the parties existing immediately prior to the nationalization and to all claims and counterclaims of both parties"; 2) suitable representatives to be appointed to represent the Iranian Government and the Anglo-Iranian Oil Company in negotiations for making arrangements for the flow of oil from Iran to world markets; 3) should the Iranian Government agree to the proposals in the above two provisions, it would be understood that (A) representatives of the AIOC would seek arrangements for the movement of oil already stored in Iran, and as the oil moved, for appropriate payments to be made; (B) the British Government to relax restrictions on exports to Iran and on Iran's use of sterling; (C) the United States Government to make an immediate grant of $10 million to the Iranian Government to assist in budgetary problems.[79] Premier Mossadegh rejected the proposals.[80]

Secretary of State Dean Acheson appealed publicly to the Premier and the Iranian Parliament to reconsider the rejection.[81] On September 7 Mossadegh attacked the Anglo-American proposals as the worst Iran ever received. About ten days later, while outlining to the Majlis his solution for the oil problem, he thretened to break off diplomatic relations with Great Britain; the Majlis voted unanimously to support the Premier in his rejection of the Truman-Churchill proposals.

On September 24, Premier Mossadegh handed to the American and British diplomatic representatives the formal rejection of the Truman-Churchill proposals. The main argument was that they were inconsistent with the laws of nationalization of the oil industry. The Premier analyzed each of the items and repeated Iran's basic claims against the Anglo-Iranian Oil Company. He then made counterproposals. He declared that the Iranian courts were the only competent channels for investigating the former Company's claims. However, should the Company want to go to the International Court, it would have to be on the basis of an agreement between the two parties, and there must be no recognition of the existence of the dispute between the two Governments. He proposed the following four articles as the basis of agreement:

1) *Compensation.* Determination of the amount of compensation to be paid for property belonging to the former oil company at the time of the nationalization of the oil industry, and arrangements for paying such compensation by installments based on any laws carried out by any country which nationalized its industries which might be agreed to by the former oil company.

2) *Basis of Examination of Claims.* Examination of the claims of both parties on the basis of one of three possibilities to be recognized by the International Court of Justice as fair and just for settling the parties' claims and used by it as basis for judgment.

3) *Determination of Damages.* Examination and determination of the amount of damages caused to the Iranian Government as a result of the difficulties and obstacles put in the way of the sale of Iranian oil by direct and indirect activities of

the former oil company, as well as the losses resulting from the delay in payment of funds, which were definitely debts owed by the Company.

4) *Payment in Advance and on Account.* Payment in advance and on account of £49 million shown on the former oil company's 1950 balance sheet as increases in royalty, taxes and dividends due to Iran from the reserves.

The Premier declared that his Government, after agreement on the four articles above, would be "prepared to agree to the judgment of the International Court, and in this case the International Court will be requested to issue its final verdict as soon and as far as possible within six months." [82] He felt that the readiness to refer to the judgment of the Court on the basis of the four articles was a sign of extraordinary concessions on the part of the Iranian Government. The proposal was valid for ten days from delivery.

Early in October, British Foreign Secretary Anthony Eden, in a message to Premier Mossadegh, refuted the implication which the latter had read into the joint Anglo-American proposals. A message from Secretary Acheson expressed President Truman's disappointment at the rejection, and he too declared that Premier Mossadegh had completely misread and misinterpreted the joint proposals. [83] However, the Iranian Premier remained unconvinced. The only concession he was ready to make was to reduce the £49 million on account, which he asked for in Article 4, to £20 million, and to defer payment of the balance until the termination of the negotiations, which were expected to last three weeks. In his letter to Acheson of October 7, he emphasized the economic hardships which Iran had experienced as a result of the unwillingness of the British to come to terms, and the danger to security and order inherent in such a situation. "I am certain you will agree that the prompt and immediate settlement of this matter would be a great and important contribution towards insuring the peace and public security of one of the sensitive areas of the world."

In his letter to Eden of the same date, the Premier reiterated his proposals of September 24 and declared that representatives of the former AIOC, invested with full powers, were invited to come to Teheran within a week from date for discussions within the limits of the Iranian Government's counterproposals. Before the departure of the Company representatives from London, £20 million, convertible into dollars, of the £49 million mentioned in Article 4 of the counterproposals, was to be put at the disposal of the Iranian Ministry of Finance, and the remainder of the £49 million was to be placed to the credit of the Iranian Government at the end of the negotiations. In this letter, as in that to Acheson, Premier Mossadegh clearly indicated the danger to the West from serious disturbances in Iran, for he concluded: "I once again remind you of the impossibility of the continuation of this state of affairs and any eventuality arising from pursuit of this policy is not the responsibility of the Iranian Government." [84]

A week later the British Government rejected the latest Iranian counterproposals, and on October 16 the Premier announced over Radio Teheran that because of the British refusal to accept his proposals, he was forced to sever diplomatic relations. The official act took place on October 27.

MOSSADEGH'S LAST EFFORTS

Since the Iranian Senate was often reluctant to support Premier Mossadegh's policies and hesitated to grant him dictatorial powers, the Majlis voted the Senate out of office on October 23, 1952. On November 16, Premier Mossadegh ordered the dissolution of the Supreme Court. Early in January 1953 he obtained from the Majlis a 64–0 vote approving the new electoral law, and on the 8th of January he asked the Majlis to extend his extraordinary powers to govern by decree for a full year. The Majlis was at first unwilling to grant these powers, and the religious leader, Ayatollah Kashani, the titular Speaker of the Majlis, strenuously opposed Mossadegh's demands. However, after a wave of demonstrations had swept the country, the Majlis, by a vote of 59–1 with 6 abstentions, granted the Premier the dictatorial powers for a year which he demanded.

Even after the breaking off of diplomatic relations between Iran and Great Britain, the United States did not completely give up the possibility of working out some solution to start the oil flowing again. During the meeting of the Foreign Ministers in Paris in the middle of December 1952, Secretary of State Dean Acheson and Foreign Secretary Anthony Eden discussed the Iranian issue, and subsequently the United States Ambassador in Teheran, Loy Henderson, began a series of conversations with Premier Mossadegh which extended into 1953. At the end of December, the Assistant Secretary of State for Near Eastern, South Asian and African Affairs, Henry A. Byroade, opened discussions in London with British representatives. By the middle of January 1953 it looked as if a solution were near, for Premier Mossadegh declared, though cautiously and with reservations, that the oil question might be resolved within two or three days, and on February 12 he appointed a committee to prepare a full list of Iran's claims against the AIOC. This was obviously connected with the proposed plan discussed by the Premier and the Ambassador. On the 20th, the British transmitted, through the American Ambassador, a solution worked out in conjunction with the United States, which provided that: 1) the International Court to decide the issue of compensation arising from the nationalization of the enterprise of the AIOC in Iran on the basis of equitable compensation for the loss of the Company's business. Iran, of course, could present counterclaims; 2) an American corporation to purchase from the NIOC oil to the amount of $133 million with general discount, and to pay an advance of $50 million immediately after signing the agreement with the NIOC; as a condition to this sale, an international company to be formed in which the AIOC could also be a member, and the Iranian Government to agree that the NIOC could enter into negotiations with this international company for an agreement for the sale of large quantities of crude and refined oil; [85] 3) before the amount of compensation was determined by the International Court, Iran to pay Britain 25 per cent of her gross receipts from the sale of oil, and after the amount was determined by international arbitration, Iran to pay 25 per cent either in sterling or in deliveries of oil to the AIOC. The period of payment was to be twenty years. [86]

After a meeting between Anthony Eden and President Eisenhower, a joint

Anglo-American communiqué was issued on March 8 approving the "final offer," and the United States declared the proposal reasonable and fair. However, in an address over Teheran Radio on March 20, Premier Mossadegh rejected the British proposal. The underlying and perhaps major stumbling block was the insistence of the British that the basis of compensation be the loss of business by the AIOC as a result of the Nationalization Law; this to the Iranians meant the same as the joint proposals of August 27, 1952.

Premier Mossadegh gave two reasons for the stiffening of the attitude of the British: they expected that anti-Mossadegh elements would replace him and then come to terms with the British; and the other big oil companies in the Middle East had been persuaded by the British of the danger to themselves from Iranian nationalization.[87] He offered counterproposals: 1) the Company either to agree to determine the amount of compensation on a basis acceptable to the Iranian Government, or state the highest amount of its claims which could be considered equitable to the Iranian Government, and that amount, together with the Iranian counterclaims, to be submitted to the International Court; 2) the Iranian Government to pay in the shortest period any amount accorded by the judgment of the International Court, either from the 25 per cent net revenue deposited, or by delivery of crude or refined oil at fair international prices; 3) instead of referring the matter to the International Court, the Iranian Government was ready to enter into direct negotiations with fully empowered representatives of the AIOC and the British Government.[88] These counterproposals were not acceptable.

As a last attempt, Premier Mossadegh wrote a letter to President Eisenhower on May 28, 1953. He complained that because of the legal efforts of the British to block the sale of Iranian oil, his country had suffered great financial hardship, and furthermore, he himself had to contend with political intrigue "carried on by the former oil company and the British Government." Reminding him of the possible danger, from an international viewpoint, which might result from the economic and political difficulties facing Iran if prompt and effective aid were not given to his country, the Premier appealed to the President that he either prevail on the British to accept the Iranian terms and make possible the renewal of revenue from the oil, or advance sufficient aid to Iran to develop her other resources in order to prevent her from completely collapsing. The letter concluded: "I invite Your Excellency's sympathetic and responsive attention to the present dangerous situation of Iran, and I trust that you will ascribe to all the points contained in this message the importance due them."

Only on June 29 did the President reply to the Premier, and the correspondence was not released until July 9 when the situation in Iran was getting entirely out of hand. Stripped of all its civilities, the President's letter told Iran that the United States expected her to live up to her international obligations, and that if she arrived at an agreement in the matter of compensation she "would strengthen confidence throughout the world in the determination of Iran fully to adhere to the principles which render possible a harmonious community of free nations; that it would contribute to the strengthening of the international credit standing of Iran;

and it would lead to the solution of some of the financial and economic problems at present facing Iran." The President bluntly told the Premier of Iran that it would not be fair to spend American taxpayers' money to aid his country "so long as Iran could have access to funds derived from the sale of its oil and oil products if a reasonable agreement were reached with regard to compensation, whereby the large-scale marketing of Iranian oil would be resumed." Nor could the United States purchase Iranian oil without a settlement between Iran and Great Britain. The President minced no words in telling the Premier that compensation merely for loss of the AIOC's physical assets could not be called "a reasonable settlement and that an agreement to such a settlement might tend to weaken mutual trust between free nations engaged in friendly economic intercourse." Mr. Eisenhower politely but firmly refused to assume responsibility for the possible international repercussions to Iran from her difficulties. "I note the concern reflected in your letter at the present dangerous situation in Iran and sincerely hope that before it is too late the Government of Iran will take such steps as are in its power to prevent a further deterioration of that situation." [89]

The failure of the Premier with the President drove him to further extremes.

Tension between the Shah and the Premier mounted steadily as the latter sought to gain control over the Army, which was traditionally under the direct control of the Shah. On March 1 the Premier had the Army chief of staff arrested after he had failed to quell an anti-Mossadegh demonstration. Nevertheless, demonstrations of pro-Shah and pro-Mossadegh elements continued to take place in Teheran throughout the early part of March. The Premier had a number of Army officers arrested whom he suspected of participating in a Court-backed intrigue against him. On April 6, 1953, the Premier asked the Majlis to order the Shah, in accordance with the recommendations of the special conciliation committee, to "reign constitutionally" instead of ruling the country. Ayatollah Kashani thereupon broke completely with Mossadegh and attacked him violently on June 30. The pro-Mossadegh deputies in the Majlis resigned their seats and thus paralyzed the lower House. On July 19 the new Speaker of the House declared the end of the Majlis. Two days later the Tudeh party and the National Front demonstrated separately in Teheran on the anniversary of Premier Mossadegh's rise to power, challenging the opposition Deputies. Mossadegh then resorted to a plebiscite to determine the struggle between him and the Majlis in his efforts to dissolve the latter. The plebiscite was held early in August and Mossadegh won over 99 per cent of the votes cast in the Teheran district. Ten days later the Ministry of the Interior announced that the government had won the plebiscite to dissolve the Majlis, by 2,043,389 votes to 1,207.

The position of the Shah became precarious, and on August 16, having failed to oust Mossadegh after issuing two decrees — one naming General Fazlollah Zahedi Premier, and the other ordering the arrest of Mossadegh — he and his queen fled to Baghdad on the way to Rome. The next day supporters of Mossadegh demonstrated in the streets of Teheran and threw down statues of the Shah and his father. Many military men who were suspected of cooperating with the Shah were arrested. Two days later backers of the Shah clashed with and overcame the supporters of Mossadegh and he was swept out of office. General Zahedi took over the government after some three hundred persons had been killed and several hundred wounded. Three days later the Shah returned from Rome.

With the end of Premier Mossadegh's regime, the chapter of the story of the Iranian oil which began with the death of General Razmara came to an end.

ATTITUDES, TACTICS AND POLICIES

This oil controversy between Iran and Great Britain revealed as never before the basic differences of attitude to the underlying issues between the two peoples and governments. The Anglo-Iranian Oil Company — and with it the British Government — approached the matter as a business venture. The Iranian Government had signed a business contract and was therefore bound, regardless of the terms, to live up to it. To be sure, the Company was willing voluntarily to consider modification of some of the terms of the contract and to make some adjustments so that Iran might improve her share in the bargain, but at no time was the Company willing to surrender its basic business advantages and practice of exploiting the very fortunate venture in which it was engaged. Nor was it even willing to grant a sizeable share of its profits to the Iranians. Yes, it was a fabulous, multi-million pound venture, but that was the Company's good fortune, as well as the result of its considerable risk of capital and of its industrial achievements. [90]

Time and again the notion of sanctity of contract was pointed out as the real issue in the controversy — by Foreign Secretary Herbert Morrison in the House of Commons, by Sir Gladwyn Jebb before the United Nations Security Council, and by the representatives of the Anglo-Iranian Oil Company. [91]

Moreover, since the Iranians could not themselves have developed the oil resources of their country, because of their poverty and lack of technical knowhow, the British felt that they had rendered great services — financial, economic, health, educational, etc. — to the Iranians, and the latter should have been grateful for what they had gained from the development of the oil resources. [92]

The British, therefore, when the question of nationalization first loomed, were unwilling to agree to any concessions to the Iranians. In spite of the changing circumstances in Iran, in the Middle East, and in the world at large, the Company and the British Government refused to modify their position. Indeed, both in Parliament especially the Conservative opposition, and in the Conservative press, a vociferous cry went up for the use of military force against the Iranians to protect British interests. Herbert Morrison, then Foreign Secretary, was under constant pressure to employ force not only to protect the lives of British nationals, but also the properties of the AIOC, and he had to concede to the extent of several times sending a warship to the area.

The Iranians, on the other hand, failed to appreciate the concept of sanctity of contracts. They felt that on the whole the Anglo-Iranian Oil Company was exploiting their natural resources and reaping fabulous profits, of which they were granted only a small, miserable fraction. At least in their eyes some of the practices of the Company were more than questionable. To be sure, the AIOC had risked capital and it did possess the technical knowledge, but the returns were extremely high — in some years as much as 150 per cent — and the Iranians felt that they should have had a much larger share of their own heaven-sent abundance. Against the statistics of the Company showing how many millions of pounds sterling it had paid to the Iranian Government in royalties, taxes, etc., the Iranians produced statistics of their own, but since they did not have access to the Company's books, these were not as authentic as the Company's. But to the Iranians their own statistics made good sense. They felt that they were losing out in income taxes, in customs duties, in the secret special price charged by the Company for oil sold to the British Navy, and in the Company's income tax payments to the British Government before the share of net profits was calculated. But most of all, they bitterly resented the very low royalties which they were receiving, even under the modified 1933 agreement. [93] The Iranians were also angry at the slow progress made in the substitution of Iranians for non-Iranians in technical and managerial positions in the Company, and at the relatively high price Iranians were charged for petroleum and petroleum products. An underlying emotional factor, at least with some of the Iranian leaders, was that the Company was dominating their country economically, and that full independence could be achieved only by gaining control of the oil industry. All these grievances made them feel that, in justice, the time had come to recover their natural resources and nationalize the oil industry.

The most important advantages the British had in the dispute were their technical knowledge and equipment, the means of transportation, the marketing facilities, and the natural unwillingness of the other great oil companies in the area to acquiesce in the act of nationalization of the Iranian Government. Furthermore, the British Government counted heavily on the support of the United States, because of the latter's oil interests in Saudi Arabia, Iraq and other areas of the Middle East, despite the fact that both in Parliament and in the British press voices — sometimes very strong — were heard declaring that the source of the trouble in Iran were the American Oil companies who, in their lust for control, had instigated the Iranians to oust the British so that they, the Americans, could come in and take over; not only were private individuals mentioned, but even Assistant Secretary of State George C. McGhee — an oil man himself — was pointed out as one of the arch villains in the Iranian oil drama.

At the beginning, therefore, the British Government refused to recognize the Nationalization Law of Iran. It maintained that the 1933 agreement was of a double character: it was a contract between Iran and the AIOC, and a concessionary agreement between Iran and Great Britain, and for Iran to nationalize the oil industry would be a violation of an international agreement between two States. It was only under strong pressure from the United States that it subsequently agreed to recognize the principle of nationalization, but under no circumstances would it recognize the nine-article Enabling Law. What exactly the British understood by the principle of nationalization was never made clear, but whatever they understood it to mean, it was not what the Iranians had understood it to mean.[94]

After submitting to American pressure and recognizing the principle of nationalization, the British were ready to work out arrangements which would not interfere too much with the actual operations of the Company and its management, although granting in principle general overall ownership to the Iranian Government. Though they were at first not too happy about President Truman's offer to send Mr. Harriman as his personal representative to Iran, they subsequently accepted the President's good offices, albeit with cautious reservations and limitations, in the hope that the United States would prevail on the Iranians to submit to the British proposals. Under the Stokes proposals which followed the Harriman mission, the Company would have kept control of production and marketing, and would have obtained the oil at greatly reduced prices. At worst the result would have been a 50-50 profit-sharing arrangement, a scheme which in fact the Company had actually offered the Iranian Government early in 1951, and which had been rejected.

The actual stoppage of the oil flow from Iran caused the British great anxiety and for a time threw them almost into a panic, for it created very serious additional currency difficulties, since a considerable quantity of oil for British consumption had to be bought in the dollar market. But as other Middle Eastern countries increased production and the British no longer had to buy in the dollar market, their attitude changed; they were ready to fight it out with the Iranians to the very end.

During the early stages of the crisis, the British concentrated on the issue of violation of contract, and they tried to force the hands of the Iranians through the International Court. After Iran's rejection of the Court's interim order and the failure of the Stokes mission, they resorted to the Security Council. When their case was lost there too, they returned to the Court for a decision on jurisdiction, and it was only when the Court declared that it was not competent to deal with the issue that the British began to talk seriously of compensation. But here, too, as in so many other respects, their conception of compensation differed basically from that of the Iranians. Not only did they expect compensation for the AIOC's physical plant, but also for the profits which the Company stood to lose as a result of the nationalization of the oil industry before the expiry of the 1933 concession. Of this basic recognition of their rights in the concession, the British apparently had little difficulty in convincing the Americans; the result was the first provision of the Truman-Churchill joint proposals of August 27, 1952, as well as the February 20, 1953 proposal. Since these were both rejected by the Iranians, the only course left for the British, short of military intervention, was to starve the Iranians into submission, and that, in fact, was exactly the policy they followed.

The Iranians calculated that after the passage of the Nationalization and Enabling Laws, the United States would exercise pressure on the British to accept the inevitable. Their major assets, they felt, were American fear of Russian penetration into the Middle East in case of serious political difficulties in Iran, and the importance of Iranian oil to NATO and for the reconstruction and development of Western Europe. They knew that Iran herself was neither able to operate the oil industry, nor to supply the transportation means; neither did they have the necessary marketing facilities. But out of naiveté, or perhaps taking a gambler s long chance, they believed that the United States, in its anxiety over possible Russian occupation of Iran, would overrule the British, send American technicians to Iran, and offer American transportation facilities. Or failing that, the Iranians thought they could employ other neutral technicians and offer to sell oil f.o.b. Iran, with American approval. In both assumptions they proved mistaken. [95]

The role of the United States in the Iranian oil crisis was, to say the least, difficult and uncomfortable, perhaps even impossible. Basically, Washington had no well thought-out policy for dealing with the crisis, and perhaps could not have had one, for the State Department, like the British Foreign Office, apparently did not foresee the rapid development of events.

On the one hand, the traditional American anti-colonial sentiment and liberal point of view disposed Americans to at least a sympathetic approach to the Iranian demands. Moreover, as Iran well calculated, American anxiety over the political consequences of possible Soviet penetration into Iran as a result of the oil crisis made Americans eager to meet the Iranians at least half way. On the other hand, as the British well calculated, the Americans could not possibly overlook the consequences for their own oil interests in the area from Iranian nationalization; the United States must recognize the importance of the oil industry to the British economy, and must not impair British friendship.

The United States thus found itself playing the unenviable role of honest broker where each side expected the broker to take his part. The Iranians thought the disinterested Americans should take their part against imperialistic Britain, and American diplomatic representatives made statements which could be so interpreted. Every time the Americans asked the Iranians to make sacrifices, the cry of protest was loud. The British felt that in fact they were also fighting the American battle, for should they lose in Iran all American oil interests in the Middle East would be jeopardized, and whenever the Americans made any gesture toward Iran, the British were quick to react.

Ultimately, the basic differences in method and approach of the British and Iranians drove the Americans into the British corner. The British arguments made much better sense to the Americans than did those of the Iranians. Not only did Americans understand the validity of the sanctity of contract and the cry against expropriation or confiscation, but they also understood the technical, financial, commercial and managerial problems involved in reactivating such a gigantic industrial organization as the Iran oil industry in all its aspects. They clearly comprehended that only by some arrangement with the AIOC could Iranian oil flow again in a manner which would not only revive the industry, but make compensation possible. Indeed, in the various formulae offered by the British, the Americans could see compromise as regards both production and the sale of the oil, and good will, and at the same time possibilities for sensible businesslike operations.

The Iranians only exasperated the Americans. While they were always ready to enter into negotiations on the basis of some compromise, at the end, when the discussions were about to materialize in a concrete proposal, they invariably reverted to their previous demands, without appreciable change or compromise.

Finally, though they were still disturbed about political eventualities, with the increased oil production in the Middle East supplanting Iranian oil, and because of their desire not to antagonize the British, the Americans came around to the British point of view that the deterioration of economic and financial conditions in Iran would force Mossadegh out, rather than bring the Soviets in, and that a new government would be amenable to a sensible solution, of course, with anticipated American-promised aid.

When, therefore, Mossadegh frantically appealed to the President of the United States for help, in May 1953, the attitude of the United States became crystal clear: agree to the British proposals and financial support will be forthcoming; refuse and the American taxpayer will not want to grant you any additional aid. But Mossadegh could no longer retreat even if he wanted to, for he had become a prisoner of his own political agitation. He was determined to nationalize the oil industry not merely in principle but in actuality; not to allow any British personnel, technicians or managers, to assume responsible independent positions in the re-activation of the industry; and to compensate the Company only for its physical plant in Iran, and this subject to Iranian counterclaims. To these demands neither the British nor the Americans would agree. The inevitable result was intransigence. Mossadegh tried desperately to continue his regime without the oil revenues, and he therefore resorted to dictatorship, but this failed and he was swept out of office.

NOTES

1. See above, Chapter III.
2. Foreign Secretary Herbert Morrison stated in the House of Commons: "In 1948 changes in world conditions suggested that some modification of rates of the payments made by the Company to the Government might be justifiable." *PDC*, 487, col. 1008, May 1, 1951.
3. According to statements by Anthony Eden and Winston Churchill in the House of Commons, the Anglo-Iranian Oil Co. had earned about 150 per cent in 1949 and paid out in dividends only 30 per cent. *PDC*, 489, col. 758, June 21, 1951; *PDC*, 491, col. 990, July 30, 1951; *PDL*, 173, col. 30, July 31, 1951.
4. Anglo-Iranian Oil Co., Ltd., *The Chairman's Statements to Stockholders in the Annual Reports for 1948, 1949, 1950, 1951, 1952 regarding the Company's interests in Iran; with a chronology of main events from 1948 to May, 1953* (London, 1953), 1.
5. See above, Chapter III.
6. *Ibid.*
7. Anglo-Iranian Oil Co., *The Chairman's Statements,* 2–3. The Company calculated that under the 1933 agreement the respective sums for 1948 and 1949 would have been £9,172,245 and £13,489,271.
8. *Middle Eastern Affairs,* II, 294, Aug.–Sept., 1951 (henceforth cited *MEA*). Most of the primary documents of the Anglo-Iranian dispute of this period are

to be found in *Pleadings; Correspondence between His Majesty's Government in the United Kingdom and the Persian Government, and Related Documents concerning the Oil Industry in Persia. February 1951 to September 1951.* Cmd. 8425, 1951; *Parliamentary Debates House of Commons* and *House of Lords;* Iranian Embassy, Washington, *Some Documents on the Nationalization of the Oil Industry in Iran* (n.d.), and other Iranian publications; *Department of State Bulletin; Middle Eastern Affairs;* and *The New York Times.* There are differences due to translation, as well as between documents drawn up in the home offices and those actually presented in the foreign countries by the diplomatic agents. Henceforward, notes will refer to one or two of these sources.

9. *MEA*, II, 295, Aug.–Sept., 1951; *PDC,* 488, cols. 1665–1666, June 11, 1951.
10. Art. XXII provided: "Any differences between the parties of any nature whatever and in particular any differences arising out of the interpretation of this Agreement and of the rights and obligations therein contained as well as any differences of opinion which may arise relative to questions for the settlements of which, by the terms of this Agreement, the agreement of both parties is necessary, shall be settled by arbitration." Cmd. 8425, 1951, 17; see above, Chapter III.
11. *MEA*, II, 230–231, June–July, 1951.
12. *PDC,* 487, col. 1009, May 1, 1951.
13. *MEA*, II, 231–232, June–July, 1951; Cmd. 8425, 1951, 32–33.
14. *MEA*, II, 232–233, June–July, 1951; Cmd. 8425, 1951, 34–36.
15. The basic attitude of Ambassador Grady and his differences with the British are outlined in his article, "What Went Wrong in Iran?", *The Saturday Evening Post,* 30, 56–58, Jan. 5, 1952.
16. *Department of State Bulletin,* XXIV, 851, May 28, 1951; *MEA,* II, 234, June–July, 1951.
17. *MEA*, II, 234–235, June–July, 1951. On May 29, Secretary Morrison stated in the House of Commons that the British Ambassador in Teheran had informed the Iranian Government: "While His Majesty's Government cannot accept the right of the Persian Government to repudiate contracts, they are prepared to consider a settlement which would involve some form of nationalization, provided – a qualification to which they attach importance – it were satisfactory in other respects.' *PDC,* 488, cols. 41–42, May 29, 1951.
18. Cmd. 8425, 1951, 37.
19. *MEA*, II, 235–236, June–July, 1951; Cmd. 8425, 1951, 37–38.
20. Cmd. 8425, 1951, 38–39.
21. For the British version of the events which culminated in the decision to submit the issue to the Court, see *PDC,* 488, cols. 40–43, May 29, 1951.
22. *Art. 7 of Enabling Law:* Purchasers of the products of the Oil Fields from which the former Anglo-Iranian Oil Company has been removed can hereafter purchase annually at current world market prices the same quantities purchased by them annually during the period commencing from the beginning of 1948 up to 29th Esfand 1329 (20th March, 1951). For additional quantities they shall enjoy priority, other conditions being equal. *MEA,* II, 295, Aug.–Sept., 1951.

23. Cmd. 8425, 1951, 40.
24. *Ibid.*, 39–41.
25. *MEA*, II, 237–238, June–July, 1951.
26. This letter from the President was released in Teheran in an Iranian translation. The original American note was not released in the US, nor was the letter to Prime Minister Attlee made public. We can, however, trace the reaction of the British from various statements made by Herbert Morrison in the House of Commons. In reply to a question, Morrison stated: "Suitable conversations have taken place with the United States Government and, on the whole, we are acting in cooperation." *PDC*, 488, col. 43, May 29, 1951.
27. Cmd. 8425, 1951, 41.
28. *MEA*, II, 238, June–July, 1951.
29. Cmd. 8425, 1951, 41; *PDC*, 488, col. 687, June 4, 1951.
30. Cmd. 8425, 1951, 42.
31. Iranian Embassy, Washington, *Some Documents*, 28–29.
32. *PDC*, 489, cols. 519–526, June 20, 1951.
33. *Ibid.*, col. 1184, June 26, 1951.
34. *Ibid.*, cols. 1184–1186.
35. *Department of State Bulletin*, XXV, 73, July 9, 1951.
36. *MEA*, II, 296–297, Aug.–Sept., 1951; Cmd. 8425, 1951, 44.
37. *Pleadings*, 45.
38. International Court of Justice, "Anglo-Iranian Oil Company Case Order of July 5, 1951," *Reports 1951*, 89–94. Judges Winiarski and Badawi Pasha dissented; their opinions are stated, *ibid.*, 95–98.
39. Cmd. 8425, 1951, 51.
40. *Pleadings*, 133–134.
41. Iranian Embassy, Washington, *Some Documents*, 27; *Department of State Bulletin*, XXV, 128–130, July 23, 1951.
42. *Department of State Bulletin*, XXV, 130, July 23, 1951; *MEA*, II, 299–300, Aug.–Sept., 1951.
43. Richard R. Stokes explained the Harriman basic achievement, which made negotiations at all possible, in that he had obtained the consent of the mixed oil committee to interpret the words in the Nationalization Law that "all operations for exploration, extraction and exploitation shall be in the hands of the Government" as under the "authority of the Government," which Stokes remarked was not quite the same thing. According to Stokes, Premier Mossadegh had never really accepted the difference. *PDC*, 494, col. 241, Nov. 20, 1951.
44. Iranian Embassy, Washington, *Some Documents*, 27–28; Cmd. 8425, 1951, 52–53.
45. Iranian Embassy, Washington, *Some Documents*, 27–35.
46. *Ibid.*, 36–37.
47. Cmd. 8425, 1951, 54–55; *Pleadings*, 134–135.
48. Cmd. 8425, 1951, 55–57.
49. Stokes subsequently explained that the reason he had withdrawn the proposal was the decision of the Premier to submit it to the Majlis without any recommendation from himself, "which seemed a quite impossible thing to allow to happen." He told Mossadegh that he had come to negotiate with him and not

with the Majlis. *PDC*, 494, col. 243, Nov. 20, 1951.

50. Cmd. 8425, 1951, 57.

51. *Ibid.*, 58–59.

52. Iranian Embassy, Washington, *Some Documents*, 44–46. For Harriman's efforts on behalf of the Stokes plan and the objections of the Iranian delegation to the plan, see *ibid.*, 39–43; see also the President's statement of Aug. 23, *Department of State Bulletin*, XXV, 382, Sept. 3, 1951.

53. *Department of State Bulletin*, XXV, 547–548, Oct. 1, 1951; Cmd. 8425, 1951, 60–61; *Pleadings*, 140–142.

54. *PDC*, 493, col. 20, Nov. 13, 1951.

55. *Department of State Bulletin*, XXV, 548–550, Oct. 1, 1951; Cmd. 8425, 1951, 63–64; *Pleadings*, 142–145.

56. Cmd. 8425, 1951, 64–66.

57. *Letter dated 28 September 1951 from the Deputy Permanent Representative of the United Kingdom addressed to the President of the Security Council and the Secretary-General*, S/2357, Sept. 29, 1951. In accordance with Article 41, Paragraph 2, of the Statute of the Court, the latter notified the Security Council of the provisional measures indicated by the Court. As mentioned above, the British immediately declared that they were willing to abide by the indication of the Court, while Iran, in protest of the indication, withdrew her declaration on the optional clause of the Court.

58. S/2358, Sept. 29, 1951.

59. United Nations, Security Council, *Official Records*, S/PV559, Oct. 1, 1951, 10. (Henceforth references to Security Council records and documents cited by their symbols — S/ followed by the number and date.)

60. *Ibid.*, 26.

61. *Ibid.*, 26–30.

62. The question of the competence of the Council to deal with the issue seriously disturbed some of the members, and some who voted for the inclusion of the item on the agenda stated that they did so in order to give Iran an opportunity to prove the Court's incompetence. S/PV559, Oct. 1, 1951, 7.

63. S/2358/Rev. 1, Oct. 12, 1951.

64. S/PV560, Oct. 15, 1951, 2.

65. *Ibid.*, 7.

66. S/2379, Oct. 16, 1951; S/PV561, Oct. 16, 1951, 15–17.

67. *Ibid.*, 19.

68. *Ibid.*, 23; S/PV562, Oct. 17, 1951, 3; S/2358/Rev. 2, Oct. 17, 1951.

69. S/PV562, Oct. 17, 1951, 10; S/2380, Oct. 17, 1951.

70. S/PV565, Oct. 19, 1951, 2–3.

71. *Ibid.*, 3–4.

72. *Ibid.*, 12.

73. *Department of State Bulletin*, XXV, 864, Nov. 26, 1951.

74. International Bank for Reconstruction and Development, Press Release No. 285, Apr. 3, 1952; – *Seventh Annual Report 1951–1952* (Washington, 1952). 17–18.

75. *Pleadings*, 281–308, and annexes.

76. *Ibid.*, 437–508, 585–627.

77. *Ibid.*, 320–370, 509–542, 628–669.

78. International Court of Justice, "Anglo-Iranian Oil Company Case (Jurisdiction) Judgment of July 22, 1952," *Reports, 1952,* 93–114. The decision was given by a 9–5 vote. The dissent of the five judges – Sir Arnold McNair, A. Alvarez, Hackworth, Read, and Levi Corneiro – is given in 116–117.

79. *Department of State Bulletin,* XXVII, 360, Sept. 8, 1952; *MEA,* III, 246–247, Aug.–Sept., 1952.

80. *MEA,* III, 247, Aug.–Sept., 1952. He rejected them orally on Aug. 27, and told the British and American diplomatic representatives that the publication of the proposals would have a bad effect on the Iranian people, and that he expected to get a satisfactory reply to his Aug. 7 demand. After hearing his reasons for rejecting the proposals, the diplomatic representatives asked him to keep the contents of the proposals secret until they had submitted a reply to the Iranian demand, in a week's time. However, on the 30th they returned with the same proposals. Mossadegh rejected them and stated that he had decided to call both houses of Parliament into session so that with consultation the necessary reply to the message can be prepared."

81. *Department of State Bulletin,* XXVII, 405–406, Sept. 15, 1952.

82. *Ibid.,* 532–535, Oct. 6, 1952; *MEA,* III, 284–289, Oct., 1952.

83. *Department of State Bulletin,* XXVII, 569, Oct. 13, 1952; *MEA,* III, 289–290, Oct., 1952.

84. *MEA,* III, 289–290, Oct., 1952.

85. *Text of the Report Broadcast by Dr. Mohammed Mossadegh, the Prime Minister, over Teheran Radio to the Iranian Nation Regarding the Nationalization of Oil Industry from Its Beginning up to 20th March 1953* (Teheran, 1953), 15–16.

86. *The New York Times,* Mar. 21, 1953.

87. *Text of the Report Broadcast,* 17–20. The tone of this broadcast would indicate that Mossadegh was no longer so sure of his position, that the pressure for a solution of the oil problem and the alleviation of economic conditions was becoming ever greater. The dictatorial powers Mossadegh obtained from the Majlis were his last desperate effort to save his policy, and he pleaded for national unified support for it.

88. *Ibid.*

89. *Department of State Bulletin,* XXIX, 74–76, July 20, 1953; *The New York Times,* July 10, 1953.

90. See below, Chapter VI.

91. Herbert Morrison stated in the House of Commons, after concluding a full debate on the Iranian oil question, that the essential point was neither Iran's right to pass legislation to nationalize the oil, nor the question of compensation. "The real issue is, therefore, the wrong done if a sovereign State break a contract which it has deliberately made." *PDC,* 489, col. 824, June 21, 1951.

Sir Gladwyn Jebb told the Security Council: "Its [Iranian Government] persistent refusal to recognize the sanctity of contracts led first to the complete stoppage of oil-tanker sailings, then to the closing down of the oil industry and then finally, to the departure of those British technicians on whom Iran must depend for the efficient operation of the industry." S/PV561, Oct. 16, 1951, 13.

92. A lengthy ' Statement of Relevant Facts up to 1st May 1951" submitted by the British to the International Court of Justice on Oct. 10, 1951 concluded with the statement: "It is no exaggeration to say that, but for the Anglo-Iranian Oil Company and its activities and investments in Iran, that country would be in a far less developed and prosperous condition than it actually is at the present day." *Pleadings*, 219.

Addressing the Security Council, Sir Gladwyn Jebb, after listing all the benefits which the Company had conferred upon the Iranians and declaring that the Iranian Government's charges that the activities of the AIOC had impoverished the Iranian people were untrue, concluded: "The Anglo-Iranian Oil Company, by its skill and foresight and good management, succeeded in developing the oil of Iran. Oil in the ground is, after all, only a potential source of wealth, and since the Iranian Government could not itself develop that oil, it was essential, rightly or wrongly, for the Iranian Government to arrange for its development by foreigners. That is a fact. The terms on which the Anglo-Iranian Oil Company offered its services were always at least as favorable to Iran as any obtaining at any given time in the Middle East. In other words, without the Anglo-Iranian Company the Iranian people would be not richer but incalculably poorer today than they are and would not possess an industrial potential which, in agreement with those who built it, can and should form the basis of the future prosperity of the Iranian nation." S/PV560, Oct. 15, 1951, 29.

After the Conservatives gained power, Foreign Secretary Anthony Eden, who had previously criticized the financial policies of the Company, declared in the House of Commons: "Much can be said about the division of profits. There is a lot I could say about my thoughts in years gone by about who has been greedy, and so on, and who has taken much out of the industry in the past. But it remains true that it was the receipt of the large oil revenues which enabled Persia to contemplate the financing of the extensive Seven-Year Plan for economic development." *PDC*, 494, col. 50, Nov. 19, 1951.

93. Declared the Iranian representative before the Security Council: "In 1950 alone, the former Company derived a profit of between $500 million and $550 million, or between £180 million and £200 million, from its oil enterprises in Iran, at international market prices. Of this huge amount, Iran received only $45 million, or about £16 million as royalties, share of profits, and taxes.

"The profits of the Company in the year 1950 alone, after deducting the share paid to Iran, amounted to more than the entire sum of £115 million cited by the representative of the United Kingdom as the total sum paid to Iran in royalties in the course of the past half-century." S/PV563, Oct. 17, 1951, 15.

94. Foreign Secretary Herbert Morrison declared: "The term 'nationalization' appears to us to have been consistently misused by Persian spokesmen." *PDC*, 489, col. 823, June 21, 1951.

Said Prime Minister Clement Attlee: "I would say that the acceptance of the principle of nationalization was very strongly pressed upon us by our American friends. What is its exact connotation? I would agree that originally the oil belonged to Persia. I do not think that we can go further than saying that the conception of nationalization is that the oil shall be worked primarily in the interests of Persia . . . when an operation of this kind is carried out, either you may pay people and they go altogether — and that is quite impossible, for it would be a loss to the world and to us and not less to Persia; or you must do the right thing, which is, I think, to work for some kind of working agreement or partnership in which we supply the knowledge, the know-how and all the rest of it and the Persians manage this thing in the interest of all . . . this seems to me what we really wish to bring under this rather general phrase of nationalization." *PDC*, 491, col. 1071, July 30, 1951; *PDL*, 173, cols. 128–130, July 31, 1951.

95. The legal issue as to the ownership of Iranian oil which the AIOC tried to establish in the courts of Aden, Italy and Japan was inconclusive. These attempts could not have been of a decisive nature, for the bulk of the oil could not possibly have been moved without British and American transportation facilities.

  To be sure, there were a number of independent American oil concerns who indicated a readiness to exploit the situation in Iran and buy up quantities of oil at greatly reduced prices — at one time the offer was 50 per cent below current rates; but on the whole the American Government consistently discouraged such attempts, and the great American oil companies at no time indicated a willingness to buy oil from Iran.

Chapter VI
THE AIOC, IRAN AND THE
INTERNATIONAL CONSORTIUM

The establishment, fabulous growth and development of the Anglo-Persian (Iranian) Oil Company form one of the most fascinating commercial ventures of modern times. As was noted above, the Company was organized in 1909; its very origin is not too clearly known. In the middle 'twenties Frank G. Hanighen, in his book *The Secret War*, spun around it a veil of mystery which was elaborated by the Hearst press. This fabric of mystery is still presented as fact by such Iranian spokesmen as Nasrollah Saifpour Fatemi. [1] But sensationalism aside, in spite of the traditional British reticence about the operation and activities of their big overseas commercial corporations, the fact that the British Government owned a practical majority of the shares made the Company subject to parliamentary inquiry. This could only result in a constant supply of information on the Company's operations and profits, as well as the Government's share, and other aspects of the Company's activities. Moreover, the actions of the Iranian Government of first cancelling the concession and later nationalizing the oil industry forced the Anglo-Iranian Oil Company and the British Government to bring the issue into the international arena — politically, the League of Nations and the United Nations; judicially, the International Court of Justice — and in arguing their case they necessarily detailed the processes of the Company's growth and development. This inevitably drew the attention of world public opinion to the Iranian oil question and to the practices of the Anglo-Iranian Oil Company.

The story of the Company's growth is phenomenal. It began modestly, with authorized capital of £2,000,000 — 1,000,000 ordinary shares[2] of £1 each, and 1,000,000 preferred shares[3] of £1 each. The 1,000,000 ordinary shares were issued as fully paid, in part payment of the purchase price of the concession (£1,380,249). Although during the first few years of operation the Burmah Oil Company had to pay the prescribed dividend on the preferred shares, the risk element in the undertaking was small, if there was any at all, and the success of the undertaking, after the lapse of the necessary time for full operations, was practically guaranteed. Two years after the Company's formation, the pipeline from the Maidan-i-Naftun oil field to the refinery at Abadan was completed, and by the end of 1911 both the pipeline and the refinery were in operation.[4] So rapid was the growth that only two years later a new pipeline had to be built and the refinery had to be expanded. When the British Government became a major partner in the Company, the authorized capital was increased to 4,000,000 shares, and the year ending March 31, 1915 showed net profits of £63,720. [5]

The first year that showed any considerable profit, however, was that ending March 31, 1917: after allowing for depreciation, interest, home charges and royalty, net profits amounted to £344,109. The dividend on ordinary shares was 6 per cent; £50,000 was allotted to the reserve fund for preferred shares, and

£100,000 to general reserve, leaving a balance of £19,537 to be carried forward. Royalty to the Persian Government for that year amount to £3,829, but it was not turned over to the Persian authorities and instead was held against the Company's claims for damages resulting from the cutting of the pipeline by Persian tribesmen. [6]

Thereafter the profits increased, new oil fields were added, the pipelines lengthened, and the facilities of the refinery at Abadan expanded. In 1918 dividends on ordinary shares went up to 8 per cent, and in 1919 to 10 per cent. During the latter year the production of crude oil had to be restricted because of the limited capacity of the pipelines. The year 1920 saw greater prosperity and expansion than ever before; net profits amounted to £ 2,611,615, and dividends on ordinary shares jumped to 20 per cent. In December 1919 an issue of 3,000,000 participating preference shares and 4,500,000 ordinary shares was authorized and issued, as well as £2,600,000 of debenture stock. [7]

The dividends on ordinary shares for 1921 and 1922 were 20 per cent. In 1923 it went down to 10 per cent, but that year the Directors applied £964,000 from the general reserve to write off investments outside Persia which had already been abandoned, and £866,000 for similar investments which were to be abandoned.[8]

In January 1922, 2,000,000 preferred shares were issued and sold at one pound one shilling per share, and 600,000 ordinary shares, which sold at 3 pounds and 5 shillings per share. In January 1923, 850,000 ordinary shares were issued and sold at 3 pounds and 15 shillings a share, and the issue was very largely oversubscribed. [9]

While dividends on ordinary shares fluctuated between 10 and 12 per cent, in 1926 £4,475,000 of the general reserve was capitalized to pay for 4,475,000 new ordinary shares to be distributed as a bonus to ordinary shareholders, in the ratio of one bonus share to each two ordinary shares registered.

In 1928 the Company's second major field in Persia, Haft Kel, 55 miles south-southeast of Masjid-i-Sulaiman (the new name of Maidan-i-Naftun since 1926) was producing oil in quantity, and the Abadan refinery was again expanded. Although world oil prices had dropped sharply, the increase in turnover more than covered the decrease; profits for the year were £3,112,529, and royalty to the Persian Government £250,806. [10]

Even after the 1929 world depression, the Company made handsome profits. The year ending December 31, 1931 showed a net profit of £2,318,717; dividends on ordinary shares were 5 per cent. while the royalty payable to the Persian Government was estimated at £134,750. [11] After the conclusion of the new concession, payments to Persia increased and the Company's profits, even during the depression years of the early 'thirties, were considerable. Thus, for the year ending December 31, 1934 the profits were £3,183,195, and the dividend on ordinary shares was 12.5 per cent. That year the Company redeemed £4,850,000 debenture stock outstanding, and further expanded the Abadan refinery. From then on the fortunes of Anglo-Iranian progressed fabulously. In 1935 the dividend on ordinary shares jumped to 15 per cent and in 1936 to 25 per cent. During the same year

£6,712,500 was capitalized from reserves (£2,327,500 from debenture stock redemption reserve, and £4,385,000 from general reserve) to pay for 6,712,500 ordinary shares which were distributed as a bonus, at a ratio of one share to each two ordinary shares registered. Crude oil production was greatly increased that year, the pipeline system was enlarged, and refining capacity extended. The payment to the Iranian Government was £2,580,206. [12]

Production increased during 1937 and 1938; profits were £7,455,094 and £6,109,477 respectively; dividends on ordinary shares 25 and 20 per cent; the refinery at Abadan was again enlarged to meet growing market demands. However, with the outbreak of World War II, the fortunes of the Company declined for a time, yet even in 1941 the dividend on ordinary shares rose to 7.5 per cent from 5 per cent to which it had fallen. By 1942 it had jumped to 20 per cent, and the net profit for the year was £7,790,282. [13]

Output was consistently increased and by 1944, when the Gach Saran field was connected by pipeline to Abadan, reached the enormous amount of 11,521,555 tons. In 1943 the Agha Jari field was brought into operation and consequently Abadan was once more extended. The Abadan refinery began to produce aviation gasoline: in 1941 it produced 40,000 tons; in 1942, 360,000 tons; and in 1944, 1,250,000 tons. [14] Production of crude in Iran soared to 19,189,551 tons in 1946, and the throughput at Abadan reached 17,673,000 tons. The dividend for 1946 was 30 per cent, and the profits £9,624,938.

The report for the year ending December 31, 1947 revealed that the profits for that year were £18,190,377. The total authorized capital was £33,000,000, composed of 12,706,252 preferred shares and 20,137,500 ordinary shares. The general reserve was £14,000,000, and preferred stock reserve £5,000,000. The total amount paid to the Iranian Government for royalty and taxation was £7,101,251, while taxes to the British Government amounted to £15,266,665. Production of crude in Iran reached the 20,194,838 ton mark. [15]

The period from that date to the end of 1950, the last full year of operation before nationalization, was one of steady progress – in production; in opening up a new field (at Lali); in adding new pipelines (linking the oil fields with Abadan and with the port of Bandar Mashur), making a total of 9,188 miles; in refining capacity expansion; in profits; in the amount of taxes paid to the British Government, [16] but apparently not correspondingly in the income of the Iranian Government. Thus the year ending December 31, 1950 saw trading profits of £84,466,342; taxes paid to the British Government, £50,706,880; net profits, £33,102,572. £25,000,000 were allotted to general reserve and £1,000,000 to preferred share reserve, making a total for general reserve of £65,000,000, against 20,000,000 ordinary shares, and preferred stock reserve of £8,000,000 against £13,000,000 stock. Dividends on ordinary shares were 30 per cent, and payments to the Iranian Government £16,031,735. [17] Production of crude reached an all-time high of 31,750,000 tons, and the Abadan refinery processed 24,059,000 tons. [18]

COMPANY ACTIVITIES IN IRAN

In summarizing forty years of its activity, the Anglo-Iranian Oil Company stated that as of the end of 1950 it had paid the Government of Iran about £150,000,000 in royalties, in addition to vast sums of money it had invested in the country for various projects. As for Abadan, it had grown from a waste, uninhabited island which in 1913 had a refining capacity of 400,000 tons, to a teeming modern city with a 24 million ton capacity. In 1950 the Company's tanker fleet had picked up 1,899 cargoes, taking 21,000,000 tons for destinations throughout the world. Abadan had sixteen oil jetties and three general cargo jetties. In spite of the water problem all over Iran, and especially in Khuzistan, every house in Abadan had piped drinking water. The Company had built housing accommodations for its staff, and 11,500 of the 40,000 employees (1,750 British staff) in 1950 occupied Company accommodations. Abadan's main hospital had a 354-bed capacity and was of a high standard.

Bandar Mashur, east of Abadan, had been developed as a port, and in 1950 tankers loaded there 5,800,000 tons of crude. About 4,300 workers (104 British) were employed.

The Company had six major producing fields in Iran at the end of 1950. Masjid-i-Sulaiman, the oldest, had been producing since 1908. Two hundred fifty-seven wells had been drilled, thirty of which were in production. The 1950 output was 2,830,000 tons; the number of employees, 7,771 (including 378 British). There was a 90-bed hospital, five clinics, and eight schools.

Haft Kel, the second major field, fifty miles to the south-southeast of Masjid-i-Sulaiman, began to produce in 1928. In 1950 it produced 9,130,000 tons, and employed 1,110 people (20 British). The third major field was Agha Jari. In 1950 it produced 15,620,000 tons, nearly half the country's output, from its sixteen wells, and had 5,977 employees, 175 of whom were British. It had a hospital with a 50-bed capacity, and three clinics, seven schools, and a brick factory for the Company's building needs.

Another field, Naft Safid ("While Owl"), about twenty miles to the south of Masjid-i-Sulaiman, came into operation prior to the outbreak of World War II. In 1951 seven wells produced 1,210,00 tons, and employed 1,400 workers (including 45 British). The fifth field, that of Lali, began operations immediately after the close of the last war. In 1950 four of the fifteen wells drilled produced 730,000 tons. It employed 1,778 persons, 45 of them British.

The final major field, Gach Saran, 125 miles southeast of Haft Kel, was discovered in 1938 and began to produce in 1940. Four of sixteen wells sunk produced 2,060,000 tons in 1950; of the 750 employees, five were Britishers.

In Ahwaz, the Company built a brick plant with a 48,000,000 annual capacity. Its labor force in 1950 was 4,000; it had four clinics and two schools. The Company maintained three pumping stations on the Masjid-i-Sulaiman — Abadan pipeline manned by 2,500 workers, including 56 Britishers.

In the early 'thirties the Company established a refinery at Kermanshah which obtained its crude by pipeline from Naft-i-Shah; 1,754 employees (30 British) comprised the force in 1950. That year the Company sold a total of 945,135 tons of refined products in Iran from 144 distribution centers. Of this, 829,135 tons came from Abadan and 116,000 tons from the Kermanshah refinery.

Altogether, the Company built 21,000 housing units for its employees in its various centers in Iran: 17,000 for married men and the rest for bachelors. All the oil fields had clinics and schools. The Company built thirty-three elementary schools which it later turned over to the Ministry of Education; these were attended by 13,750 children. It built forty major bridges in Iran during the course of its operations. [19]

The Company's growth in Iran was accompanied by an expansion in the various aspects of oil production outside of Iran — exploitation, production, transportation, marketing and distribution. [20] In addition, its marketing facilities extended to many foreign countries in complex combinations with foreign companies. [21] It had first a half, and then a quarter, interest in the Iraq Petrolem Company and in the other oil companies in Iraq and those operated by IPC, as well as half a share in the Kuwait Oil Company. The annual reports of the Company give detailed information on the constant expansion of the various major subsidiaries; that for 1951 carries a complete list of principal allied and subsidiary companies. [22]

There can be no doubt that the major source of the Company's income at the beginning and for many years thereafter was derived from the oil of Persia. It was also from the great profits that the Company reaped from its venture in Persia that it invested in areas which brought no returns. Those profits made it possible for the Company not only to expand its activities in Iran, but also in the fields of transportation, marketing and distribution. It must nevertheless be recognized that a good share of its growth and profits was the result of its far-flung enterprises outside of Iran, and in evaluating the profits of the Anglo-Iranian Oil Company, one must keep in mind that they were not derived exclusively from Iranian oil. [23]

Table I (pp. 132-133) indicates that the profits of the Company were extremely large, but that the income of the Iranian Government was not nearly as great. The company constantly maintained that its profits were only commensurate with the risks it took and the enormity of its investment; that the Iranian Government and the Iranian people benefited greatly from its efforts; and that the Iranian Government, in accordance with the provisions of the 1933 agreement, also received a share of the profits derived from subsidiaries operating outside of Iran. But above all, it argued that but for the Anglo-Iranian Oil Company's enormous direct payments to the Iranian Government and the indirect benefits which accrued to the Iranian people as a result of the Company's activities, as well as the direct benefits in health, education and other services given by the Company, the general standard and conditions of living would have been much lower than they were in Iran in 1950. While the Iranians possessed great underground wealth in the oil pools of their territory, it was only by bringing the oil to the surface and refining and

Table I.   *Iranian Oil**

Year	Stock Authorized		Stock Issued		Debentures	Oil Production (in tons)	Net Profit £
	Ordinary	Preferred	Ordinary	Preferred			
1910	1,000,000	1,000,000	1,000,000	600,000	(5%)600,000		
1911							
1912				300,000		43,084	
1913						80,800	
1914						273,635	
1915						375,977	63,72
1916	2,000,000		2,000,000	100,000		449,394	85,76
1917						644,074	344,10
1918		1,000,000		1,000,000	1,800,000	897,402	1,308,55
1919						1,106,415	2,010,80
1920	4,500,000		4,500,000	3,000,000	2,600,000	1,385,301	2,611,61
1921				3,500,000		1,743,557	4,028,02
1922			600,000	2,000,000		2,327,221	3,130,38
1923			850,000			2,959,028	2,689,14
1924						3,714,216	2,507,86
1925						4,333,933	3,571,96
1926						4,556,157	4,383,23
1927	4,475,000		4,475,000			4,831,800	4,635,44
1928						5,357,800	3,112,52
1928a							2,832,95
1929						5,460,955	5,206,76
1930						5,939,302	4,648,57
1931		2,500,000				5,750,000	2,318,71
1932				2,206,252		6,450,000	2,379,67
1933						7,086,706	2,653,978
1934						7,537,372	3,183,19
1935						7,487,697	3,519,18
1936	6,712,000					8,198,119	6,123,46
1937			6,712,500			10,167,795	7,455,094
1938						10,195,371	6,109,47
1939						9,583,286	2,986,358
1940						8,626,639	2,841,90
1941						6,605,320	3,292,31
1942						9,399,231	7,790,282
1943						9,705,769	5,639,122
1944						13,274,243	5,677,142
1945						16,839,490	5,792,447
1946						19,189,551	9,624,938
1947						20,194,836	18,564,857
1948						24,871,058	24,064,920
1949						26,806,564	18,390,016
1950						31,750,147	33,102,572

a. In 1928 the accounting year was changed from Mar. 31 to Dec. 31. 1928 covers the nine month from Mar. 31 to Dec. 31 and henceforth all years terminate Dec. 31.

* Sources: Anglo-Persian (Iranian) Oil Co., Ltd., *Report of the Directors and Balance Sheet to 31s, March 1910* – (the whole series until 1950, titles differ somewhat, and beginning with 1949 it i under the name, *Annual Report and Accounts as at 31st December)*; Anglo-Persian Oil Co., *Persi Past and Present* (London, 1950), 25–27; Azami, *Le Pétrole en Perse,* 98–105, 150; League o Nations, *Official Journal,* 13th Year, 1932, 2299; *Pleadings,* 189, 200–201, 210, 214 *et passim PDC,* 139, col. 695, Mar. 10, 1921; *PDC,* 140, cols. 275–276, Apr. 6, 1921; *PDC,* 143, col. 4, Jun 13, 1921; *PDC,* 151, col. 1700, Mar. 10, 1922; *PDC,* 163, col. 2157, Mar. 8, 1923; *PDC,* 273, col. 9 Dec. 12, 1932; *PDC,* 311, col. 737, Apr. 28, 1936; *PDC,* 324, col. 1187, June 3, 1937; *PDC,* 488 col. 787, June 21, 1951; Wilson, *Persia,* 94–95.

ALLOCATION TO

Dividend on Ordinary Shares %	Payments to Iranian Government	General Reserve	Preferred Stock Reserve	Debenture Redemption	Concession Amortization	Extra Depreciation[g]
	1,326,000					
8		100,000	50,000			
8		100,000	50,000	80,000	100,000	
10		300,000	50,000	80,000	100,000	
20	469,000	575,000	50,000	80,000	100,000	
20	585,000	455,000	50,000	80,000	100,000	
20	593,000	300,000	100,000	100,000	100,000	400,000
10	533,000			100,000		495,000
10	411,000		150,000	100,000		415,000
12½	831,000		500,000	200,000	100,000	393,000
12½	1,054,000	1,500,000		200,000	100,000	228,000
7½	1,400,000	1,000,000		200,000	100,000	450,293
7½	502,000	500,000		200,000	100,000	453,267
12½	529,000	400,000		150,000	50,000	201,051 12/31
15	1,437,000	1,000,000	500,000	200,000	100,000	352,465
10	1,288,000	250,000		200,000	100,000	487,217
5	1,339,132[b]			200,000	100,000[f]	501,944
7½	1,525,383			200,000		320,829
7½	1,812,442			200,000[e]		302,183
12½	2,189,853					459,107
15	2,220,648					408,747
25	2,580,206	1,200,000				514,976
25	3,545,313	546,872				744,215
20	3,307,479	300,000				602,602
5	4,270,814[c]					571,250
5	4,000,000[c]					625,000
7½	4,000,000[c]					625,000
20	4,000,000[c]		500,000			2,000,000
20	4,000,000[c]		500,000			
20	4,464,438		500,000			
20	5,624,308		500,000			
30	7,131,669	2,000,000	500,000			
30	7,101,251	10,500,000	1,000,000			
30	9,172,269[d]	16,000,000	1,000,000			
30	13,489,271[d]	10,000,000	1,000,000			
30	16,031,735[d]	25,000,000	1,000,000			

b. This amount is the result of the adjustment under the 1933 agreement. The original amount was £307,000.

c. Additional payments were made in 1939: £1,500,000, making a total of £4,270,814; and for 1940–43 additional sums to make up £4,000,000. The original amounts for the years 1939–43 were: £2,770,814; £2,786,104; £2,025,364; £3,427,933; £3,617,917 respectively.

d. The Company calculated that had the Iranians accepted the supplemental agreement, the amount for 1948 would have been £18,677,822; for 1949, £22,890,261; and for 1950, £22,888,557.

e. The balance of debenture stock outstanding, amounting to £4,850,000, was redeemed on Dec. 31, 1934.

f. "The Reserve for Amortization, £1,250,000, has been transferred in reduction of the assets on the balance sheet, formerly entitled 'Purchase Price of Concession, Land, Refinery, Etc.' " Report, as of 31st December, 1932, 6.

g. This was in addition to regular depreciation which was deducted prior to the calculation of net profits.

marketing it that made benefits from their treasure actually possible for the Iranians. Without the Company the Iranians could never have reaped the fruits of their oil resources. [24]

THE LABOR ISSUE

One of the major issues between the Company and the Government was the employment of Iranians, especially in the upper administrative and technical echelons. Even in the concession granted to D'Arcy, as mentioned above, it was provided (Article 12) that the workmen, with the exception of the technical staff such as managers, engineers, borers and foremen, should be Persians. The Government was naturally anxious to increase the number of Iranians on the technical staff, and the 1933 Agreement (Article 16) provided that the Company should recruit artisans as well as technical and commercial staff from among Persian nationals. However, the qualification "to the extent that it shall find in Persia persons who possess the requisite competence and experience" seriously reduced the possibilities, if it did not nullify them altogether, of employing Iranians in the higher echelons. Both sides agreed to study the situation and prepare a general plan for progressive and yearly reduction of non-Persian employees and for their progressive replacement in the shortest possible time by Persians. Moreover, the Company agreed to grant £10,000 annually to enable Persian nationals to obtain in Great Britain the necessary professional education for the oil industry. A committee was organized and a general plan worked out in 1936 which envisaged the progressive replacement of non-Iranians by Iranians.

The question of training, however, was the chief difficulty between the two sides. To what extent did the Company make a serious effort to train Iranians to replace non-Iranians on the technical staff? The Company maintained that it had met its obligations; that the basic problem lay in the fact that the Iranian trainees soon left the Company's employ to go to other parts of the country, climatically more agreeable than the oil-producing provinces, where their technical skill was eagerly sought. [25] Nevertheless, the Company pointed out that the general plan of 1936 had worked satisfactorily. In 1934 the total number of Iranians employed by the Company was 7,174, and of non-Iranians, 1,799; in 1950 there were 43,080 Iranians and 4,503 non-Iranians. The Company stated that the actual reduction in the number of foreign employees in relation to total staff, apart from unskilled labor, during the period 1936–1950, was practically that envisaged by the general plan. [26]

The International Labor Office mission reported that as of November 30, 1949 (statistics supplied by the Company) the number of salaried employees of the Company was: graded foreigners 2,756 and Iranians 765; non-graded foreigners 696 and Iranians 3,544. As of December 31, 1949, wage earners, practically all Iranians, were 42,614; of these, 35,267 received wages above the minimum rate, while 7,347 received the minimum rate.[27]

Since the Iranian Government did not feel that the progress was satisfactory and blamed the Company, the question of the replacement of non-Iranians by Iranians became one of the issues in the negotiations for the supplemental agreement. In the debate following nationalization, the Iranian Government implied that the Company, in spite of its forty years presence in Iran, had intentionally not trained Iranians who could take over and manage the oil industry. The International Labor mission, however, declared: "It will be difficult to increase the rate at which Iranian nationals are recruited for employment in the higher categories of wage earners and as members of the supervisory staff. There is no reluctance on the part of the Company to recruit and promote Iranians for those categories." [28]

The real issue remained as to who should be responsible for preparing Iranians to be able to take over the management of the industry. The Company argued that it would have employed Iranians in all categories if any were to be found, while the Government felt that it had been the Company's duty during its long stay in Iran to have prepared Iranians for employment in the higher grades.

As the tension between the Company and the Government rose, the very treatment of labor in general became an issue. In March 1948 the Iranian Government asked the International Labor Office to send a mission to report on the social conditions in the oil industry in Iran. In January of 1950 a three-man mission, including an Iranian of the Labor Office staff, arrived in Iran. It examined the problems of the industry under three headings: conditions of work; social, community and welfare services; and industrial relations.

After studying the various aspects of the problems and the immediate grievances, the mission came to the general conclusion that, taking into consideration the unusual conditions of the petroleum industry — the characteristics of the country (size, isolation, climate, etc.), the general widespread poverty, malnutrition and disease, the low standard of living, inadequate educational facilities, and the need for improved health and medical services (water supply, sewage disposal and local transportation) — "against this background the working and living conditions of the oil workers appear as an encouraging example of what can be done." [29] On the whole, the report was favorable to the Company. It stated that the Government felt that the Company should pay more attention to the problems of housing, health, food supplies and education; that although these services are governmental responsibilities rather than employer obligations, the Government maintained that in the petroleum areas the Company should be responsible for supplying the needs of the workers and the general population since the problems in those areas had been created by the operations of the Company. [30]

In comparison with labor conditions in other industries in Iran, as well as with labor conditions in the petroleum industry in other parts of the Middle East, the complaints of the workers and of the Government against the Company were not completely justified. The Company was rendering important and far-reaching services in housing, education, health and sanitation, which are not directly required of employers. The entire aspect of technical training in Iran, which percolated into areas other than petroleum, was the direct result of the Company's training efforts, especially through its technical school in Abadan. It is a moot

question, however, whether all these services redounded to the benefit of the
general population, or were limited only to the workers in the Company's plants.
One may also wonder whether Anglo-Iranian's training practices and policies were
such as ultimately to make it possible for Iranians to be capable of running the
industry. The policies and practices followed by the Company over the years obviously did not produce the necessary Iranian personnel able to do so. It is also a
moot question whether the Company's practice of supplying health, educational
and sanitary services — a governmental task — did not weaken the powers of both
central and local authorities, and as a result made the population locally more
dependent on the Company, thus aggravating the relations between the Company
and the central government, and at the same time relieved the Iranian authorities of
their obligations to supply such services.

IMPACT OF OIL ON IRAN

The influence of the exploitation of the oil resources of their country on the
Iranians could be discerned in a number of manifestations. Most obvious was the
financial income, both in terms of direct payments to the Iranian Government
(royalties and taxes) and their role in the governmental budgets as well as in
development projects, and in the indirect spending by the Company for its operation in Iran. It could also be traced in the general technical impact of the industry
on the population, on housing, health, education and other social services which the
Company introduced in the limited areas of its operations, as it affected the standards of the rest of the country. It could be noted in the very availability and usage
of petroleum in the industrial and other phases of the nation's life. And, finally, it
could also be seen in the impact which the whole aspect of petroleum has had on
the political leaders, not only in their relations with the British oil concern, but also
in their relations with foreign governments, and the extent to which they believed
their petroleum wealth determined the attitude and policies of foreign governments
towards Iran.

The income from royalty and other payments made by the Company played an
important role in the life of the country, and created a dependence on this income
which affected the entire financial structure of Iran. To be sure, the percentage of
the income from the oil in the total Government budget, whenever this was revealed, was not very high. [31] However, the very fact that of all items of income the
surest and safest one was that derived from petroleum exploitation was in itself a
decisive factor in the economic and financial life of the Government of Iran. If one
studies carefully the tax paying — or rather non-paying — habits of the Iranians, [32]
especially the rich landowning classes, if one takes into account the complete lack
of reliable statistics, and appreciates the Government's fiscal operations and
methods of calculations and budget estimates, one may very well conclude that the
revenue from the oil industry was of a much more significant character than would

appear from the official percentage it represented in the annual budgets.

The large amounts of money which the Company expended on its operations supplied work, directly and indirectly, to thousands of Iranians (nearly 70,000 in 1950); it went for social services and for salaries to the British staff; for the exchange of rials for sterling at an official rate lower than the free market rate, which made available much of the necessary foreign exchange for Iran's needs. United Nations economic experts calculated that the foreign exchange received from the petroleum industry provided 70 per cent of the foreign exchange disbursements of the various banks of the country in 1948/49, 54 per cent in 1949/50, and 65 per cent in 1950/51. Petroleum production constituted in recent years about one-third of the total product of the country, and the oil industry accounted for about one-tenth of the national income. [33]

Although during the debate on nationalization some Iranians attempted to minimize the role of the petroleum income in the economic life of their country, and although they were willing, in the face of Great Britain's determination not to allow the industry to operate under nationalization, to build their economy without the oil income, [34] the responsible majority of the leaders recognized that the oil resources were, in fact, the main and perhaps the major wealth that they possessed, and the more enlightened Iranians especially looked upon it as the means of extricating their country from its social and economic backwardness. Declared Nasrollah Entezam, the renowned Iranian who served as President of the General Assembly and headed his country's delegation to the United Nations, before that body at its sixth session in 1951: "As you know, oil is the main source of our national wealth. It is therefore proper that we can countenance its exploitation only as a way of ensuring the general welfare of our people." [35]

The importance of the oil revenue to the Government of Iran increased with the growing realization of the backwardness of the country, as well as with the absolute increase in the revenue resulting from the vast jump in production. Indeed, in the spirit of the times Iranian leaders began to talk of a plan for developing their country within the framework of a limited number of years. Premier Ghavam es-Sultaneh announced right after the war a very ambitious seven-year development plan which would have involved an expenditure of $1,840,000,000. [36]

At the beginning of 1947, the Iranian Government invited a group of American experts to prepare a plan, which was submitted in July as the Morrison-Knudson report. The program recommended was of a much more modest character than the Ghavam project and involved an expenditure of $250,000,000; it strongly advocated that the stress be on greater efficiency and productivity in agriculture rather than a hasty development of industry. [37]

In November, the Premier appointed a Supreme Planning Board to prepare an economic program based on the Morrison-Knudson findings. The following September the Board submitted a scheme to be executed in seven years at an approximate cost of 21 billion rials (at the official rate, about £160,000,000). The money was to be derived from the following sources:

Royalty and other payments from the
    Anglo-Iranian Oil Company   –   7,800,000,000 rials
Credits from Bank Melli Iran   –   4,500,000,000
Sale of Government property   –   1,000,000,000
Loan from International Bank   –   6,700,000,000
Private contributions   –   1,000,000,000

The bill introduced with the report provided not only for the approval of the program, but also for the establishment of a Plan Organization to supervise and execute it. [38] Article 5 of the bill stated:

> In the year 1327, 600,000,000 rials, and in the following years till the end of the seven-year period, the entire revenues of the Government derived from the payments of the Anglo-Iranian Oil Company shall be specifically assigned for the execution of the Plan.
>
> After the expiration of the seven-year period the revenues described in this article shall be specifically assigned to service the interest and amortization payments of the loans, which by law may be obtained for the execution of the Plan. [39]

Thus, the income from the oil revenue was not only to supply the major portion of the cost of the plan, but was also to service the interest and amortization of the loans which made up the major portions of the cost.

While considering the Iranian Government's application for a loan, the International Bank for Reconstruction and Development requested that further study be undertaken, especially of the social, health and educational conditions, and it recommended the engagement of the Overseas Consultants, Inc., a group of eleven New York management, engineering and administration firms, which had been organized to advise on the rehabilitation of Japan. Early in February 1949 the Iranian Government engaged the Overseas Consultants, [40] to assist in the implementation of the Seven Year Plan. On August 22 the group submitted a five-volume report covering (in general within the limits of the Seven Year Plan law) public health, education, agriculture, water resources, meteorology, surveying and mapping, town improvement and housing, transportation, communications, industry and mining, electric power, petroleum, statistical organization, distribution, and the general issue of organization for the plan's implementation. [41] The conclusions and recommendations of the Overseas Consultants seemed to be sound and within the realm of realization; the obstacles and dangers inherent in the conditions of the country as well as in the stage of general development of the Iranians were clearly and pointedly stated.

Yet, despite all the preparations and planning, and no doubt the sincere desire of some of the Iranian leaders to utilize the oil revenues for general social and economic improvement, the undertaking of the Overseas Consultants turned out to be a dismal failure. It would seem that while some of the Iranian leaders had reached the stage of realizing the need for foreign experts and consultants, they had not yet reached the stage of self-discipline which would enable them to change their basic attitudes and habits in order to bring about the desired reforms. They apparently became impatient with the foreign experts, and in their eagerness to effect the changes practically overnight, they decided to get rid of the foreigners and return to their old methods. As regards the Iranians' share in the failure of the Seven Year Plan, Clifford S. Strike, president of F.H. McGraw and Company, one of the constituent companies of Overseas Consultants, listed the following: politics as usual; an unscrupulous minority with the cunning and influence to sabotage the plan so that they could preserve their parasitical way of life; ignorance of the plan and therefore lack of broad acceptance by the masses; and trouble over oil as it developed at the very beginning of the implementation of the plan. [42]

GENERAL CONDITIONS IN IRAN

We must search a little deeper for an explanation of the failure of the Overseas Consultants with the Seven Year Plan, an explanation which would also apply to the previous instances when foreign experts were invited to rehabilitate even limited areas of the Iranian regime and then were abruptly dismissed. We should begin with the general stage of progress of the Iranian people. Even as late as 1948 there were no reliable statistics bearing on all the important aspects of the life of the country. The state of education can be discerned from the fact that illiteracy was estimated to be as high as 85 per cent of the total population, and the character of the existing schools was far from satisfactory. [43] As for agriculture, the methods of cultivation were antiquated and hence the yields were far from adequate, and the system, equivalent "to sharecropping," reduced the income of the peasant to about one- to two-fifths of the total crop. [44] Furthermore, only 16,760,000 hectares were cultivated, at a time when there were 33 million hectares of unused, potentially productive land in a country where, out of an estimated population of about 16 million, 85 per cent were rural, and those engaged in industry, including petroleum, did not exceed 200,000. [45]

In spite of the fact that the AIOC had been in Iran for over forty years, and in spite of the very strenuous and dictatorial efforts of Riza Shah to industrialize the country, the total number of persons employed in modern enterprises in 1950 was only about 120,000. This figure comprised 52,000 directly engaged in the oil industry, 15,000 employed by contractors engaged in construction work for the oil company, 20,000 in Government factories, 30,000 in private industry, and 5,000 in mining. [46]

As for the physical condition of the Iranians, the United Nations reported that the average daily calorie content of the food supply available for human consumption for the period 1934—1938 was 2,010, and for the period 1946—1949, 1,811. Both these were respectively lower than in Turkey, Egypt and Iraq. [47] In 1948 the total number of registered doctors was 1,500; of these about 750 were in Teheran. [48] There were about 4,000,000 cases of malaria annually, [49] as well as a large incidence of trachoma, tuberculosis, venereal and intestinal diseases. The infant mortality rate was 500 per 1,000 live births. It was estimated that there were in 1949 about 1,000,000 opium addicts. The country had no adequate insect control program, and cities and towns were without sanitary water supply and sewerage systems. [50]

The attitude of the Iranian leaders to social services is best illustrated by the percentage allotment for such services in the annual government budgets. [51]

Table II. *Social Services in Iranian Budget*

Year	Education	Health	Social Welfare
1939	6	3	—
1947	8	3	—
1948	7	3	—
1949	9	3	—
1950	9	2	—

These few facts on economic and social conditions suffice to illustrate Iran's general backwardness. If one adds to this the corruptibility of Government officials, the conslusion would easily be drawn that the Iranians were not ready for the self-imposed social reform and economic rehabilitation to be brought about by the revenue from the oil exploitation.

The oil industry, in spite of its gigantic commercial operations and fabulous profits, employed a relatively small fraction of the labor force, far out of proportion to the wealth produced, and its influence on the total population and on Iran's industrial development was not impressive. Anglo-Iranian used very little of Iran's products, which would have helped in the country's economic growth. The impact of the income of the Iranian oil employees and the social services supplied by the Company was not felt to any marked degree by the rest of the population. The revenues derived from oil helped the Government with its budget, and many of the officials who were connected with the industry in one form or another no doubt enriched themselves directly and indirectly from it, but the majority of the population, the poor peasants, remained as backward as ever. Iran's great need, as the Overseas Consultants pointed out throughout their report, was the special knowledge and skills required for modern industrial life and for all the aspects of the Seven Year Plan. [52] This need was filled to a limited extent, as was mentioned above, by the training which the Company supplied to some Iranians who later left its employ and offered themselves in other industrial sections of the country. But on the whole the forty years of the Company's operations produced very little of specialized knowledge or enough skilled workers to bring about the necessary changes in the industrial structure of the country. The limitations of the technical impact of the oil industry are illustrated by the consumption of petroleum in the country, including that used by the refinery at Abadan: [53] in 1950, 2,895,000 tons; 1951, 2,060,000 tons; 1952, 1,200,000 tons. [54]

During the Company's forty years of operations, the Iranian Government received a total of some $450 million [55] in direct payments, plus indirect payments in salaries and wages, in social services, in transportation facilities, and it was assisted considerably with the foreign exchange needs of the country. The basic social and economic ills, however, were not greatly eased as a result of the oil exploitation; neither did the social and other services which the Company supplied to its workers carry over to the rest of the community. Nor could it be said that the technical impact of the petroleum industry on the general industrial development was very great. The comsumption of petroleum products by Iranians was considerably lower than that consumed by Egypt, which does not possess the great oil reserves of Iran, and a goodly portion of whose oil must be imported.

The Iranians, because of their backwardness, were not able to utilize the revenue from the oil for necessary social services and for the economic development of their country. They themselves were not capable of executing economic plans and social projects, nor were they capable of entrusting such plans and projects to foreigners. The problem was very diplomatically stated by the Overseas Consultants when they laid down as principles for the execution of the Seven Year Plan: ''National effort towards social and economic advancement must be applied first to fundamentals''; ''capital is not a substitute for skill and experience''; and ''coordination is an essential ingredient of the Plan.'' [56] These principles and the corresponding qualities were completely lacking in the Iranians. The Overseas Consultants soon realized through their own experience that the plan could not be materialized.

The income which the oil industry brought to the Government and the social services which it supplied to its employees, the schools it built, the clinics it equipped and the hospitals it presented to the Iranian Government, also created discontent. This was heightened as oil production increased and Company profits soared. The feeling grew among a considerable number of Iranians that those fabulous profits rightfully belonged to Iran, and that if their country had all the income from the oil, its ills could be remedied. Politically this became the great issue, especially after World War II, and it was out of these feelings that the various development projects grew and ultimately emerged as the Seven Year Plan.

Throughout the period of the oil exploitation, the Iranians felt that the policies of Great Britain and Russia towards Iran were always conditioned by the existence of oil and its exploitation by Great Britain. Moreover, they felt that the influence of Britain in the international arena determined the policies of other nations, and even international bodies, towards Iran. [57]

During the nationalization period the Seven Year Plan receded to the background and in the heat of the debate there emerged a school of thought that maintained that the economic development of Iran could be achieved without the oil revenue. However, with the fall of Mossadegh, the income from oil revenue became more than ever a decisive aspect of the national economic policy, and the Plan Organization was revived, though this time under purely Iranian management. It was decided that the future income from the oil would be applied to various projects aimed at social and economic reforms which would, with Western encouragement, prevent Iran from falling into the Communist camp.

THE CONSORTIUM ACCORD

The fall of Mossadegh and the rise of Fazlollah Zahedi must be considered a victory not only for the Anglo-Iranian Oil Company, but also for the British policy in Iran — in which England was eventually followed by the United States — of bringing Iran to terms through economic pressure resulting from the stoppage of revenue from the oil, as well as through completely sealing off Iran from the international market. The crisis demonstrated the decisiveness of the oil revenue in the government financial structure; and the implied promise of the United States to help, should the policy of the Government change, operated effectively to bring down Premier Mossadegh's regime.

To exist at all, as well as to receive United States' aid, Premier Zahedi had to show how completely and utterly his predecessor had failed, and thus pave the way for a solution which would meet with British-United States approval. The first major act of the new government was to announce to the Iranian people on August 27 — only eight days after it took office — that the previous government had left a debt of about $210 million, and that a comparable amount had been lost in oil revenue; that Premier Mossadegh had not only increased the paper money by fifty per cent, but had used up some $50 million in dollars and sterling currency held as reserve for the original issue of the paper money, [58] and that the country's economic rehabilitation depended on the income from oil production.

However, the basic issues involved were still to be solved. It was understood in London, and even more so in Washington, that whatever solution was found would have somehow to be within the framework of the nationalization law, even if only formally. Even Zahedi could not ignore public sentiment by repudiating that law. Moreover, it was generally agreed that while the oil industry could not be reactivated without foreign technicians and experts, not only in marketing but also in producing and refining, it would not be possible to bring back the AIOC as exclusive producer, refiner and marketer even within the framework of the nationalization law.

As far as the oil companies were concerned, the major problem was the integration of Iranian oil again in the world market, for since nationalization the former Iranian supply had gradually been replaced by increased production in other areas of the Middle East, and this had become an important factor in the financial, and therefore political, relations between the concessionaires and the countries and governments granting the concessions. Restoration of the Iranian supply would therefore have meant a curtailment in production in other parts of the region. Finally, whatever solution was found, it must not grant the Iranians better terms than those enjoyed by other Middle East governments, else the whole oil concession structure in the area would collapse, with a universal rush for nationalization.

Though the United States followed the British — somewhat reluctantly at first — in forcing Iran to come to terms, after the fall of Mossadegh it was anxious, much more so than Britain, to make it possible for Zahedi to arrive at a satisfactory solution in order to ensure that there would be no possibility of Iran's falling into the grip of Communism and the Soviet Union. The United States, therefore, in addition to promising direct financial aid, undertook to bring both reluctant parties in the dispute to a speedy solution.

Herbert Hoover, Jr. was named petroleum adviser to Secretary of State Dulles and dispatched to Iran, where he and the American Ambassador in Teheran, Loy Henderson, began to work for an agreement. The first hurdle to overcome was the renewal of diplomatic relations between the disputants. Here a very serious difference of opinion developed. The Iranians maintained that negotiations for a solution of the oil dispute should take place first; after they could bring to their countrymen a successful solution, they would be ready to resume diplomatic relations. The British took the view that before they sat down to negotiate the oil problem they must be on speaking terms with the Iranians. On October 20, 1953, Foreign Secretary Anthony Eden declared in the House of Commons: "I hope that a new chapter has opened in Persia. There is a new government there and to them, and to the Persian people, Her Majesty's Government wish sincerely to extend once more the hand of friendship. The Persian Government are aware that we are ready to resume diplomatic relations; if this can be done it will then be easier for us to discuss together the complex problem of Persian oil. I should like to say that the United States Government are working very closely with us in these matters." [59]

The Iranians were at first adamant, but after Secretary of State Dulles indicated his support on November 3, 1953 of the British point of view, [60] negotiations between the British and the Iranians, through the agency of the Swiss Government, progressed rapidly, and on December 5 an official communiqué issued both in London and Teheran announced the resumption of diplomatic relations. [61] Premier Zahedi carefully explained to the Iranians that since the British Government had given assurances of willingness to work out a solution, taking into account Iran's national aspirations and the oil nationalization law, the Iranian Government had decided to establish normal friendly relations. [62]

On December 8, 1953 it was reported in Teheran that Premier Zahedi had ordered a four-man special committee to begin arrangements for negotiations to settle the oil dispute with Great Britain. [63] In the meantime Herbert Hoover, Jr. had arrived in Teheran, on October 17, and he and Ambassador Henderson had immediately begun to confer with Government leaders, including the Premier and the Shah, as well as with the national commission appointed by the Premier to formulate a new oil policy. Hoover's task was neither to offer a solution, nor to negotiate in the name of the United States Government, but to give information to the Iranians on the general world oil situation and to point out how Iranian oil could be fitted once again into that situation; to review the actual conditions of the refinery and the measures and personnel necessary to reactivate it. His efforts were apparently successful, for by the end of the month the oil commission presented to Premier Zahedi a report of their conferences with Hoover, who was to make a report to London. [64]

From then on Hoover practically shuttled back and forth between Washington, Teheran and London. He became the instrument through which the proposals and counterproposals of the British and Iranians, sanctioned or rejected by the United States Government, passed from one side to the other. As the discussions progressed, the solution seemed to be veering in the direction of an international consortium involving the major oil companies operating in the Middle East. These were in fact the companies which controlled the greater proportion of the world supply. Such a solution had the advantage of removing the objections of the Iranians to the return of the Anglo-Iranian Oil Company, and providing for the absorption of the newly produced oil in world markets, as well as protecting the existing concessions in the area. The discussions therefore became triangular: the Iranian Government, the British Government and Anglo-Iranian Oil Company, and the American companies.

According to the president of the Anglo-Iranian Oil Company, he sent out invitations early in December 1953 to seven companies: five American (Standard Oil of New Jersey, Socony-Vacuum Oil Co., Standard Oil of California, Gulf Oil, and Texas); Royal Dutch-Shell and Compagnie Française des Pétroles, for a meeting in London to examine the various problems involved should Iranian oil flow again into world markets. The result was that the companies indicated readiness to form a provisional consortium for the solution of the Iranian oil dispute. The subsequent discussions between Dennis Wright, the British chargé d'affaires in Teheran, and the Iranian authorities convinced the Anglo-Iranian Oil Company that a consortium was the best possible answer to the problem. Talks with the seven companies were therefore resumed in February 1954. They hinged on two issues: the proposals to be submitted to the Iranian Government, and the arrangements to be worked out by the members of the consortium themselves. Meanwhile, a technical mission representing the eight companies visited Iran at the invitation of the Iranian Government.

After considerable bickering and harsh charges and countercharges about the control of the Iranian oil industry, which were echoed in editorials in *The New York Times*, the companies arrived at an understanding, on April 9, on the provisional formation of the consortium.[65] The Anglo-Iranian Oil Company was to have a 40 per cent share; the five American companies 8 per cent each; Royal Dutch-Shell, 14 per cent; and Compagnie Française des Pétroles, 6 per cent. The permanent formation of the consortium would of course depend on the agreement with Iran. This understanding was reached after consultation with and consent of both the United States and British Governments. The Iranian Government immediately issued an invitation for a delegation of the consortium to commence negotiations. Orville Hardin of Standard Oil of New Jersey, J.H. Loudon of Royal Dutch-Shell, and H.E. Snow of Anglo-Iranian, accompanied by technicians, arrived in Teheran in the middle of April to begin negotiations with the Iranian Government.[66]

The United States Government was anxious to find a solution to the Iranian dispute as quickly as possible, but the five American companies were not overly eager to join the consortium without some special inducements. Since they represented the major Middle East oil producers, they were reluctant to accept additional oil to market. J.H. Carmical, the petroleum expert of *The New York Times*, reported early in February that the United States Government, in its eagerness to get the Iranian oil matter settled, had put pressure on the American companies to join the consortium;[67] as an inducement, Attorney General Herbert Brownell, Jr., under the direction of the National Security Council, had granted the companies immunity from anti-trust prosecution.[68]

Simultaneously, while representatives of the consortium were negotiating with the Iranian Government, representatives of the British and Iranian Governments began to negotiate in Teheran the question of compensation. In the meantime, however, that issue had lost its acuteness due to the impending formation of the consortium, since the question of future profits would be solved partly by AIOC's retention of a 40 per cent share in the consortium, and partly by payments made by the seven other companies to AIOC for their shares in the consortium. Differences soon developed as to the price to be charged the consortium by the National Iranian Oil Company, and on the operation of the Abadan refinery. After a short interruption, negotiations were continued in the latter part of June [69] under the leadership of Howard Page, vice-president of Standard Oil of New Jersey, for the consortium, and Dr. Ali Amini, Finance Minister, for Iran, advised by Torkild Rieber, who had been engaged as special oil adviser by the Iranian Government. [70] After prolonged negotiaions, the Iranian Government announced that it would pay less than £30 million in settlement of all claims advanced by the AIOC as a result of nationalization. [71]

On August 15 the agreement between the Iranian Government and the consortium covering the production and refining of oil, and the agreement between the Iranian Government and the AIOC covering compensation for the latter's properties, were completed, and on the 31st they were initialed by Ali Amini and Howard Page. After the first agreement had been approved by the heads of the eight companies, it was to be submitted to the Majlis for ratification. [72]

On October 10 the Iranian Government submitted to Parliament a bill to ratify the agreement. Eleven days later the Majlis ratified the agreement, 113 to 5, with one abstention. For this the Deputies received from Premier Zahedi the commendation: "You have truly carried out your obligations to the nation." [73] A week later the Senate also ratified the agreement, 41—4, with 4 abstentions. The next day the Shah gave the royal assent, and on the 30th Iranian oil began flowing again at Abadan into British, French, American and Dutch tankers, to be carried to the markets of the world.

AGREEMENTS OUTLINED [74]

After setting up in the preamble the needs for additional captial, experienced management and transport and marketing facilities which the eight companies could supply, and their willingness to share equitably the profits resulting from their operation and management of some of the oil properties of the Iranian Government and the National Iranian Oil Company, as the reasons for concluding the agreement, the document contained the following major provisions:

1. An Exploration and Producing Company and a Refining Company, known as the Operating Companies, to be incorporated by the consortium members under the laws of the Netherlands, these companies to sign the agreement; they are to be registered in Iran, and the Board of Directors of each to consist of seven members, two of them nominees of the National Iranian Oil Company (Art. 3).

2. The Exploration and Producing Company to be endowed with rights and powers to explore for, drill for and produce, extract and take crude oil and natural gas, operate field topping plants, process oil and gas produced by it to the extent necessary for its operations, store such oil and derivatives, and transport and deliver them on board ship.

   The Refining Company to have rights and powers to refine and process the crude oil and natural gas produced by the Exploration and Producing Company. The rights and powers of the Operating Companies not to be revoked or modified at any time during the term of the agreement. "The Operating Companies shall exercise their respective rights and powers . . . in behalf of Iran and the National Iranian Oil Company." The Operating Companies to minimize the employment of foreign personnel by ensuring "so far as reasonably practicable, that foreign personnel are engaged only to occupy positions for which the Operating Companies do not find available Iranians having the requisite qualifications and experience." In consultation with the NIOC, the Operating Companies to prepare plans and programs for industrial and technical training and education, and cooperate in their execution, to replace foreign personnel by Iranians. The NIOC to have means for financial and technical supervision over the Operating Companies (Art. 4).

3. All stocks, stores and materials, movable plant and equipment, mechanical transport and drilling plants and tools, except items for use in internal distribution, to be delivered to the Operating Companies by the NIOC (Art. 6).

4. The Operating Companies to be granted exclusive rights within the delimited area, except for internal distribution (practically the same as enjoyed by the AIOC). New land within the area covered by the agreement necessary for the operation of the Companies, if not used, to be given gratis; if used, a rental fee to be paid by the Companies (Art. 7).

5. The Operating Companies to be entitled to the following fees from the Trading Companies for export and from the NIOC for internal consumption: Exploration and Producing Company – one shilling per cubic meter of crude oil delivered; Refining Company – one shilling per cubic meter of crude oil refined. Both the Trading Companies and the NIOC also to pay a proportionate share of operating costs and expenses (Arts. 13 and 14).

6. The NIOC to continue to perform and carry out non-basic operations, such as provision, maintenance and administration of housing, public roads, medical and health; food supply systems, canteens, restaurants and

clothing stores; industrial and technical training and education; welfare facilities, and similar services (Art.17).

7. Title to crude oil sold by NIOC to the Trading Companies to pass at well head (Art.18).

8. The consortium members to guarantee production: for 1955, 17,500,000 cubic meters; 1956, 27,500,000; 1957, 35,000,000. After the third year, the quantity produced to be in line with the trend of supply and demand for Middle East crude oil (Art. 20).

9. Each Trading Company (a subsidiary of a consortium member) to pay the NIOC for crude oil purchased 12½ per cent of the applicable posted price of such crude oil; the NIOC may select to take crude oil in lieu of all or part of the stated payments for crude oil (Arts. 22 and 23).

   Except for definite imposts — such as income taxes payable by the Trading Companies and the Operating Companies, customs duties on certain items, and other payments specifically outlined in the agreements — the consortium members, Trading Companies, their respective affiliates, and the Operating Companies, to be "free from all taxation by any governmental authority in Iran," and no taxes to be imposed on dividends paid by these companies (Art. 28).

10. All materials and equipment necessary for the Operating Companies, as well as drugs and equipment for hospitals, to be imported without any license and to be exempt from import and other customs duties, taxes or charges (Art. 34). No export taxes or other charges or payments to be made on the products and purchases of the Trading Companies, affiliates, etc. (Art. 35).

11. Power and authority to execute the contract to be vested in the hands of the Ministry of Finance (Art. 37).

12. Article 41, Paragraph B, stated: "No general or special legislative or administrative measures or any act whatsoever of or emanating from Iran or any governmental authority in Iran (whether central or local) shall annul this Agreement, amend or modify its provisions, or prevent or hinder the due and effective performance of its terms. Such annulment, amendment or modification shall not take place except by agreement of the parties to this Agreement."

13. Differences as to the execution or interpretation of any provisions of the agreement to be handled first by direct discussions between the parties; if that fails, by a conciliation committee; and if that fails, by arbitration (Arts. 42 and 45).

14. Both the English and Persian texts of the agreement to be valid, but in case of divergent interpretation, English to prevail (Art. 48).

15. The term of the agreement to be for twenty-five years, with the right of renewal for three additional periods of five years each (Art. 49). [75]

The agreement between the Iranian Government and the Anglo-Iranian Oil Company on compensation for the latter's loss of its properties as a result of nationalization stated in the preamble that in consideration of the agreement with the consortium, the AIOC had agreed to relinquish all claims in respect to its assets in southern Iran. In consideration, on the one hand, of the assests of the Company in the internal distribution equipment, the Kermanshah refinery and the Naft-i-Shah oil field and the resultant loss from the disruption of the AIOC enterprise, and, on the other hand, the disruption of Iran's economy and the sums that would have been due Iran had the supplemental agreement been ratified, the agreement provided:

1. That after offsetting the £51 million due Iran against the Company's claims, Iran to pay £25 million, interest free, to the AIOC in ten equal annual installments beginning January 1, 1957, in the form of oil to be given to a subsidiary of the AIOC, which would not be subject to income tax (Art. 1).[76]

2. This payment to be accepted by both parties "in full and final settlement of all claims and counterclaims by Iran and the NIOC on the one hand, and the Anglo-Iranian Oil Company, Limited, on the other" (Art. 2).

Immediately after the completion of the agreements, Premier Zahedi announced an ambitious five-year development plan which was to be financed from the revenue derived from the reactivation of the oil industry, together with loans from the United States and the International Bank for Reconstruction and Development. It called for an expenditure of $500 million, and was aimed at a large increase in agricultural and mineral production and improvements in transportation and communications. Supervision of this program was to be entrusted to a five-man supreme planning council. [77]

On the other hand, it has been calculated that the Iranian Government needed for its annual operations about $100 million, most of which would be derived from the oil revenue. [78] However, the Government's estimated income from the oil during the first three years was to be about £150 million. This would hardly suffice for current expenditures, let alone long-range development projects. Moreover, except for their bitter economic and financial experience during the nationalization period, one wondered why the Iranians should be more capable and ready to undertake long-range economic projects and social reforms, without the benefit of foreign advisers and experts, than they were, with the aid of foreign experts, in the Seven Year Organization Plan prior to nationalization. [79]

The AIOC came out of the dispute victorious. When nationalization was enacted in 1951, the Company's compensation claims were estimated at a billion dollars. Its concession was to run until 1993, and it actually offered Iran a fify-fifty profit-sharing proposal. The consortium agreement, so far as Iran was concerned, would not net her more than 50 per cent of the profits; the agreement potentially ran for about forty years thus actually covering the life of the AIOC concession. While theoretically the Company properties now belonged to the NIOC, the compensation which the AIOC received was in fact more than had been estimated at the time of nationalization. The Company received £25 million in direct compensation; in addition, it retained the approximately £51 million in the special contingencies account which would have been due the Iranian Government under the supplemental agreement. The seven consortium companies agreed to pay AIOC £32,400,000 in three equal installments over the next twelve months; and the Company was to receive the equivalent of ten cents per barrel of crude oil and products exported from Iran by these companies, until a total of $510 million had been paid. [80]

On November 3, 1954 the Company announced to its shareholders that £80,550,000 from the general reserve would be used to pay for the purchase of new ordinary stock to be distributed as a bonus, at a ratio of four new shares to every registered share. This the Company described as "a measure that is in accordance with the growth and wealth of the Company now that it is free from the uncertainty of the Iranian involvement." [81] Thus the fabulous venture begun by D'Arcy in 1901, taken over by the Anglo-Persian Oil Company in 1909, and joined in by the British Government in 1914 had, in spite of two world wars, in spite of a cancellation and nationalization by the Iranian Government, progressed to the point where it had amassed such enormous profits that it was able to distribute such a bonus to its shareholders at this late stage in its development. Both the British Government and the Company had come out satisfied.

The United States Government did not conceal its satisfaction with the conclusion of the agreement, and the President deemed it necessary to congratulate personally not only the Shah, but also the American Ambassador in Teheran, Loy Henderson, and the special petroleum adviser, Herbert Hoover, Jr., for their roles in the successful settlement of the dispute. In his letter to Mr. Hoover, the President emphasized that the conclusion of the agreement not only promised further progress in Iran, but also the promotion of the American "objective of maintaining peace in the area." [82]

It would seem that the major parties in the three-and-a-half-year-old dispute — Iran and Britain directly, and the United States indirectly — had arrived at an amicable and satisfactory solution to ensure the successful continuation of the production of oil in Iran. The fourth party, the Soviet Union, which always remained in the shadows in all the negotiations, had kept exceptionally, and no doubt intentionally, quiet. Russia neither interfered directly nor offered advice to or threatened the Iranians. It would seem that the United States, with some persuasion by the British, had outmaneuvered the Soviet Union and that the revenue from the oil and other financial aid which were to come to Iran strengthened the hands of her government and gave it some stability. [83]

NOTES

1. Nasrollah Saifpour Fatemi, *Oil Diplomacy. Powderkeg in Iran* (New York, 1954), 11–17. For an official Company point of view see Sir John Cadman, "Great Britain and Petroleum," *American Petroleum Institute Bulletin*, December 21, 1921, 21.
2. The Company's shares are always £1 each.

3. Ordinary shares carry two votes each, and preferred shares one vote to every five shares. *PDC*, 139, col. 1061, Mar. 14, 1921.

4. Anglo-Persian Oil Co., *Report of the Directors and Balance Sheet to 31st March 1912* (henceforth cited *Report*, with date).

5. *Report to 31st March 1915.*

6. *Report to 31st March 1917.*

7. *Report to 31st March 1920.*

8. *Report to 31st March 1923.*

9. As regards the cost of production of Persian oil, the Chairman of the Company, Sir Charles Greenway, stated at the fifteenth annual meeting of the Company held in 1924, that it was always much lower practically than anywhere else, "because in this respect we stand, as I have previously pointed out, in an exceptionally favored position, owing to the unique productivity and life of our wells and to the very high quality of our crude."

As to the Company's growth and development in the thirteen years since it began operations, Sir Charles stated: "It is a record of development which, I am confident, has not been equalled in the same short space of time by any other concern in the commercial history of the world." He reported that during the previous three years "we had provided out of Revenue no less than £12,000,000 for new capital outlay. I will now add to these figures by telling you that since we first became a revenue-producing concern in 1914 — only ten years ago — we have provided out of our earnings a total of no less than £19,000,000 for expenditure of a captial nature in addition to paying out the sum of £9¼ millions in dividends and debenture interest." Anglo-Iranian Oil Co., *Fifteenth Ordinary General Meeting*, 1924, 4.

10. *Report to 31st March 1928.*

11. *Report to 31st December 1931.* The final payment offered for the year was £306,872. See Chap. III above, page 38.

12. *Report to 31st December 1936.*

13. *Report at 31st December 1942.* During the summer of 1940 the Company agreed to an arrangement guaranteeing the Government £4,000,000 for the years 1940 and 1941; this arrangement was continued until the termination of hostilities with Germany and Italy.

14. *Report and Balance Sheet at 31st December 1944.*

15. *Report and Balance Sheet at 31st December 1947.*

16. It was stated in the House of Commons that in 1948 the Company had paid in taxes to the British Government around £28,000,000, and in 1949, £22,000,000, while the Iranian Government received in royalty and taxes about £8,000,000 and £12,000,000 for the respective years. *PDC*, 489, col. 787, June 21, 1951.

17. The Company calculated that had Iran accepted the terms of the supplemental agreement, the amount for 1950 would have been £22,888,557.

18. *Annual Report and Accounts as at 31st December 1950.*

19. Anglo-Iranian Oil Co., Ltd., *AIOC Operations in Iran* (London, [1951]).

20. In Apr. 1929 the list of the Company's subsidiaries was given in the House of Commons: — Scottish Oils Ltd.; British Tanker Co., Ltd.; First Exploitation Co., Ltd.; National Oil Refineries, Ltd.; British Petroleum Co., Ltd.; Tanker Insurance Co., Ltd.; D'Arcy Exploitation Co., Ltd.; Khanaqin Oil Co., Ltd.;

North Persian Oil Co., Ltd.; Britannic Estates Co., Ltd.; and Anglo-Persian Oil Co. (India) Ltd. *PDC,* 227, col. 16, Apr. 15, 1929.

    A list given in the House of Commons in Mar. 1953 also included the Aden Petroleum Refinery, Ltd.; Australasian Petroleum Co., Ltd.; and Kent Oil Refinery, Ltd. *PDC,* 512, col. 93, Mar. 10, 1953.

21. In reply to a question as to the area supplied by the AIOC, Kenneth Younger of the Foreign Office stated: "The Anglo-Iranian Oil Company, either directly or indirectly from other countries, supplies practically all the major markets in Europe, the Middle East and the Far East, excluding Russia, her satellites and China. Supplies also go to the Western hemisphere." *PDC,* 489, col. 2, Jan. 18, 1951.

22. Anglo-Iranian Oil Co., Ltd., *Annual Report and Accounts as at 31st December 1951,* 18–19. The list gives sixty-five companies covering Great Britain, the British Commonwealth, Algeria, Austria, Belgium, Denmark, Egypt, France, French West Africa, Germany, Holland, Iceland, Italy, Morocco, Norway, Sweden, Switzerland and Tunisia.

23. See the Company's *Annual Report and Accounts for the Year ended 31st December 1952,* and *Annual Report and Accounts for the Year ended 31st December 1953.* During these two years the Company operated entirely without Iranian oil, yet its profits were larger than ever, and both the crude oil and refining capacity of Iran were slowly being supplanted by other sources.

24. In summarizing the achievements of the Company in Iran, United Nations experts stated that cumulative production of crude oil in the 40-year period 1912–1951 was 333 million tons, of which 32.3 million were produced in 1950. Exports of crude oil and petroleum products from Iran during the period 1912–1951 were estimated at 290 million tons with an approximate value of £1,200,000,000. Direct payments to the Iranian government in the form of royalties, taxes and share of profits amounted to £122,000,000 in the same period. United Nations, Department of Economic and Social Affairs, *Economic Developments in the Middle East. 1945 to 1954,* 71.

25. This fact was confirmed by the International Labor Office mission report International Labor Office, *Labor Conditions in the Oil Industry in Iran* (Geneva, 1950), 72.

26. *Pleadings,* 208. The Company reported in 1952 that the actual figures of its total personnel in Iran as of Mar. 1951 were:

Staff	Iranians	Non-Iranians	Total
Senior	30	89	119
Other	5,492	3,534	9,026
Labor			
Top grade	17,550	896	18,446
Middle grade	12,225	1	12,226
Lower grade	4,411	—	4,411
Unskilled	13,925	—	13,925
Apprentices and trainees	3,392	—	3,392
Contractors' employees	13,603	—	13,603
Total	70,628	4,520	75,148

"Anglo-Iranian Answers Iran with Facts," *The Oil Forum*, VI, xi, April, 1952. See also UN, Dept. of Economic Affairs, *Summary of Recent Economic Developments in the Middle East 1950–1951*, 50; Anglo-Iranian Oil Co., *The Anglo-Iranian Oil Company and Iran* (London, 1951), 14.

27. International Labor Office, *Labor Conditions in the Oil Industry in Iran*, 16, 22.
28. *Ibid.*, 72.
29. *Ibid.*, 83.
30. *Ibid.*, 85 *et passim.*
31. In 1924 it was about 7 per cent; in 1931 about 18 per cent, Wilson, *Persia,* 297, 298; in 1947, 12 per cent; in 1948, 12 per cent; in 1949, 6 per cent, and in 1950, 12 per cent, UN, Dept. of Economic Affairs, *Review of Economic Conditions in the Middle East 1951–52*, 77. Oil royalties were not included in the general budget up to 1942/43 and were kept outside the budget in the reserve accounts. In 1937/38 they were about 13 per cent; 1942/43, 12 per cent, UN, Dept. of Economic Affairs, *Public Finance Information Papers Iran* (New York, 1951), ST/ECA/SER, A/4, 34.
32. For a full discussion of the system of taxation and collection methods and results, see UN, Dept. of Economic Affairs, *Public Finance Information Papers Iran*, 27–29 *et passim.* As a drastic measure to force the richer classes to pay their taxes, the government of Ali Razmara resorted to broadcasting over the national radio the names of those who had not paid, and threatening them with confiscation of their property. *The New York Times*, Oct. 1, 1950.
33. UN, Dept. of Economic Affairs, *Summary of Recent Economic Developments in the Middle East 1950–51*, 49–50.
34. UN, *Official Records of the General Assembly, Seventh Session Plenary Meetings* (Oct. 14, 1952–Aug. 28, 1953), 278. Statement of the Iranian delegate, Nasrollah Entezam, on Nov. 13, 1952.
35. UN. General Assembly, Sixth Session, *Official Records*, Nov. 14, 1951, 128; A/PV, 344.
36. Kirk, *The Middle East 1945–1950*, 92; Great Britain, Board of Trade, *Iran Economic and Commercial Conditions* (London, 1948), 30 (henceforth cited as Great Britain, Board of Trade, *Iran*); UN, Dept. of Economic Affairs, *Public Finance Information Papers Iran*, 22.
37. Great Britain, Board of Trade, *Iran* 30.
38. *Ibid.*, 32; US, Dept. of Commerce, *International Reference Service*, IV, July, 1949, 4.
39. *Report on Seven Year Development Plan*, I, Exhibit 2, p. 5 of 9; UN, Dept. of Economic Affairs, *Review of Economic Conditions in the Middle East 1949–50*, 28–29.
40. UN, Dept. of Economic Affairs, *Public Finance Information Papers Iran*, 22.
41. *Report on Seven Year Development Plan*, I, 1–2.
42. *New York Herald Tribune*, Oct. 28, 1951. The UN reported that the Seven Year Plan Organization absorbed 2,234 million rials (approximately $70 million) of oil revenues during the period July 1949–Nov. 1950; this was about 84 per cent of the total income of the Organization, including a loan of 195 million rials from Bank Melli. UN, Dept. of Economic Affairs, *Summary of Recent Developments in the Middle East 1950–51*, 55–56. On Jan. 8, 1951,

when the Iranian Government cancelled its contract with the Overseas Consultants, Inc., Max W. Thornburg, vice-president of the group, in discussing the cancellation, stated: "The plan was doomed from the start when politicians moved in and took over. Important appointments were made on the basis of political and personal interest instead of competence and experience. Utilization of the money available was determined by personal and political interest. In the absence of modern accounting methods it is impossible to know what really happened to the money." *The New York Times,* Jan. 9, 1951.

43. Said the Overseas Consultants after carefully investigating the school system: "The instruction in the existing curricula was very poor; and the programs of study bore little relationship to the apparent needs of society." *Report on Seven Year Plan,* I, 21.

44. The average farmer must live and support his family on approximately $100 per year. Lyle J. Hayden, "Living Standards in Rural Iran," *The Middle East Journal,* III, 143, Apr., 1949.

45. Great Britain, Board of Trade, *Iran,* 42, 49; UN, Dept. of Economic Affairs, *Review of Economic Conditions in the Middle East 1951–52,* 16. In 1945 it was estimated that only about 10–15 per cent of the total land area was under cultivation and that another 20–30 per cent could be cultivated if irrigation were available. B.A. Keen, *The Agricultural Development of the Middle East* (London, 1946), 6.

46. UN, Dept. of Economic Affairs, *Review of Economic Conditions in the Middle East 1951–52,* 40–41.

47. *Ibid.,* 13.

48. Great Britain, Board of Trade, *Iran,* 42.

49. About 82 per cent of the population had the malaria parasite in their blood. Hayden, "Living Standards in Rural Iran,' *Middle East Journal,* III, 145, Apr., 1949.

50. *Report on Seven Year Development Plan,* I, 13.

51. UN, Dept. of Economic Affairs, *Review of Economic Conditions in the Middle East 1951–52,* 79. It should be noted that the Iranian system is highly centralized and that the years 1949–50 were included in the Seven Year Plan in which education and health were major items. Another study of the UN stated that the percentages in the total Government budget for the social services (health and education) were in 1937/38 – 7; 1942/43 – 8; 1945/46 – 12; 1946/47 – 12. UN, Dept. of Economic Affairs, *Public Finance Information Papers Iran,* 25.

52. *Report on the Seven Year Plan,* I, 10.

53. Excluding the refinery, internal consumption figures were: 1937, 172,000 tons; 1949, 780,000; 1950, 842,000. UN, Dept. of Economic Affairs, *Summary of Recent Developments 1950–1951,* 27; including refinery consumption, figures were: 1929, 793,000; 1937, 1,545,000. UN, *Review of Economic Conditions 1951–1952,* 62.

54. UN, Dept. of Economic Affairs, *Summary of Recent Economic Developments in the Middle East 1952–53,* 57. The Company reported that it had sold 945,135 tons of refined products in the country in 1950. As for the price of petroleum, the *Report on the Seven Year Plan* (IV, 253) noted: "Taxes account for about 44 per cent of the retail selling price and transportation

about 43 per cent, a total of about 87 per cent of the price paid by the consumer. The remainder is divided between marketing costs, profits and the base cost of the product at the refinery." It further stated: "It is deplorably true that most of Iran's population today is inadequately supplied with oil products and that this situation promises rapidly to become worse." *Ibid.*, 254.

55. Iranian Embassy, Washington, *Press Release,* Nov. 23, 1952. The Company gives £115,000,000. Anglo-Iranian Oil Co., Ltd., *AIOC Operations in Iran,* 1.

56. *Report on Seven Year Plan,* I, 12.

57. In addressing the General Assembly of the UN on Nov. 13, 1952, the Iranian delegate unequivocally implied that the International Bank had refused Iran a loan because of Britain's objection to the nationalization of the oil industry. UN, *Official Records of the General Assembly, Seventh Session Plenary Meetings* (Oct. 14, 1952–Aug. 28, 1953), 277–278. The Iranians were under the impression that the US put obstacles in the way of the Export-Import Bank granting them a loan because of Britain's oil difficulties with Iran. *The New York Times,* June 23, 1951.

58. *The New York Times,* Aug. 28, 1953. The full report as it was read the following day over the Teheran radio by the Deputy Premier and published fully in the press contained this significant policy statement: "In the present situation without oil revenues, most of the development plans cannot be carried out." *The New York Times,* Sept. 30, 1953.

59. *PDC,* 518, col. 1812, Oct. 20, 1953.

60. *Christian Science Monitor,* Nov. 4, 1953. It is interesting to note that in the speech from the throne on the occasion of the opening of Parliament on Nov. 3, the Queen stated as regards Iran: "My Government hope for a renewal of those friendly relations which have been traditional between this country and Persia and for an early resumption of normal diplomatic relations between the two countries." *PDC,* 520, col. 4, Nov. 3, 1953. Nothing was said about the oil dispute.

61. *The New York Times,* Dec. 6, 1953.

62. *Ibid.*

63. *New York Herald Tribune,* Dec. 9, 1953; *MEA,* V, 37, Jan., 1954.

64. *MEA,* IV, 422–423, Dec., 1953.

65. A remark made two years previous to the formation of the Consortium about the companies comprising the Consortium is interesting: "The outstanding characteristic of the world's petroleum industry is the dominant position of seven international companies. The seven companies that conduct most of the international oil business include five American companies – Standard Oil Co. (New Jersey), Standard Oil Company of California, Socony-Vacuum Oil Co., Inc., Gulf Oil Corp., and the Texas Co. – and two British-Dutch Companies – Anglo-Iranian Oil Co., Ltd., and the Royal Dutch-Shell group."

"Control of the industry by these seven companies extends from reserves through production, transportation, refining and marketing. All seven engage in every stage of operations, from exploration to marketing." US Senate, *International Petroleum Cartel,* 21–23.

66. *PDC,* 526, col. 796, Apr. 12, 1954; Anglo-Iranian Oil Co., Ltd., *Annual Report of Accounts for the Year Ended 31st March 1953* (London, 1954), 16–18;

The New York Times, Apr. 18, 1954. While it is not quite clear from the report of the president of the Anglo-Iranian Oil Co. who was the father of the consortium idea, it would appear that it emerged as a result of Hoover's discussions with the Iranians on the one hand, and with the British on the other. See above, Chap. V.

67. *The New York Times*, Feb. 7, 1954.

68. *The New York Times*, Feb. 1, 1954; *MEA*, V, 103–104, Mar., 1954.

69. *PDC*, 529, col. 1780, July 5, 1954.

70. See above, Chap. V. At that time he functioned as the representative of the International Bank.

71. *The New York Times*, Aug. 2, 1954; *MEA*, V, 335, Oct., 1954. In addition, the Company was to retain £49,987,440, which it held in a special contingencies account, which would have been due to the Iranian Government had the supplemental agreement been ratified *(Manchester Guardian*, Aug. 6, 1954), plus the amounts paid by the seven companies for their shares in the consortium. This was given by Herbert Hoover, Jr. as $600 million.

72. *The New York Times*, Sept. 1, 1954; *MEA*, V, 336, Oct., 1954.

73. *The New York Times*, Oct. 22, 1954.

74. From a draft of the agreement supplied by one of the five American companies in the consortium.

75. This agreement differed basically from all previous concessions and agreements between the Anglo-Iranian Oil Co. and the Iranian Government in that it went into great and very fine detail on the various aspects of the actual operations of the industry, methods of accounting and calculating profits, process of arbitration, and other such issues. From a technical legal point of view, differences between the parties would become a rather complex problem for arbitrators or others assigned the task of settling such differences.

76. These payments were regularly made, and in 1967 the Company reported having received the last installment of £ 2½ million. BP (successor to AIOC), *Annual Report and Accounts . . . 1966*, 10.

77. *The New York Times*, Aug. 11, 1954. On Oct. 30 it was reported from Iran that some $700 million of oil revenue would be spent under the seven-year development plan being worked out by Abolghassam Ebtehaj, who was to name a seven-man supreme council to help him integrate the plan. In a sense this would be the continuation of the Seven Year Plan Organization, under complete Iranian control and management. Ebtehaj stated that the program's aim was to eliminate Iranian unemployment and give the living standard a major upward swing; he gave no indication, however, of specific projects. *New York Herald Tribune*, Oct. 31, 1954. Further interesting discussion of the five-year plan and a comparison with the original seven-year plan is to be found in *Le Commerce du Levant-Beyrouth Express*, Oct. 13, Oct. 23, 1954.

78. *The New York Times*, Apr. 18, 1954.

79. As soon as the oil dispute settlement was anticipated, Iran went on a spending spree, buying on credit all over the world. In the middle of October Ebtehaj, the Governor of the Central Bank, tried to stop the spree by sharply reducing imports by Government departments; he also planned to curb commercial imports. Ebtehaj stated that Iran's foreign credit commitments, estimated at

$50,000,000, would prevent large-scale application of oil revenues to development projects for two years. *The New York Times,* Oct. 14, 1954.

80. *The New York Times,* Oct. 31, 1954. An interesting table presenting the "credit" and "debit" of the Company's compensation question is given in "Anglo-Iranian Nationalization Losses Cut,' *The Oil Forum,* VIII, 298, Sept., 1954. Anthony Nutting stated in the House of Commons that the AIOC would be paid £214 million by the other seven companies for their 60 per cent share in the consortium. *PDC,* 531, col. 30, Nov. 1, 1954. The payments of the companies were annually reported by the British Petroleum Company. The final payment of £6.4 million was made in 1971. BP, *Annual Report and Accounts for 1971,* 10.

81. Anthony Nutting told the House of Commons that in October 1951 the Anglo-Iranian £1 shares stood on the stock market at £5 each, but as soon as the consortium agreement was signed they went up to £18 each. *PDC,* 531, col. 31, Nov. 1, 1954.

82. *Department of State Bulletin,* XXXI, 266, Aug. 23, 1954.

83. On April 29, 1955, the American members of the Consortium, in accordance with a prior agreement with the Iranian Government, allocated 1/8 of their holding to a group of nine other American oil companies. They were: American Independent Oil Company, Atlantic Refining Company, Getty Oil Company, Hancock Oil Company, Richfield Oil Corporation, San Jacinto Petroleum Corporation, Signal Oil and Gas Company, Standard Oil Company (Ohio) and Tidewater Oil Company. These have constituted the Iricon Agency, and have had a 5% interest in the Consortium. Iran Oil Operating Companies, *1959 Review,* 4.

Chapter VII
POST-NATIONALIZATION ERA: 1954–1971

The nationalization of the petroleum industry which was later proclaimed by Iranian leaders as the far-reaching achievement which they subsequently called the "White Revolution," the product of the courage and wisdom of the Shahanshah Aryamehr, but which was in reality the product of former Premier Mohamed Mossadegh, who forced the Shah to flee the country at the height of the nationalization controversy, produced, with the extraordinary efforts of the United States, the International Consortium. Theoretically the entire petroleum industry in the country became the property of the National Iranian Oil Company, and the Consortium was merely servicing the national concern; it was actually, at first, a mere face-saving device which made possible the continuation of the old operations: exploration, development, production, refining, transportation and marketing under the guise of a new organization. The specific functions given to the National Iranian Oil Company, formerly performed by the Anglo-Iranian Oil Company, were the non-basic operations and local distribution of oil products. In order to enable the national company to function it was assigned considerable portions of the oil revenue.[1] However, in July 1957, the Iranian Majlis enacted a new petroleum law which gave a new turn to the National Iranian Oil Company, to oil developments in Iran, and perhaps to the entire oil industry in the Middle East. The new enactment gave NIOC new broad powers for expanding exploration for, and production of, oil as rapidly as possible throughout Iran, outside the Consortium area, and for its refining, transportation and marketing at home and abroad.

The act authorized the national concern to divide the country, including off-shore areas, outside the Consortium agreement limits into districts of about 80,000 square kilometers each. It was empowered to declare any district open, and to enter into joint-partnership agreements with foreign companies for exploitation.[2]

From that time on Iran has made continuous, steady strides in her relations with the Consortium almost diametrically reversing the roles of the early relationship between Government and Company — in the magnitude of production and revenues; and — as a result of great economic development — in the amounts of home consumption of petroleum products; in the development and utilization of natural gas, for domestic power and home consumption and for export both as raw and liquefied; in the resultant rapid development and expansion of a petrochemical industry; in the evolvement of new and highly favorable concessions and contract agreements with foreign companies for the exploitation of oil, natural gas and petrochemicals of the country on- and off-shore; in building — to be sure, limited — a refining system in various parts of the country to meet local needs; in building, operating and successfully maintaining a complex pipeline system of various diameter-sizes and lengths carrying crude oil, products and natural gas; in obtaining a

157

slow but steadily expanding international oil market for crude, natural gas, and ultimately, oil products; and finally in engaging in oil industry ventures in foreign countries.

This change in the role of the Iranians in the growth and acceleration of the oil industry was brought about by a number of factors, some characteristic of all the Middle Eastern oil-producing countries, and some unique to Iran. First are the general changed conditions in international relations in the world which began right after World War II and which have been accelerated ever since under the impact of the United Nations. This was enhanced by the growing role of the smaller developing countries in the global East-West struggle, and by the great number of new independent nations which emerged in Asia and Africa and have created an international strong anti-colonial mood.

As decisive at least was and is the constantly increasing demand for oil in Europe, Asia and potentially America, and the realization that the cheapest and most obvious, if not the major, single source to meet the demand was the Middle East – a fact which escaped no Middle Eastern producing country.

Since the end of World War II the Iranians have developed a greater economic, political and technological sophistication in dealing with foreign companies and with the industrial complexities as well as potentialities of the oil industry. They have indeed travelled a very long way from the days when they granted the D'Arcy Concession.

An important contribution to Iran's stronger and advanced position was the Organization of Oil Producing Countries, of which she was a founding member. Iran had taken full advantage – when it suited her purposes – of the possiblities which were afforded her by OPEC, although it always maintained friendly – rather than antagonistic – relations with the Consortium.

The evolvement of new types of concession-agreements with foreign companies other than the major international companies, with very favorable terms, gave the Iranians a sense of pioneering and achievement and a determination to press ahead with newer and better agreements, with better terms from the Consortium and with developing the oil industry themselves.

Perhaps a determining factor in the new Iranian position was the fact that of all the major oil-producing countries in the Middle East Iran had the largest population, and she had a longer tradition of political independence and relative stability under the leadership of the Shah, and the broader political base of a parliamentary regime.

Similarly, because of the large population, Iran had continuously and persistently worked on plans and projects for economic development in order to establish an economy independent of oil revenue. The Iranians, at least theoretically, looked upon the oil income primarily as the financial means to build their future economic independence.

Finally, although Iran is a Muslim state and located in the Middle East on the Persian Gulf, she is not an Arab country, and therefore played, plays and no doubt hopes to play a special independent role in the international politics of the Middle

East in general and of oil in particular. Iran had exploited in the past, and would no doubt do so again in the future, her unique position in the Middle East to advance the importance of her oil resources in times of crisis and guaranteed steady supply in normal times.[3] All these factors plus the fact that after the withdrawal of the British from the Persian Gulf area Iran emerged — not without the resentment and bitterness of her Arab neighbors and the assistance and support of the West, especially the United States — as the greatest military power in the region, contributed to her favorable position in the oil developments of the Middle East and in her relations with the Consortium, to her own oil resources development and to her dominance in the Middle East.

PRODUCTION

The most outstanding fact about Iran during the period under discussion is the phenomenal growth of production of crude oil as illustrated by Table III. Both the percentage increase between 1954 and 1971 and the cumulative total are indeed staggering. This becomes even more impressive when the well-productivity in Iran is realized. Thus, for instance, in the year 1961 a total of 113 wells in the Consortium area produced an average daily output of 1,171,000 barrels, making an average per well output of 10,362 barrels.[4] Ten years later, in 1971, a total of 193 wells produced an average daily output of 4,183,933 barrels, making an average per well ouput of 21,678 barrels.[5] In ten years the output per well more than doubled.[6]

After production, or indeed the direct motive for it, came the question of the Government's oil revenues. Table IV presents the direct payment revenue in British pounds and United Stated dollars. The overall drive of the Iranian Government — regardless of international market conditions, the production levels of neighboring states and the objectives of OPEC — in its relations with the Consortium was for greater production, which meant greater income. It is obvious that increased production will naturally augment revenue; however, as staggering as production figures were, the increase in revenue was disproportionate, especially in the last few yars, to the increment in output. In the year 1956, for instance, the average payment to the Government for a metric ton amounted to roughly $6.00; in 1971 it amounted to roughly $8.00. This jump in per ton income from 1956 to 1971 was due to the constant improvements in the companies' payments, most of which were the results of the OPEC efforts.

These consisted first in the elimination of marketing allowances. The concessionaire companies deducted about ½% of posted prices as the producing countries' contribution towards marketing expenses. The Fourth Conference of OPEC meeting in Geneva in April, 1962, passed a resolution asking member states to adopt measures to eliminate any contribution to marketing expenses. The Iranian Government thereupon undertook negotiations with the Consortium. Early in the following year the latter officially notified the former that the major part of the Governmental contribution would be reduced, from about 1½ cents to ½ cent per barrel. In 1962 Iran's share of marketing expenses had amounted to $3 million;

Table III. *Iran Oil Production, 1954–1971*

Year	Cubic Meters	Metric Tons	Barrels
1954	3,480,000	3,000,000	21,900,000
1955	18,920,000	16,100,000	117,530,000
1956	31,095,000	26,400,000	192,720,000
1957	41,500,000	35,600,000	259,880,000
1958	48,000,000	40,400,000	294,920,000
1959	54,000,000	46,400,000	337,625,000
1960	61,000,000	52,000,000	385,749,000
1961	68,000,000	59,400,000	431,491,000
1962	76,000,000	66,000,000	481,947,000
1963	85,000,000	73,100,000	538,108,000
1964	98,000,000	85,400,000	618,730,000
1965	109,000,000	95,000,000	688,321,000
1966	123,000,000	105,200,000	771,234,000
1967	151,000,000	129,300,000	947,678,000
1968	166,000,000	142,200,000	1,042,223,000
1969	196,000,000	168,100,000	1,231,707,000
1970	222,000,000	190,700,000	1,403,558,000
1971	263,000,000[a]	226,200,000[b]	1,700,701,000[c]

a.   NIOC, 1971, 48. The breakdown for 1971 was: Consortium 240,500,000, LAPCO 7,801,000, IPAC 7,246,000, SIRIP 3,335,000, IMINOCO 3,835,000, NIOC 700,000. *Ibid.*, 7, 44.

b.   British Petroleum Company, *Statistical Review of the World Oil Industry 1964*, 18; – *1969*, 18; – *1971*, 18; *Petroleum Press Service*, XXXIX, 10, January, 1972.

c.   *Petroleum Press Service*, XXXVI, 40, January, 1969; – XXXVII, 236, June, 1970; – XXXVIII, 400, October, 1971; *World Oil*, CLXX, 96, February 15, 1970; – CLXXI, 168, August 15, 1970; NIOC, *Iranian Oil A Decade of Success*, 9; – *Oil and Economic Development*, 67.

under the new arrangement between $2 and $2.5 million of this sum had been eliminated.[7]

The expensing of royalty payments was a much more difficult problem to solve and took a much longer time. The 50% tax on profits payment as the Government's share consisted of two parts: 12½% of profits as royalty payment, and the balance up to 50% of the profits, in other words the royalty was part of the 50% arrangement. The same Fourth Conference of OPEC adopted a resolution requesting that royalty payments be considered an expense and that the Government receive 50% of the balance of the profits after the deduction of the royalty payments.[8]

Iran and Saudi Arabia were asked, because of their good relations with the concessionaire companies, to undertake negotiations. However, Aramco for Saudi Arabia and the Consortium for Iran proved very resistant and the negotiations made

Table IV. *Iran, Consortium Revenue, 1954–1971*

Year	£	$
1954/55	35,429,652	99,200,000
1956	54,792,420	153,400,000
1957	76,038,349	212,900,000
1958	88,339,354	247,300,000
1959	93,745,172	262,500,000
1960	101,794,000	285,200,000
1961	107,554,655	301,100,000
1962	118,770,000	333,700,000
1963	134,990,000	384,700,000
1964	171,500,000	480,200,000
1965	183,300,000	513,200,000
1966	207,400,000	580,700,000
1967	265,600,000	711,500,000
1968	336,300,000	810,500,000
1969	378,500,000	924,000,000
1970	437,600,000	1,070,800,000
1971	745,100,000[a]	1,750,000,000[b]

a. Iranian Oil Operating Companies, *1960 Annual Review*, 2; – *1962 Annual Review*, 2; – *1963 Annual Review*, 2; – *1964 Annual Review*, 5; – *1965 Annual Review*; – *1966 Annual Review*; – *1967 Annual Review*, 6; – *1968 Annual Review*, 4; – *1969 Annual Review*, 5; – *1970 Annual Review*, 7; – *1971 Annual Review*, 6; NIOC, *The Role of the Oil Industry in Iran's Economy*, 14; – *Oil and Economic Development in Iran*, 34, 71; – *Annual Report 1968*; – *NIOC in 1969*; – *NIOC in 1970*, 1; – *1971*, 3. The companies' *1971 Annual Review*, 27, gave the total direct payments for the period 1954–1971 as £3,542,000,000. Total capital investment for the same period amounted to £358,400,000.
b. NIOC, *Economic Impact of Petroleum Industries on Iran*, 23; *Petroleum Press Service*, XXXVIII, 326, September, 1971.

no progress: OPEC went from one conference to another without achieving results. However, the Consortium and some of the other companies offered counter proposals in the form of a compromise. They were ready to recognize the principle of expensing royalty payments, but the Governments would have to allow, for tax calculation purposes, a discount from the posted prices which, they claimed, were unrealistic. After almost two years of negotiations – for Iraq was holding out against the compromise, while Iran was determined to accept the Consortium offer which was made in the middle of 1964 – a solution was found.[9]

Royalty was to be expensed, and during the first year the governments would allow an 8½% discount from posted prices; the second year 7½%, and the third year 6½%. It had been estimated that on the average this would have given the governments an extra 3½ cents per barrel in the first year, 4 cents the second year and 4½ cents the third year. [10]

In order to benefit from the formula as applicable to 1964 as the first year, the governments would have had to agree to it before the end of the year. When the OPEC Seventh Conference convened in Djakarta, Indonesia, Iran and other governments had already decided to sign agreements with their concessionaires, [11] and as a result the Conference dropped the item of royalty expensing from its agenda. On December 1, 1964, OPEC decided on an extraordinary meeting to be held in Geneva. At the end of the month it was announced that five Middle Eastern countries, except Iraq, accepted the companies' proposals.

The above formula was operative through 1966, and the pressure for the elimination of the discount allowance was continued by OPEC. Meanwhile the 1967 Arab-Israel war broke out, which resulted in the closure of the Suez Canal, and which brought about radical changes in the production patterns of Middle East oil and in companies – governments relations.

On January 6, 1968 the companies agreed to gradually eliminate the allowance off posted prices for tax purposes until 1971. The schedule proposed was: 1967 – 6½%, 1968 – 5½%, 1969 – 4½%, 1970 – 3½%, 1971 – 2%, and nothing thereafter. The extraordinary conference of OPEC meeting in Beirut two days later recommended that the governments accept the companies' proposals. [12]

Yet, the major purpose for which OPEC came into being – restoration of posted prices to the pre-August 1960 level – was not accomplished until 1970. All the efforts from 1961 to the end of the decade failed. [13]

Negotiations between the Persian Gulf OPEC states and 22 concessionaire companies at the end of 1970 and early in 1971 resulted in the February 14, 1971 Teheran Agreement. Iran, Iraq, Saudi Arabia, Kuwait, Abu Dhabi and Qatar were to receive on the average a 35-cent increase per barrel. It eliminated all the remaining OPEC allowances to the companies. The agreement was for a period of five years with annual increases until 1975. In addition to the above increase the companies were to increase the posted prices by 5 cents per barrel on June 1, 1971, and thereafter 5 cents each on January 1, 1973, 1974 and 1975. The countries agreed in turn not to seek any further increases in the terms of the settlement during these five years. It was estimated that the new increases in the posted prices and the improved terms would give the Gulf states and extra $1.2 billion in 1971 over 1970, and would probably rise to $3 billion in 1975. The above increase in Iranian revenue from 1970 to 1971 indicates the accuracy of this estimate. [14]

In spite of the satisfaction of the companies that they had assured themselves a stable five-year supply without further demands from the producing countries, at the end of 1971 OPEC advanced two new requests. Because of the devaluation of the United States dollar, the governments asked for an adjustment in the calculation of the posted prices. The OPEC Middle Eastern countries requested participation in the concessionaire companies; while some countries asked for outright nationalization, the majority was willing to begin with 20%.

Iran was not interested in participation since she already had nationalized the oil industry. But she was interested in the first request. On January 20, 1972, representatives of the Persian Gulf states, including Iran, and ten representatives of 16 Western companies signed an agreement in Geneva for an immediate increment of 8.49% in posted prices to compensate for the devaluation of the United States dollar. [15]

Thus, through the utilization of OPEC, Iran succeeded in increasing the direct payments. But in addition to direct payments the Consortium companies have made payments for custom duties and import taxes, for wages and salaries, for contributions to the workers' social insurance organization, for contractors and local purchases. These payments were mostly met by bringing in money from abroad; and Iran thereby increased her foreign exchange resources. The magnitude of the other payments, though by no means proportionate to the direct payments, was nevertheless considerable, as evidenced by Table V.

Table V. *Iran Consortium–Other Payments, 1959–1971*

Year	£
1959	29,690,000
1960	30,565,000
1961	37,423,000
1962	35,105,500
1963	31,800,000
1964	32,700,000
1965	33,800,000
1966	40,400,000
1967	42,100,000
1968	47,800,000
1969	56,300,000
1970	62,700,000
1971	55,600,000[a]

a. Iranian Oil Operating Companies, *1959 Annual Review*, 32; – *1960 Annual Review*, 36; – *1961 Annual Review*, 32; – *1962 Annual Review*, 5; – *1963 Annual Review*, 5; – *1964 Annual Review*, 7; – *1965 Annual Review;* – *1966 Annual Review*, 36; – *1967 Annual Review*, 36; – *1968 Annual Review*, 42; – *1969 Annual Review*, 40; – *1970 Annual Review*, 44; – *1971 Annual Review*, 32.

CONCESSIONS AND CONTRACTS

The most impressive development of Iranian oil has no doubt been, after the Consortium, the achievements of the National Iranian Oil Company since 1957. During the period under review NIOC evolved three types of oil-exploitation undertakings: the joint-venture, the contractual-service agreement, and the modified joint-venture accord.

On August 3, 1957 NIOC signed an agreement with the Italian state company Aziende Generale Italiana Petroli, AGIP, to form the company Société Irano-Italiènne des Pétroles (SIRIP), to be owned equally by both parent companies. SIRIP was granted an area of 22,900 square kilometers in the Zagros area and offshore in the Persian Gulf. Both parent companies were to share equally in the profits after deduction of Iranian taxes. AGIP undertook to bear all the expenses of SIRIP. When oil in commercial quantities was discovered NIOC was to put up half the captial needed for development. [16]

The general terms of the joint-venture company were limited in any one district to not more than 16,000 square kilometers, drilling was to begin within four years from date of signature, duration of the concession was limited to 25 years with provision for renewals of three terms of five years each, and at the end of 10 years half of the area not developed had to be relinquished.

The actual terms for AGIP were: to spend at least $22 million for exploration, $6 million within the first four years, and $16 million within the next eight years. After the discovery of oil in commercial quantities, SIRIP was to repay AGIP its exploration expenditures, at an annual rate of not less than 10 cents per barrel produced. A noteworthy provision was that AGIP might suspend explorations at the end of four years, or any year thereafter, but in doing so it would have to pay NIOC one half of the unexpended balance of the $22 million exploration obligation. SIRIP was not required to pay royalty payments; it was to pay only income tax on its profits from its share of the oil. [17]

Early in 1961 oil was discovered in Bahregan Sar in the SIRIP concession area. From that time onward production grew steadily, as is indicated in Table VI. The later discovery of the Nowruz field made it necessary for SIRIP to complete a 58-mile, 36-inch underwater pipeline connecting the new field to the storage facilities on shore near Imam Hassan, from which crude was shipped from a tanker mooring-buoy, nine miles out in the Persian Gulf. [18]

Early in June 1958 NIOC approved a similar agreement with Pan American Petroleum Corporation, a subsidiary of Standard Oil, Indiana. The area was 16,000 square kilometers offshore in the northern part of the Persian Gulf. The joint company formed was Iranian Pan-American Oil Company, IPAC. The terms were the same as those of AGIP, except that Pan American was to pay a $25 million cash bonus. [19]

IPAC discovered three offshore fields: Darius in 1961, Cyrus in 1966 and Fereydoun in 1970. The last one is located on the Iranian side of the Persian Gulf median with Saudi Arabia. [20] In October 1964 IPAC completed its 100,000-barrel

Table VI. *Iran, SIRIP Production in Cubic Meters, 1961–1971*

1961	202,536
1962	362,590
1963	862,286
1964	1,334,275
1965	1,387,238
1966	1,382,326
1967	1,197,592
1968	1,048,843
1969	1,547,334
1970	1,853,913
1971	3,335,000[a]

a. NIOC, *National Iranian Oil Company,* 6; – *Annual Review 1968,* 10; – *NIOC in 1969;* – *NIOC in 1970,* 7; – *1971,* 44; *World Oil,* CLVII, 174–6, August 15, 1963; – CLIX, 164, August 15, 1964; *World Petroleum Report,* XI, 70–72, March 15, 1965; *Petroleum Press Service,* XXXIII, 175–177, May, 1966.

Table VII. *Iran, IPAC Production in Cubic Meters, 1965–1971*

1965	2,679,439
1966	3,968,621
1967	5,823,748
1968	5,961,790
1969	6,022,699
1970	5,371,649
1971	7,246,000[a]

a. *Petroleum Press Service,* XXXIII, 175–7, May, 1966; *World Petroleum Report,* XIII, 68, 1967; *World Oil,* CLXIII, 173, Aug. 15, 1966; NIOC, *Annual Report 1968,* 10; – *NIOC in 1969;* – *NIOC in 1970,* 7; – *1971,* 44.

per day capacity storage facility at Kharg Island, and oil production began in earnest in 1965 and continued since. The growth is shown in Table VII.

That same month, Sapphire Petroleum Company of Canada signed an agreement with NIOC for the exploration and exploitation of oil in a 10,000-square-kilometer offshore area lying between the sections granted SIRIP and the Consortium. Terms were similar to those of IPAC, except that Sapphire was not to pay a bonus, and was to spend $18 million on exploration. [21]

NIOC itself entered the production phase of the industry, primarily in two fields, Naft-i-Shah and Albroz near Qum. As Table VIII indicates the quantities produced were not of a large magnitude. Nevertheless, as mentioned above, at first NIOC received a goodly portion of the Consortium payments; since the beginning of 1963 NIOC had been able to support all its activities from the income of oil products sold for domestic consumption. NIOC analysts reported that up to the end of 1962, NIOC had received £69.5 million from oil revenues. From the beginning up to 1965 NIOC invested £81 million, of which £38 million was spent on the construction of various pipelines, on the purchases of machinery and on various projects, and on the investment in the joint-companies such as SIRIP and IPAC. [22]

Table VIII.   *NIOC Production Barrels, 1961–1971*

1961	2,900,000
1962	5,029,769
1963	6,074,777
1964	2,597,607
1965	2,757,000
1966	3,125,537
1967	3,234,077
1968	3,422,932
1969	3,403,864
1970	3,735,675
1971	4,403,000[a]

a.   *World Petroleum Report,* VIII, 44, February 15, 1962; – XI, 70–72, March 15, 1965; – XV, 83, 1969; *World Oil* CLIX, 164, August 15, 1964; – CLXI, 174, August 15, 1965; CLXV, 148, August 15, 1967; – CLXVII, 205, August 15, 1968; NIOC, *Annual Report 1968,* 10; – *NIOC in 1970,* 7; – *1971,* 7.

In spite of the efforts of NIOC, no foreign companies responded to the invitation to explore for oil in the various districts. However, in 1965 the ice was broken, when Teheran announced the signing of agreements with six groups for concessions in the offshore area. The six groups were to pay bonuses amounting to $190 million immediately, and $51 million partly after 5 years and partly after commercial exploitation; and they were to spend $164 million in exploration. The terms were similar to those of IPAC. [23]

As of the end of 1971, only two of the above six joint-venture companies were producers: LAPCO and IMINOCO. LAPCO has been producing since 1968; its major field was Sassan. In 1968 it produced 801,878 cubic meters; 1969 – 6,924,281; 1970 – 8,266,784 and 1971 – 7,801,000 cubic meters. At the end of 1968, completed projects on Lavan Island were a 22-inch-diameter submarine pipeline to transport the crude oil from the Sassan field, six crude oil storage tanks of ½ million barrels each, and an oil jetty capable of handling ocean-going tankers of 200,000 tons. [24]

IMINOCO began producing in 1969 from Rostam field. In the first year it produced 887,594 cubic meters; 1970 – 3,209,597 and 1971 – 3,835,000 cubic meters. An 18-inch submarine pipeline connecting the Rostam field with Lavan Island was inaugurated in November 1969. The Company shared the storage facilities of LAPCO. In 1970 a second offshore field, Rakhsh, was discovered; it began producing in 1971, and it was connected with Lavan Island. [25]

CONTRACTUAL-SERVICE AGREEMENTS

In 1966 NIOC departed from the joint-venture concession and entered into a contractual agreement with the French state group ERAP, Enterprise de Recherches et d'Activities Petrolières. The French group was to provide capital, technical services and management skills to expand production and export of Iranian oil. ERAP was to lend Iran the money to cover the cost of exploring a 200,000 square kilometer area onshore and about a 20,000 square kilometer area offshore. ERAP was to organize a subsidiary SOFIRAN, Société Française des Pétroles d'Iran, under the laws of Iran. At the end of one year SOFIRAN and NIOC were to select smaller areas for intensive exploration. At the end of six years the areas to be exploited were to be reduced to 5,000 square kilometers onshore and 3,333 square kilometers offshore. When oil of commercial quantities was discovered NIOC was to be entitled to set aside as a national reserve 50% of the oil. The other 50% were to be exploited. SOFIRAN was only the general contractor, and all the production equipment used by the subsidiary and the oil produced belonged to NIOC.

The contract was for 25 years, and NIOC was to sell ERAP, during the contractual period, 35 to 45% of the output of the balance of 50% of the oil at cost plus 2%. From the money to be realized NIOC was to repay ERAP for the exploration expenses at the rate of about 10 cents per barrel of oil produced, per year. The profits of ERAP from the oil bought were subject to Iranian 50% income tax.

ERAP also obligated itself to find markets for the oil it was to produce, between 3 and 4 million tons of crude annually, for 10 years. ERAP was to invest between $35 and 40 million during the first six years. Iranian analysts declared that under the most favorable circumstances, Iran's share of the profits could be as high as 90%.[26]

Early in 1969 NIOC concluded its second service-contract agreement with a group of five West European oil companies for the exploration for oil in a 27,260 square kilometer area relinquished by the Consortium in Southern Iran, including territorial waters in the Persian Gulf up to the three-mile limit. The terms were similar to those of ERAP. The European group obligated itself to buy up to 2 million tons a year of NIOC's share, were the total annual production under 5 million tons, and half of the total if production reached over 5 million tons.

A third service-contract was concluded at the same time with the American Continental Oil Company to explore 12,860 square kilometers in Southern Iran. The terms were the same as with ERAP. After 5 years 50% of the area was to be relinquished, and two years after that 50% of the remainder was to be returned to NIOC.[27]

NEW JOINT-VENTURE AGREEMENTS

Soon, however, NIOC reverted to the joint-venture pattern. The Iranians maintained that the new joint-partnerships were better and more favorable to them than the earlier types. The question nevertheless remained, why did they abandon the contractual-service agreements, which, they asserted, gave them between 89 and 91% of the profits? We should perhaps examine, at his juncture, the role which NIOC saw for itself. In its *Oil and Economic Development of Iran* (p. 33), NIOC declared that it had three objectives: "To expand the country's petroleum industry and increase the income derived from oil; to provide for internal requirements, and to expand the country's influence abroad." In another of its studies, *Iranian Oil, A Decade of Success* (p. 2), NIOC's goal is stated in somewhat more ambitious terms: 'To secure for itself reliable and long-term outlets in the international market by participating directly in downstream operations, including refining and distribution of petroleum products in the major consuming countries of the world, so that eventually it will be supplying its products under its own trade mark at petrol stations." The reason for its service agreement with the West European group, as given in *The Economic Impact of Petroleum Industries on Iran* (p. 8), was "the fact that its members lacked access to sufficient quantities of crude oil could therefore be depended upon to facilitate the entry of Iranian crude into foreign markets."

It would seem, however, that the results were disappointing. Indeed from the point of view of returns, the income from NIOC's undertakings in general were very low in comparison with the returns from the Consortium. Production outside of the Consortium area was about 10% of the total in 1970, yet the returns were less than

5% of the revenue from the Consortium that same year. Total revenue from the Consortium was £443.3 million; LAPCO, £11.23; IPAC, £3.96; IMINOCO, £2.56; and SIRIP, £1.04 million. [28]

The Petroleum Press Service calculated that the 1970 return to the Iran Government per barrel produced was from the Consortium 83 cents, SIRIP 21 cents, IPAC 28 cents, IMINICO 30 cents; only from LAPCO were the returns somewhat higher. [29]

NIOC, therefore, and perhaps because of the unwillingness of foreign companies to conclude contractual-service agreements, reverted to the joint-venture type with some modifications. In 1971 NIOC entered into three joint-venture undertakings covering the relinquished areas of the Consortium. The total cash bonuses paid by the three foreign groups amounted to $47 million; bonuses conditional on oil production amounted to $26 million, and the minimum exploration expenses were set at $85 million. The basic differences between the old and new type partnership were: duration of agreement — after six years of exploring, 20 years, and only two five-year extension periods; throughout the agreement period, Iran had the right to receive crude oil required for domestic consumption at cost price, on condition that the quantity did not exceed 10% of production. NIOC had the right to assume possession of any gas produced if it were not informed by the partners of a program for its utilization within six months of discovery. Iranian tankers, under equal conditions, were to be given priority. Expensed royalty payments were to be from 12½ to 16% of posted prices and income tax payments were to be based on the OPEC rate. There were also provisions for the establishment of petrochemical installations and for building refineries for export of products. [30]

REFINING

After production, refining is the most important stage in the oil industry. It increases the value of the crude as far as the company is concerned, and it provides additional employment, economic wealth and cheap oil products for the domestic market as far as the country is concerned. Generally speaking, the concessionaire companies preferred to lift their oil as crude and refine it in their own refineries or in the refineries of their home countries, which would be the beneficiaries of the added value, employment and economic activity. [31] In Iran, however, from the very beginning the Anglo-Persian Oil Company built its great refinery at Abadan, which also served as the port-terminal for loading both crude and refined products for export.

When the industry was nationalized, the refinery became the property of NIOC. The Consortium agreement provided for a separate company for refining the crude oil and natural gas produced by the Exploration and Producing Company. The Refining Company had the unrestricted use of the refinery, for the operation of which it was paid operating costs plus a fee of one shilling per cubic meter. It also operated the topping plant at Masjid-i-Sulaiman.

In addition a topping plant at Naft-i-Shah and a small refinery at Kermanshah were operated by NIOC in accordance with the terms of the Consortium agreement. With the growth in producion and in domestic needs all the existing refining facilities were expanded and new local refineries were established.

The growth of the Abadan refinery is illustrated by Table IX. The daily throughput of the refinery was increased from 351,000 barrels in 1960 to 422,348 barrels in 1971. [32]

Table IX.   *Abadan Refinery Throughput, 1955–1971. Long Tons*

1955	7,000,000
1956	11,000,000
1957	15,000,000
1958	15,000,000
1959	15,500,000
1960	20,156,000
1961	15,390,000
1962	17,315,000
1963	17,566,000
1964	18,067,000
1965	18,269,000
1966	18,608,000
1967	19,278,000
1968	20,257,000
1969	20,148,000
1970	20,675,000
1971	20,675,000[a]

a.   Iranian Oil Operating Companies, *1960 Annual Review*, 34; – *1961 Annual Review*, 15; – *1962 Annual Review*, 17; – *1963 Annual Review*, 21; – *1964 Annual Review*, 21; – *1965 Annual Review; – 1966 Annual Review; – 1967 Annual Review*, 11; – *1968 Annual Review*, 46; – *1969 Annual Review*, 39; – *1970 Annual Review*, 15; – *1971 Annual Review*, 18; NIOC, *Annual Report 1968*, 18; – *NIOC in 1969; – NIOC in 1970*, 8; – *1971*, 8.

The first local refinery was built by NIOC in Teheran at a cost of $92 million. It was completed and inaugurated in March 1968, and its daily capacity expanded from 53,836 barrels in 1968 to 85,022 barrels in 1971. [33]

Capacity in both the Naft-i-Shah topping plant and the Kermanshah refinery was steadily augmented. In September 1971, both plants were closed and the new expanded plants of the Kermanshah refinery were on stream in the same month. While the Kermanshah refinery daily average throughput in 1966 was 4,255 barrels, its average daily throughput in 1971 was practically doubled to 8,222 barrels. Its daily capacity was increased to 15,000 barrels. In 1971 a new refinery was being built in Shiraz, and plans were adopted to build other local refineries at Tabriz, Meshed and Isfahan. [34]

One of the major objectives of NIOC was fully realized. Local consumption of oil products grew by leaps and bounds. Table X tells that story.

Table X. *Iran, Domestic Consumption, 1955–1971. Cubic Meters*

1955	1,200,000
1956	2,000,000
1957	2,400,000
1958	3,000,000
1959	3,400,000
1960	3,900,000
1961	4,500,000
1962	4,200,000
1963	4,100,000
1964	5,500,000
1965	5,900,000
1966	6,600,000
1967	7,500,000
1968	8,539,000
1969	9,560,000
1970	10,400,000
1971	11,630,000[a]

a.   Iranian Oil Operating Companies, *1962 An-nual Review*, 34; – *1964 Annual Review*, 39; – *1969 Annual Review*, 42; NIOC, *NIOC in 1969; – NIOC in 1970*, 12; – *1971*, 12.

NIOC–CONSORTIUM RELATIONS

The relations between Iran and the Consortium may be generally characterized as friendly. The main issues between them were, of course, as mentioned above, the restoration of cuts in posted prices to the pre-August 1960 level, disallowance of marketing expenses, expensing of royalties, increases in posted prices and an increase in the Government's share of the profits. But the major issue which commenced from the very beginning and, in a sense, is still a bone of contention between the parties was the rate of production. Iran was determined to recoup her losses of oil production during the nationalization period years, to regain her position as the leading producer in the area and to achieve her economic development program, which was primarily dependent on oil revenue.

Annual large increases in production meant ever greater investments for the Consortium companies in exploration, in development and production, in pipelines, in refining capacity, and in loading stations and terminals. The Consortium members maintained that they could not produce more oil than the market could normally absorb; should it become glutted with oversupplies, prices would drop and profits diminish. Iran on the other hand argued that the Consortium was intentionally holding down production in order to reap greater profits at the expense of the producing country. Above all Iran needed the revenue, and the Consortium companies should not dictate the economic policies of a sovereign power.

Under constant pressure, the Consortium tried to accommodate the Iranians; although every once in a while the atmosphere would become tense, neither side permitted the situation to get out of hand, and reason seems to have prevailed. As the position of the Iranians grew stronger and that of the Consortium weaker, the Iranians in the end always had their way.

Motivated by the same desire to increase revenue, Iran asked the Consortium to relinquish areas which were not actually being exploited, although the agreement did not provide for such relinquishment until 1979. Moreover, since NIOC did not itself produce sufficient crude to meet its commitments to some of the Eastern European countries, the Consortium was requested to deliver extra crude for that purpose.

To all of these demands the Consortium responded in time. In 1959 it built six new oil-storage tankers at Bandar Mashur and increased capacity to 4,500,000 barrels. In the same year it built pipeline facilities from Gach Saran field and developed Kharg Island as an oil-loading terminal. A 26/28-inch pipeline, 71 miles long, was laid from Gach Saran to Ganaveh on the coast; from there, a 30-inch submarine pipeline, 28 miles long, was constructed to take the oil to Kharg. The initial terminal capacity of Kharg Island accommodated at the same time one 100,000-ton tanker, one 45,000-ton tanker, and two 35,000-ton tankers. The daily capacity of the pipeline was 230,000 barrels, and the tank-farm capacity at the terminal was 2,278,000 barrels.[35]

On September 19 and 20, 1961, high-level talks between the Consortium and Iran were held in London, and the Iranian chief delegate, Abdullah Entezam, Chairman of NIOC, warned that the increase in production would necessitate increases in facilities and further Iranization of the industry.[36] In 1962 the Consortium declared that "in order to be able to meet the expected increase in demand for Iranian crude oil, both short-term and long-term, extensive exploration activites in search of new oil reserves went on throughout the Agreement Area." On July 10, 1962, Premier Ali Amini declared in Teheran, while negotiating with the Consortium for the requests made by OPEC, that "the Government of Iran has no differences with the Consortium," and the issues on the agenda could be resolved by negotiations, and he added emphatically: "We have, however, no intentions of revising the Consortium agreement." [37]

To facilitate deliveries to Kharg Island, the Consortium built — two years later — a second 28-mile, 30-inch submarine pipeline from Ganaveh to Kharg, and the export capacity of the terminal was increased by 500,000 barrels per day. [38]

In the following year facilities were further expanded. A 106-mile, 42-inch pipeline was built from Agha Jari field to Ganaveh, and two additional 30-inch, 28-mile submarine lines from Ganaveh to Kharg were completed. In March 1965 a 67-mile, 26/30-inch loopline from Gach Saran to Ganaveh was built. [39]

Dissatisfied with the rate of production, the Iranian Government demanded in the fall of 1966 greater increases in production, in facilities, and relinquishment. On October 24, Premier Abbas Hoveida issued a very grave warning to the Consortium: either meet the Government demands for increased production or suffer the consequences. Two months later an amicable agreement was concluded. The Consortium undertook to increase offtake during the next two years, to install additional crude production facilities, to relinquish about 25% of its area and to furnish NIOC with 20 million tons of crude during the period 1967–1971 to meet NIOC's commitments to Eastern European countries. The results were fast in coming. Exports in 1967 rose by almost 23% over those in 1966; in March 1967 the Consortium relinquished ¼ of its area and undertook to expand its export facilities.

It reported that it spent £30 million during 1966 for facilities from well-head to loading jetty, including eleven new half-million-barrel tanks, and for loading techniques on Kharg Island which were capable of accommodating tankers of 205,000 tons. Early in 1966, the export-port for main products was shifted from Abadan to Bandar Mashur, and Kharg became the largest single crude oil export terminal. [40]

The Arab-Israel war gave the Iranian oil industry a new turn. Although the Suez Canal was blocked, because of the Arab stoppage of the flow of oil to the Western countries, Iran increased her exports, and newer facilities were needed.

In order to raise the necessary funds needed for the Fourth National Development Plan (1968–1972) Iran requested the Consortium to increase output in order to raise revenue to $1 billion a year. The Consortium was given an ultimatum: unless the member companies came up with a program that would raise production during the current year of 1968 to the level of revenue required, Iran would take drastic measures to force the rate of production. After these negotiations the Consortium offered a compromise. It would increase production by 10%, and the difference between actual realized revenue and the goal of $1 billion would be advanced on account of next year's production. The proposal was accepted. [41]

The Consortium reported that in 1969 a large number of new oil and gas handling facilities was installed during the year. These included production units, pipelines, booster stations, recovery facilities for natural gas liquids and export terminal facilities to cater to requirements in the future as well as for 1969. In the same year the third expansion phase of Kharg Island was completed at a cost of $41 million. Storage tankage was raised to 11,180,000 barrels; a one-million-barrel tank was erected and the loading rate was raised to 60,000 tons an hour. [42]

During 1970–1971 further expansion took place at Kharg. Jetties were made accessible to tankers up to 250,000 tons. Construction was undertaken for a two-berth sea-island type of loading facility to accommodate tankers of up to 500,000 tons. A 42/48-inch pipeline from Ahwaz to Ganaveh was built, and two additional 1,000,000-barrel crude oil storage tanks were nearing completion, increasing the total storage capacity to 14,000,000 barrels. A 72-inch pipeline linked the three-million-barrel tanks. [43]

With the explanation that Eastern Europe was never a market for Western companies and in line with the 1966–67 agreement, the Consortium delivered at Kharg terminal to NIOC for export to Eastern European countries, beginning with 1968, 2,400,000 barrels; in 1969 – 11,300,000; 1970 – 18,000,000; and 1971 – 16,000,000 barrels. [44]

The magnitude of the Consortium's expansion through the phenomenal output of oil was indicated above. The growth of the number of oil fields and their productivity should also be pointed out. In 1950–1951, the last year before nationalization, Anglo-Iranian Oil Company had six major producing fields, which were described above in Chapter VI. By 1971 the Consortium had 18 major producing fields; a comparison of productivity would illuminate the almost sensational advance. Agha Jari/Karanj/Marun/Paris fields produced, as a group, from 81 wells, 850,000,000 barrels; Gach Saran, from 28 wells, 322,000,000; Ahwaz, from 19 wells, 18,400,000; Bibi Hakimeh, from 13 wells, 162,000,000; Pazanan, from one well, 9,800,000; Masjid-i-Sulaiman/Lali fields, from 19 wells, 3,400,000; Naft Safid, from 11 wells, 11,800,000; Kharg, from 4 wells, 8,100,000; Ramshir, from one well, 1,600,000; Rag-e-Safid, from 2 wells, 13,200,000; Binak, from one well, 19,100,000; and Par-e-Siah produced, from one well, 4,800,000 barrels. The total number of wells was 193, and the total number of barrels produced was 1,522,300,000; the average annual production per well was about 7,887,564 barrels. [45]

NIOC ACTIVITIES AND EXPANSION

The agreement between Iran and the Consortium provided NIOC with two major functions: the responsibility for the non-basic operations of the oil industry and the supply of domestic oil and gas requirements. In 1959, the operating companies transferred to NIOC in Abadan the administration and maintenance of company housing, construction of non-basic houses and buildings, and the administration of the Abadan Institute of Technology. By the end of 1961, the transfer of all non-basic operations was complete. In 1963, NIOC organized a centralized unit, the Non-Basic Operations Organization, which worked in close cooperation with the operating companies in the oil fields, terminals and refineries. Twelve thousand staff and employees were exclusively occupied in providing the social and welfare needs of the oil workers. [46]

In line with the agreement provision for domestic needs, NIOC launched a refining-capacity expansion program as outlined above. But the new and projected refineries had to be provided with means of transportation to receive the supplies of crude and send the refined products to the consuming centers. NIOC therefore built a network of pipelines of different lengths and diameters which carried crude to the refineries and products from the refineries to the distribution outlets. From 1955 to 1965 NIOC spent £38 million on the construction of a total of 1,904 kilometers of oil pipelines: Ahwaz-Teheran, Teheran-Kazvin, Teheran-Sharud-Meshed, Azna-Isfahan and Naft-i-Shah-Kermanshah. In 1970 NIOC reported that its oil pipeline network consisted of 3,698 kilometers, and in 1971 an additional 72 kilometers of pipeline were added. [47]

NATURAL GAS AND PETROCHEMICALS

The most striking if not spectacular accomplishment of NIOC had been in the area of natural gas utilization and in developing a variegated petrochemical industry. Natural gas is found in Iran partly as caps above the oil columns in the wells and partly as dissolved in the oil itself, and in huge natural gas deposits as such. An Iranian source stated that with every ton of oil produced from the Consortium oilfields, 144 cubic meters of natural gas comes to the surface. Ever since oil was produced in Iran on a commercial basis, billions of cubic meters of associated gas have been flared.

In 1960 it was reported that while exploring for oil NIOC discovered a huge natural gas deposit in Sarajeh in the extreme Western part of the country. Two years later it was reported that NIOC planned to spend $25 million on building a 20-inch, 90-mile gas line to supply the capital with natural gas from Sarajeh. [48]

The first major step towards utilization of natural gas as a power source in industry was the construction of the Gach Saran-Shiraz pipeline in 1961, which supplied gas for a cement factory, two sugar factories and the Shiraz fertilizer plant. [49]

On April 22, 1966, NIOC organized a subsidiary, the National Iranian Gas Company (NIGC) for the purpose of dealing with all natural gas exploitation issues. It also entered into an agreement with the Soviet Union to supply natural gas in 1971 at an annual rate of 6.2 billion cubic meters, increasing the rate to 10.85 billion cubic meters in 1977. NIOC faced two major tasks: 1) to obtain the gas resources with which to meet its commitment, and 2) to find the means of transporting it.

For the latter a project called the Iranian Gas Trunkline (IGAT) was launched. For the former NIOC reached an agreement with the Consortium at the end of 1966 on the treatment and supply of gas for the planned gas trunkline, and for the installation of facilities for the manufacture and export of liquefied petroleum gas (LPG) by the trading companies. [50] Construction of the 40/42-inch, 1,104-kilometer, gasline began in September 1967. The contracts were awarded to different companies. Soviet and German concerns were awarded the building contracts while British, French and Japanese companies were awarded the contracts to supply the great variety of installations. The project was the greatest that NIOC had ever undertaken. According to a NIOC analyst, *The Times* of London called it "the most significant undertaking in the Middle East."

The line began at the Bid Boland gas treatment center near the Agha Jari and Marun oil fields, from which it was carried to Saveh in central Iran, and from there to Astara on the Russian frontier. The trunkline was connected, by spur lines 675 kilometers long, with the major cities Shiraz, Isfahan, Kashan, Qum, Saveh, Kazvin, Resht and Astara; a branch line was built to Teheran. The system also contained 103 kilometers of gas-gathering pipeline, compression facilities, and metering and control instruments for the transportation of the gas.

When the line, which cost some $750 million, had been completed it was to supply the industrial needs of the towns through which it passed as well as Teheran, and later to service domestic needs. [51]

On October 1, 1970, the IGAT line went into operation. This inaugurated the first export of natural gas from a Middle Eastern producing country. The initial daily capacity of the line was 830 million cubic feet, of which 200 million was delivered to the Iranian towns and 630 million was exported to the Soviet Union.

One week after the opening of the line a new agreement was signed with the Soviet Union which provided for a wide range of projects, including an increase in gas export in coming years which would necessitate the construction of a parallel trunkline. An Iranian analyst declared: "An important facet of this project is that gas exports will mainly pay for a steel mill currently being constructed in Isfahan." [52]

Meanwhile three developments took place in the natural gas area; new gas fields were discovered, more facilities for the production of gas and the manufacture of liquid gas were being constantly added by the Consortium and other companies, and the domestic gas and liquefied consumption was rapidly increasing.

NIOC reported that a new gas field was discovered in 1968 in Khangiran near Sarakhs in the northeast. [53] A year later gas deposits were discovered on Qeshm Island.

The Consortium complied with its agreement with NIOC and installed facilities for supplying both raw and liquefied gas for internal consumption and for export. [54]

Early in 1970 the Natural Gas Liquids Refinery at Bandar Mashur was commissioned, and exported during the year 1,328,379 barrels of propane, 1,514,700 barrels of butane and 1,353,934 barrels of liquid gasoline. In 1971 it produced 1,417,000 cubic meters of liquid gas. [55] In October 1971, a fifth Natural Gas Liquids plant with a daily capacity of 225,000,000 cubic feet was commissioned by the Consortium.

NIOC supplied natural gas and products for local needs at an ever increasing rate. In 1969 a total of 82,780 tons of liquid gas was delivered to 17 distribution companies, while 162 million cubic meters of natural gas was distributed through pipelines in the Shiraz, Ruddasht, Kavar and Gooyam areas for industrial and home consumption. [56]

In 1970 NIOC delivered a total of 112,020 cubic meters of liquid gas to 16 gas distributors and a total of 180,561,489 cubic meters of natural gas for commercial and home use. [57]

In 1971 total production of natural gas was 1,305 billion cubic feet, of which domestic consumption accounted for 551.5 billion. The Consortium delivered a total of 547.584 billion cubic feet; 374.014 billion went to NGL plants, and 137.57 billion to other consumers. An additional Natural Gas Liquids plant with a daily processing capacity of 225 million cubic feet was completed in the same year. [58]

The great advance in the export of natural gas came in 1971 when two letters of intent were signed by Iran: one for the sale of 3 million tons of LNG annually for a period of 20 years to the French Company Elf–ERAP and the Japanese firm of C. Itoh, and the other for the sale of 4 to 6 million tons of LNG annually from Qeshm Island for a period of 20 years to two other Japanese firms. It was also reported that provisional agreement had been signed with the International Systems and Controls Corporation of the United States for the liquefaction and export of 4 to 7 million tons LNG annually, mainly to the United States, for a period of 20 years. [59]

In 1972 it was reported that Iran was disposing daily of more than 1½ billion cubic feet of natural gas. But, curiously, in all the different accounts of Iranian sources about the phenomenally rapid development of the natural gas industry no mention is made of the reportedly very huge gas fields of Sarajeh and Khangiran. [60]

PETROCHEMICALS

The growth of the petrochemical industry was the direct consequence of the utilization of natural gas, The National Petrochemical Company, a subsidiary of NIOC, had two fully owned companies: the Iran Fertilizer plant established in 1963 with $24 million in capital and the Pazargad Chemical Company established in 1966 with $2 million in capital, both producing various petrochemical products.

In May–July of the same year the NPC signed three agreements with foreign companies for developing the petrochemical industry of the country: 1) Abadan Petrochemicals was a partnership in which NPC owned 74% and B.F. Goodrich 26%; the initial capital was $8 million and the projected captial was $30 million; it commenced operations in 1969. 2) Shapur Chemical Company was a 50/50 partnership between NPC and Allied Chemicals; the initial capital was $25 million and the projected capital $200 million; it commenced operations in 1970. 3) Kharg Chemical Company, a joint-partnership between NPC and Amoco International; the initial captial was $7 million and the projected capital $40 million; it commenced operations in 1970. [61]

Late in 1971 Iran signed a 50/50 joint-venture agreement with a Japanese group for building a petrochemical complex at Bandar Shapur for the manufacture of various petrochemical products. Part of the products was to be taken by NPC for local disposal and the balance was to be marketed in Japan. Cost of construction was to be $358 million, of which $100 million were to be paid in cash by the two partners, with the remaining $258 million raised in Japanese capital markets. [62]

NIOC MARKETING

In its attempt to penetrate foreign markets, NIOC was not very successful. These were tightly controlled by the international companies which produced their crude, refined it, transported it and marketed it. Yet NIOC did succeed in disposing of its crudes in the so-called fringes of the international markets which were closed to the international companies, or in which the latter were not interested. NIOC concluded barter agreements with Eastern European countries, and, as mentioned above, not producing itself the quantities to which it committed itself, pressed the Consortium for extra crude deliveries, which the latter obliged.

In 1965 NIOC concluded a barter agreement with Argentina for 2.5 million barrels of crude for Argentine wheat. In the same year it signed an undertaking to supply Rumania with $100 million worth of oil over a 10-year period in exchange for industrial products including drilling and refining equipment, port facilities and machinery. In 1968 this was expanded into a $235 million agreement. Iran was to deliver about 10 million tons of crude over a period of 5 years ending in 1975; the schedule for the three immediate years was set: 1968 — 400,000 tons; 1969 — 1,200,000; 1970 — 1,900,000 tons. [63]

A second barter agreement was signed with Bulgaria for a total of $10 million and covering deliveries of 2,375,000 tons during 1967–1970: 75,000 tons in 1967, 400,000 in 1968, 700,000 in 1969 and 1,200,000 in 1970. [64]

The third barter agreement was arranged with Czechoslovakia in 1968 for a $200 million credit for 20 million tons of crude to be delivered during the 1971–1980 period.

In 1970 a five-year trade agreement was signed with the Soviet Union for $1 billion in credits for Iranian exports of natural gas and petrochemicals from the Abadan and Bandar Shapur complexes. [65]

The desire of NIOC to participate in downstream aspects of the oil industry prompted it among other things to enter the transportation phase of the industry. The National Iranian Tanker Company, in which NIOC had a 50% interest, was reorganized in 1969 and had at that time four ocean-going tankers, two with a tonnage of 55,000 tons each, and two of 35,000 each, and several small-size vessels for oil-products transportation to Iranian ports. In 1970, NIOC became the sole owner of the Tanker Company. An Iranian analyst gave the following two objectives of the transportation company: "Firstly ... it will reduce the country's dependability on others, and secondly, and perhaps more profoundly, it will provide Iran with a link to the consuming nations. A link that is essential if a producing country is ever to deal directly with a consuming area" [66]

From marketing Iranian crude and natural gas NIOC went into foreign downstream operations, one of its basic aims as outlined by its managers; these ventures involved disposing of some of the crude produced outside the Consortium area. In

1968 NIOC became a 17.5% owner of the Madras refinery being built in India; the capacity of the refinery was to be 2.5 million tons annually and NIOC was to furnish 70% of the crude for a period of 15 years. NIOC helped with the actual building of the new facility, which went on stream in 1969; IPAC supplied the crude. In 1970 IPAC exported 2.9 million cubic meters, and 5.3 million tons in the following year.

NIOC also became a partner, together with Total, a subsidiary of Compangie Française des Pétrole, and the South African Coal Oil and Gas Corporation in building the National Petroleum Refineries of South Aftica at Sasolburg. It went on stream in 1971 with an initial capacity of 2.5 million tons annually; 70% of its crude requirements were supplied by LAPCO. [67]

NOTES

1. At the time, the Consortium was the only source of oil revenue for Iran. Of the total revenue received in the period 1954/5, the Plan Organization was assigned 59%, NIOC 29% and the Ministry of Finance, for the current government-budget needs, 12%; in 1956 the Plan Organization was assigned 53%, NIOC 25% and the Ministry of Finance 22%. From then on, the percentage of NIOC went down until 1962, when it received 2%, the Ministry of Finance 58% and the Plan Organization only 40%. NIOC, *Economic Impact*, 23.

2. For details, see "Petroleum Law of Iran," *World Trade Information Service*. Pt. I, No. 58–37, Apr., 1958.

3. Until recently practically all producing governments urged the companies to produce greater quantities of oil in order to obtain greater revenues. The companies on the other hand would have wanted to limit production to market demands, and the Consortium resisted the pressure for higher production. The Shah as spokesman for Iran resented the commercial company's dictating to a sovereign state the rate at which its resources should be developed and used. A 1970 report of NIOC argued for special consideration for Iran. It declared: "Quite apart from the economic need for development in Iran, Iranian opinion is convinced that the country deserves some special acknowledgement of its success and stability, for the way in which it stands as a pillar of strength in a crucial and strategic part of the world, and the good sense and fortitude shown in maintaining supplies of a vital commodity to the western world at a time when all other sources of supply in the area were summarily and arbitrarily turned off. This reciprocal gesture of good will is not, however, to be one sided.

   "There is a very large measure of self-interest on the part of the commercial companies operating in the region to increase their offtake from demonstrably their most reliable source of supply. There is an even more practical reason why governments, business interests and private individuals in North America, in Western Europe and in Japan should see that to recognize such stability and reliability is in their best interest too." NIOC, *Oil and Economic Development of Iran* (Teheran, 1970), 6. Henceforth, cited NIOC, *Oil and Economic Development*.

4. Iranian Oil Operating Companies, *1961 Annual Review,* 13. Henceforth citeᴜ IOOC, *Annual Review* and year.

5. IOOC, *1971 Annual Review.*

6. Professor M.A. Adelman of Massachusetts Insititue of Technology calculated that in the period 1962—1964 the production cost per barrel in Iran was 7 cents, compared with $1.56 in Texas, 62 cents in Venezuela, 15 cents in Libya and 10 cents in Kuwait and Saudi Arabia. He also established that the development investment per initial barrel per day was $130 in Iran compared with $3,250 in Texas, $863 in Venezuela, $167 in Kuwait, $160 in Saudi Arabia and $149 in Libya. *Petroleum Press Service,* XXXIII, 178, May, 1966.

7. *World Petroleum Report,* X, 81—84, Mar. 15, 1964; *Petroleum Press Service,* XXXIV, 262, Oct., 1967. This rate of ½ cent per barrel Government contribution became general in the Middle East until 1971.

8. Thus, if the profit per barrel had been, for the sake of argument, $2.00, the Government's share would have been $1.00 — 25 cents in royalty and 75 cents in share of profits. However, if the 12½% royalty had not been deducted from the 50% income tax paid to the Government, the latter would receive the 25 cents in royalties. The balance of $1.75 would then have been equally divided between the Government and the Companies. This would have given the Government a total of $1.125; that is 12½ cents more than under the existing system. Moreover, there were at the time indications that OPEC would press for a 20% royalty.

9. *Petroleum Press Service,* XXXI, 425—426, Mar., 1964.

10. "Djakarta's Green Light," *Petroleum Press Service,* XXXII, 2—4, Jan., 1965.

11. Reported the *Petroleum Press Service:* "Before the delegates had gathered, tension had been built up by press reports that the Iranian Government had accepted the Consortium's latest offer and was on the point of signing an agreement with the Company." XXXI, 470, Dec., 1964.

12. OPEC, *1968 Review and Record,* 4—5; *Petroleum Press Service* XXXV, 44, Feb., 1968.

13. See below,Chapter XXV.

14. British Petroleum Company, *Annual Report and Accounts for 1970,* 5; henceforth cited BP, *Annual Report* and year; *Petroleum Press Service,* XXXVIII, 82, Mar., 1971, 175, May, 1971; *World Oil,* CLXXII, 7, Mar., 1971; United Nations, Economic and Social Office in Beirut, *Studies on Selected Development Problems in Various Countries in the Middle East* (United Nations, 1971), 102; henceforth cited UN, *Studies* and year; *Petroleum Times,* LXXV, 30, Feb. 12/26, 1971. The companies justified the increases by the stability of supply which they hoped to achieve during the next five years. The Chairman of British Petroleum, in his statement to the shareholders, saw a new role for the companies and declared somewhat philosophically: "The discussions in Teheran have convincingly demonstrated that the industry has an important international role to play between the governments of the countries which produce oil and those of the countries which are the main consumers." *Annual Report 1970,* 5.

15. *Petroleum Times,* LXXVI, 37, Feb. 11, 1972.

16. The Iranians claimed that through this type of concesssion they changed the

50/50 division of profits pattern into 75/25 in favor of Iran. Actually the 50/50 principle remained intact, for Iran was a 50% partner in the venture, except that it was to contribute its share only after oil was discovered in commercial quantities.

17. "New Oil Agreements in the Middle East," *The World Today*, XIV, 135–140, Apr., 1958; George W. Stocking, *Middle East Oil. A Study in Political and Economic Controversy* (Nashville, 1970), 169. AGIP was not required to pay any bonuses either at signature or at later dates. *The New York Times'* Washington correspondent reported on Sept. 1, 1958 that the State Department was worried and that the big American oil companies were angry about the probability that Iran was about to break the solid pact of the 50/50 profit-sharing agreements. He reported that some American oil executives had called the Iranian-Italian agreement "blackmail." Moreover, during the negotiations preceding the conclusion of the agreement, the State Department had made unsuccessful representations both in Teheran and in Rome.

18. *Petroleum Press Service*, XXXVII, 430, Nov., 1970.

19. *Teheran Economist*, May 12, 1958.

20. Late in 1968 Iran reached an agreement with Saudi Arabia on the demarcation of the median line in the Persian Gulf. In the middle of this area the Fereydoun field (Iran) and the Marjan field (Saudi Arabia) is located with a 10 billion barrel estimated reserve. Under a new agreement signed in 1969 both countries were to share 50/50 in the combined field. *World Oil*, CLXVII, 205–6, Aug. 15, 1968; – CLXIX, 193, Aug., 1969; *World Petroleum*, XL, 29–30, Jan., 1969; *World Petroleum Report*, XV, 83, 1969.

21. *Foreign Commerce Weekly*, 24, Aug. 25, 1958; *International Oilman*, XII, 258, Sept., 1958; *The New York Times*, June 22, 1958. The Canadian venture did not materialize.

22. NIOC, *Oil and Economic Development of Iran*, 27–28.

23. These groups were: LAPCO, Lavan Petroleum Company, covering an area of 8,000 square kilometers, a joint-partnership with Atlantic Group; IROPCO, Iranian Offshore Petroleum Company, covering an area of 2,250 square kilometers, a joint-partnership with Tidewater Group; IMINOCO, Iranian Marine International Oil Company, covering an area of 7,960 square kilometers, a joint-partnership with AGIP/Phillips/Hydro Carbon Company, and India; FPC, Fars Petroleum Company, covering an area of 5,750 square kilometers, a joint-partnership with French Group; DOPCO, Dashtestan Offshore Petroleum Company, covering an area of 6,036 square kilometers, a joint-partnership with Shell; PEGUPCO, Persian Gulf Petroleum Company, covering an area of 5,150 square kilometers, a joint-partnership with German Group. NIOC, *Oil and Economic Development of Iran*, 43; – *The Economic Impact*, 7; *World Oil*, CLXI, 170–171, Aug. 15, 1965. In 1970 Shell surrendered its rights in the joint-venture. *World Petroleum Report*, XVIII, 66, 1972.

24. *World Petroleum*, XL, 34, July, 1969; *World Petroleum Report*, XV, 83, 1969; NIOC, *Annual Review 1968*, 10, 12; – *NIOC in 1969*; – *NIOC in 1970*, 7; – *1971*, 44.

25. *World Petroleum Report*, XVI, 72, 1970; NIOC, *Annual Review 1968*, 12; – *NIOC in 1969*; – *NIOC in 1970*, 7; – *1971*, 44; – *Economic Impact*, 7.

26. NIOC, *Economic Impact*, 7; — *Oil and Economic Development of Iran*, 43—44; Stocking, *Middle East Oil*, 178—179. In 1968 it was reported that SOFIRAN had discovered oil offshore near Siri Island in the eastern part of the Persian Gulf, but until the end of 1971 no production was reported. *Petroleum Press Service* XXXV, 385, Oct., 1968. Addressing a seminar organized by the Kuwait Institute of Economic and Social Planning in the Middle East, which examined the terms of oil concessions, Nadhim al-Pachachi, oil adviser to Adu Dhabi, emphasized that the contract signed between Iran and ERAP represented a definite departure from existing agreements since the Government retained overall responsibility for decision-making and the sole ownership of all oil produced at wellhead.*Ibid*, 464, Dec., 1968.

27. *World Oil*, CLXIX, 193, Aug. 15, 1969; NIOC, *Oil and Economic Development of Iran*, 46; — *Economic Impact*, 7—8; *Petroleum Press Service*, XXXVI, 451—453, Dec., 1969. It was reported in 1970 that the Russians were exploring the offshore waters in the Caspian sea on a contract basis for NIOC. *World Petroleum Report*, XVII, 65—66, 1971.

28. NIOC, *NIOC in 1970*, 1.

29. *Petroleum Press Service*, XXXVIII, 384, Oct., 1971.

30. The partnerships were INPECO, Irano-Nippon Company, covering an area of 8,000 square kilometers in the Luristan province, and a group of four Japanese companies and Mobil Oil Corporation; BUSCHO, Bushir Oil Company, covering an area of 3,715 square kilometers of Iran's continental shelf opposite Bushir, and the Amerada-Hess Corporation; HOPECO, Hormuz Petroleum Company, covering an area of 3,500 square kilometers of Iran's continental shelf in the Straits of Hormuz, and Mobil Oil Corporation. NIOC, *Economic Impact*, 9; — *1971*, 1; *Petroleum Press Service*, XXXVIII, 303, Aug., 1971; Saudi Arabian Monetary Agency, Research and Statistics Departments, *Statistical Summary*, 18—19, Sept., 1971. Henceforth cited SAMA, *Statistical Summary* and date.

31. The *Aramco Handbook: Oil and the Middle East* (Dhahran, 1968), 82, stated: "For a number of reasons the newer, modern refineries mostly are constructed near the market for their products. Before the war the tendency was to locate the refinery near the crude oil source and to ship the products to market. But these factors have resulted in a shift in location of new refineries: It is cheaper to transport crude oil than refined products by tanker. Nations can buy more crude oil than refined products with their foreign exchange. Manufacturers in industrial centers now obtain feedstock, formerly considered worthless, from nearby refineries and use them in making petrochemicals. Consuming nations obtain the economic benefits of refinery construction and operation." What the handbook fails to point out is the fact that the changed relations in terms of payments and the basis for calculating these payments reversed the position of the companies as to the location of the refineries.

32. NIOC, *Economic Impact*, 11; IOOC, *1960 Annual Review*, 17; — *1971 Annual Review*, 18.

33. NIOC, *Economic Impact*, 11; *Petroleum Press Service*, XXXVI, 453, Dec., 1969.

34. NIOC, *Economic Impact*, 10—12; — *NIOC in 1970*, 8; — *Oil and Economic*

Development, 28, 68; *Petroleum Press Service,* XXXIII, 176, May, 1966; — XXXVI, 453, Dec., 1969; *World Petroleum Report,* VIII, 44, Feb. 15, 1962; — XIII, 67—68, 1967; — XVII, 65—66, 1971.

35. IOOC, *1959 Annual Review,* 1; — *1960 Annual Review,* 17, 34; BP, *Annual Report 1969,* 30; *Petroleum Press Service,* XXVII, 381, Oct., 1960.
36. *New York Times,* Sept. 17, 1961.
37. IOOC, *1962 Annual Review,* 2.
38. IOOC, *1964 Annual Review,* 4; *World Petroleum Report,* XI, 70—72, Mar. 15, 1965.
39. IOOC, *1965 Annual Review.*
40. IOOC, *1966 Annual Review,* 4, 6; — *1967 Annual Review,* 6—7; BP, *Annual Report 1966,* 25; — *Annual Report 1967,* 27; NIOC, *Oil and Economic Development,* 25; *World Oil,* CLXV, 142, Aug. 15, 1967; *World Petroleum Report,* XIII, 67—68, 1967; *World Petroleum,* XXXVIII, 49, July, 1967.
41. *World Oil,* CLXXI, 168, Aug. 15, 1970; *World Petroleum,* XL, 20, July, 1969. Remarked a NIOC analyst: "This year, as before, wisdom and good sense prevailed at the negotiations between those guiding the policies of the National Iranian Oil Company and members of the Consortium." *Oil and Economic Development,* 15.
42. IOOC, *1969 Annual Review,* 18; *World Petroleum Report,* XVI, 72—73, 1970.
43. IOOC, *1970 Annual Review,* 6; — *1971 Annual Review,* 6; NIOC, *Economic Impact* 16.
44. IOOC, *1968 Annual Review,* 7; — *1969 Annual Review,* 39; — *1970 Annual Review,* 43; — *1971 Annual Review,* 27. These deliveries fell, of course, far short of the 20 million tons agreed upon, because the closing of the Suez Canal made it very difficult for some of the Eastern European countries to lift their quotas of oil originally planned for.
45. IOOC, *1971 Annual Review,* 34. By the end of 1971, NIOC produced from Naft-i-Shah field; SIRIP from Bahregan Sar and Nowruz; IPAC from Darius and Cyrus; LAPCO from Sassan; and IMINOCO from Rostam and Rakhsh fields. NIOC, *1971.* 7; *Petroleum Press Service,* XXXVI, 451—453, Dec., 1969.
46. IOOC, *1959 Annual Review,* 19, 26; — *1961 Annual Review,* 13; *Petroleum Press Service,* XXX, 103—105, Mar., 1963; NIOC, *Oil and Economic Development,* 26. The Iranian source stated: "The operating companies agree to refund to NIOC the costs of providing the services in the Consortium Agreement area."
47. NIOC, *Oil and Economic Development,* 28; — *NIOC in 1970,* 9; — *1971,* 9; *World Petroleum Report,* XVII, 65—66, 1971.
48. *Petroleum Press Service* XXX, 270, July, 1963. NIOC apparently had second thoughts about the pipeline. For in 1972 it was reported in Teheran that in order to meet the gas requirements of the capital, a 30-inch branch line 111 kilometers long has been built from Saveh, on the trunkline, to Rey, the gas terminal for Teheran. From Rey the gas was conveyed to the consumption centers in Teheran through three main branch lines. *Iran Oil Journal,* 5, Sept., 1972.
49. NIOC, *Economic Impact,* 12; — *Iranian Oil. A Decade of Success,* 5; — *Role of Oil in Iran's Economy,* 11.

50. IOOC, *1966 Annual Review*, 4; BP, *Annual Report 1966*, 25.
51. NIOC, *Oil and Economic Development of Iran*, 54–59. Without giving any details as to the supplier, builder or scope of production, this source stated: "A noteworthy project, closely linked with IGAT, is the construction of the Pipe Rolling Mills at Shiraz where pipes for the gas trunkline are being made." 58; – *Economic Impact*, 12–13. It noted: "The NIOC was responsible for the construction of the gas trunkline from the southern oil fields to Saveh in central Iran, and the Soviet Union participated in the laying of the Saveh-Astara stretch by supplying the necessary equipment and manpower." 13. See Chapter IV above.
52. *Petroleum Press Service*, XXXVII, 421, Nov., 1970; NIOC, *NIOC in 1969;* – *Iranian Oil. A Decade of Success*, 5.
53. A NIOC report stated that it was the greatest gas discovery in Iran, and that Iran was "currently estimated to have about 60% of the Middle East's total proven gas reserves." *Economic Impact*, 13.
54. The extent to which natural gas was wasted even after the 1966 agreement is illustrated by the following table:

Gas Production and Utilization 1966–1968. Million Cubic Feet

Year	Production	Utilization	% Utilized
1966	664,782	48,959	7.4
1967	733,556	51,784	6.9
1968	851,708	55,450	6.5

NIOC, *Oil and Economic Development*, 67.

55. IOOC, *1970 Annual Review*, 17; – *1971 Annual Review*, 5; *World Petroleum Report*, XVII, 65–66, *1971;* NIOC, *1971*, 8.
56. NIOC, *NIOC in 1969*. In 1968 LAPCO and IPAC made no use of their natural gas. Of NIOC's production of 76,375,072 cubic meters of natural gas only 23,721,270 were utilized; of SIRIP's production of 176,162,731 cubic meters only 40,600,000 were utilized; and of the Consortium's 22,644,956,235 cubic meters produced only 4,103,220,516 were utilized. Even as late as 1970, LAPCO flared all the gas it produced. NIOC, *Annual Review 1968*, 10; – *Economic Impact*, 12.
57. NIOC, *NIOC in 1970*, 17. The Consortium reported that during 1970 it delivered 426.070 billion cubic feet of natural gas; of these 270.830 billion went to NGL plants and 155.240 billion to other consumers. The corresponding figures for 1969 were 163.916, 36.825 and 127.091 billion cubic feet, respectively. IOOC, *1970 Annual Review*, 43.
58. IOOC, *1971 Annual Review*, 7, 27; NIOC, *Iranian Oil. A Decade of Success*, 9; – *1971*, 16. The last source gives the total export figure to the Soviet Union as 5.546 billion cubic meters, and internal consumption in Iran as 363 million cubic meters.
59. NIOC, *1971*, 16; – *Iranian Oil. A Decade of Success*, 6.
60. *Petroleum Press Service*, XXXIX, 225, June, 1972.

    IOC, *Oil and Economic Development*, 70; – *Economic Impact*, 13–14; –
Annual Review 1968, 48–50; – *NIOC in 1969;* – *NIOC in 1970*, 18; *World
Petroleum Report*, XVII, 65–66, 1971; SAMA, *Statistical Summary*, 16, Dec.,
1969–Jan., 1970.
.. SAMA, *Statistical Summary*, 17, Dec., 1971; NIOC, *Economic Impact*, 15.
Exports of petrochemicals in 1971 totalled 590,870 tons. NIOC, *1971*, 1.
63. NIOC, *Oil and Economic Development of Iran*, 47; – *Annual Report 1968*, 39;
Petroleum Press Service, XXXIV, 194, May, 1967; – XXXV, 74, Feb., 1968.
64. *Petroleum Press Service*, XXXIV, 276, July, 1967; *World Petroleum Report*,
XVI, 73, 1970; – XVII, 66, 1971.
65. *World Petroleum Report*, XV, 83, 1969; – XVII, 66, 1971.
66. NIOC, *NIOC in 1969;* – *NIOC in 1970*, 2: – *Iranian Oil. A Decade of Success*,
3.
67. NIOC, *Annual Review 1968*, 40; – *NIOC in 1969;* – *NIOC in 1970*, 10; –
1971, 10; Compagnie Française de Pétroles, *Annual Report 1970*, 53, 69; –
Annual Report 1971, 72. Henceforth cited CFP, *Annual* and year.

Chapter VIII
IRAN–DEVELOPMENT AND OUTLOOK

As was pointed out in Chapter VI, the Iranian leaders realized the potentialities of oil revenue for the economic development of their country. They agreed to adopt a development plan which would utilize the income from the oil industry for the rehabilitation of Iran. Perhaps out of different motives and the contradictory aims of the various forces operating in the country, Iran adopted the Seven-Year Development Plan for the years 1948–1955. The reasons for its ineffectiveness were intrinsic: it was a first effort, awkward and untried, and it was in a sense an outside imposition, and as such was bound to be resisted, as social practices and attitudes operated against it; but the greatest cause for its failure was the onset of nationalization, which deprived it of all income.[1]

Since the adoption of the First Plan many changes had taken place both in Iran and abroad. The Iranians, at least the articulate leadership, had come to look upon oil not as an exhaustible source of wealth to be exploited sparingly in order to spread the benefits therefrom over as long a period as possible, but as an exhaustible source of wealth which must be used to gain economic independence. They argued that the sooner the oil was produced in ever greater quantities the sooner would the country achieve prosperity and economic independence.[2]

Like many other developing countries, Iran believed that the only answer to all its economic and social ills lay in transforming a predominantly backward agrarian economy into a dynamic industrially advanced society. This could be achieved through the oil industry, technologically as well as financially. Some of the social and economic patterns of Iran had undergone serious changes after the nationalization of the oil industry, which were greatly hastened by the persistent efforts of the Shah and his administrators, especially in agrarian reform. Moreover, the increase of oil consumption throughout the world and particularly in Europe and the United States made the Middle East, of which Iran was the top producer, the major source of supply. This changed the basic pattern of relations between the Consortium and the Iranian Government, and strengthened the latter at the expense of the former. As a result, in any struggle between Iran and the foreign companies, the will of Iran generally always prevailed.

With the resumption of oil production by the Consortium in 1954 and the renewal of the flow of revenue, the Second Seven-Year Plan, for 1955–1962, was adopted. In the main it was a resumption of the objectives and projects of the First Plan, except that it was entirely in the hands of Iranians. Some of the basic difficulties and weaknesses described in Chapter VI were plaguing the Second Plan. It called for a total investment of £414 million; actual investments were £65 million short of that figure. The system of priorities was: transport and communications 35%, agriculture 22%, regional development 14%, social welfare 13%, industries and mines 8% and other items 8%. The structure of the Gross National Product for the last year of the plan (1962) was given as: agriculture 28.2%, industry and mining 34.8% and services 37%.[3]

The Third Plan was for the five-year period 1962–1967. It called for an expenditure of £592 million. NIOC stressed that the Third Plan's aim was to speed up the achievement of the projects undertaken during the First and Second Plans, and limited itself to such new projects as could not or would not be undertaken by the private sector. The system of priorities shifted somewhat but not radically: transport and communication, still first with 25%, agriculture 22.5%, education and health 16%, fuel and energy 13.5%, industry and mines 16% and others 12%.

In spite of the reported improvement in the economic growth of the country, amounting to an average annual increment of 6% in the mid-50's and 8.4% in the early 60's, the Plan became progressively more dependent on the oil revenue from the Consortium, both in actual amounts and in percentages. In 1957 it received $106.4 million, which was 50% of the total; in 1967, the last year of the Third Plan, it received $569.2 million, which amounted to 80% of the revenue.[4]

The Fourth Five-Year Development Plan covered the years 1968–1973. Its expenditure was to be $10.854 billion. Of this amount about $6 billion or 54.7% was to be contributed by the oil revenue and $4.8 billion or 45.3% was to be raised from the private sector through foreign loans and foreign investments. NIOC stated that the "Fourth Development Plan surpasses anything in the field of economic planning which has so far been attempted by this country."[5]

Table XI gives the sector apportionments and percentages of the Fourth Plan. The overall aim of the plan was "to provide a better standard of living for the people and help to create a just society." The more particular objectives and the changed basic approach are reflected in the priorities of the plan. The lion's share went to industries and mines, while transport and communications was still

Table XI. *Fourth Five-Year Plan, 1968–1973*

Sector	$	Percentage
Agriculture and Irrigation	885,996,000	8.2
Transport and Communication	1,858,304,000	17.2
Industries and Mines	2,773,632,000	25.5
Regional Development	416,712,000	3.8
Social Welfare	1,105,440,000	10.2
Fuel and Energy	2,602,128,000	23.9
Manpower	24,120,000	.2
Housing	1,159,056,000	10.7
Others	33,504,000	.3
	$10,853,892,000	100%

NIOC, *Economic Impact,* 25.

awarded 17.2%; fuel and energy, which is directly connected with industry, came second with 23.9%; social welfare which was steadily climbing from 3.9% in the First Plan to 13.7% in the Second Plan and to 15% in the Third Plan, was apportioned only 10.2%; while housing, which was granted in the Third Plan only 6%, was allotted 10.7%.

The target for the annual increase in the GNP during the five-year period of the Fourth Plan was 9.4%. NIOC claimed that it was not an optimistic aim, for the increase in 1971 was 12%. The main thrust was of course industry and mining; the objective in this sector was "to achieve a 100% increase, equivalent to an annual average increase of about 15%, in the total value of production of the industrial and mining sector (excluding oil), whose share of total GNP is expected to rise from its 1968 level of 32% to 40% by the end of the Fourth Plan, with the share of the agriculture sector decreasing to 16% of GNP during the same period."[6]

What have the three first development plans achieved for Iran in the field of elementary education? Iran's leaders have recognized their basic problems, and NIOC reported in 1972 that "successive governments, supported by His Majesty, have shown a readiness to challenge head-on the greatest obstacles facing the country, such as the formal feudal basis of agriculture, illiteracy, disease, lack of social services and general lack of knowledge concerning economic development." Yet, even after 17 years of development efforts and spending considerable amounts, the education objective was formulated: "To expand educational activity by increasing the literacy level of the 10—45 age group from 35% to 60% by the end of the Fourth Development Plan period."[7]

The oil revenue financed not only the development plans but the regular Government budget as well. During the period 1960—1971, oil revenue accounted for about 50% of the budget. A NIOC report stated: "Perhaps the most important role of the oil industry in the Iranian economy is as an earner of foreign exchange." It pointed out that between 1956 and 1971 the oil industry earned an annual average of 75% of all foreign exchange earnings.[8]

If past experience is any guide, in spite of the most optimistic reports of NIOC about the economic progress of the country, after the Fourth Plan there will no doubt come a Fifth Plan, which will be of even greater magnitude in objectives and spending. The only source for the plan's spending will still be the Consortium. The means to obtain the resources will again be greater production and higher percentages in the profits for the Government. It has already been mentioned in Chapter VII that in 1971 the total production of oil of all the other foreign companies and NIOC was less than 10% of the Consortium production, and revenue was less that 5%. Table XII, which is the NIOC summary of Iranian Government revenue for 1971, reveals that the Consortium's direct payments and the profits of NIOC from the crude supplied by it, amounted to over 90%.

Thus, in spite of all the efforts and impressive achievements of NIOC as outlined in the previous chapter, Iran must still depend very heavily on the Consortium member companies. To meet its needs the Iranian Government will no doubt press the Consortium for the above-mentioned objectives. This raises two basic questions:

Will the Consortium member companies yield on both? Or if the Consortium resists one or both what will the Iranian Government do?

Table XII. *The Iranian Government Oil Revenues, 1971,*
In 1000 Rials

1.	Income tax paid by the Consortium's trading companies	109,869,246
2.	Income tax paid by the Iranian oil operating companies	1,529,117
3.	Stated payments (less 2 per cent general reserve)	29,690,823
4.	Income tax paid by SIRIP	396,034
5.	Income tax paid by IPAC (2nd party share)	709,857
6.	Income tax paid by IPAC (NIOC share)	1,054,094
7.	Income tax paid by LAPCO (2nd party share)	2,202,730
8.	Income tax paid by LAPCO (NIOC share)	1,127,144
9.	Income tax paid by IMINOCO (2nd party share)	258,166
10.	Income tax paid by IMINOCO (NIOC share)	455,687
11.	Tax on NIOC net profits	1,947,966
12.	NIOC dividend payments	1,523,352
13.	Taxes, dues and product price increase	10,823,686
14.	Employee income tax	1,094,775
15.	Contractors' income tax	805,141
16.	Bonus payment after deduction of 2% general reserve	74,725
	TOTAL	163,562,543

NIOC, *1971*, 4.

Before attempting to answer these questions we should clearly understand the fundamental elements of the situation. In order for the Consortium to increase production it would have to augment vastly its investments in exploring, development, production, transportation and marketing facilities. As the rate of increase rises the ratio of investment jumps much higher than the increase in production; this means enormous investments and reduced returns. Assuming that the additional huge sums of investment were available, the Consortium companies would want to be assured of the safety of their investments and a reasonable return for them. In view of the experiences of the foreign companies in the Middle East and the Consortium experience with Iran these would seem to be doubtful.

Should the Consortium refuse to comply with the above two demands of Iran, the Iranian Government would have the alternative of taking over the total operation of the industry. From the technological point of view and the international-political climate, Iran would be in a far better position to take over the industry than it was in 1951. But Iran would still have to face almost insuperable difficulties; in addition to technological and administrative obstacles, it would not have the investment resources necessary to operate the existing industry and certainly would not have the means for increased production. Iran would not have the means of transporting the oil nor the markets to sell it in.

In determining their position in regard to the Iranian Government's demands, the Consortium member companies would have to weigh two very grave consequences: 1) partial if not total loss of their present investments in Iran, and 2) the loss of supply of oil for their home and international markets.

One is inclined to think, guided by past experience, that both sides would realize the consequences to both sides from a break between them, and would follow a reasonable course of cooperation. The pattern of relations described in Chapter VII would continue, and the Iranian Government would gain in strength and position.

At a press conference in London in the middle of June 1972, the Shah outlined the future of the Iranian oil industry. It was to be of maximum output, maximum security of supply, stability of conditions for the period of the agreement, and after that tougher terms.[9]

At the end of the month a far-reaching agreement was concluded with the Consortium. It was expected that the increase in production might reach as high as 8 million barrels a day at the end of the seventies. NIOC was given a much more expanded role in taking over from the Consortium some parts of production, refining and marketing; the Consortium was to enter into joint-venture projects with NIOC in areas of exploration, production, processing of natural gas and LPG manufacture and export, and for the manufacture of petrochemicals. [10]

Later it was reported that the Abadan refinery was to be handed over to NIOC, and a new joint-refinery, probably on Kharg Island, was to be built with a tentative 15 to 18 million ton capacity. Plans also called for the expansion of the Kharg Island terminal to accommodate 500,000-ton tankers. Of course, the takeover of Abadan would make NIOC a giant exporter of products. On January 6, 1973, *Le Monde* of Paris reported that the Shah notified the Consortium that Iran intended to take over the entire oil industry, and would not extend the agreement after its expiration in 1979.

To guarantee continued marketing of Iran's output, the Shah decided that the Consortium companies would buy the oil produced after 1979 and thus assure the companies of continual supply. According to *Le Monde* the decision of Iran was not negotiable. [11]

NOTES

1. The Plan envisaged an expenditure of £125 million; the total income was practically to have been derived from the oil industry. It spent only £24 million. NIOC, *Oil and Economic Development*, 27.
2. A NIOC spokesman saw the future of Iran in industrial exports, and declared: "With a deeper penetration of Eastern Bloc markets and the establishment of huge export-oriented industries such as gas, aluminum and petrochemicals, it is

expected that non-oil exports will increase from about $300 million in 1971 to $566.7 million in 1973." NIOC, *Economic Impact,* 4.

3. *Ibid.* The original total investment is given in United States dollars as 1.120 billion. For the financial difficulties of the Second Plan see United Nations, *Economic Development 1958–59,* 37–38.
4. NIOC, *Economic Impact,* 1, 23.
5. *Ibid.,* 1; – *Oil and Economic Development,* 30.
6. NIOC, *Economic Impact,* 4. The same source asserted that while the per capita GNP in 1960 amounted to $181 it had passed the $400 mark in 1972. *Ibid.,* 1.
7. *Ibid.,* 3, 20.
8. *Ibid.,* 18; Sam H. Schurr and Paul T. Homan, *Middle East Oil and the Western World* (New York, 1971), 103; Gholam Reza Nickpay, *The Role of the Oil Industry in Iran's Economy and Future Programmes* (Teheran, 1962), 14.
9. *Petroleum Press Service,* XXXIX, 259, July, 1972.
10. *Petroleum Press Service,* XXXIX, 259, July, 283, Aug., 341, Sept., 1972.
11. *Ibid.,* 366, Oct., 1972; *New York Times,* Dec. 17, 1972; – Jan. 7, 1973.

Section Three
IRAQ

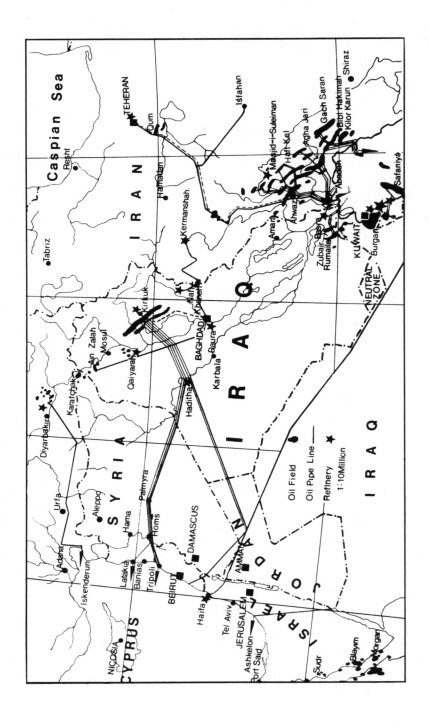

Chapter IX
EARLY ATTEMPTS TO OBTAIN CONCESSIONS

THE TURKISH PETROLEUM COMPANY

The knowledge of the existence of petroleum in what is now Iraq goes back perhaps to ancient times, but it was not until the latter part of the last century that it became known that the vilayets of Baghdad and Mosul contained oil fields which might have commercial potentialities, and the Ottoman authorities became aware of the possibilities of exploiting these fields. In 1888 and 1898 Sultan Abdul Hamid conferred, by special firmans,[1] the concessions of the vilayets of Mosul and Baghdad, respectively, on his Civil List, thus making the concessions his private business.[2]

In 1904 the Civil List signed a contract with the Anatolian Railway Company — nominally Turkish but actually German — acting for the Deutsche Bank, to carry out a preliminary survey of the fields in the two vilayets, with an option, within two years after completing the survey, to enter into a contract with the Civil List for joint operation of the fields. Two years later the Civil List considered the agreement with the Anatolian Railway terminated and began negotiations with William Knox D'Arcy, who had the active support of the British Ambassador at Constantinople. While these negotiations were going on, the Young Turk Revolution took place and the Mesopotamian oil fields were transferred from the Civil List to the Ministry of Finance. D'Arcy continued to negotiate with the Ministry, but with no tangible results.

In 1912 the German interests of the Anatolian Railway Company attempted to obtain confirmation of their old concession. Simultaneously, the Royal Dutch-Shell group, through its subsidiary, the Anglo-Saxon Oil Company,[3] was trying, with the assistance of an Armenian, Sarkis Colouste Gulbenkian, to obtain a concession in Mesopotamia, and the American Chester group (details about which later) was also seeking a concession. By 1912 there were therefore four different groups seeking concessions in Mesopotamia: 1) German—Deutsche Bank; 2) British—D'Arcy (APOC);[4] 3) Dutch—Anglo-Saxon Oil Company; and 4) American—Chester group.

Two considerations operated in uniting three of the four groups in their efforts to gain the Mesopotamian oil concession. The Europeans were determined to keep the Americans out. At the same time the British were approaching the stage where they were ready to cooperate with the Germans in the Baghdad railway project,[5] and as the Dutch group was to a considerable extent a British company, the best way of obtaining the concession was by putting combined pressure on the Turks. Negotiations were therefore carried on, not between the Turkish Government and the groups seeking the concessions, but between the Turkish Government and the British and German Governments.

In 1912 Sir Ernest Cassel, a German-born English b
Bank of England, formed a British joint stock comp
Company, with a capital of £80,000, for the purpos
Mesopotamian oil fields, as well as prospecting for oi
Empire. He was represented in Constantinople b
president of the Turkish National Bank (a British-o
per cent of the shares of the Turkish Petrole
confidence of the Turkish Government and the
Anglo-Saxon Company and the Deutsche Bank
and each became a 25 per cent owner of the '
Anglo-Persian Oil Company balked.[6] The nef
on March 19, 1914 an agreement was signr
popularly known as the Foreign Office A
British and German Governments, the T'
Shell Company, and the D'Arcy group, :

This agreement, which united all th
oil concessions, gave the British absc
Turkish Petroleum Company from £8
new certificates to the D'Arcy group
Turkish National Bank in the Turki
per cent, and the Anglo-Saxon C
beneficiary interests, without voti   :r
cent from Anglo-Saxon) to S.   ne
Company was to be composed (and
two each by the other two gr   was
the ban on concession-seeking   rtners
without the participation of t

    The three groups   any shall
give undertakings on   ies associated with them not   duction or
manufacture of cru   a, except in
that part which is   rnment or of
the Sheikh of Kc   Turco-Persian
frontier, otherwis   y.[8]

    Only five day   Great Britain, in
identical notes to   urkish Petroleum
Company a conce   e vilayets of Mosul
and Baghdad. C   lim, replied to the
German Ambas

    "Mr. Anc
    "In response to ...   ellency had the kindness

to address to me under the date of the 19th instant, I have the honor to inform you as follows:

"The Ministry of Finance being substituted for the Civil List with respect to petroleum resources discovered, and to be discovered in the vilayets of Mosul and Baghdad, consents to lease these to the Turkish Petroleum Company, and reserves to itself the right to determine hereafter its participation as well as the general conditions of the contract.

"It goes without saying that the Society must undertake to indemnify, in case of necessity, third persons who may be interested in the petroleum resources located in these two vilayets."[9]

Because of the outbreak of World War I, no further steps were taken and this was the last and final act of the negotiations between the Turkish Petroleum Company, through the British and German Ambassadors, and the Turkish Government.

The Grand Vizier's letter raised a number of questions. Did it grant a legal right which, in accordance with the provisions of the peace treaties with Turkey, had to be recognized and respected by the successor states? Or did it only promise a contract or a lease,[10] and as no subsequent action was taken, the Turkish Petroleum Company had no valid claim which had to be recognized by the victors or successor states or governments?

Before considering the controversy which arose between the United States and Great Britain as to the validity of the Company's rights, we shall outline the American concession claim as well as Anglo-French relations during World War I and during the peace negotiations.

THE CHESTER CONCESSION

Early in 1908 the American Consul at Aleppo sent a report to the State Department pointing out the opportunities for American firms to construct and operate railways in parts of the Ottoman Empire. Rear Admiral Colby M. Chester, who as captain of the *U.S.S. Kentucky* had been sent to Constantinople in 1899 to give support to the American Minister in obtaining indemnity for American missionary property destroyed during the Armenian massacres of 1896, apparently saw the Consul's report. That same year, 1908, he went to Constantinople for the purpose of obtaining railway and mining concessions. He had the backing of the New York Chamber of Commerce and the New York State Board of Trade, and was supported by President Theodore Roosevelt and Secretary of State Elihu Root.[11]

In 1909 Admiral Chester entered into an agreement with the Turkish Government for a concession to construct a port and three railway lines: from Harput to Lake Van, via Arghana, Diarbekr and Bitlis; from Yourmourtalik to some point on line "a"; from some point on line "a" to Sulaimaniah and the Persian border via Mosul and Kirkuk. The concession granted mineral rights, including oil, for twenty kilometers on both sides of the line, and Chester deposited "caution" money to the amount of 20,000 Turkish pounds.[12] To implement the concession, Chester organized the Ottoman-American Development Company under the laws of the State of New Jersey, with an authorized capital of $600,000.[13]

On March 9, 1910 the concession was signed by the Turkish Minister of Public Works and the following year was sent by the Grand Vizier to Parliament for ratification. The outbreak of the Turco-Italian war and of the Balkan wars prevented the ratification, and renewed efforts in 1914 to have it ratified were interrupted by the onset of World War I. Since the concession was not ratified, Admiral Chester had to admit to Secretary of State Charles Evans Hughes, in April 1921, that it had never actually been granted.[14]

The Anatolian Railway Company offered the main opposition to the Chester concession at the time, and Marshall von Bieberstein, the German Ambassador at Constantinople, protested to the Turkish Government that the proposed Chester concession violated some of the rights of the German companies. British interests, through the British Ambassador, also acted against the project, but the American Embassy in Constantinople gave Chester its full diplomatic support.[15] The fact remains, however, that the Chester concession was not ratified. At best it could be considered as having the same status as the Turkish Petroleum Company's concession, but in some respects it was even inferior since it had been sent to Parliament in 1911 but was not ratified, whereas the TPC concession was signed by the Grand Vizier in 1914, fully three years later.[16]

The final chapter of the Chester concession, however, was not written until the early 1920's and we must pick up the story of the relations between the French and the British.

ANGLO-FRENCH RELATIONS

The question of the future of the territories of the Ottoman Empire after the outbreak of the war was a subject of considerable discussion between the members of the Triple Entente; every one of the three was striving to make past ambitions come true, as well as to obtain guarantees for future positions. Russia was determined to gain control of Constantinople and the Straits; France saw in the partition of Turkey's territories a chance to realize her old romantic ambition of becoming the ruler of the Levant; while England aspired to build, under her protective wings, a new Islamic empire which would guarantee her position in the Middle East and guard her line of empire to India. The upshot was the Constantinople Agreement of March 1915 among the three allies; and the Sykes-Picot agreement,[17] between England and France, relating to the Arab territories of the Ottoman Empire.

The latter agreement divided most of Asiatic Turkey into zones: an international, brown zone covering most of Palestine; a French, blue zone covering parts of Syria; a British, red zone covering parts of Palestine and parts of Mesopotamia; between the blue and the red zones an area of "a" and "b" zones was marked off for the establishment of Arab states or a confederation of Arab states; in "a" the

French were to be predominant, and in "b" the British. The vilayet of Mosul was included in the French "a" zone.

In a secret letter dated May 15, 1916, from the British Foreign Secretary, Sir Edward Grey, to the French Ambassador at London, Paul Cambon, it was stated that the British would be grateful, before confirming the Sykes-Picot agreement, if they could be assured that in the regions which, under the conditions of the proposal, would become "entirely French, or in which French interests are recognized as predominant, any existing British concessions, rights of navigation or development, and the rights and privileges of any British religious, scholastic or medical institutions will be maintained." To which Cambon agreed, in a letter of the same date, with the understanding that the same would apply to French rights in the regions assigned to Great Britain. *"J'ai l'honneur de faire connaître à votre Excellence que le Gouvernement français est prêt à sanctionner les diverses concessions britanniques ayant date certaine antérieure à la guerre dans les régions qui lui seraient attribuées ou qui relèveraient de son action."* [18]

Lloyd George, however, was never too happy with the Sykes-Picot agreement and sought to introduce two important modifications: the status and boundaries of Palestine, and the inclusion of Mosul in the British zone. The opportunity presented itself when Clemenceau visited London in December 1918. The French demands on Germany at the impending peace negotiations placed England in a good bargaining position. In return for British support of French demands in the Ruhr, Clemenceau made concessions on the boundaries of Palestine and ceded to the British Mosul, where they were in occupation after the armistice of Mudros. The French were promised a share in the oil of Mosul. [19]

LONG-BERENGER NEGOTIATIONS

Subsequently, Walter Long, British Minister in charge of Petroleum Affairs, and Senator Henry Berenger, the French Commissioner General of Petroleum Products, began negotiations on coordinating Franco-British oil policies in Rumania, Asia Minor, the French colonies and protectorates, and the British crown colonies, and on April 8, 1919, the two initialed an agreement in Paris, subject to confirmation by their respective Foreign Ministries. It provided for a common British-French oil policy for the Near East and the countries adjacent to the Mediterranean, and for the exploitation of the oil fields in those regions. As regards Mesopotamia, the agreement provided that should Great Britain receive it as a mandate, it would obtain from the Mesopotamian government — for the Turkish Petroleum Company or a new company to be formed — the rights acquired by the Turkish Petroleum Company. The French Government was to have a share in the capital of the Company, with all rights of representation. That capital was to be composed as follows: British interests, 70 per cent; French interests, 20 per cent; native government interests, 10 per cent. The French, on their part, were to facilitate by every means at their command the construction of two separate pipelines for the transport of oil from Mesopotamia and Persia to a port or ports on the eastern Mediterranean. If the pipelines crossed territory over which France was mandatory, she was "to use her good offices to secure every facility for the rights of crossing without any royalty or wayleaves on the oil transported." [20]

There was apparently a serious difference of opinion between Lord Curzon, Foreign Secretary, and Prime Minister Lloyd George on the one hand, and the other members of the cabinet, especially the Petroleum Executive, on the other. The French began to press for a share in the Mesopotamian oil as early as January 6, 1919, when the French Ambassador to London wrote to Lord Curzon to ask for discussions on the oil issue. Lord Curzon was inclined to refuse on the ground that no discussions should be held before the territorial decisions of the Peace Conference were arrived at. Lloyd George, on his part, suspicious of Clemenceau and experiencing serious difficulties with him in connection with Syria, was determined to keep oil out of the discussions. The other members of the cabinet, however, felt that the only way to keep the French from forming a united front with the United States against British oil interests, was to negotiate with them. In spite of Lord Curzon's strong opposition, it was decided on January 15, 1919 "that His Majesty's Government should at once signify their willingness to co-operate before the French secured American assistance, and before this country was forced by decisions at the Peace Conference to adopt in self-defense and practically under compulsion the policy of co-operation to which it was now invited." Negotiations were immediately initiated between Sir John Cadman, representative of the Petroleum Executive in the British peace delegation, and Henry Berenger. In a note from the British Ambassador of February 1, 1919 the hope was expressed that the results of the negotiations "were likely to have a favorable issue." Two days later Sir John Cadman asked Arthur James Balfour, the executive head of the British peace delegation, for his sanction to inform Berenger that the British Government was ready to admit a twenty–thirty per cent French participation in Mesopotamia.

While Lord Curzon was working to prevent an oil agreement with the French in Mesopotamia, he was handed, on March 15, 1919, a copy of the provisional agreement between Long and Berenger. In an Inter-Departmental committee meeting on Eastern affairs on April 29, 1919, it was pointed out that the Long-Berenger agreement was "an important part of the more important negotiations by which H. M. Government hoped to secure control of the Royal Dutch-Shell combine." Subsequently, on May 16, in a letter from the Foreign Office to the French Ambassador, the Long-Berenger agreement was confirmed. [21]

In the meantime, Lloyd George was encountering opposition from Clemenceau on the pipelines and railways from Mosul to Tripoli. Learning about the Long-Berenger agreement, Lloyd George informed the French Premier on May 21, 1919: "Inasmuch as you regard the British proposal for railways and pipelines from the Mosul area to Tripoli as a departure from the Agreement which we entered into in London in December last, I do not propose to proceed further with the proposed ıich I hereby withdraw." [22] Lord Curzon was not at all reluctant to ench Ambassador on July 22, 1919 that as a result of conversations bь.   ,d George and Clemenceau, the agreement was annulled. [23] Berenger, on his paı., in a memorandum to M. Philippe Berthelot and Clemenceau, prepared at the latter's request, after analyzing the British oil policy, [24] urged the French Government to come to an understanding with the British and work out a petroleum policy to assure France a proper participation in oil development; he advocated the ratification of the Long-Berenger agreement.

Sir John Cadman resumed negotiations with the French, and in a note submitted by Berthelot to Lloyd George on December 12, France demanded an equal share with Britain in the oil of Mesopotamia and Kurdistan in return for giving up Mosul. [25]

THE SAN REMO AGREEMENT

On December 29, 1919, Sir Hamar Greenwood, British Minister in charge of Petroleum Affairs, and Senator Henry Berenger initialed a memorandum of agreement. As regards Mesopotamia, the French were granted 25 per cent in the Turkish Petroleum Company. France agreed "to the construction of two separate pipelines and railways necessary for their construction and maintenance and for the transport of oil from Mesopotamia and Persia." [26] This memorandum led ultimately to the San Remo agreement.

In April 1920 the San Remo conference prepared a draft peace treaty with Turkey; it awarded the detached territories of the Ottoman Empire as mandates to England and France. On the 25th the British and French signed a special oil agreement. The paragraph covering Mesopotamia read:

> The British Government undertake to grant to the French Government or its nominee twenty-five per cent of the net output of crude oil at current market rates which His Majesty's Government may secure from the Mesopotamian oil fields, the British Government will place at the disposal of the French Government a share of twenty-five per cent in such company, the price to be paid for such participation to be no more than that paid by any other participants to said petroleum company. It is also understood that the said petroleum company shall be under permanent British control. [27]

The original objective of the Turkish Petroleum Company was thus achieved. The British held a 75 per cent interest (for in the meantime the Royal Dutch-Shell combine had for all intents and purposes become a British concern), [28] while the 25 per cent interest of the Germans was now assigned to France. The validity of the original concessionary claims of the Turkish Petroleum Company was guaranteed since the Treaty of Sèvres, signed by the Turks on August 10, 1920 (though under protest), confirmed Allied nationals in acquired rights in territories detached from Turkey.

ENTER THE UNITED STATES

The United States was, as regards the territories detached from the Ottoman Empire and to be placed under mandate, in a rather peculiar position. It had never declared war on Turkey; it had not participated in the negotiations on the territorial disposition of the Ottoman Empire, and therefore took no part in the peace negotiations with Turkey. Nevertheless, it felt very strongly that since it had contributed materially to the Allied victory over Turkey, it was entitled to the fruits of victory, and that Americans should be given an equal opportunity with the other victors in developing the economic possibilities of the mandated territories. The Americans in Paris for the peace conference soon learned about the Long-Berenger negotiations and on May 13, 1919 Leland Summers of the American commission wrote to Sir H. Llewellyn Smith, head of the economic section of the British peace delegation, for information about the reported Anglo-French agreement on oil in Rumania and Galicia, and activities "in connection with the oil properties throughout the Orient." Smith admitted that negotiations were going on, but stated that he was not aware of any concluded agreement. Summers became even more curious and told Smith that it would be highly desirable if he could give him "some idea of the negotiations which have taken place as we should like to know that American oil interests are not excluded from participation, and of course the preliminary stages are the ones that would be the most important to America." [29]

Balfour in Paris prepared a note outlining the reply to Mr. Summers, detailing the agreement and containing an assurance that it was not the intention of either government to exclude the United States from adopting a similar policy. Before the note could be presented to the French for their approval, Lloyd George withdrew the agreement. On June 25 Summers was advised that Anglo-French negotiations were in abeyance and that if the United States had any proposals to make, they should be communicated to the Foreign Office. [30]

The news of the reported Anglo-French negotiations caused even greater agitation in the United States over the exhaustion of the country's oil reserves. The cry was raised that the flow of oil from the United States during the war had taxed the oil resources to the limit, and that unless foreign oil concessions could be obtained, within a short time — about twenty years — the United States would be at the mercy of foreigners (meaning, of course, the British) for the country's oil needs. On July 29, 1919, Senator James D. Phelan delivered a speech in the Senate pointing out the critical supply situation; he fortified his assertions with statistical data supplied by the Federal Bureau of Mines. [31] The clamor was especially vociferous against the British restriction that the search for and operation of the oil fields in their territories be confined to British nationals, and the demand was made that the United States retaliate in kind, especially against the Shell Oil Company in the United States. [32]

Reacting to the reported Anglo-French agreement on oil in Mesopotamia, the American Ambassador in London submitted a memorandum, on May 12, 1920, to the Secretary of State for Foreign Affairs, in which he stated the American grievances about the British oil policy in Mesopotamia and Palestine. He outlined the general principles of equal opportunity of the Open Door policy as especially applicable to areas to be placed under mandate, and pointed out the discriminatory action against American citizens taken by the British authorities in Palestine and Mesopotamia. [33]

The three underlying issues in the dispute between the United States and Great Britain over the exploitation of oil resources in Mesopotamia were: Was the United States as an ally entitled to share with the other allies in the mandated territories since it had not declared war against Turkey? Could United States citizens claim a share in the economic exploitation of the oil resources of the area when a concession had already been granted to the Turkish Petroleum Company? And, finally, was the San Remo agreement discriminatory and in violation of the principles of the Open Door? The major aspect of the dispute, and the one which became a determining factor, was the comparable oil resources and supply of Great Britain and the United States.

On July 28, 1920, the American Ambassador, John W. Davis, acting on instructions from Secretary of State Bainbridge Colby, presented a very strong note to the British Foreign Office. The United States maintained that although it had not formally declared war on Turkey, it had materially contributed to the Allied victory and was therefore fully entitled to a share in the fruits of the victory. As to the Turkish Petroleum Company, the United States maintained that its concession had no legal validity, and that American citizens must therefore be given an equal opportunity to exploit the Mesopotamian oil fields. As regards the San Remo oil agreement between Great Britain and France, the note stated that it was not clear to the United States how such an agreement could "be consistent with the principles of equality of treatment understood and accepted during the peace negotiations in Paris" − an implication of bad faith on the part of both Great Britain and France. [34]

Two weeks later Lord Curzon replied to the American notes of May 12 and July 28. He emphasized that the feeling in the United States against Great Britain was completely unjustified since the output of oil in the British Empire and Persia was only about 4.5 per cent of world production, while the United States produced 70 per cent, and in addition American companies owned at least three-quarters of the Mexican output, which accounted for another 12 per cent, thus giving the United States about 82 per cent of world production. Under the circumstances the allegation that Great Britain threatened the oil supremacy of the United States could hardly be taken seriously. Lord Curzon categorically denied that there was anything improper in the San Remo agreement; the Turkish Petroleum Company possessed a legally valid concession. As for the question of mandates, he argued that the only place to discuss mandates was the Council of the League of Nations. [35]

In reply, Secretary of State Colby wrote on November 20, 1920 that the United States, as one of the Allied and Associated Powers, had every right to discuss any issue connected with the mandates, and that the question was not confined to League Council discussions; that as to oil shortages, the question was

not one of production but of supply, and that the United States must replenish its
supply by drawing on the latent resources of underdeveloped regions. The Secretary
took sharp issue with Lord Curzon's statement that the concessions in the territories of the former Ottoman Empire would have to be considered by the Arab
states which were to be organized, while at the same time the British Government
claimed that the Turkish Petroleum Company's concession must be honored and
other claims (American in Palestine) must await further clarification. He reiterated
that legally the United States could not agree that the Turkish Petroleum Company
possessed rights to petroleum concessions or to the exploitation of oil. Moreover, in
view of the official British disclaimer that as the mandatory power it had no
intention of establishing on its own behalf any kind of monopoly, how could it
justify the provision in the San Remo agreement that any petroleum company
which might develop the Mesopotamian oil fields "shall be under permanent British
control"?

The implication in Lord Curzon's note that the United States Government's
advocacy of the Open Door policy was motivated by the desire for oil control by
American interests elicited from the Secretary of State a rather strong reaction. "I
should regret any assumption by His Majesty's Government or any other friendly
power that the views of this Government as to the true character of the mandate
are dictated in any degree by consideration of the domestic need or production of
petroleum, or any other commodity," although the fact was that the United States
possessed only one-twelfth of the world petroleum resources, and during the war
had supplied more oil to the war effort than had any other nation. [36]

The American oil companies, particularly the Standard Oil Company of New
Jersey, in the person of A. C. Bedford, chairman of its Board of Directors, had
evinced an interest in Near East oil. As indicated above, Sir John Cadman intimated
that Mr. Summers' inquiries about the Anglo-French oil negotiations had been
prompted by Mr. Bedford. Bedford had also made inquiries of Berenger as to the
nature of the agreement between the British and the French, and had even asked
for the text of the agreement. To what extent, however, the American oil
companies were the instigators of the Government's demands for the Open Door
policy in Mesopotamia is difficult to determine with certainty on the basis of the
available material. It would seem that while the Standard Oil Company, which
claimed concessionary rights in Palestine, had complained to the State Department
about British interference and preferential treatment of British nationals in Mesopotamia, the Department took greater initiative there.

At a conference between representatives of Standard Oil and the Foreign Trade
Adviser's Office of the State Department, apparently called at the initiative of the
Department, the former stated that if the Mesopotamian oil fields were opened to
American capital, Standard would take steps to participate. The State Department
had to be sure that in advocating the Open Door policy for American citizens, there
would be American oil companies of the proper calibre which would be willing to
take advantage of the opportunity. [37]

In order to apply the Open Door policy on a broad basis, it was felt that other companies besides Standard of New Jersey should participate. It would appear that the then Secretary of Commerce, Herbert Hoover, took it upon himself to invite the big American oil companies to Washington to interest them in Mesopotamian oil and work out a plan of action. On November 3, 1921 the representatives of these companies reported to Hoover that after they had conferred with him about entering into Mesopotamian oil exploitation, the group would be ready to send a party of geologists and engineers to the region to make preliminary surveys. They also informed him that since reaching this decision they had drafted a letter to the Secretary of State which would be dispatched immediately on receipt of a telegram from Hoover. The letter was sent to Secretary of State Hughes on the same day. [38] It stated that the group desired to investigate prospective areas in Mesopotamia with a view to undertaking petroleum production; that a party of geologists and engineers representing the group was ready to start out as soon as the group had the assurance that it was accorded the privilege of making the survey; that the group would appreciate any further information and instruction, as it was aware of the exchange of communications on the subject between the United States and British Governments.

Secretary Hughes replied on the 22nd that as soon as the Department was advised that permission for prospecting in Mesopotamia was being or might be granted by the authorities in the territory, it would at once notify the group, and he thanked the group for its readiness to participate in the exploitation of Mesopotamia. [39]

At about the same time, the American Ambassador in London, on instruction from the Secretary of State, delivered a note on November 17, 1921 to Foreign Secretary Lord Curzon, which declared that the United States could not accept the claim of the Turkish Petroleum Company except in accordance with the principles which had been accepted by the British Government as applicable to mandated territories, and on the basis of a satisfactory determination of the character and extent of the Company's rights. Since the views of the two Governments were at variance as to the validity of the concession, the note proposed suitable arbitration to determine the character of the concession. [40]

While the negotiations were going on, the State Department was also involved in the question of the north Persian oil fields. Through the efforts in London of A. C. Bedford of Standard Oil of New Jersey, and of Sir John Cadman of Anglo-Persian in the United States, Standard joined the APOC in its attempt to obtain the concession from Persia; [41] the representatives of these two companies also held discussions on Mesopotamia and the Turkish Petroleum Company. [42]

On June 22, 1922, A. C. Bedford called on the State Department on behalf of the seven American oil companies to ascertain the Department's attitude to negotiations between American and British interests on the question of Mesopotamian oil. The Department insisted on the principle of equal opportunity, and on the invalidity of the Turkish Petroleum Company's concession. However, the Department, recognizing that its primary purpose was not theoretical principles, added: "It is not the desire of the Department, however, to make difficulties or to prolong needlessly a diplomatic dispute or so to disregard the practical aspects of the situation as to prevent American enterprise from availing itself of the very opportunities which our diplomatic representatives have striven to obtain." It declared that it would have no objection to American and British interests entering into final negotiations, provided that any reputable American oil company which was ready and willing to participate was not excluded from the arrangement decided upon, and that the legal validity of the Turkish Petroleum Company's claims should be determined by the methods suggested by the American Government.

In answer to the first condition, Bedford assured the Department that the seven companies included in the group were all the American companies that were likely to be interested. To meet the second condition, the Department itself suggested that it might be possible at the proper time to obtain a new or confirmatory grant of the concession to the Turkish Petroleum Company. To this Bedford agreed. [43]

Four days later he cabled Sir Charles Greenway, chairman of the Anglo-Persian Oil Company, negotiating for the Turkish Petroleum Company, the position of the State Department and the suggestions it had made. However, he raised another issue: "The seven American companies interested have considered views of State Department and questions concerning American participation and their views are that percentage you indicated to me would not be adequate from the point of view of what would be an equitable proportion to allocate to American interests." [44]

What this percentage had been was not officially disclosed, but on August 1, 1922 the American Ambassador in London, George Harvey, notified the Secretary of State that the partners in the Turkish Petroleum Company had decided to offer the American group 12 per cent participation. "Naturally the offer appeared entirely too low to Mr. Teagle and there was a rupture of official negotiations." [45] It would thus appear that the difficulties based on principles which had been raised by the State Department were practically overcome, but that the percentage share of American participation was the main stumbling block which caused the negotiations to break down. On August 26, 1922, Bedford, in a memorandum to the State Department detailing the negotiations between the American group and the Turkish Petroleum Company during which a formula had been worked out to meet State Department insistence on the Open Door policy, reported that the Turkish Petroleum Company was willing to limit its operations to specific areas which would be selected two years after the concession was granted, and that the remainder of the concession areas would be subleased at public auction, to which the Turkish Petroleum Company would not be admitted as a bidder. The principle against monopoly in the Open Door policy would thus be met more than half way. On the percentage share of the American companies, however, no agreement had been reached. Bedford intimated to the Secretary of State, when he left the memorandum with him, that the partners of the Turkish Petroleum Company were willing to offer 20 per cent, and he asked for the Department's opinion.

The British, on their part, not giving up their original contention that the

Turkish Petroleum Company concession was legally valid, but eager to avoid a serious clash with the United States, were willing to grant the American group 20 per cent in the concession, and since the American companies were ready to accept the offer, the State Department was obliged to face the problem realistically. On August 22, 1922, the Secretary of State wrote to W. C. Teagle, who was acting on behalf of all the American companies: "In its support of the Open Door policy it is not this Government's desire to set up impractical and theoretical principles or to place obstacles in the way of the participation of American companies in foreign enterprises, but rather to open to American companies the opportunity for such participation if they desire it. It rests chiefly with American commercial interests themselves, once the opportunity is offered, to determine the extent and terms of their participation and to decide whether, under existing circumstances, an adequate opportunity is offered." The Secretary added that if all the American companies interested in sharing in the development of Mesopotamian oil resources were invited to participate, and if there was no attempt to establish a monopoly in favor of the Turkish Petroleum Company, the Department would not consider the proposed sub-leasing scheme contrary to the spirit of the Open Door policy. [46]

While these negotiations were progressing and the dispute between the United States and Great Britain was nearing a solution, the Near East witnessed a war between Turkey and Greece, the emergence of a new Turkey, the repudiation of the Treaty of Sèvres; and the raising of a question concerning the very status of one of the important vilayets in the oil concession — Mosul. These new conditions brought back into the struggle the old Chester concession and delayed for several years longer the solution of the Mesopotamian oil question.

IN SUMMARY — A STALEMATE

The early attempts to obtain concessions from Turkey clearly demonstrated the pre-World War I methods of governmental pressure — German and British — on the Turkish Government to gain concessions for their nationals. It also demonstrated that in spite of their serious national differences, by June 1914 the British and Germans were ready to cooperate in the Baghdad railway project and the exploitation of the Mesopotamian oil fields. As to the concession of the Turkish Petroleum Company, it cannot be claimed that the Grand Vizier's letter of June 28, 1914 was a legally valid concession which could have been protected by the peace treaty with Turkey. Neither could it be claimed that the Chester concession was legally valid.

The British found themselves after the war in a very uncomfortable position. The Americans, who had not declared war on Turkey, and who had refused to join the League of Nations, would not accept a mandate over former Turkish territory, and yet, having supplied a very large part of the oil consumed for war operations, were demanding a share in the economic exploitation of the mandated territories. The British idea of the mandate system, much as it may have differed from that of the French, was far from conforming to that held by the United States, and the

British stubbornly insisted that the Turkish Petroleum Company concession was legally valid. They were not unmindful of the consequences to their entire future in the Middle East should they admit the invalidity of the concession. They were also not unmindful of the methods employed in obtaining the concession and they probably knew that they could not get an exclusive new concession either from the new Turkish regime or from the authorities in Mesopotamia unless it were based on the old concession. Moreover, the British were never enthusiastic about the American "Open Door" policy, especially when it involved oil, and they rendered lip service to it only to avoid clashes with the United States. [47] Furthermore, they were aware that unless the oil could be brought by pipeline to a port on the eastern Mediterranean, the Mesopotamian fields could not be exploited, and the road to the Mediterranean was in the hands of the French. The French were prevailed upon to give up Mosul — especially when it was militarily occupied by the British — and to permit the oil to flow to the Mediterranean in return for a share in the oil. However, unless the Turkish Petroleum concession was conceded to be valid and Great Britain was the possessor of the concession, the French certainly would not get their promised share.

Thus, all through the long negotiations, the British maintained that the Turkish Petroleum Company concession was legally valid; as the mandatory over Iraq they were "convinced" they could manage to obtain confirmation of the concession from the new Mesopotamian government. However, in order to avert conflict with the United States in the oil area, they were willing to allow American companies to share in the concession. The American State Department created serious difficulties.

In the United States the hue and cry over exhaustion of oil reserves as a result of the vast amount of oil taken from the country during the war reached a point which the administration could no longer ignore. The British methods in Persia, Mesopotamia and other oil areas aroused suspicion, which was followed by demands that the United States Government enter directly into the oil industry, in the same manner as the British Government, to guarantee the country's oil supply. [48]

As a consequence, the State Department began to question British actions in Mesopotamia and to press the Open Door policy to assure American participation in the economic exploitation of the mandated territories. The big American oil companies seemed at first reluctant to take advantage of the opportunity the State Department was seeking for them; no doubt they feared Government interference and participation modeled on the British Government's system, and they were not quite happy about the Open Door policy, which could also work against them. However, after several conferences with Secretary of Commerce Herbert Hoover, the companies indicated their readiness to participate in the exploitation of the Mesopotamian oil fields and so informed the Secretary of State on November 3, 1921. The American companies subsequently began direct negotiations with the partners of the Turkish Petroleum Company. This was in a sense a recognition by the United States Government and the American oil companies of the Turkish

Petroleum Company concession, and complicated negotiations between the two governments.

Here the issues were reduced to three: the Open Door; the validity of the Turkish Petroleum Company concession; and the share of the American companies. The first two issues were soon overcome by the State Department and the representatives of the American group. The Open Door principle could be satisfied if all the American companies who wished to participate had the opportunity to do so, and by arriving at some arrangement that would not grant the Turkish Petroleum Company monopolistic rights of exploitation. After the President of the Standard Oil Company of New Jersey assured Secretary of State Charles Evans Hughes that the seven companies which composed the group were the only American companies ready and willing to participate, the Department was satisfied.

The second issue was to be met by the device of having the new government of Mesopotamia confirm the old Turkish Petroleum Company concession, or grant a new concession.

The only stumbling block was the percentage share the American group was to receive. At first 12 per cent was offered; later this was raised to 20 per cent, and the American companies were inclined to accept. The Turkish-Greek war upset the plans and as the British were going to the Lausanne conference at the end of 1922, where the fate of the Mosul vilayet was to be decided, they were looking for strong American support. On December 12, 1922, Teagle of Standard Oil of New Jersey reported to the State Department that the Turkish Petroleum Company was willing to grant 24 per cent of the Company's shares to the American group, and in consideration of the surrender of this 24 per cent by the Anglo-Persian Oil Company which originally possessed 50 per cent of the concession, Anglo-Persian was to receive free of charge 10 per cent of the crude oil produced by the concession. In return the State Department was to agree not to question the title of the Turkish Petroleum Company and to advise the American representatives at Lausanne to give strong support to this arrangement, to the exclusion of any other interests, American or otherwise. Secretary of State Hughes rejected the last two conditions. The Department would not recognize the Turkish Petroleum Company and again advised confirmation or a new concession from the Mesopotamian government. Nor would the Department support this arrangement to the exclusion of other American interests. "This Department can never take the position that it will support at any time any arrangement to the exclusion of American interests." [49] It was on this stalemate that the Lausanne conference opened.

NOTES

1. For the Turkish text of these firmans and their English translation, see "The Mesopotamian Oil Fields," *Oil Engineering and Finance*, Feb. 17, 1923, 198–201.
2. *Papers Relating to the Foreign Relations of the United States* (Washington, 1936), II, 81 (henceforth cited as *Foreign Relations* and year); Great Britain,

Foreign Office, *Correspondence between His Majesty's Government and the United States Ambassador respecting Economic Rights in Mandated Territories,* Cmd. 1226, 1921, 11.

3. The Royal Dutch Oil Co., organized at The Hague in 1890 with a capital of 1,300,000 florins, obtained a concession from the government of the Dutch East Indies in Sumatra. In 1892 the capital was increased to 1,700,000 florins to enable the Company to build a refinery. Three years later the capital was further increased to 2,300,000 florins; dividends that year were 44 per cent. In 1897 the capital was increased to 5,000,000 florins, and dividends were 52 per cent. In its expansion in the Far East, Royal Dutch met with keen competition from the Shell Transport and Trading Co., a British firm which was not only engaged in transporting oil but also in producing it in Borneo. In 1902 Shell and Royal Dutch organized the Asiatic Petroleum Co. as the marketing agency for the products of both concerns, and ten years later the two amalgamated. Earlier, on Jan. 1, 1907, the two groups transferred their assets to two companies, Dutch and British — Bataafsche Petroleum Maatschapfij and the Anglo-Saxon Petroleum Co. Pierre l'Espagnol de la Tramerye, *The World-Struggle for Oil* (New York, 1924), 59—65. For a detailed analysis of the companies, subsidiaries and other group arrangements of the Royal Dutch-Shell combine, see United States, Federal Trade Commission, *Report on Foreign Ownership in the Petroleum Industry* (Washington, 1923), 21—69.

4. In 1909 the D'Arcy interests were transferred to the newly formed Anglo-Persian Oil Co. (See above, Chapter II.)

5. Edward Mead Earle, *Turkey, The Great Powers and the Baghdad Railway: A Study in Imperialism* (New York, 1923), 261 *et passim.*

6. According to Longrigg, the British Government favored Anglo-Persian participation in the Turkish Petroleum Company. In 1913 both the National Bank and Gulbenkian received summonses from the Foreign Office that their shares should be made available for re-allotment. "The intrusion of Anglo-Persian was, in fact, unwelcome to Deterding, though not unacceptable to the Germans as providing a broader basis for Anglo-German co-operation." Stephen Hemsley Longrigg, *Oil in the Middle East: Its Discovery and Development* (London, 1954), 31.

7. Joseph Berteloot in "Aux Lueurs du Pétrole Gênes, La Haye, Lausanne," *Etudes,* 175, April—May—June, 1922, declares that the Anglo-Persian Oil Co. was granted 30 per cent of the shares by this agreement.

8. *Foreign Relations 1927,* II, 821—822; Edward Mead Earle, "The Turkish Petroleum Company: A Study in Oleaginous Diplomacy," *Political Science Quarterly,* XXXIX, 277—279, June, 1924. On Oct. 19, 1912 the Deutsche Bank wrote to the Turkish Petroleum Co.: "We hereby undertake that we will not directly or indirectly be interested in the production or manufacture of crude oil in the Turkish empire in Europe and Asia apart from our interests in the Turkish Petroleum Company Ltd." Similar letters dated Oct. 22 and Oct. 23 respectively were written to the Turkish Petroleum Co. by the National Bank of Turkey and the Anglo-Saxon Co., Ltd. *Foreign Relations 1927,* II, 80.

9. *Foreign Relations 1920,* II, 662.

10. In reply to questions, Bonar Law stated in the House of Commons that no financial terms nor time period had been fixed in the grant to the Turkish

Petroleum Co. *PDC*, 131, col. 1012, July 5, 1920. The British Government was hard pressed by various members of Parliament to produce the grant of the concession to the Turkish Petroleum Co. or state clearly its terms; the Government refused persistently to do either. *PDC*, 131, col. 219, June 29, 1920; col. 629, July 1, 1920; col. 1031, July 5, 1920; col. 1960, July 12, 1920; col. 2575, July 15, 1920.

11. Henry Woodhouse, "American Oil Claims in Turkey," *Current History*, XV, 953–954, Mar., 1922.
12. State Dept. file 867. 77/285, 1914.
13. In 1912 the Ottoman-American Exploration Co., organized under State of Delaware laws, succeeded to all the assets of the Ottoman-American Development Co. State Dept. file 867. 77/81.
14. *Foreign Relations 1921*, II, 920.
15. *Ibid.*, 918–920; – File 867. 77/81.
16. A memorandum in the State Dept. of a meeting between the Secretary of State and Admiral Chester on April 18, 1921 contains the following:
    "*Comment:* In view of the objections now felt by the Department to certain monopolistic oil concessions proposed by Great Britain in her own mandated area of Mesopotamia, it would seem impractical at the present time for the Department to make a definite statement or to give definite encouragement to an American concession in Mesopotamia."
    Foreign Relations 1921, II, 920.
17. So named after Sir Mark Sykes of the British Foreign Office, and Charles François Georges Picot of the French Ministry of Foreign Affairs, who early in 1915 continued the negotiations which had been begun the previous year and arrived at the plan which formed the final agreement on May 16, 1916. David Lloyd George, *The Truth about the Peace Treaties* (London, 1938), II, 1087 *et passim.*
18. E. L. Woodward and Rohan Butler (eds.), *Documents on British Foreign Policy 1919–1939* (London, 1952), First Series, IV, 244–245 (henceforth cited *Documents*). It is interesting to note that in a letter dated 13/26 April, 1916 from Maurice Paleologue, the French Ambassador at Petrograd, to the Russian Foreign Minister, Sazonov, relating to French and Russian interests in the partitioning of Turkey, the Ambassador stated: "*En ce qui concerne les institutions, administrations, établissements religieux, scolaires, hospitaliers, etc., relevant des deux nations, ils continueront à jouir des privilèges qui leur étaient accordés jusqu'ici par les traités, accords et contrats conclus avec le Gouvernement ottoman.*" *Ibid.*, 243. It would seem that Grey had specifically in mind the Turkish Petroleum Co. The full correspondence forming the Sykes-Picot agreement is in *Documents*, 245–247.
19. David Lloyd George, *The Truth*, II, 1038–1060, 1090, 1145, 1155; *Foreign Relations 1919*, V, 3, 760, 809–810 – VII, 217; *Documents*, First Series, IV, 596. See also *Foreign Relations 1919*, V, 809–810, and VIII, 217; *The Times* (London), June 12, 1920.
20. *Documents*, First Series, IV, 1089–1091.
21. *Documents*, First Series, IV, 1090–1095.
22. *Ibid.*, 1092.
23. *Ibid.*, 1101.

24. See above, Chapter II.
25. *Documents,* First Series, IV, 583; Lloyd George, *The Truth about the Peace Treaties,* II, 1100–1101.
26. *Documents,* First Series, IV, 1115. See also Temperley, *History of the Peace Conference,* VI, 181–183. The 25 per cent in the Company which the French were to receive was the 25 per cent which had belonged to the Deutsche Bank; for it they were to pay the amount paid by the British Government to the Public Trustee, plus 5 per cent interest since the date of payment. Lloyd George insisted on railways as necessary for the construction of the pipelines.
27. *Memorandum of Agreement between M. Philippe Berthelot, Directeur des Affairs Politiques et Commerciales au Ministère des Affaires Etrangères, and Professor Sir John Cadman, Director in charge of His Majesty's Petroleum Department,* Cmd. 675, 1920.
28. At the beginning of World War I, the Royal Dutch-Shell submitted to British control in order to continue to operate on the high seas. Henry Deterding, the manager of the Company, became a British subject and the Company's head-quarters were moved from The Hague to London. For his services to the British Empire, Deterding was knighted in 1920.
29. *Documents,* First Series, IV, 1095. Sir John Cadman informed the Foreign Office that A. C. Bedford of the Standard Oil Co. of N. J. "evidently prompted this request of Mr. Summers." *Ibid.,* 1098.
30. *Ibid.,* 1096.
31. *Congressional Record. Proceedings of Debates,* First Session of the Sixty-Sixth Congress, 58, part 4, 3304–3310, July 29, 1919.
32. In a letter to Viscount Grey, British Ambassador in Washington, dated Dec. 20, 1919, Secretary of State Lansing summarized the situation thus: "The best technical authorities seem to believe that the peak of petroleum production in the United States will soon be reached, and that the reserves will be practically exhausted within a measurable period. The situation in the United States will be more serious because of its enormous domestic consumption, and because in the past there has been relatively little investment of American capital in important foreign producing fields.

    "These facts, together with the exclusion of American citizens either in law or in fact, from commercial production in other countries, have given rise in this country to agitation for some form of government action." *Foreign Relations 1919,* I, 171. This agitation was not limited to the daily press, but serious professional journals followed the same line, or perhaps formulated the line, as is evidenced by such articles as those of Van H. Manning, "International Aspects of Petroleum Industry," *Mining and Metallurgy,* 1–10, Feb., 1920 (the author was a director of the US Bureau of Mines); David White, "Our Future Oil Supply," *Engineering and Mining Journal,* 951–955, June 4, 1921. An explosive article written by a Britisher in support of American fears and confirming American suspicions of British policy, and which was subsequently constantly cited by Americans, was: E. Mackay Edgar, "Britain's Hold on the World's Oil," *Sperling's Journal,* V, 38–49, Sept., 1919. For additional correspondence between the two governments on the general issue of petroleum policy, see *Foreign Relations 1919,* I, 163–171, and *Foreign Relations 1920,* I, 350–370 *et passim.*

33. *Foreign Relations 1920,* II, 651–655.
34. *Foreign Relations 1920,* II, 658–659.
35. *Foreign Relations 1920,* II, 663–667 *et passim.*
36. *Foreign Relations 1920,* II, 669–673.
37. In a letter to W. C. Teagle, President of the Standard Oil Co. of N. J., under date of Nov. 22, 1921, Secretary of State Charles Evans Hughes stated: "It is helpful to know that American oil companies are prepared to take advantage promptly of the opportunities which are expected to be presented in that region and, accordingly, your courteous and timely statement of the position and plans of your Company is thoroughly appreciated." *Foreign Relations 1921,* II, 87–88.

    Testifying before the Senate Committee Investigating Petroleum Resources in 1945, Charles W. Hamilton of the Gulf Oil Corp. recalled the activities of the Government in 1920, thus: "Representatives of the industry were called to Washington and told to go out and get it." US Senate, *American Petroleum Interests in Foreign Countries* (Hearings before the Special Committee Investigating Petroleum Resources, June 27–28, 1945, Washington, 1946), 55.
38. The letter was signed by E. L. Doheny, Pres., Mexican Petroleum Co.; Amos L. Beaty, Pres., Texas Co.; George S. Davison, Pres., Gulf Refining Co.; G. W. Van Dyke of the Atlantic Refining Co.; H. F. Sinclair of the Sinclair Consolidated Oil Corp.; C. F. Meyer, Vice-Pres., Standard Oil Co. (N. Y.); and W. C. Teagle, Standard Oil Co. (N. J.).
39. *Foreign Relations 1921,* II, 87–88.
40. *Ibid.,* 89.
41. See above, Chapter IV.
42. Sir John Cadman, in an article discussing the world oil question after the Standard-APOC agreement on the north Persian oil fields had been reached, said about Mesopotamia: "There is good reason to hope that the same spirit of cooperation will finally solve the tangled problem of the Mesopotamia oil-fields." "The World's Unexplored Resources," *The Petroleum Times,* Dec. 30, 1922, 954.
43. *Foreign Relations 1922,* II, 337–338. The sense of this understanding was communicated to the American Embassy in London on June 24, 1922.
44. *Ibid.,* 339.
45. *Ibid.,* 340.
46. *Ibid.,* 333–345.
47. See below, Chapter X.
48. David White, "Our Future Oil Supply," *Engineering and Mining Journal,* 953, June 4, 1921; *The New York Times,* June 15, 1920.
49. *Foreign Relations 1922,* II, 348–349. The reference to the exclusion of American interests was to the Chester concession which, in view of the new situation, was being pushed again in Turkey.

Chapter X
FROM THE LAUSANNE CONFERENCE TO THE IRAQ PETROLEUM CO.

The armistice with Turkey was signed at Mudros on October 30, 1918, and considerable portions of Turkey were subsequently occupied by the Allied forces. With the reluctant consent of France, the Allies permitted the Greeks to occupy Smyrna, and Greek forces marched into the city on May 15, 1919. One month later Mustafa Kemal Pasha began his nationalist movement in Anatolia. While the Allied powers were discussing peace terms with the Turkish Government in Constantinople, the Nationalists organized the Grand National Assembly, which met in Ankara (then called Angora) in April 1920, with Mustafa Kemal Pasha as president. But even as early as February of that year the Nationalists under Kemal operated with considerable success against the French who were in occupation in Cilicia.[1]

The policy which the French were pursuing in the Near East differed basically from that of the British. The latter were determined to reduce Turkey to complete impotency as a force in international affairs, but in order to assuage the millions of their Moslem subjects, they hoped to build up a new Islamic empire centered on the Sherifian family: Hussein in Mecca, his younger son Feisal in Damascus, and his older son Abdullah in Iraq. They therefore supported the Greeks against the Turks. The French, on the other hand, although they had reluctantly consented to the Sykes-Picot and other agreements on the Near East which aimed at establishing an Arab state or confederation of Arab states, resented the actions of the Arabs, especially the proclamation of Feisal as king of Syria, as interference in French schemes, and they looked on the Arab nationalist leaders as agents of the British. As soon as Clemenceau encountered difficulties with Feisal and met with military pressure from the Turkish nationalists in Cilicia, French policy began to aim at restoring Turkey as a power, and through her rebuilding French prestige in the Near East and among the Moslems, as well as acquiring economic advantages in the new Turkey. Circumstances within France — financial hardship, manpower shortage, war weariness — all operated to make France come to terms with the Turkish Nationalists and through them strike back at British policy in the area. Thus, M. Franklin-Bouillon, a former Minister, paid a number of visits to Ankara, ostensibly in a private capacity, and on October 20, 1921 he signed an agreement with Yussuf Kemal Bey, Minister of Foreign Affairs in the government of the Grand National Assembly.[2]

The treaty, which provided for French evacuation of Cilicia and for special privileges for the Turks in Alexandretta, also gave the Turks control of some four hundred miles of the Baghdad railway, as well as transportation privileges — civilian and military — over the railway going through Syrain territory. The significant aspect of the treaty was the recognition by France of the Nationalist regime and the termination of the state of war between the two countries. In a letter accompanying the treaty, Yussuf Kemal Bey promised France economic concessions.[3]

This so-called Franklin-Bouillon-Angora agreement seriously disturbed the British; they accused the French not only of violating specific treaties by concluding a separate peace treaty with Turkey and by so doing undermining the entire Allied position, but of actually fomenting anti-British riots in British Mesopotamia; and they asserted that through the border changes between Syria and Turkey and the railway privileges granted to Turkey, their position in Mesopotamia was endangered.[4] The vilayet of Mosul was at stake.

The French denied all the accusations: they claimed that the agreement was nothing but a local arrangement which the military conditions of the area made necessary, and that it had no other significance.[5] The fact nevertheless remained that the French and British were working at cross purposes.[6]

In the war between Turkey and Greece the French supported the Turks while the British supported the Greeks, so that in a sense the Greeks and the Turks were fighting the battle of the British and the French for influence and prestige in the Near East. The Greeks lost, and Turkey emerged as a new force in Europe and the Middle East. In November 1922 the Lausanne Conference was called to negotiate a peace treaty between Greece and Turkey, as well as between Turkey and the Allies. It was recognized by then that the Treaty of Sèvres was a dead letter.

One of the most disturbing elements in Anglo-French relations, Anglo-Iraqi relations, and Anglo-American relations, and a disturbing aspect of the Lausanne Conference itself was the uncertainty of the frontier between Turkey and Iraq. On January 28, 1920, Kemal Pasha renounced, in his National Pact of Independence, all claims to former Ottoman territories inhabited by Arab majorities, but the Kurds were specifically claimed as Turks.[7] Turkey maintained that Mosul was overwhelmingly Turkish and therefore a part of Turkey over which she had never given up her sovereignty.

The situation arising from the Turkish victories placed the Turkish Petroleum Company in a precarious position, for it had put all its hopes on Great Britain being the mandatory over Iraq. In view of the fact that the United States Government persistently questioned the validity of the Company's concession, there was hardly the slightest possibility that the new Turkish Government — which the British had been striving to defeat — would recognize the concession if Mosul should be awarded to Turkey. The first effect was for the Turkish Petroleum Company to offer the American group a 24 per cent share if the United States supported the Company claim at Lausanne.

The United States on its part had to be very cautious. It had never declared war on Turkey and consequently did not actively participate at the Lausanne Conference, and more that once it had indicated that it was not interested in the territorial issues to be discussed at the Conference. Should Mosul fall into British hands as part of the Iraq mandate, American companies were assured of a share of the Turkish Petroleum Company, but if it fell to Turkey, what then?

CHESTER CONCESSION RENEWED

A partial answer to this question, and one which also indirectly helped the American group in its negotiations with the Turkish Petroleum Company, came in the renewal of the Chester concession.

Admiral Chester called on the State Department in June 1920, and many times again in 1921. He was always sympathetically received, but it was generally understood by all concerned, according to a Department of State memorandum, that while he perhaps had moral claims, he did not have a legal concession. Early in 1922 new interest in the United States and Canada was aroused in the Chester concession, and in March of that year the Ottoman-American Development Company was reorganized, under a charter from the State of Delaware, with a total of 5,000 shares of no par value; it acquired all the assets of the Ottoman Exploration Company.

Representatives of the new company went to Turkey to negotiate with the Turkish Government, through the Minister of Public Works, at Ankara, for the renewal of the Chester concession. On December 5, 1922 the US acting High Commissioner at Constantinople reported to the State Department that the Council of Ministers and the Public Works Commission of the National Assembly had been kept informed of the negotiations, which were nearing conclusion.[8]

This development, at the very time the Lausanne Conference was in session, could not but upset the British and the American group negotiating with the Turkish Petroleum Company. In a letter from Walter C. Teagle to Secretary of State Hughes in the latter part of October 1922, it was stated that recent developments in the Near East, and particularly the statement which Mustafa Kemal Pasha had made about oil concessions in his instructions to the Turkish delegation to the Lausanne peace conference, made the American group negotiating with the Turkish Petroleum Company feel that it should reconsider the entire situation. Especially disturbing was Kemal's insistence that the Kurdish territories which the Treaty of Sèvres included in the new kingdom of Iraq be returned to Turkey.

On November 17, 1922, A. C. Bedford cautiously asked the State Department, whether in view of the negotiations between the American group and the Turkish Petroleum Company, the Department had any information regarding Admiral Chester's concession or any recent grant obtained by Chester from the Turkish Government, or any opinion as to the validity and standing of the reported concession. The reply, on November 22, clearly stated that the Department was not in a position to give the group information on the negotiations which might be carried on by other American concerns with regard to concessions in the former Ottoman Empire.

After due reflection, the American group concluded that its best chances were with the Turkish Petroleum Company, and on November 29, 1922 it intimated to the State Department that even under the changed conditions, and even after the group had been informed that Ambassador Child had reported to the Department that there were indications that the entire or part of the vilayet of Mosul might be transferred to Turkish sovereignty, the most satisfactory way of securing for the American oil industry an opportunity to participate in the development of the Mesopotamian petroleum resources was still through participation with the British and French associated in the Turkish Petroleum Company. It urged the State Department to recognize the validity of the TPC concession so that even if Mosul were transferred to Turkey, it would still be within the area covered by the concession and would have to be so recognized by the new Turkish Government.

The State Department refused to commit its policy exclusively to the Chester concession, and on January 15, 1923 the Secretary of State advised the American acting High Commissioner in Constantinople, who was pressing for full support of the Chester group, that while giving support to all Americans seeking concession, the Department could not support one against the other. There were at the time three groups in the running: the Chester group; the group negotiating with the Turkish Petroleum Company; and a new group working on behalf of the heirs of Abdul Hamid.[9] "In dealing with these three specific proposals the Department made it clear that it would accord proper diplomatic support to American citizens and interests but that it played no favorites and would grant no special privileges to any one American company which it would deny the other." [10]

The French protested vigorously to the Turkish Government against the adoption of the Chester project on the ground that portions of the railway lines included in the concession had been granted before the war to French interests. Nevertheless, on April 10, 1923 the Turkish National Assembly accepted the Chester project by a vote of 141 to 16.[11] Secretary of State Hughes declared that the granting of the Chester concession should be understood to mark the triumph of the Open Door policy as against such policies as inspired the tripartite agreement of the Baghdad railway concession. Yet the Department was not ready to support the claim that Chester had obtained valid legal rights under the negotiations which took place prior to the existing concession. On April 29 the Chester agreement was signed in Ankara. [12]

The Ottoman-American Development Company failed to obtain the necessary financial resources to effect its concession. By August, after the Lausanne Treaty had been signed and it had been decided that the vilayet of Mosul was to remain under British-Iraq control for a least one year, the financial interests in the United States no longer displayed any willingness to invest in the Chester concession, and the American High Commissioner reported to the Secretary of State that the Turkish Government was becoming gravely concerned over the Company's inactivity. To this communication the Secretary replied, on November 12, 1923: "The information which the Department has does not indicate that any American company or group has the slightest interest in advancing funds to make it possible for the Ottoman-American Development Company to begin work on the Samsoun-Sivas line." [13] On December 18, 1923 the Turkish Government annulled the Chester concession.

THE LAUSANNE CONFERENCE

The fluctuations in the fortunes of the Chester group were in great measure determined by the developments at the Lausanne Conference. The position of the Turks, based on their National Pact, soon made it clear that the question of Mosul would be one of the major issues of the Conference. The United States instructed its Ambassador to Italy, Richard Washburn Child, and its Minister to Switzerland, Joseph Grew, to attend the Conference as observers. On November 26, 1922 the American delegates reported that there was reason to believe that Great Britain's policy might be to seek withdrawal from Mesopotamia in return for concessions, especially in the matter of petroleum. Secretary of State Hughes outlined for them the American policy and stated bluntly: "The position taken by the Department is that in view of American contributions to the common victory over the Central Powers, no discrimination can rightfully be made against us in a territory won by that victory." He told the American diplomats: "You should proceed openly and candidly, in view of the delicacy of oil questions, and in view of the danger that Turks may attempt to raise dissensions over oil among the powers represented at Lausanne." [14]

On November 26, 1922, as the Conference was about to take up the question of the boundaries between Turkey and Iraq, Ismet Pasha asked Lord Curzon, who was presiding, to defer public discussion of the subject pending private negotiations between them. [15] This was agreed to and oral discussions between the two commenced; these were soon converted to an exchange of written memoranda. [16] These private negotiations resulted in no agreement, and on January 23, 1923 the issue came before the Conference.

The bases for the decision as to whom the vilayet of Mosul should be awarded were to be ethnographical, political, historical, geographical, economic and military. Although these were questions of fact, the two sides presented conflicting data. Basically, however, both agreed that the Arabs were not a majority of the population in the area; moreover, that the Kurds were the single major element of the population. But while the Turks claimed the Kurds as part of the Turkish population, the British asserted that they were anti-Turkish. Contradictory figures were presented on the composition of the population, the racial origin of the Kurds, the nature of the language spoken by them, the economic realities of the vilayet and its military significance. The longer the debate continued the more uncompromising became the attitude of the disputants, especially the Turks. Under such circumstances it was inevitable that the question of oil should be brought up, even if not directly at the conference; it was loudly publicized in the press of the world. Lord Curzon declared that although Ismet Pasha had not alluded to oil, the press of the world had, and he therefore found it imperative to state: "It is supposed and alleged that the attitude of the British Government with regard to the retention of Mosul is affected by the question of oil. The question of the oil of the Mosul vilayet has nothing to do with my argument. I have presented the British case on its own merits quite independent of any natural resources there may be in the country. I do not know how much oil there may be in the neighborhood of Mosul or whether it can be worked at a profit, or whether it may turn out after all to be a fraud. During the time I have been connected with the foreign affairs of my country I have never spoken to or interviewed an oil magnate. I have never spoken to or negotiated with a single concessionaire or would-be concessionaire for the Mosul oil or any other oil." [17]

It would seem, as was asserted by the opposition members of the British Parliament, that Lord Curzon was protesting too much. It might also be, as had happened with the Long-Berenger negotiations, [18] that Lord Curzon was not fully informed of the petroleum policies and negotiations of the British Government and therefore could make the assertions that he did. Be that as it may, Curzon reviewed the story of the Turkish Petroleum Company concession and declared: "The British Government, after full examination, were convinced, and remain convinced, of the validity of this concession. They felt and feel bound to uphold it." Not to frighten the Americans, and perhaps to gain their support, the British Foreign Secretary added: "Both the British Government and the Company itself recognize that oil is a commodity in which the world is interested and as to which it is a great mistake to claim or to exercise a monopoly. Accordingly the Company with the full knowledge and support of the British Government took steps, and negotiations have ever since been proceeding, to associate the interests of other countries and other parties in this concession, so that all those who are equally interested may have a share." [19]

The United States Government, aware of the Chester negotiations proceeding at the time, was not to be appeased by this offer, and the American representative, Ambassador Richard Washburn Child, handed Lord Curzon a statement which reiterated the American position questioning the validity of the Turkish Petroleum Company concession and demanding equal opportunity for American citizens in the exploitation of the oil resources. [20] On February 1, 1923 Child wrote to Lord Curzon, at the instruction of Secretary Hughes, again repeating the American position and calling his attention to previous American notes on the subject. [21]

Lord Curzon clearly intimated to the Conference that on the question of Mosul the British could not yield. Realizing the uncompromising position of both sides, he offered to refer the issue to the League of Nations for an independent inquiry and decision. The Turkish delegation refused to submit to arbitration, Turkey maintaining that she still had sovereignty over Mosul, and she would not submit her natural inheritance to arbitration. Thereupon Curzon threatened that Turkey's refusal might mean war: he declared that he had information that there might be movements of Turkish troops from Ankara in the direction of Mosul should an adverse decision be taken. As the Turkish delegation adamantly continued to reject arbitration, he announced that he would be obliged to invoke Article 11 of the Covenant of the League of Nations. [22]

This placed the French in a most delicate position. They could neither side openly with the Turks, nor would they allow a renewal of hostilities between Turkey and Great Britain. The French delegate, M. Bompard, declared that the gravity of the situation imposed on him the duty of intervening with a view to averting the disaster which threatened, and he urged the Turkish delegation not to reject outright the British proposal. [23] Nonetheless, the Turkish delegation persisted in rejecting the proposal; Lord Curzon then announced that he would proceed immediately with action against Turkey before the Council of the League. [24]

On January 31, 1923 the Turkish delegation was handed a draft treaty. Article 3, which dealt with boundaries, stated about Iraq: "From the point on the Tigris which constitutes the terminal part of the frontier referred to in paragraph (1) of the Article [with Syria]:

"A line to be fixed in accordance with the decision to be given thereon by the Council of the League of Nations."

The same day Curzon announced that the British Government had already applied to the Council of the League. [25]

For a number of reasons the Turks were determined not to permit the issue to reach the League, and on February 4 Ismet Pasha proposed that the Mosul question be excluded from the program of the Conference in order that it might, within a period of one year, be settled by common agreement between Great Britain and Turkey To this Lord Curzon answered that it was out of the question since the League had already been notified of the issue, but that he would be willing to accept the proposal for a year of direct negotiations between the parties, for which period the British would ask the Council not to proceed with the case; in the meantime the status quo in the region would be preserved. [26]

The Conference broke up on that note. When it reconvened in April, new difficulties arose, for Sir Horace Rumbold, representing Great Britain, insisted that a provision guaranteeing the status quo during the year of negotiations be inserted into the treaty. The Treaty of Lausanne, which was finally signed on July 24, 1923, provided in Article 3, paragraph 2, the following about Mosul:

With Iraq:

The frontier between Turkey and Iraq shall be laid down in friendly arrangement to be concluded between Turkey and Great Britain within nine months.

In the event of no agreement being reached between the two Governments within the time mentioned, the dispute shall be referred to the Council of the League.

The Turkish and British Governments reciprocally undertake that, pending the decision to be reached on the subject of the frontier, no military or other movement shall take place which might modify in any way the present state of the territories of which the final fate will depend upon that decision. [27]

The British failed in their effort to have inserted in the treaty a specific recognition by the Turkish Government of the Turkish Petroleum Company's concession, [28] but with the signing of the treaty and the question of Mosul to be amicably solved between the Turks and the British or ultimately by the League of Nations, the Ottoman-American Development Company could not attract the necessary capital and it became obvious that the only way Americans could participate in the exploitation of Mesopotamian oil resources would have to be through the Turkish Petroleum Company. The British attitude now stiffened.

Before returning to the negotiations between the American group and the Turkish Petroleum Company, we should follow the course of the disposition of the question of Mosul.

MOSUL BEFORE THE LEAGUE OF NATIONS

In accordance with Article 3 of the Lausanne Treaty, negotiations between the British, represented by Sir Percy Cox, and the Turks, represented by Fethy Bey, opened in Constantinople on May 19, 1924; they continued until June 5, but without any positive results. [29] At the conclusion of the negotiations the British representative attempted to formulate a joint application to the Council of the League for a decision, but the Turks naturally refused, and on August 6 the British Government asked the Council to put on its agenda: "Iraq Frontier: Article 3 (2) of the treaty signed at Lausanne on July 24th, 1923." [30]

With the consent of Turkey, the Council put the issue on its agenda on August 30. About three weeks later hearings began, with Fethy Bey representing Turkey and Lord Parmoor representing Britain. The arguments, both in the documentary evidence and in the oral presentation, were repetitions of those in the memoranda first exchanged between Lord Curzon and Ismet Pasha, debated before the Lausanne Conference, and restated in the direct negotiations in Constantinople. Even the legally technical issue — whether the vilayet of Mosul was the question under consideration, as the Turks argued, or only the question of the frontier between Turkey and Iraq, as the British maintained, an issue which loomed high in the debate before the Council — actually went back to the Lausanne Conference. Even the proposal by the Turkish representative that the question be decided by a plebiscite was a repetition of that offered at the Lausanne Conference. Of more immediate concern for the Council were the charges and countercharges by both parties of violation of the status quo since the signing of the Lausanne Treaty.

The British delegate pursued Lord Curzon's original purpose at the Lausanne Conference by asking that the Council, after it had determined the scope of the question, appoint "a commission of disinterested and unbiased persons to settle the dispute after studying the documents that have already been prepared and any further evidence which they may consider necessary." [31]

A stumbling block to action by the League was the status of the Council under the Treaty of Lausanne. The British declared that they regarded the Treaty as placing the Council in the position of an arbitrator, whose ultimate award must be accepted in advance by both parties; the British Government would therefore consider itself bound by the Council's decision. The Turks, however, referred to Article 15 of the Covenant of the League.

On September 29 the Rapporteur stated that agreement between the parties on interpretation as to the issue before the Council had been arrived at, and after a statement by Fethy Bey indicating that Turkey would be ready to abide by the

Council's decision, it was decided to set up a commission of three members to examine the existing documents, obtain additional documents, conduct an on-the-spot investigation, and advise the Council on a possible solution. [32]

Before the commission could enter on its duties, however, a serious difficulty arose as to the definition of the boundary line constituting *status quo* at the time of the signing of the Treaty of Lausanne. The result was the emergence of the so-called "Brussels Line" (the Council was then meeting at Brussels) formulated by a committee composed of representatives of Sweden, Spain and Uruguay. On October 31, 1924 the Enquiry Commission was appointed: Count Paul Teleki, former Premier of Hungary, M. de Wirsen, Swedish Minister at Bucharest, and Colonel Paulis, a retired officer of the Belgian Army. The Commission met in Geneva on November 13, 1924 and elected M. de Wirsen chairman. [33]

The report of the Commission submitted to the League Secretariat on July 16, 1925 went into a detailed analysis of the various problems involved, and after advising against partitioning the area, it recommended that the whole territory south of the "Brussels Line" be united with Iraq subject to the following conditions:

> (1) The territory must remain under the effective mandate of the League of Nations for a period which may be put at twenty-five years.
> (2) Regard must be paid to the desires expressed by the Kurds that officers of Kurdish race should be appointed for the administration of their country, the dispensation of justice and teaching in the schools, and that Kurdish should be the official language of all these services. [34]

The Council opened debate on the Commission's report on September 3, 1925. Great Britain accepted the two recommendations. The Turkish delegate, Tevfik Rouschdy Bey, argued that legally the vilayet of Mosul was under Turkish sovereignty, that since Turkey had never recognized the mandatory system, and since the Commission recommended a mandate under League control, the Council could not judge in its own case. [35]

The Council itself raised the following questions: What was the character of the decision to be taken by the Council in virtue of Article 3, paragraph 2, of the Treaty of Lausanne — an arbitral award, recommendation, or simple mediation? How was the decision to be arrived at — unanimously, or by majority? Should the parties to the dispute vote? A subcommittee of the Council recommended that these questions be submitted to the Permanent Court of International Justice for an advisory opinion. The British reluctantly agreed; the Turkish representative, however, maintained that the Council's function, as stipulated in the Lausanne Treaty, was no more than that of good offices, and he stated: "The Turkish Government sees no necessity on the present occasion for referring anything to the Permanent Court of Justice at the Hague — the less so in view of the fact that the questions put are essentially extremely political questions. The advisory opinion of the Court could not, therefore, in any way affect the rights which the Turkish Government holds under the Acts of Lausanne, or modify the role of the Council laid down by those Acts." [36] The Council voted, over the objection of the Turkish representative, to submit the question to the Permanent Court for an advisory opinion.

The first public meeting of the Court was held on October 26; both Turkey and Great Britain were invited to supply information on the question before the Court. The Turkish Government sent a telegram declaring that it did not think it necessary to be represented. [37] The British submitted documents [38] and sent a group of representatives headed by Sir Douglas Hogg, Attorney General, who addressed the Court on October 26 and 27. [39] On November 21, 1925 the Permanent Court gave the following opinion:

> (1) That the "decision to be taken" by the Council of the League of Nations in virtue of Article 3, paragraph 2, of the Treaty of Lausanne will be binding on the Parties and will constitute a definitive determination of the frontiers between Turkey and Iraq;
> (2) That the "decision to be taken" must be taken by a unanimous vote, the representatives of the Parties taking part in the vote, but their votes not being counted in ascertaining whether there is unanimity. [40]

On December 8, 1925, the Council adopted the advisory opinion of the Court, despite bitter opposition from the Turkish representative, and thereafter the Turkish representative refused to attend any further meetings of the Council. [41] In a letter dated December 16, Rouschdy Bey wrote the Council: "I desire further to declare that the sovereign rights of a State over territory can only come to an end with its consent, and that, therefore, our sovereign rights over the whole of the vilayet of Mosul remain intact." [42]

On the same day the Council adopted the recommendations of the Enquiry Commission. The territory south of the "Brussels Line" was to be allocated to Iraq; in view of the twenty-five year provision in the Commission's recommendation, the Council asked that a new treaty of alliance between England and Iraq be concluded which would make that recommendation possible. After detailing the boundary line, the Council's resolution provided that:

> As soon as, within a period of six months from the present date, the execution of this stipulation has been brought to the knowledge of the Council, the Council shall declare that the present decision has become definitive and shall indicate the measures required to ensure the delimitation on the ground of the frontier line.
> The British Government, as mandatory power, is invited to lay before the Council the administrative measures which will be taken with a view to securing for the Kurdish populations mentioned in the report of the Commission of Enquiry the guarantees regarding local administration recommended by the Commission in its final conclusions.
> The British Government, as mandatory power, is invited to act, as far as possible, in accordance with the other suggestions of the Commission of En-

quiry as regards measures likely to ensure pacification and to afford equal protection to all the elements of the population, and also as regards the commercial measures in the special recommendations of the Commission's report. [43]

The new treaty between Great Britain and Iraq was signed at Baghdad on January 18, 1926 and was sent to the Council; on March 11, 1926, the Council declared the decision of December 16, 1925 definitive. [44] In the meantime, the British were carrying on negotiations with the Turkish Government, and on June 6, 1926 Sir Austen Chamberlain informed the Council that a treaty between Great Britain, Iraq and Turkey had been signed, with some minor adjustments in the boundary in favor of Turkey. [45]

ANGLO-AMERICAN ISSUE

While the dispute with the Turkish Government over Mosul was satisfactorily settled, the negotiations with the American companies and the State Department were far from concluded. After the State Department's refusal to support the claim of the Turkish Petroleum Company at Lausanne in return for the 24 per cent share offered to the American group, W. C. Teagle informed the Company that its proposal, with the further proviso that 10 per cent of the oil was to be delivered to the Anglo-Persian Oil Company, was impracticable as a business proposition. [46] Conferences in New York and London continued the discussions, and on October 25, 1923 Teagle submitted to the State Department a draft agreement, which eliminated the 10 per cent oil delivery provision to the Anglo-Persian Oil Company, but included a provision that the Turkish Petroleum Company select some definite limited areas for exploitation within two years after the signing of the concession with the Iraq Government, and that the remainder be subleased under public auction at stated periods; [47] and that the Government of Iraq grant the Company a new concession. On November 8, the Secretary of State informed Teagle that the Department would be prepared to give appropriate diplomatic support to the American interests concerned. [48]

In the ensuing negotiations between the American group and the Turkish Petroleum Company, the question of subleasing seemed likely to prevent agreement. The Iraq Government in its discussion with the TPC on a new concession insisted on having the final say in the selection of the sublessors. The American companies were expecting to obtain the subleases for themselves and so get from Iraq much more than their percentage share in the Turkish Petroleum Company, but if the Iraq Government — which they felt would be under British influence, if not actual control — should have the final say in the selection of the sublessors, they would certainly be systematically eliminated. This became all too apparent when the Iraq Government (perhaps with British advice) began to reject the subleasing scheme altogether and demanded that the Turkish Petroleum Company itself develop the entire area of the concession. [49]

Another obstacle to an understanding between the British and the Americans was S. C. Gulbenkian. The primary purpose of the American group was to obtain crude oil; the Turkish Petroleum Company was to be merely a producer of crude, which it was to sell at seaboard to the respective participating groups, according to their shares. Gulbenkian declared that he was not an oil trader and that he did not want crude oil; he was determined that the Turkish Petroleum Company should be a fully operating company so that he would benefit to the full extent from his five per cent share of the profits. In order to meet the American demands he advanced conditions that neither the Americans nor the other partners of the Turkish Petroleum Company could possibly accept. The American group's attempt to deal directly with Gulbenkian proved of no avail. It seemed, therefore, that unless the State Department intervened the Turkish Petroleum Company might obtain the concession from Iraq without American participation. [50]

On September 18, 1924 the American group asked the State Department to make diplomatic representation to the British Foreign Office — Gulbenkian was a naturalized British subject — to compel Gulbenkian to accept, on a reasonable basis, the principle that the oil produced in Iraq should be divided among the partners, rather than the profits of any joint enterprise. In a letter dated September 20, 1924 to Ambassador Frank Kellogg at London, Secretary of State Hughes clearly implied that since Britain was firmly established in Iraq, the British partners might attempt to obtain the concession from Iraq without sharing it with the Americans; Gulbenkian's uncompromising attitude might be inspired. The Ambassador was requested to speak sternly with the Foreign Office. [51] In the middle of October Kellogg reported to the Secretary of State that the Foreign Office held that Gulbenkian's arguments were sound legally and practically and that it would not intervene in business negotiations or disputes, but that it was ready to use all good offices to compose, if possible, the differences between the parties. [52]

The determined attitude of the Americans, combined with the British success in the League against Turkey, persuaded the British Government to proceed with the arrangements with the Government of Iraq to grant the concession to the Turkish Petroleum Company.

IRAQ GRANTS THE CONCESSION

As mentioned above, the armistice of Mudros was signed on October 30, 1918, and the occupation was extended to Mosul proper, according to British claims in order to meet military imperatives. The British Sherifian policy saw Emir Abdullah as the future ruler of Iraq. [53] The San Remo Conference assigned Mesopotamia to Great Britain as an "A" mandate. Soon, however, difficulties arose in Syria for the French, and in Mesopotamia for the British. In May, 1920 an uprising began in Iraq which in July became a full-scale revolution. By October the insurrection was quelled, but not before the British alone had suffered 2,500 casualties.

British Middle Eastern policy was consequently reorganized. The Colonial Office under Winston Churchill took over the Middle East Department form the

Foreign Office, [54] and in 1921 the Cairo Conference was convened to revamp Britain's entire policy in the area. The result was a switch — Feisal was to become king of Iraq and Abdullah was relegated, at the advice of Colonel Lawrence, to Transjordan. [55] On July 11, 1921, after the necessary preliminary measures had been taken, [56] the Emir Feisal was elected king and installed. Great Britain although still mandatory could no longer continue the former relationship with the new kingdom of Iraq, and a treaty of alliance between the two countries was signed on October 10, 1922. [57] The treaty was to come into force as soon as it was ratified and accepted by the Iraq Constituent Assembly. It was to remain in force for twenty years unless Iraq was admitted to membership in the League of Nations before that date, when it would automatically terminate. However, in April of the following year, in an attempt to meet public opinion in Iraq, the British agreed to a protocol to the treaty which reduced the twenty-year period to four years from the date of ratification of the peace treaty with Turkey, or the admission of Iraq to membership in the League, whichever came first. [58]

After prolonged delays, due primarily to internal dissension in the newly formed Kingdom of Iraq where the major religious group was Shiite, while the newly arrived King belonged to the Sunni sect, the Constituent Assembly opened on March 27, 1924. [59] The Assembly moved slowly, and the British threatened that unless the treaty were ratified without further delay, they would have no alternative but to bring the issue of the Iraq mandate before the Council of the League. On June 11, 1924, when only 69 of the 100 delegates were present, the Assembly approved the treaty by a vote of 37 for, 24 against, and 8 abstentions. [60] Regarding the boundary between Iraq and Turkey, the resolution affirming the treaty declared: "This treaty and its subsidiary agreements shall become null and void if the British Government fail to safeguard the rights of Iraq in the Mosul Vilayet in its entirety." [61]

While the British Government was persistently pursuing its efforts to have Mosul included in the Iraq kingdom, the Iraq Government signed an agreement with the Turkish Petroleum Company, on March 24, 1925, granting it a concession for seventy-five years.

IRAQ-TURKISH PETROLEUM COMPANY AGREEMENT

The major provisions of the agreement were:

1. Exclusive oil rights to the Turkish Petroleum Company in all of Iraq except the vilayet of Basrah and the "transferred territories" for a period of seventy-five years, at the end of which all the Company's property to revert to the Government of Iraq free of charge.

2. The Company to prepare a geological survey within eight months; within thirty-two months it was to select twenty-four rectangular plots, each of an area of eight square miles; it was to commence drilling within three years.

3. Article 6 read: "The Government shall, not later than four years after the date of this Convention, and annually thereafter, select not less than 24

rectangular plots, each of an area of eight square miles, and the Government shall offer the same for competition, by sealed tender, between all responsible corporations, firms and individuals, without distinction of nationality, who desire leases." [62]

4. The Company to pay to the Iraq Government royalties of 4 shillings (gold) per metric ton on net oil production for twenty years after the completion of its pipeline; after that time payments to be based on the market value of oil averaged over ten-year periods.

5. The Company must remain a British company registered in Great Britain and its chairman to be at all times a British subject. [63]

MOTIVATION

The three-cornered, though unequal, struggle among Turkey, Great Britain and the United States over Mosul revealed the different objectives and motivations of each of the three for the final determination of the issue.

The Turks were the least motivated by the petroleum possibilities of Mosul in their persistent efforts to re-occupy the vilayet as part of the new Turkey. National prestige and military-strategic considerations determined the attitude of the Turks from the moment the Lausanne Conference opened down to the decision of the Council of the League to include Mosul in Iraq under British mandate. Moreover, not yet a member of the League of Nations, and suspicious that the League was a mere guise for an Allied front, Turkey refused not only to give up Mosul, but even to recognize the authority of the League. As for the oil potentialities of Mosul, the Turks were willing to grant the British oil concessions, practically on the latter's own terms, if their own sovereignty over the vilayet was recognized. [64]

Once the Turks realized that the British were victorious as a result of the decision of the Council of the League and that they were committed to the Iraqi Government to include Mosul within its territory, and since they themselves had meanwhile become reconciled to the League in anticipation of membership, they came to terms with the British. With minor territorial adjustments and a 10 per cent share in the royalty payments to Iraq from the oil of Mosul, they signed the treaty with Britain and Iraq in June 1926.

The motivation of the British was more complex. To begin with, there were strategic considerations of Empire defense; not only Mosul, but Iraq as a whole with Mosul as a decisive part of it, was an important link in the lifeline of Empire, as well as an integral part of the new British expansion in the Middle East. Moreover, Iraq had cost the British many millions of pounds and thousands of lives, first to conquer from the Turks and then to hold and administer. Lloyd George, as Prime Minister, pointed out — when his former chief, Herbert H. Asquith, advocated British withdrawal from Mosul — that he could not understand withdrawing from the more important and more promising part of Mesopotamia. "Mosul is a country with great possibilities. It has rich oil deposits." It was through these rich oil deposits that the British hoped to recover their heavy investments in Iraq. Lloyd George left the House of Commons in no doubt that when the question of mandatories was to be decided, the British would "claim the right as the mandatory power of Mesopotamia, including Mosul." [65]

Later, Lord Curzon, at the Lausanne Conference, as well as Colonial Secretary Leopold S. Amery and others in Parliament and in the press, protested righteously that oil had not been a consideration in the determined efforts of the British to include Mosul in Iraq; that their only consideration had been their obligation to the League of Nations which had entrusted them with the mandate, and their promise to King Feisal and to the people of Iraq to include Mosul in their country. [66]

The Labor opposition was not ready to accept the protestations of the Government ministers, and various members of Parliament pointed out that in the San Remo agreement the Mesopotamian oil had loomed very large indeed. [67] Some bluntly stated that if it had not been for the oil, the British would never have battled so persistently for Mosul. The Government's theme was that it could not possibly have been influenced by oil since there was no certainty of the existence of oil resources in the territory. To this, Arthur Ponsonby, in 1926, made the following significant statement: "I should be very much surprised, in spite of the disclaimers we have heard today, if in the course of the next few years we do not hear of the discovery of oil in Iraq and of fortunes which may be made out of it, and this will always be wrapped up by our being told that the prosperity of the people of Iraq will also increase as the result of the discovery of oil. What I object to in the sequence of events by which we undertake these fresh obligations is the vein of hypocrisy which runs through them." [68]

Apparently because of the anti-oil atmosphere at home as well as because of the negotiations with the Americans, the British Government was quite sensitive about the oil motivation in its Mosul policy and labored hard to show that it had been actuated by the noblest of intentions. To prove that the British had not been interested in the oil of Mosul, Sir Austen Chamberlain, who represented Britain in the Council of the League, recalled to the House of Commons in 1926: "I was approached by the Turkish representative with the proposition in March of last year to settle this question apart from the League of Nations. What was the basis of the proposition? It was that Turkey should have so much as she desired of the vilayet of Mosul. That was one side of the bargain. The other side was that a British company, approved by His Majesty's Government, should have the exploitation of all the oil. Pipe lines were to be necessary, and a British company should have the construction of the pipe lines. A port or two ports would be required, and a British company should have the concession for the ports. I think — I speak from memory — 300 kilometres of railway was offered. If we were after oil, we could have had a concession for all the oil in Mosul and concessions for anything else we liked. The reply of His Majesty's Government was that they were trustees for Iraq; that they were not possessors but mandatories, and that as mandatories and trustees they could not bargain away the rights and interests of Iraq and her people in exchange for concessions to British capitalists." [69]

Sir Austen must have been aware that his revelation proved nothing. In March 1925, after the position taken by the British Government at the Lausanne Conference, its treaty commitments to the Iraqis and the strong stand it took before the Council of the League for the whole world to hear, if the British Government had been willing to bargain away Mosul for the oil concessions, it would certainly have been condemned by the entire world. Once it took the position that Mosul must be included in Iraq and, as the opposition charged, because of the oil there, it could no longer retreat even if Turkey had offered much more attractive terms.

When one recalls the Anglo-French oil negotiations on Mesopotamia which had preceded the San Remo agreement, the Anglo-American correspondence in which Lord Curzon pointed to the mere 4.5 per cent of the world's petroluem supply which England controlled compared to the 82 per cent controlled by the United States, the constant bickering in the negotiations with the American group over the latter's prospective percentage share in the Turkish Petroleum Company, and the entire oil policy of the British Government, one cannot possibly accept without seriously questioning the assertion that oil had not "had the slightest influence" in determining British policy in Iraq. Sir Harry Brittain came perhaps nearer the truth when he declared: "I am not afraid at all to take up the subject of oil. Whenever oil is mentioned there are in some quarters of the House sure to be sneering observations in regard to it. Whether you like it or not we have arrived at the age of oil. We live in a country in which there is plenty of coal and no oil. We have to get oil with which to run our ships where shipowners insist upon burning oil . . . If oil can be obtained from Iraq, then Iraq will gain just as much as any commercial company will gain." [70] As a matter of fact, when the Labor Party was in control and pursued the same policy before the League of Nations as the Conservative government, the Colonial Secretary, J. H. Thomas, admitted openly that it was true to say that boring for oil in Mosul had had some bearing on the question of the frontier between Turkey and Iraq. [71]

It would therefore seem that oil had been of very considerable importance in the determination of British policy as regards Mosul. A configuration of factors made it possible for the British to harmonize their own interests with those of the Iraqis. The inclusion of Mosul in Iraq was a condition for the approval of the Anglo-Iraqi treaty by the Constituent Assembly, and this greatly helped the British in their efforts to retain Mosul for their Empire defense and for its oil fields, as well as to fulfill their obligations to the League as mandatory for Iraq.

The motivation of the United States was wrapped up in the abstract principles of equality of opportunity for American nationals, in accordance with the Open Door policy to be applied to mandated territories. While the Open Door policy was also advocated by the United States in other areas of the world — China, for example — and had an aspect of international peace worthy of respect, one cannot altogether close one's eyes to the underlying causes which brought that policy in connection with the oil in Mesopotamia into full blossom, and its subsequent untimely withering.

The general disillusionment in the United States after the high idealism of the war generated in Wilson's regime, created a suspicion about British motives and actions, and the Anglo-French San Remo oil agreement could not but have inflamed this suspicion. The cry over the exhaustion of American oil resources, the attack on the British oil exploitation practices, and the demand for retaliatory action against British oil companies in the United States forced the American Government to try to obtain outside oil resources, especially those in territories under British control. The most likely areas were the territories detached from the Central Powers from Turkey in particular, for not only did the United States feel it was entitled to share in the fruits of the victory, but practically all the known potential oil fields were located in the former territories of the Ottoman Empire. The answer to the supposed American oil shortage lay in the Mesopotamian oil fields. The Open Door policy provided an ideal opening. The British saw clearly what was behind the demands of the United States. When, however, Lord Curzon implied that the motivation for the vigorous prosecution of the Open Door policy was the desire for oil control by the American companies, Secretary of State Colby was outraged.

But in order to achieve the desired end — outside oil resources — the State Department had to find practical solutions to overcome the abstract principles of the Open Door policy. The monopolistic character of the Turkish Petroleum Company was overcome by a scheme of subleasing; the free and equal participation of all American companies and individuals who were willing to venture into the Middle East was overcome by the assertion that no other companies except those in the American group were ready and willing to participate; the objection to the validity of the Turkish Petroleum Company's concession was overcome by the device of making the Iraqi Government grant the TPC a new concession. The State Department completely condoned the self-denial provision of the agreement between the American group and the Turkish Petroleum Company [72] despite the fact that this was a restrictive measure on the members of the Company and a complete repudiation of the basic principles of the Open Door.

The American Government, not being sure of the outcome of the Mosul issue between Great Britain and Turkey before the Lausanne Conference, played safe by supporting the Chester concession; it was thus in a position to assure the American interests of participation in the exploitation of the Mosul oil fields no matter what the final disposition of Mosul would be. The British emerged from Lausanne at least partially successful, the Chester concession faded from the scene, and the State Department began to work more intensively than ever before with the American group to obtain a proper share in the Turkish Petroleum Company.

NOTES

1. "The Near East," *The Round Table*, XII, 319–337, Mar., 1922.
2. Great Britain, Foreign Office, *Despatch from His Majesty's Ambassador at Paris enclosing the Franco-Turkish Agreement signed at Angora on October 20, 1921*, Cmd. 1556, 1921. Turkey No. 2 (1921).
3. Yussuf Kemal wrote: "The government of the Grand Assembly, desirous on its part to promote the development of the material interests common to the two

countries, authorizes me to inform you that it is disposed to grant the concession for the iron, chrome and silver mines in the Karshut valley, for a period of ninety-nine years to a French group, which, within a period of five years from the date of the signature of the present agreement, must begin to work this concession through a company constituted in accordance with Turkish law, in which Turkish capital shall participate to the extent of 50 per cent.

"In addition the Turkish Government is prepared to examine with the utmost good will other requests for concessions for mines, railways, ports and rivers which may be put forward by French groups, on condition that these requests are in accordance with the reciprocal interests of Turkey and of France." Cmd. 1556, 1921, 8.

4. Great Britain, Foreign Office, *Correspondence between His Majesty's Government and the French Government respecting the Angora Agreement of October 20, 1921.* Cmd. 1570, 1922. Turkey No. 1 (1922), 6, 22, 26 *et passim.*

5. *Ibid.,* 18 *et passim.*

6. For Kemal's interpretation of this agreement, see *A Speech Delivered by Ghazi Mustapha Kemal* (Leipzig, 1929), 523–527.

7. Great Britain, Foreign Office, *Lausanne Conference on Near Eastern Affairs 1922–1923,* Cmd. 1814, 1923, 370; Ahmed Amin, *Turkey in the World War* (New Haven, 1930), 276–277.

8. *Foreign Relations 1922,* II, 966–983.

9. In the general state of uncertainty about the future of Turkey and especially the vilayets of Baghdad and Mosul, the heirs of the Sultan Abdul Hamid claimed the two vilayets as the personal property of the Sultan and hence belonging to them as his heirs. A group of American concerns supported them in expectation of the concession; they were represented by Samuel Untermeyer who approached the State Department in their behalf.

10. *Foreign Relations 1923,* II, 1199.

11. *Foreign Relations 1923,* II, 1202. The full text of the agreement is reproduced in 1220–1240; Benjamin Gerig, *The Open Door and the Mandate System* (London, 1930) gives the date as April 9, 1923, and the vote as 185 to 21, 147.

12. The British acting High Commissioner in Constantinople registered an emphatic protest against the grant of rights within the borders of Iraq and declared that the validity of any such grant would not be recognized by the British Government. *PDC,* 163, cols. 1345–1346, May 2, 1923.

13. *Foreign Relations 1923,* II, 1251.

14. *Foreign Relations 1922,* II, 346.

15. Cmd. 1814, 1923, 339.

16. The memoranda exchange appears in an appendix, 363–393 in the English version of the Conference, Cmd. 1814, 1923; it is not included in the French version, France, Ministère des Affaires Estrangères, *Documents Diplomatiques Conférence de Lausanne* (Paris, 1923), 2 vols.

17. Cmd. 1814, 1923, 360. The Turkish delegation was apparently beset by concession hunters and Ismet Pasha reported that since its arrival at Lausanne the Turkish delegation had been confronted with requests and openings from certain groups which wished to know whether Turkey would recognize their rights to work the Mosul oil fields after Mosul had been reoccupied. *Ibid.,* 397.

18. See above.
19. Cmd. 1814, 1923, 361.
20. *Ibid.,* 405.
21. United States Senate, *Oil Concessions in Foreign Countries,* 68th Congress, First Session, Document No. 97, 1924, 56 (henceforth cited Document No. 97); *Foreign Relations 1923,* II, 241.
22. Cmd. 1814, 1923, 402.
23. For the bitter resentment of the British against the development of events at the Conference, see Lloyd George, *The Truth about the Peace Treaties,* II, 1351 *ff.;* for a different point of view, see Henry H. Cumming, *Franco-British Rivalry in the Post-War Near East* (London, 1938), 186–213.
24. Cmd. 1814, 1923, 404.
25. *Ibid.,* 688, 433.
26. *Ibid.,* 846–851.
27. Great Britain, Foreign Office, *Treaty of Peace with Turkey and Other Instruments Signed at Lausanne on July 24, 1923.* Cmd. 1929, 1923, 15.
28. It has been ascertained that this was due, at least in part, to the open opposition of the American delegation to the concession, and Sir Horace Rumbold even accused the American delegation of urging the Turks not to recognize the concession. Gerig, *The Open Door,* 138–139.
29. The full reports of these meetings are in *La Question de Mossoul de la Signature du Traité d'Armistice de Moudros (30 Octobre 1918) au 1er Mars 1925* (Constantinople, 1925), 180–200, (Turkish Red Book); important excerpts of these reports are reproduced in Permanent Court of International Justice, *Treaty of Lausanne, Article 3, Paragraph 2,* Series C, No. 10, Acts and Documents Relating to Judgments of Advisory Opinions (Leyden, 1926), 166–177.
30. League of Nations, *Official Journal,* 1924, 1465.
31. *Ibid.,* 1319.
32. *Ibid.,* 1337–1339, 1358–1363.
33. *Ibid.,* 1648–1653, 1659–1662; – 1925, 145.
34. League of Nations, *Question of the Frontier between Turkey and Iraq. Report to the Council by the Commission Instituted by the Council Resolution of September 30th, 1924* (Geneva, 1925), C. 400. M. 147. 1925. VII, 88–89.
35. League of Nations, *Official Journal,* 1925, 1313–1327.
36. *Ibid.,* 1377–1382.
37. Permanent Court, Series C. No. 10, 9.
38. *Ibid.,* 198–258.
39. *Ibid.,* 18–54.
40. Permanent Court of International Justice, *Collection of Advisory Opinions No. 12 Article 3 Paragraph 2, of the Treaty of Lausanne* (Leyden, 1925), Series B. No. 12, Nov. 21, 1925, 33.
41. League of Nations, *Official Journal,* 1926, 121–128.
42. *Ibid.,* 187.
43. *Ibid.,* 187–192.
44. *Ibid.,* 503.
45. *Ibid.,* 858–859. In addition to the boundary adjustments, the treaty provided that Iraq was to pay Turkey, for a period of twenty-five years, 10 per cent of the oil royalties from the area of Mosul. It also contained an option that

Turkey might accept £500,000 in lieu of such royalties. *Treaty between the United Kingdom and Iraq and Turkey regarding the Settlement of the Frontier between Turkey and Iraq,* Cmd. 2679, 1926.

46. *Foreign Relations 1923,* II, 242.
47. On July 21, 1922, the American companies submitted a memorandum to the State Dept. containing the provision that the Turkish Petroleum Co. was to "select for their own exploitation within two years from the date of the confirmation of the concession by the Iraq Government a total not to exceed 12 blocks, the area of each block not to exceed 16 square miles.

    "The oil rights of the balance of the territory covered by the concession, totaling some 150,000 square miles, to be opened for sublease to any responsible individual; firm or corporation who may be interested in developing oil production in Iraq." This would eliminate the monopolistic possibilities. *Foreign Relations 1922,* II, 342. See also the modified proposals in *Foreign Relations 1923,* II, 247–257.
48. *Foreign Relations 1923,* II, 259. It was about this time that the Department recognized that the Chester concession was dead. See the Secretary's statement of Nov. 12, 1923 mentioned above, p. 218.
49. *Foreign Relations 1924,* II, 222–226.
50. *Ibid.,* 229–230.
51. *Ibid.,* 232–234.
52. *Ibid.,* 235–236.
53. The Iraq Covenanters, a nationalist society, passed a resolution offering the throne of Iraq to Abdullah on the same day, March 8, 1920, that the Syrian Arab Congress offered the throne of Syria to Feisal.
54. *PDC,* 138, col. 1409, Feb. 28, 1921.
55. *PDC,* 301, col. 1114, Mar. 24, 1936; Philip W. Ireland, *Iraq: A Study in Political Development* (New York, 1938), 328–329.
56. Ireland, *Iraq: A Study in Political Development,* 311 *ff.*
57. Great Britain, Foreign Office, *Iraq Treaty with King Feisal,* Cmd. 1757, 1922; – *Treaty Between Great Britain and Iraq October 10, 1922,* Cmd. 2662, 1922.
58. *PDL,* 53, col. 1057, May 3, 1923. The full protocol is produced there.
59. Great Britain, Colonial Office, *Report of His Britannic Majesty's Government on the Administration of Iraq for the period April 1923–December 1924,* Colonial No. 13 (London, 1925), 17.
60. *Report 1923–1924,* Colonial No. 13, 22.
61. *Ibid.,* 23; *PDC,* 176, col. 869, July 21, 1924, and cols. 193, 1964–1965, July 29, 1924. For the interpretation given by the British Government and by the opposition for this reservation of the Iraq Constituent Assembly, see the entire debate on Middle Eastern Services, *Ibid.,* cols. 1919–1995.
62. Arnold J. Toynbee in *Survey of International Affairs 1934* (London, 1935), 189, remarks about this provision: "This arrangement was the Company's own choice; for they had been deterred, by American insistence, upon the maintenance of the 'open door' in Iraq, from applying for exclusive exploitation rights throughout the area over which their concession extended."
63. Great Britain, Colonial Office, *Special Report by His Majesty's Government in the United Kingdom of Great Britain and Northern Ireland on the Progress of Iraq during the Period 1920–31,* Colonial No. 58 (London, 1931), 303–315.

64. The Turks tried to offer the British oil concessions during the early stages of the Lausanne Conference in 1922, Cmd. 1814, 1923, 360 *et passim;* and again when the issue was to be decided before the Council of the League in 1925. *PDC,* 191, cols. 2276–2277, Feb. 18, 1926.

65. *PDC,* 127, cols. 662–664, Mar. 25, 1920. *The Times* (London) commented that Lloyd George seemed to think that the oil would pay the cost of administration. "We doubt it, for oil profits generally seem to find their way by some invisible pipeline into private pockets." *The Times,* Mar. 27, 1920.

66. *PDC,* 189, cols. 2140–2141, Dec. 21, 1925. Said Amery: "I need not repeat that no interest of any sort directly concerned with oil has influenced the policy of the British Government, or of any British Government." Again, in Feb. 1926, in urging the House of Commons to ratify the new Anglo-Iraq treaty, Amery declared: "Oil has not had the slightest influence one way or another, in determining our policy with regard to Iraq." *PDC,* 191, col. 2178, Feb. 18, 1926.

67. *PDC,* 191, col. 2178, Feb. 18, 1926.

68. *Ibid.,* 2235.

69. *Ibid.,* cols. 2276–2277.

70. *Ibid.,* 2253–2254.

71. *PDC,* 176, col. 1942, July 29, 1924.

72. For further details concerning this see below, Chapter XI.

Chapter XI
THE CONCESSION

THE "RED LINE" AGREEMENT

In the negotiations between the American group and the Turkish Petroleum Company, the question of the self-denial clause, in which the participating companies obligated themselves not to seek oil concessions in the territories of the Ottoman Empire except through the Turkish Petroleum Company, was one of the major difficulties. In August, 1922, W.C. Teagle, president of the Standard Oil Company of New Jersey and spokesman for the American group, submitted a long memorandum to the State Department outlining the bases on which his group was negotiating with the partners of the Turkish Petroleum Company. In his reply Secretary of State Hughes wrote on August 22, in reference to the self-denial clause: "While American petroleum interests are at liberty to take such action as they may see fit to limit their activities throughout the Turkish Empire, no such undertaking would affect the attitude of this Government in the future consideration of claims in this area or its support of American interests there, and the Department will reserve its entire freedom of action in dealing with such a future contingency."[1]

Since the American group could not arrive at a satisfactory understanding with the partners of the Turkish Petroleum Company, the latter proceeded on its own and obtained the new concession from the Government of Iraq on March 24, 1925, as outlined above. The difficulty created by Gulbenkian, which the United States diplomatic agent felt was British-inspired, combined with the State Department's refusal to approve the self-denial clause, endangered the position of the American group. At the end of 1925, the latter reported to the State Department that the failure of the partners of the Turkish Petroleum Company to come to an agreement with Gulbenkian might bring the negotiations to an impasse. Secretary of State Frank Kellogg thereupon warned the British Government that should the American group withdraw because of failure to obtain participation in the Turkish Petroleum Company on a fair basis, the United States Government would reserve its entire freedom of action should reasonable and proper efforts be made by the interested American company to secure the right to a fair share in the development of oil resources of Mesopotamia "through other means than the Turkish Petroleum Company."[2]

The State Department sought to overcome the difficulty with Gulbenkian by questioning his legal status in the new concession. Since the original position of the United States Government was that the Turkish Petroleum Company's concession had no legal validity, it would follow that Gulbenkian, whose claim was based on that concession, had no legal status. Acting Secretary of State Joseph Grew cabled Ambassador Houghton on January 14, 1926 that "it would be inconsistent with

237

our earlier correspondence with the British Government to admit legal foundation of Gulbenkian's claim."[3]

The American group, however, represented by Teagle, continued to negotiate with the partners of the Turkish Petroleum Company in an effort to arrive at some compromise with Gulbenkian. These negotiations were prolonged into 1927, when a formula was worked out between Gulbenkian[4] and the other groups which made it possible for the American group to participate, but it involved acceptance by the American group of the self-denial provisions of the early negotiations of 1912 and 1914. This was, of course, contrary to the principles of the Open Door, as Secretary of State Hughes had clearly stated in 1922. Nevertheless on April 9, 1927 the Secretary of State informed Guy Wellman, representing the American group, that the State Department would not object to the formula.[5] The Department thus condoned, if it did not actually sanction, the self-denial provisions as stated in Paragraph 10 of the "Foreign Office" Agreement of March 1914, cited above.[6]

On July 31, 1928 all the participants in the Turkish Petroleum Company signed the Group agreement which limited the activities of each participant in the specified area, which was marked out on a map attached to the Agreement by a red line, and hence became known as the "red line agreement."[7]

The percentages of shares in the Turkish Petroleum Company were: Anglo-Persian Oil Company,[8] Royal Dutch-Shell Company, Compagnie Française des Pétroles,[9] and the American Group, each 23 3/4 per cent, and S.C. Gulbenkian, 5 per cent. The American group, composed of five companies, was organized in February 1928 as the Near East Development Corporation; of the seven original companies, Sinclair and Texas withdrew.[10] Subsequently, the name of the Turkish Petroleum Company was changed to Iraq Petroleum Company.[11]

IRAQ PETROLEUM CONCESSION

According to the agreement, the Iraq Petroleum Company was to choose 24 plots, each eight square miles, and the Government of Iraq was to offer the remaining territory for competitive bidding.[12] Instead, a new agreement was signed on March 24, 1931. The new concession gave the Company the sole right to exploit all lands situated to the east of the Tigris River, covering an area of 32,000 square miles. The Company was to construct a pipeline with a total capacity of not less than 3,000,000 tons per annum. Royalties to be paid to Iraq were four shillings (gold) per metric ton, with a minimum required payment of £400,000 (gold) for the first twenty years beginning with the first exports. Until the commencement of export, the Company was to pay the Government the same sum, of £400,000 (gold) annually, of which half, £200,000, would be recoverable by the Company from future royalties in excess of £400,000; while £200,000 was to be dead rent. The Company was exempt from taxation in consideration of yearly payments to the Government of £9,000 (gold) up to the time of commercial export, and thereafter £60,000 (gold) on the first 4,000,000 tons produced and pro rata, and £20,000 (gold) on each additional million metric tons produced and pro rata.[13]

THE MOSUL PETROLEUM COMPANY CONCESSION

That the British had never meant to abide by the system of subleasing and the provision eliminating the Turkish Petroleum Company as a bidder at public auctions (a provision inserted in the concession simply to meet the stubborn insistence of the State Department), [14] was evidenced not only from the negotiations leading up to the revised agreement of March 24, 1931, but also from the negotiations carried on by the British Oil Development Company for concessions in the rest of the area. When the Iraq Government was about to begin negotiations with BOD, Secretary of State Henry L. Stimson instructed Charles Dawes, the American Ambassador to Great Britain, in the middle of April 1931, to remind the British Foreign Office that according to the tripartite convention of January 9, 1930, the Government of Iraq was not to grant a concession covering any of the oil lands in question without affording Americans, individuals and corporations, an equal opportunity to bid, and that the United States expected the British Government to intervene with the Iraq Government to protect their rights. The British Government replied on July 17, 1931 that while it had brought the view of the United States to the attention of the Iraq Government, it took the position that the provisions of the tripartite treaty barred discrimination between the different nationals when bids were up, but that there was no obligation on the part of the Iraq Government to put every concession up for public tender before granting it. [15]   However, to meet the objections of the United States, the Iraq Government went through the motions of accepting four tenders for the concession of oil-bearing fields in the rest of the country, and at the end of 1931 announced that since the bid made by the British Oil Development Company was the best, the others had been rejected. [16]

On April 20, 1932, the British Oil Development Company obtained a 75-year concession covering all the lands in the vilayets of Mosul and Baghdad west of the Tigris River and north of the 33rd parallel (about 46,000 square miles). Until commercial quantities of oil were found, BOD was to pay dead rent: £100,000 (gold) in 1933, increasing this by £25,000 (gold) annually up to £200,000. It was to construct a pipeline with a minimum annual capacity of 1,000,000 tons, or to make arrangements for the transfer for export of that minimum quantity. The Iraq Government was entitled to take, free of charge, up to 20 per cent of the oil for local consumption and for resale to the Company. Royalties were to be four shillings (gold) per metric ton; the Company was to be tax exempt in consideration of £1,000 paid to the Government annually until commercial production, and then to be the same as that paid by the Iraq Petroleum Company. [17] Ten years later this concession was transferred to the Mosul Petroleum Company, a subsidiary of the Iraq Petroleum Company. [18]

THE BASRAH PETROLEUM COMPANY CONCESSION

On December 4, 1938, the Basrah Petroleum Company, also a subsidiary of the Iraq Petroleum Company, obtained a 75-year concession covering all lands not included under previous concessions (about 93,000 square miles). Except that Basrah was to pay dead rent of £200,000 (gold) annually until commercial quantities were exported, all the other provisions were the same as those of the Mosul Petroleum Company agreement.[19] Basrah began producing and exporting oil in 1951.[20]

As discussed in Chapter II above, after the Persian Government granted the concession to D'Arcy, Great Britain succeeded in obtaining recognition by the Turkish Government that the concession covered a strip of territory on the Turco-Persian border which had been transferred from Persia to Turkey in 1913. This strip, known as the "transferred territories," was recognized as part of the Anglo-Persian concession by all the parties involved in the discussions, negotiations and prolonged deals over Iraqi oil. On August 30, 1925, an agreement was concluded between the Government of Iraq and the Anglo-Persian Oil Company, granting the concession in the "transferred territories" for thiry-five years and establishing the royalty payments on a percentage of profits basis. The agreement also called for the formation of a subsidiary company to operate within Iraq. The subsidiary company was to erect a refinery for oil products to be consumed in Iraq; all the oil produced was to be for local consumption. The following year, on May 24, 1926, the agreement was revised. Royalties were to be paid on a per metric ton basis, the same as IPC; the subsidiary company was also to become a marketing agency throughout the country; and the concession was extended for an additional thirty-five years.[21]

The subsidiary organized was the Khanaqin Company. It produced oil from the Naft Khaneh field, and its refinery at Alwand, near Khanaqin, terminus of a twenty-five mile pipeline, was completed in May 1927. Early in 1932 a marketing company, the Rafidain Oil Company, was organized for the purpose of supplying Iraq's petroleum requirements. In May of the same year Khanaqin made over to Rafidain that portion of its business which was concerned with the marketing of oil and oil products in Iraq.[22]

PRODUCTION

Oil in commercial quantities was discovered as early as 1927.[23] It was not until 1934, however, with the completion of the Company's pipeline, that large quantities of oil were produced.[24] Since the major fields were inland, at Kirkuk, the output was inevitably limited by the capacity of the pipelines. Table XIII, on the following page, gives total production from 1927 up to and including 1951.

PIPELINES AND REFINERIES

The production of oil in Iraq was conditioned, at least in large measure, as was noted above, by transportation, hence the provisions in the original concessions for the transport of the oil.

The first major pipeline built by the Iraq Petroleum Company was a parallel

Table XIII. *Iraq Oil Production, 1927–1951*

In Metric Tons[a]

1927	45,000	1940	2,514,000
1928	95,000	1941	1,566,000
1929	116,000	1942	2,595,000
1930	121,000	1943	3,572,000
1931	120,000	1944	4,146,000
1932	115,000	1945	4,607,000
1933	123,000	1946	4,680,000
1934	1,031,000	1947	4,702,000
1935	3,729,000	1948	3,427,000
1936	4,011,000	1949	4,067,000
1937	4,255,000	1950	6,479,000
1938	4,298,000	1951	8,351,000
1939	3,963,000		

In Barrels[b]

1927	338,000	1940	24,225,000
1928	713,000	1941	12,650,000
1929	798,000	1942	19,726,000
1930	913,000	1943	24,848,000
1931	830,000	1944	30,943,000
1932	836,000	1945	35,112,000
1933	917,000	1946	35,665,000
1934	7,689,000	1947	35,834,000
1935	27,408,000	1948	26,115,000
1936	30,406,000	1949	30,957,000
1937	31,836,000	1950	46,760,000
1938	32,643,000	1951	61,534,000
1939	30,791,000		

a. United Nations, *Statistical Yearbook, 1949–1950,* 147, and *1951,* 141. Production in metric tons covering the years 1928, 1933, and 1938–1950 is given in UN, *Review, 1949–50,* 60; figures in metric tons for IPC in Walter E. Skinner's *Oil and Petroleum Year Book* (London), for the years 1934–1951 in the volumes for 1936, 1940, 1946, 1949, 1951 and 1952. IPC production for 1934, 1938 and 1946–1951 appears diagrammatically in *Iraq Oil in 1951,* 8; the cumulative production from the beginning of operations to the end of 1951 is given as 68 million tons, 9.

b. *World Oil,* July 15, 1952, 66; see also the graph in *Iraq Oil in 1951,* 8, covering two pre-war years and 1946–1951 – limited, however, to IPC.

12-inch line from Kirkuk to Haditha, a distance of 150 miles, where it branched off in two directions, one line going to Haifa (Palestine) for a distance of 470 miles, and one to Tripoli (Lebanon) for a distance of 380 miles.[25] The line, which had an annual capacity of 4,000,000 metric tons, was completed in 1934. The Anglo-Iranian Oil Company and the Dutch-Shell Company received their shares of oil at Haifa where, after the Consolidated Refineries were completed, it was refined. The American and French groups received their crude at Tripoli.

Immediately after World War II (October 1946) a second parallel pipeline to terminate at Haifa and Tripoli was started, but because of the Arab-Israel war the branch to Haifa stopped at the boundary, about 42 miles from the terminus; the branch to Tripoli was completed in July 1949. A 30-inch pipeline from Kirkuk to Banias (Syria) covering a distance of 556 miles and with an annual capacity of 12,000,000 tons, was completed in April 1952 and officially inaugurated on November 18. This large new pipeline made possible a 95 per cent increase in the oil production from Kirkuk as compared with that of 1951.[26]

For the Basrah area a 75-mile 12-inch pipeline, begun in 1948, was completed in October 1951 from the Zubair field to Fao on the Shatt-al-Arab, on the Persian Gulf; a parallel 24-inch line was commissioned on January 1, 1954. The new oil field of Rumaila, 20 miles west of Zubair, was connected with the last-named field by a 20-mile 20-inch pipeline.[27]

In July 1952 a 135-mile 12-inch pipeline from Ain Zalah, in the Mosul area, to Station K2 on the main Kirkuk-Mediterranean pipeline, was completed.[28]

The question of refineries was one of the major issues between the Iraq Government and the Iraq Petroleum Company. The former was anxious to have refineries on its territory; they would not only supply refined petroleum products necessary for home consumption, but would also increase the country's income, employ a greater number of workers, and enable the Iraqis to learn the necessary technique in the advanced stages of the oil industry.

A small refinery to meet local needs — but perhaps more because the AIOC would not build a long pipeline to carry the limited quantity of oil produced at Naft Khaneh in the "transferred territories" — was erected at Alwand in 1927 by the Khanaqin Oil Company, as mentioned above.[29] It produced about 70 per cent of the petroleum products consumed in Iraq; the remainder was supplied by the Abadan refinery in Iran.[30]

In spite of persistent agitation by the Government, until 1951 the Iraq Petroleum Company refused to build more refineries in Iraq; it had two very small refineries at Kirkuk and Haditha, and a processing plant at Kirkuk. The great refineries were in Haifa and in Europe.

Closely connected with the question of refining was that of marketing of oil and oil products for internal consumption. As mentioned above, this was concentrated in the Rafidain Company. On December 25, 1951, the Iraq Government and the Khanaqin and Rafidain Companies signed an agreement providing for the purchase by the Government, at an agreed amount, of the Alwand refinery and the Rafidain facilities. Rafidain, acting on behalf of the Government, was to operate the refinery and, until 1961, was also to act as sole distributing agent for certain oil products for local consumption. [31]

To satisfy all local consumption demands, the Government decided to build a large refinery in the vicinity of Baghdad. Early in July 1951, it announced that it had accepted a bid from W.W. Kellogg[32] of New York to erect a refinery south of Baghdad, with an annual capacity of 1,000,000 tons. Work on the refinery at Daura started in 1952 and was finished in 1955, at a cost of £10,000,000. A 135-mile 12-inch pipeline brought crude from pumping station K2. As a temporary measure to meet the emergency created by the shutdown of Abadan, a small refinery with a 175,000-ton annual capacity was erected at Muftiya, outside of Basrah. In addition, the stabilizing and topping plants at Kirkuk helped to meet immediate local needs. [33]

In July 1952 the Iraq Government set up a five-man Refineries Board to administer the refineries, distribution and prices.

The problem of lubricating oil was not met by all these refinery projects, and in the middle of May 1954 the Government allotted 3,000,000 dinars (decided upon in September 1953) for the erection of a lubricating oil refinery, in the vicinity of Daura, which would have an annual capacity of 25,000 tons. This project was to be administered by the Refineries Board. [34]

Iraq was well on the way to meeting home requirements for petroleum products; in fact, when the general refinery in Daura was completed, the refined products from Alwand had to be exported. Such a possibility was anticipated and provisions were made with the Khanaqin Oil Company to operate Alwand on an export basis. [35]

THE 1952 AGREEMENT

The Red Line Agreement provided, in addition to the restrictions mentioned above, that in case a group was unable to take its share of oil, that share would be forfeited without compensation. The French group could not take its share during the Vichy regime, nor could C.S. Gulbenkian; Standard of New Jersey could not take deliveries while the Mediterranean was closed to tankers.

When the war was over the Iraq Petroleum Company again expanded operations, and each group took its proportionate share of the crude oil produced. The Compagnie Française des Pétroles and Gulbenkian attempted to obtain a reaffirmation of the restrictive provisions of the Red Line Agreement. This would have guaranteed to the French and Gulbenkian not only increased production in Iraq, but also a proportionate share in any acquisition which Standard of New Jersey and Socony-Vacuum might have acquired in the Arabian American Oil Company. The American companies were unwilling to reaffirm the restriction; on January 29, 1946 Eugene Holman, president of Standard of New Jersey, advised CFP that "there had been a substantial change in the attitude of the American public and Government toward restrictive agreements and under current conditions, reaffirmation of the agreement seemed inadvisable." In October 1946 the American group declared the Red Line Agreement to be dissolved; it cited the increased need for crude oil from the Middle East to meet the greater demand of its expanded markets as the major reason.

Standard, in association with Socony-Vacuum, began negotiations with the Anglo-Iranian Oil Company on the possibility of constructing a jointly owned pipeline from the Persian Gulf to the eastern Mediterranean and on Standard buying from the AIOC and Kuwait — neither of which was included in the "red line" area — considerable quantities of oil over a twenty-year period; on December 26, 1946 an agreement between the three companies was signed. [36] However, according to Standard, even this additional supply would not have met its great demands for crude oil. [37]

Whether Standard ever actually meant to go through with this pipeline scheme, or whether it was just a maneuver to break the Red Line Agreement so that it could buy a share in Aramco, is hard to tell. AIOC, however, proceeded with the project, organized the Middle East Pipelines Ltd., [38] and entered into negotiations with Iraq and Syria for wayleaves. [39]

While the negotiations with AIOC for the new pipeline were going on, Standard of New Jersey started negotiations with the Arabian American Oil Company for a supply of crude oil. In December 1946 it was reported that Standard and Socony-Vacuum had reached an agreement in principle with Standard of California and the Texas Company for buying shares in the Arabian American Oil Company. The Compagnie Française des Pétroles thereupon brought suit in February 1947 in the British courts against the American companies and the other partners in the Iraq Petroleum Company. Four months later Standard of New Jersey and Socony-Vacuum filed a statement of claim with the British High Court of Justice denying CFP's allegations about the validity of the restrictive provisions of the Red Line Agreement. In spite of the fact that the American companies had been abiding by the provisions of the Red Line Agreement ever since they had signed it in 1928, they now declared: "Any agreement contained in the said declaration was in restraint of trade and contrary to public policy and void and unenforceable in law."

The partners of IPC were determined to avoid a public airing of the Company's affairs and after a tentative agreement was arrived at in May 1947, negotiations were continued for a permanent solution that would reconcile the conflicting interests of the various groups. On November 3, 1948 a new Heads of Agreement was signed and the suit and countersuit were withdrawn. Standard of New Jersey and Socony-Vacuum were freed to complete their acquisition of an interest in Aramco; the overriding 7.5 per cent royalty to the Anglo-Iranian Oil Company in the Iraq Petroleum Company concession was reaffirmed, and the Compagnie Française des Pétroles was assured increased supplies of crude from Iraq through the provision of the necessary additional facilities, especially pipelines, to export Iraqi oil in conformance with a planned program of production. [40]

This dispute [41] highlighted one of the most controversial issues between the Iraq Petroleum Company and the Government of Iraq. Although the increase in production of crude oil in Iraq was steady, it was not so spectacular as in Iran, and certainly not as in Saudi Arabia. A number of Iraqis advanced the argument that the British Government, which controlled the Anglo-Iranian Oil Company, was more interested in expanding production in Iran than in Iraq. The French, on their part, while agreeing with this argument, feared an additional curtailment of Iraqi production – their major supply – if the participating American companies were permitted to enter into other oil undertakings in the area. To the French, therefore, the Red Line Agreement was a guarantee, because of the limitations it placed on the American companies, of a steady, and even of an increased, supply of oil from Iraq.

The Iraqis were dissatisfied with the royalties they were receiving; nor were they satisfied with the number of their nationals in the higher positions of oil production operations. They complained of the lack of training facilities for Iraqis for the conduct of the industry. A constant source of irritation between the Government and the Company was the lack of refining facilities in the country. Another issue, of a technical nature, was the evaluation of gold shillings paid as royalties. Difficulties developed when the Company's second payment fell due in January 1932, after the British pound went off the gold standard. A temporary arrangement, without prejudice to either party, was made after Premier Nuri es-Said Pasha went to London and discussed the issue with the Company; Iraq was paid £578,000 instead of the £409,000 provided by the agreement. Subsequently, it was agreed that payments be made in sterling at the current price of gold. [42] Later a serious difference arose as to the interpretation of the "current price of gold." The Company maintained that the gold shilling must be based on the official London rate of exchange of gold, while the Iraqis argued that it should be based on the free market value, which was, of course, considerably higher.

Against the Iraqis' contention that production was being methodically kept down by the Company, the latter argued that the relatively low rate of production of Iraq oil was due to transport difficulties; that the Company was investing huge sums for pipelines, but the process of building the pipelines was conditioned by the exigencies of war which limited the supply of steel. It pointed out that immediately after the war, in spite of a serious steel shortage, the parallel 16-inch pipeline to Haifa and Tripoli had been built, alongside the existing 12-inch lines, but that Iraq's refusal to allow the oil to flow to Haifa (Israel) had cut down production to practically half the potentialities of the pipelines.

In the middle of 1948 – despite the fact that the original concession provided that only after 1954 should royalty payments be reconsidered – the Iraq Government asked IPC for an increase in royalties. After prolonged negotiations, the Government announced, on November 11, 1950, that the discussions with the Company had resulted in an increase from four to six shillings (gold) per metric ton of oil. No agreement could be reached, however, on the question of the gold evaluation of the shilling, and the Iraq Government decided to take the issue to court.

Two important events which had occurred in the meantime completely upset Iraq's royalty increase and necessitated a different arrangement based on the new realities in the Middle East. Early in 1951, Saudi Arabia signed an agreement with Aramco replacing its existing royalty payment with a 50-50 profit-sharing arrangement (to be dealt with in the next section), and agitation in Iran against the Anglo-Iranian Oil Company culminated in the nationalization of that country's oil industry in April.

Two weeks after the nationalization of oil in Iran, Premier Nuri Said, addressing the Finance Committee of the House of Deputies, threatened the oil companies that if they did not meet Iraq's demands for higher royalties — equal to those paid to Saudi Arabia and Iran, and payment in the value of gold on a free market basis — they might lose their concessions.

It would appear that this threat was the result of a stalemate in negotiations, and was an effort to intimidate the oil companies in view of the agitation by some political parties for the nationalization of the oil industry. It apparently attained its objective, for on the following day the Iraq Government announced in Baghdad that the companies had agreed to increase royalties to levels similar to those received by neighboring countries, and that negotiations for a detailed agreement were to begin immediately. At the end of May, Nuri Said announced in the Senate that the Iraq Government and IPC had agreed in principle on a new level of royalty payments; this was confirmed by the Company in London at the end of July.

A hitch, however, developed when the Iraqi representative insisted on, and the Company refused to agree to, the insertion of a clause providing for the reopening of the contract to adjust royalty payments to any changes in the neighboring countries. Finally, on August 13, 1951, it was announced that IPC and its subsidiaries had reached agreement with the Iraq Government; on February 3, 1952, the new agreement was signed. It was approved by the House of Deputies on the 14th, and by the Senate on the 17th.

AGREEMENT PROVISIONS

The major provisions of the agreement and its enclosures were:

1.   The Government to share on a fifty-fifty basis in the profits resulting from the operation of the companies before deduction of taxes. In addition, each of the companies to pay annually £20,000 for tax commutation (Arts. 2 and 9).

2.   Should the Government so decide, it could take in kind up to 12.5 per cent of the net production of each of the companies as part of its share of the profit and dispose of it in the open market or sell it back to the companies at current world prices (Art. 3).

3.   The companies to guarantee the Government's share in each calendar year to be not less than 25 per cent of the net production value at posted prices of the Iraq Petroleum Company and the Mosul Petroleum Company, and 33 1/3 per cent of the Basrah Petroleum Company (Art. 4).

4.   The Iraq Petroleum Company to guarantee a production minimum of 20.75 million tons of crude oil annually beginning January 1, 1954; the Mosul Petroleum Company to guarantee an annual minimum of 1.25 million tons

beginning January 1, 1954; and the Basrah Petroleum Company to guarantee an annual minimum of 8 million tons beginning January 1, 1956. These guarantees to be effective for the life of the agreement (Art. 5).

5. All the companies jointly and severally to guarantee that the Government's share be not less than £20 million annually for the years 1953 and 1954, and not less than £25 million for 1955 and each year thereafter (Art. 6).

6. The agreement, signed in the name of the Government by Abdul Majid Mahmoud, Minister of Economy, and for the companies by Horace Stephen Gibson, Managing Director, was drawn up both in Arabic and English; both to be operative, but in the event of discrepancies between the two, the English to prevail (Art. 15).

7. The companies to supply the Government refineries with all the crude oil necessary for local consumption at 5.5 shillings per ton.

8. If, in future agreements between oil companies and neighboring Near Eastern countries, royalties were increased, Iraq to have the right to request similar increases.

9. Each of the three companies to appoint two Iraqi directors.

10. The companies to send at their own expense fifty Iraqi students annually to British universities for specialized studies in the oil industry, and IPC to establish a school in Kirkuk to train Iraqi workers.

11. Should production be halted because of *force majeure* or "an act of God," the companies to pay the Government a minimum of £5 million a year for two years. [43]

With the signing of the new agreement, Iraq entered a new era. The complete shutdown of the Iranian oil sources, combined with the terms of the new agreement, increased production beyond normal proportions and made the oil revenue a decisive factor in the development of Iraq. Oil production in 1952 jumped to 18,066,797 tons, in 1953 to 27,220,199 tons, and in 1954, to 29,550,000 tons, considerably above the estimates of the agreement. Royalty income in 1951 was £13,700,000, and in 1952 £33,000,000, plus the £5,000,000 paid in settlement of past claims; in 1953 it jumped to £51,400,000, and to £68,390,000 in 1954. [44]

The expansion in production was in all three areas. The Kirkuk field (IPC) produced 15,522,715 tons in 1952, and 22,865,538 tons in 1953 and 23,719,672 tons in 1954. The Zubair field (BPC) produced 2,238,177 tons in 1952 and 3,077,522 tons in 1953, and together with the new field Rumaila produced 4,584,358 tons in 1954. The Ain Zalah field[45] (MPC), which began production in October 1952, produced 281,977 tons that year and 1,277,239 tons in 1953 and 1,281,827 tons in 1954. [46]

REFLECTION

The development of Iraq oil from 1925 when the Iraq Government granted the first concession to the Turkish Petroleum Company, to 1953 was one of considerable progress. The Iraq Petroleum Company was composed of four different national groups and was primarily a producer of crude oil which was to be apportioned among the partners, rather than a full operating company as were the other concessionaires in the area. Production of oil was of course conditioned on the pipelines, and after the first pipeline was completed the growth was not spectacular. This might perhaps be attributed to the British policy of conserving their oil resources — for the British were practically a one-half shareholder in the Company — as well as to the overall international policy of limiting production in order to maintain high prices.

In their dispute with the Americans, the British came out rather well. While they granted the American group 23 3/4 per cent, the AIOC collected a royalty on all the oil produced. The monopolistic character of the Company was retained through the revision of the agreement in 1931, and the concession to the British Oil Development Company in 1932. The State Department's insistence on equal opportunity for American companies to exploit Iraq's oil fields was thus vitiated. The consent of the American group to the Red Line Agreement annulled any limitations of the Open Door policy. It should be noted, however, that the two remaining American companies of the original seven were not at all disturbed at the time by the restrictive aspects of the agreement. It was not until the oil fields of Saudi Arabia loomed on the horizon that the American companies began to oppose the Red Line restrictions and finally succeeded in removing them. [47]

With the building of additional pipelines, production increased, and although World War II and then the war with Israel upset the normal growth, the serious disturbance in the production of Middle East oil through the Iranian attempt at nationalization and the terms of the 1952 agreement caused production to jump to spectacular heights, and with it the revenue of the Iraq Government.

These same factors made possible the expansion of refinery facilities in Iraq and their ownership by the Government; internal distribution was also taken over by the Government, and at least part of the industry was to be run by the Iraqis themselves.

NOTES

1. *Foreign Relations, 1922*, II, 344.
2. *Foreign Relations, 1925*, II, 239—245. The British were apparently not impressed with this warning, and on December 9, 1925 Prime Minister Stanley Baldwin declared that the British Government regarded as "undoubtedly valid the concession made on 14th March 1925," by the Iraq Government to the Turkish Petroleum Co. They welcomed the inclusion of French interests in the Company, "and have watched with sympathy negotiations for the inclusion also of American interests. If these negotiations result in American interests acquiring an interest in the Turkish Petroleum Company, such a result will be welcomed by His Majesty's Government." *PDC*, 189, col. 427, Dec. 9, 1925; see also *PDC*, 188, cols. 2016-2017, Dec. 1, 1925.
3. *Foreign Relations, 1926*, II, 364.
4. For the arrangements with Gulbenkian, see US Senate, *International Petroleum Cartel. A Staff Report to the Federal Trade Commission Submitted to the*

Subcommittee on Monopoly of the Select Committee on Small Business (Washington, 1952), 63. Henceforth cited US Senate, *International Petroleum Cartel.*

5. *Foreign Relations, 1927,* II, 823.

6. The final action was taken on Apr. 16, 1928 when the State Dept. wrote to Guy Wellman: "With reference to your inquiry on April 10, 1928, concerning the attitude of the Department of State, I take pleasure in informing you that, in the light of the information at hand, the Department considers that the arrangement contemplated in view of the special circumstances affecting the situation are consistent with the principles underlying the open door policy of the Government of the United States." *Ibid., 824.*

7. Art. 10 of the Agreement provided: "All the parties hereto agree that the Turkish Company or a nominee of the Turkish Company shall, except as hereinafter mentioned, have the sole right to seek or obtain oil concessions within the defined area, and each of the Groups hereby covenants and agrees with the Turkish Company and with the other Groups that excepting only as herein provided or authorized such Groups will not nor will any of its Associated Companies either personally or through the intermediary of any person, firm, company, or corporation seek for or obtain or be interested, directly or indirectly, in the production of oil within the defined area or in the purchase of any such oil otherwise than through the Turkish Company or an Operating Company under the Turkish Company." US Senate, *International Petroleum Cartel,* 66.

8. For surrendering 25 per cent of its holding to give the Americans a share, all the other partners agreed to give Anglo-Iranian a 10 per cent overriding royalty on all the oil obtained from the 24 plots reserved for exclusive exploitation by the Turkish Petroleum Co. In 1931, however, when the concession was revised to cover 32,000 square miles of the area east of the Tigris River, the issue of the 10 per cent overriding royalty was reopened; in November 1934 a compromise was worked out, substituting a 7.5 per cent royalty for the 10 per cent and covering the entire area of the new concession. The royalty was to be paid in crude oil delivered to the Anglo-Iranian Oil Co. free of cost at the field, with IPC paying the royalty to the Iraq Government. US Senate, *International Petroleum Cartel,* 64, 89; *Standard Oil Company (N. J.) and Middle Eastern Oil Production* (New York, 1947), 6.

9. The Compagnie Française des Pétroles was organized in 1923 by leading French bankers and by the marketing and refining subsidiaries of the international oil companies. Early in 1924 an agreement was concluded with the French Government under which the CFP was granted exclusive rights to the French share of Iraqi oil. In 1929 the French Government took a 25 per cent share interest in CFP, which two years later was increased to 35 per cent, with a 40 per cent control in the Company. US Senate, *Internaitonal Petroleum Cartel,* 54.

10. A few years later Gulf, Atlantic and Mexican sold their interests to Standard Oil of N.J. and Standard Oil of N.Y.; the latter became Socony-Vacuum Oil Co. in 1931, and later Mobil Oil Corporation.

11. The prolonged negotiations between the US and Great Britain regarding the

rights of American nationals in Iraq were finally formalized in a treaty signed on Jan. 9, 1930. Art. 2 provided: "The United States and its nationals shall have and enjoy all the rights and benefits secured . . . to members of the League of Nations and their nationals, notwithstanding the fact that the United States is not a member of the League of Nations." *Rights of the United States of America and of Its Nationals in Iraq. Convention and Protocol between the United States of America, and Great Britain and Iraq* (Treaty Series No. 835, Washington, 1931), 2.

12. Under the Mar. 14, 1925 concession, TPC was to make its selection within thirty-two months from the date of the concession, and the Iraq Government was to make its selection of 24 plots, to be offered for competition by sealed bids, within four years. Selection by the Company was therefore to have taken place by Nov. 1927, and the Government's selection by Mar. 1929. Instead, the Company asked for more time, and in Aug. 1927 the Iraq Government granted a year's extension. In May 1928 the Company sought a further extension.

In the meantime, an independent British-Italian syndicate, the British Oil Development Co. (BOD), approached King Feisal to obtain oil leases outside the 24 plots selected by TPC, for which it would build a railway from Iraq to the Mediterranean. To counter this, TPC offered to make a survey for the pipeline and a railway to the Mediterranean if it were granted a two-year extension. At the same time, the Company was also negotiating for a modification of the 1925 concession.

Discussions for modification took place in June 1929, while the Company prepared to make a tentative selection of its plots to comply with the provisions of the agreement and thus obviate the necessity of getting a time extension. In 1931 the revised concession was signed. US Senate, *International Petroleum Cartel*, 68–70.

13. *Special Report,* Col. No. 58, 316–325; *PDC*, 282, col. 2400, May 22, 1931; *PDC*, 267, col. 1787, June 29, 1932.

14. See above, Chap. X, footnote 62; also footnote 12 of this chapter.

15. *Foreign Relations, 1931*, II, 607–608.

16. *Ibid.*, 610.

17. Great Britain, Foreign Office, *Report by His Majesty's Government in the United Kingdom of Great Britain and Northern Ireland to the Council of the League of Nations on the Administration of Iraq for the period January to October, 1932* (London, 1933), 43–57.

18. In Nov. 1932 the Mosul Oil Fields, Ltd., with British, Italian, German, Dutch, French-Swiss and Iraqi holdings, was organized for the purpose of acquiring the share capital of the British Oil Development Co. In 1935, the Italians became the majority shareholders, with a 52 per cent interest. In order to regain control over the British Oil Development Co., the Mosul Holdings Ltd., a subsidiary of the Iraq Petroleum Co., was registered in 1936 for the purpose of acquiring the shares of Mosul Oil Fields. By 1937 the Iraq Petroleum Co. interests were in effective control, and in 1941 Mosul Holdings, renamed the Mosul Petrolem Co., received the assignment from the British Oil Development Co. of its 1932 concession. Longrigg, *Oil in the Middle East*, 79–80; US Senate, *International Petroleum Cartel*, 85; Mohammed al Naqib, "How Iraq's Oil Production Began and Grew," *Al Aswaq al Tijariya*, February 6, 1954.

19. The full text of the concession is reproduced in *Iraq Government Gazette*, Dec. 4, 1938, 686—706; the law ratifying the agreement, 672.

20. Iraq Petroleum Co., *Iraq Oil in 1951* (London, 1952), 21.

21. *Special Report*, Col. No. 58, 1931, 222. Six hundred eighty-four square miles were covered by this concession.

22. *Report 1932*, 40. A short summary of the various concessions is given in *The Middle East. A Political and Economic Survey* (Royal Institute of International Affairs, London, 1954), 281—282, and in *An Introduction to the Past and Present of the Kingdom of Iraq* (by a Committee of Officials, Baltimore, 1946), 43—45. A generalized outline of the terms of the concessions, almost on a schematic basis, appears in Raymond F. Mikesell and Hollis B. Chenery, *Arabian Oil: America's Stake in the Middle East* (Chapel Hill, N.C., 1949), 44—48; a sketchy outline is also given in *Review of Economic Conditions in the Middle East 1949—1950*. United Nations. E1910/Add.2/Rev.1, ST/ECA/9 Add. 2 (New York, 1951), 58. Henceforth cited UN, *Review 1949—50*; Longrigg, *Oil in the Middle East*, 66—76.

23. Drilling began in April and in October Baba Gurgur No. 1 well came into production with 600,000 barrels a day. Mohammed al Naqib, ' How Iraq's Oil Production Began and Grew," *Al Aswaq al Tijariya*, Feb. 6, 1954; see also *PDC*, 207, col. 658, June 13, 1927.

24. The Company spent about $62,000,000 from the time of the signing of the concession in 1925 until the oil began to flow through the pipeline for commercial marketing. US Senate, *American Petroleum Interests*, 228.

25. A detailed description of this pipeline is given in *American Petroleum Interests*, 212; UN, *Review 1949—50*, 61. The building of this parallel pipeline was provided for in the agreement signed on Mar. 24, 1931 and the British Government had previously obtained agreements from the Governments of Palestine and Transjordan, *PDC*, 248, col. 212, Feb. 10, 1931; *PDC*, 250, col. 1080, Apr. 1, 1931.

26. UN, *Review 1949—50*, 61; *Iraq Oil in 1952*, 2. It took seventeen months to complete and cost £41 million.

27. *Iraq Oil in 1953* 27; *Iraq Oil in 1954*, 14.

28. *Iraq Oil in 1952*, 27—30.

29. *PDC* 207, col. 1467, June 20, 1927.

30. International Bank for Reconstruction and Development, *The Economic Development of Iraq. Report of a Mission* (Baltimore, 1952), 150—151. Henceforth cited as International Bank, *Iraq Mission Report*.

31. Iranian Oil Co. Ltd., *Annual Report and Accounts as at 31st December 1951*, 21.

32. *The Christian Science Monitor*, July 6, 1951.

33. *Iraq Oil in 1952*, 16—17; *The Middle East. A Political and Economic Survey*, 282; United Nations, Dept. of Economic Affairs, *Summary of Recent Economic Developments in the Middle East 1952—53* (New York, 1954), 49—50. Henceforth cited as UN, *Summary 1952—53*.

34. *Al Zaman*, Sept. 7, 1953. For a description of the Government refinery projects see United Nations, Department of Social and Economic Affairs, *Economic Developments in the Middle East, 1945 to 1954* (New York, 1955), 117.

35. Longrigg, *Oil in the Middle East*, 193.
36. *PDC*, 432, cols, 74–75, Jan. 23, col. 361, Feb. 1, 1947; George Kirk, *Survey of International Affairs. The Middle East 1945–1950* (London, 1954), 84.
37. *Standard Oil Company (New Jersey) and Middle Eastern Oil Production*, 12. The AIOC was apparently more interested in the project than Standard. Sir William Fraser, chairman of AIOC, after fully describing the arrangement between AIOC and the two American companies, declared: "I have no doubt that these arrangements, when they are brought into effect some years hence, will be of great advantage to the company." Anglo-Iranian Oil Company Ltd., *Report and Balance Sheet at 31st December 1946*, 14.
38. This Company was owned 60.9 per cent by AIOC, 24.7 per cent by Standard of N.J., and 14.4 per cent by Socony-Vacuum. US Senate, *International Petroleum Cartel* 28.
39. Anglo-Iranian Oil Company Ltd., *Report and Accounts as at 31st December, 1948,* 16; – *Annual Report and Accounts as at 31st December, 1949,* 17–18.
40. US Senate, *International Petroleum Cartel,* 96–106.
41. For a discussion of the Red Line Agreement, see John A. Loftus, "Middle East Oil: The Pattern of Control," *The Middle East Journal*, II, 21–22, Jan., 1948; K. Grünwald, "The Oil Fields of the Middle East," *Economic News* (Tel Aviv), IV, 9, Oct. 1951; US Senate, *International Petroleum Cartel* 56–84 *et passim*.
42. *Report 1932*, 38; *Survery of International Affairs 1934*, 192–193.
43. Iraq Petroleum Company Ltd., Mosul Peteoleum Company Ltd., Basrah Petroleum Company Ltd., *Agreement with the Government of Iraq Made on the Third Day of February, 1952 Together with Law No. 4 of 1952 Ratifying the Agreement* (English and Arabic Texts); *Iraq Oil in 1951*, 15–18. The companies also agreed to pay a lump sum of £5 million in settlement of past disputes. United Nations, Dept. of Economic Affairs, *Summary of Recent Economic Developments in the Middle East 1950–1951* (New York, 1952), 36. Henceforth cited as UN, *Summary 1950–1951*.
44. UN, *Summary 1952–53,* 60; UN Department of Economic and Social Affairs, *Economic Developments in the Middle East, 1945 to 1954* (New York, 1955), 97, henceforth cited as UN, *Economic Developments 1945 to 1954;* "Oil and Social Change in the Middle East," *The Economist,* CLXXVI, 3, July 2, 1955.
45. Ain Zalah, which is in the Mosul area and is operated by the Mosul Petroleum Co., was developed in 1940, but early in 1941, because of the war, operations were suspended; they were resumed in 1952. After the completion of the pipeline from Ain Zalah to K2 station on the Kirkuk-Mediterranean pipeline, oil began to flow. *Iraq Oil in 1952*, 25.
46. *Iraq Oil in 1952*, 30; *Iraq Oil in 1953*, 21–30; *Iraq Oil in 1954* 6, 11, 14.
47. In 1947, Standard Oil Company (New Jersey) issued a pamphlet: *Standard Oil Company (New Jersey) and Middle Eastern Oil Production*, which declared that the Company was considering buying a share in the Arabian American Oil Company because it needed more oil for the Eastern Hemisphere markets and that in view of developments, it no longer considered the "red line" restrictions of the 1928 agreement valid; 3–15.

Chapter XII
OIL AND DEVELOPMENT OF IRAQ: 1927–1955

Iraq may serve as an excellent example for the study of the relationship between oil production and the social and economic patterns of an underdeveloped country. Production of oil began almost simultaneously with the emergence of the independent state. Though Iraq's social and economic traditions were fixed when it was part of the Ottoman Empire and seriously bedeviled the growth and development of the new kingdom, the special Iraqi national patterns were being molded during the very period when oil was flowing in ever-increasing quantities through pipelines hundreds of miles long to world markets. Moreover, Iraq had the benefit of Western guidance, at first British administration and afterwards advice; yet during the first twenty-five years of oil production the country did not progress as spectacularly as the oil figures (from 45,000 tons in 1927 to some 28,000,000 tons in 1953). The basic obstacles to the rapid social and economic development of the people and the transformation of the country could apparently not be overcome by the mere availability of oil, some technical knowledge, and royalties from the oil industry.

Before drawing any conclusions, we should examine some fundamental facts about the country, the people, and the factors which operated in Iraq during this period to create the conditions that prevailed there.

LAND AND PEOPLE

The area of Iraq is about 175,000 square miles, and it had a population of approximately 5,000,000 people in 1947.[1] Because of the lack of sufficiently reliable data, as well as the cultivation practices of the Iraqi peasants, no exact figures were available as to the cultivable and cultivated areas or the extent of the uncultivable part. The first comprehensive survey of land tenure, methods of cultivation, and other related questions was made by Sir Ernest Dowson in 1930. He estimated the cultivable area as 92,000 square kilometers — 41,000 in the rainfall zone and 51,000 in the irrigation zone — and that only a fraction of the zones, possibly one-fifth to one-tenth, was cultivated in any one year.[2] In 1945 it was estimated that out of the total of some 112,000,000 acres, the potentially cultivable area was about 30,000,000, and that only about 6.5 million were irrigated.[3] The United Nations reported in 1949 that only 3,200,000 hectares — 1,300,000 under irrigation — were cultivated, out of 6,000,000 cultivable.[4] In the same year the British estimated the cultivable area as 35,000 square miles and a small portion as utilized in any one year.[5]

The International Bank Mission to Iraq in 1951 could not state with any certainty the cultivated and cultavable areas. According to the Mission, the official estimates indicated that the amount of cultivable land was about three times that under cultivation, but the definition of cultivability was not established and no accurate figures were available.[6]

Regardless of the exact statistics, it was universally agreed that although about 80 per cent of the population derived its living directly or indirectly from the soil, a significant portion of the cultivable land, perhaps more than half, was not cultivated; that with a little effort and systematic organization, the area under cultivation could have been rapidly expanded; and that with better and more intensive cultivation, the wealth and earning capacity of the country could have been substantially increased. Of all the Middle Eastern countries, Iraq has all the necessary requisites for sound and rapid economic development, especially in the agricultural sector. It has a very considerable quantity of land ready for immediate cultivation, as defined by the most conservative standards, and with modern methods very great stretches defined as uncultivable could have been improved and brought into the cultivable class; it has an abundance of water for irrigation as well as for power, and it has at its disposal enormous quantities of cheap fuel both in petroleum and natural gas.

What, then, have been the economic, educational, health and social conditions of the country? As regards the size and growth of the population, it would be very difficult to state categorically the percentage increase of the population. The accuracy of the figure advanced by the October 1947 census has been questioned by the International Bank Mission.[7]

Despite substantial strides in public health – an increased number of hospitals, dispensaries and doctors – during the first thirty years of the existence of the kingdom, the Bank Mission reported in 1951 that the infant mortality rate was very high, "perhaps as much as 250 per 1,000 births," and that there was a very high incidence of such endemic diseases as trachoma, hookworm, bilharzia, malaria and dysentery. Malaria was reported to cause 50,000 deaths every year. In 1948 no less than 6,657,474 visits to medical institutions for treatment were reported, and 1,291,225 cases of infectious diseases were treated. Of these, 603,698 were malaria and 524,740 trachoma.[8] In 1944 the total number of doctors in the country was 528; in 1951 it was reported that there was one doctor to every 7,000 in the population, which would make a total of about 915 doctors; but a large percentage of them were concentrated in Baghdad.[9]

There can be no doubt that educationally the country progressed considerably from the stage in which it was in 1921, but all in all the sum total was not very impressive; the rate of progress, discouraging both to the educators and to the Iraqi people, reflected on the possibilities of development, which in the final analysis depended on the general educational level of the population. To be sure, in the period 1930–1950 the number of schools rose from 262 to 1,100, the number of pupils from 32,750 to 175,000, and the number of primary school teachers from 1,325 to 6,588; nevertheless, out of about 750,000 children of school age only 175,000 were in schools; the percentage of dropouts as the children advanced to the higher grades was so large as to cause real alarm. The facilities and quality of technical and vocational education were woefully deficient, and adult education was almost completely neglected. As to the quality of the educational program, the Mission stated: "Educational concepts are far too narrow and too little related to everyday problems of living." But what was most alarming was the fact that as late as 1951 about 90 per cent of the total population was still illiterate. [10]

Though possessing so much land, so much water, and considerable revenue from oil, Iraq's standard of living was fairly low. Annual per capita income was estimated in 1949 at about $84 and not much higher in 1953. [11] This compared with $140 for Lebanon, $100 for Syria, $125 for Turkey, $100 for Egypt, and $389 for Israel.

Industrial growth advanced at a very slow pace. In order to stimulate the establishment of industry, the Government passed a law in 1929 granting tax relief and other concessions to those who would develop industrial projects, but almost twenty years later the United States Department of Commerce reported that the development of manufacturing industries was still in the embryonic stage. [12] In 1949 it was reported that in spite of the impetus given to local production by the shortage of import goods during the war, Iraq's industries were still underdeveloped. [13] In 1951 the International Bank Mission reported that industry was little developed; that exclusive of those employed by the oil companies, 60,000 workers were engaged in industrial and handicraft production, but most of those were employed in small undertakings, with only about 2,000 working in what might be classified as modern industrial plants. [14] The United Nations reported in 1951 that industry continued to play a very minor part in the Iraqi economy and that the gross output was estimated at not more than ID 3,000,000. [15]

The general industrial and economic situation could perhaps be gauged by the consumption of oil and oil products, especially since the country had these supplies in such abundance. In 1929, internal oil consumption, including refinery needs, was 70,000 metric tons; in 1937 it had risen to 132,000 tons, but in 1950 it was only 929,000 tons. Excluding refinery needs, consumption in 1937 was only 88,000 tons; in 1949, 518,000; in 1950, 619,000; in 1951, 635,000; and in 1952, 690,000. [16]

The level of civic responsibility and national consciousness was illustrated by the taxpaying practices of the Iraqis. The first income tax law was enacted in 1927, but not before considerable modifications had been made and a long time had elapsed did the Government begin to receive any substantial revenue. Table XIV shows the annual revenue from income tax, the total revenue, and the percentage of income taxes in the total revenue.

The methods of collection revealed the attitude of the general public to personal taxes. In 1945 the total number of individuals who paid income taxes was 18,482; one-third of these were Government employees, ten per cent were from other salaried groups. [17]

These random facts about Iraq and its population indicate, on the one hand, a low standard of living and general underdevelopment; and, on the other hand, excellent potentialities for expansion and improvement in agriculture and industry. Th importance of the income from oil and its utilization for capital improvement

Table XIV. *Iraq, Percentage of Income Tax in Total Revenue*

Year	Income Tax (in ID)	Total Revenue (in ID)	Percentage
1943	1,016,082	18,810,000	5.4
1944	1,293,089	20,700,000	6.2
1945	1,149,394	20,790,000	5.5
1946	1,032,054	25,520,000	4.0
1947	1,081,379	23,510,000	4.6
1950	2,354,000	31,579,000	7.5[a]

a. Great Britain, *Iraq*, 22; UN, *Review 1949–50*, 78; *Annual Report of the National Bank of Iraq, July 1949–December 1950*, 53.

were realized from the very beginning. In 1931 the Five-Year Capital Works Program was enacted in order to hypothecate the revenues from advance oil royalties and dead rent. The monies earmarked for the Capital Development Works Scheme Account were never, however, completely used for capital works projects, and from time to time the Government used the funds to cover the accumulated deficit.[18] The Iraqi Parliament subsequently adopted two yearly budgets, ordinary and extraordinary.[19] The latter, for development schemes, was derived mainly from oil revenues and income from the Iraqi Currency Board, but its revenues were frequently diverted to other than the intended purposes.[20]

The various Iraqi governments — as a rule they had a high frequency of change and a low rate of duration — not only realized the importance of the oil revenue to the general development of the country, but also the need for expert surveys and development plans. From time to time specialists were invited to prepare reports for the Government. The most important of these, prior to the International Bank Mission, was perhaps the Haigh Irrigation Commission, which worked from 1946 to 1949 and recommended a £100 million development plan which would have increased the irrigated area from 8 to 17 million acres.[21] Yet in 1949 the United Nations Survey Mission reported that the financial resources available to the Iraq Government, not only from oil royalties but also from the accumulated wartime sterling surplus, had been largely diverted to financing consumption imports and the operating expenses of the State.[22]

In this statement the Mission touched on the basic problem of the area, even of those sections which have available financial resources for development projects. The realization of the importance of investment in development projects which would bear returns in the future necessitates self-denial of immediate consumption goods. Such long-range vision, especially when resources for gratification of immediate needs are readily available, is, however, a quality with which neither the masses nor many of the political leaders were endowed. It is perhaps in the lack of this quality that the explanation for the failure of numerous development plans of the Middle Eastern oil-producing countries may be found. The ability to curtail current spending for the sake of investments which will increase not merely the individual's returns but the wealth-producing potential of the nation as a whole would be a *sine qua non* for any growing and even highly developed economy, and certainly for underdeveloped areas. But the temptation to utilize the "external" income – from oil – for current needs which could never before be afforded, for alleviating the tax burden, for improving political standing with the electorate, proved too strong, and sooner or later all succumbed.

THE DEVELOPMENT BOARD

To return to Iraq. In May 1950 the Government promulgated a law (1950 No. 23) creating a Development Board to which was to be assigned the revenue derived from the oil companies; in 1952 the law was amended assigning only seventy per cent of the revenue to the Board. The Board, with the Premier as chairman, the Minister of Finance as member *ex officio,* and six non-political full-time executive members, was charged with preparing an overall development plan and given the authority to execute major capital development projects. On October 10 of that year Iraq officially requested the International Bank for Reconstruction and Development to send a survey mission to review the economic potentialities of the country and make recommendations for a development program. [23] The Mission arrived in Iraq on February 25, 1951, and a year later, on February 11, 1952, the report was submitted to Premier Nuri as-Said. It proposed a development plan for the years 1952–1956 involving a total expenditure of some $470,000,000, to be derived from the oil revenue for the same period. Basing itself on the Mission's report, the Development Board worked out the following plan:

Table XV. *Iraq Development Board Program*

Projects	Cost (in ID)
Irrigation schemes	53,400,000
Roads	26,700,000
Other communications	4,100,000
Buildings (including hospitals and schools)	18,000,000
Public utilities	4,100,000
Agriculture	7,500,000
Industry (including erection of a chemical factory)	31,000,000
Miscellaneous	10,600,000
Total	155,400,000[a]

a. This differs in some respects from the recommendations of the Survey Mission both in total expenditure as well as in the allotments for the various items. See International Bank, *Iraq Mission Report,* 73.

As far as the financing was concerned, the Development Board had no difficulty. As mentioned in the previous chapter, oil production in 1952 and 1953 ran ahead of schedule, and the Board's share of revenue in those two years alone was about ID 30,000,000.

The general progress of the Development Board under the existing conditions was tolerably satisfactory, but the long-range objectives and the conditions necessary for their realization were not appreciated either by the people or by the politicians. In 1953 the Development Board Law was amended and a new Ministry of Development was created. Though the Development Board was retained, the execution of the development plan was placed under both the Board and the Ministry. The Minister of Development became an additional *ex officio* member of the Board, responsible to Parliament for the conduct of his Ministry and of the Board. [24] The original intent of the law to remove the Board from the vagaries of political change was thus negated. Moreover, because of the general educational and technological level of the population, the Development Board found itself in the unusual, if not unique, position of not being able to spend the constantly increasing revenue. [25] The Board consequently resorted to lending its funds to other Government agencies not concerned with long-range development, and a significant portion of its income was diverted to other regular expenses. [26]

Despite inherent limitations and basic difficulties, the awareness of the people and the Government that the oil revenue could be used as a means of advancing the development of the country had had some positive and some negative effects on Iraq's growth. One might note a general tendency to depend exclusively on oil income, to the exclusion of all other resources of governmental and economic operations. For obvious reasons the Government adopted a policy of reducing taxes [27] in view of the gigantic income from oil. [28] Such a policy, combined with enormous spending, brought in its wake a host of problems and dangers: inflation, a great strain on a primitive economy, shortages of human resources (civil servants, teachers, doctors, engineers), and the overburdening and eventual breakdown of transportation and communications. Nevertheless, progress had been registered. Many irrigation and drainage projects were undertaken, and some were completed; a beginning of agrarian reform and rehabilitation through such projects as Diyala was made. [29] In 1952 the United Nations reported that a slight expansion in industry had been achieved and a basis for more rapid development laid. Capital investments in major industries rose from a cumulative total of ID 3.18 million in 1948–1949 to 5.84 million in 1951–1952; the value of output rose from ID 1.4 million to 4.4 million, and the number of workers increased from 1,650 to 4,550. [30]

In its report for the fiscal years 1952–1953 the Development Board stated that considerable progress had been made on some of the construction and irrigation projects although it had not used up its revenues and had granted ID 6.3 million to municipalities and Government departments to carry out new programs. [31]

The impact of the oil industry on Iraq was not as great as it could have been. In spite of Government effort, and in spite of the belief held by some Government leaders that it was an unconditional, integral part of the overall development program, industry made very little progress. There had been some advances in agriculture, but not as much as was hoped for. There existed a serious danger that after the irrigation projects had been completed and the land available for cultivation had increased, unless the Iraqi peasant was educated — psychologically as well as technically — to utilize the additional acreage effectively and constructively and to keep the irrigation machinery in working order, the enormous expenditures would be wasted. Nor were the new hospitals and the new school buildings able to fulfill their proper functions, for there were not enough trained doctors, nurses and teachers with a social consciousness and a deep-seated and genuine desire to help the impoverished, uneducated masses. [32] The investment in education would have been a complete waste if children continued to drop out of schools in great numbers as they progressed to the higher grades in the primary schools.

The consumption of petroleum and petroleum products, which cost less in Iraq than anywhere in the world, according to the International Bank Mission, did not indicate that the Iraqis had taken advantage of the opportunities presented by cheap fuel for the general development of their country. The difficulties which the Development Board experienced in getting ahead with its work, the political interference with its efforts, and the inability to spend its income on long-range approved projects were indications of the inherent obstacles to transforming a "backward, semi-feudal state into a prosperous, progressive modern nation." [33]

It might therefore be concluded that with the basic conditions of backwardness, illiteracy, inadequate health conditions, lack of educational (vocational and professional) facilities, a country — even when endowed with a plentiful supply of land, abundant water resources, cheap fuel and tremendous financial resources — cannot by its own efforts utilize these advantages to rehabilitate itself at a pace which seems warranted by the favorable conditions. The will to achieve realization of the development plans was no doubt there, at least on the part of the leadership; the intentions were no doubt good; the various plans and projects prepared by foreign experts would no doubt have been operative under normal conditions. But for any of these plans to succeed two basic conditions were prerequisite. The masses for whom the projects were intended would have had to cooperate fully with the authorities; they would have had to understand the objective, they would have had to be prepared to work diligently for its achievement, and they would have had to be prepared to wait for its maturity; above all, they would have had to identify themselves and their effort as part of the plan and its realization. Those in charge of the execution had to have not only good intentions and willingness to work with foreign experts both on the preparation and execution levels, but had to possess deep understanding and have faith in the social and national objective.

Unfortunately, the Iraqis were not prepared for long-range development projects; neither were they educated to understand them. Nor had the leadership — the implementers of such projects — displayed the qualities necessary to convince and educate the masses not only of the good to be derived but of the difficulties to be overcome. Nevertheless, the oil industry and the revenue derived from it did one thing for the Iraqis. It made them realize the possibilities of improving their conditions; it made them lose their fatalistic acceptance of their lot and created dissatisfaction with their situation as it was. The oil industry also opened up to some Iraqis, even though to a very limited extent, the world of modern Western technology which was based on petroleum. The result was a stirring, a disturbance, in the social, cultural and economic patterns of the country. The fabulous revenue from the oil could direct this discontent, and in some instances even impatience, into constructive and coordinated channels.

THE COMPANY'S POLICIES

Table XVI shows the income contributed to the national budget from oil royalties up to the establishment of the Development Board.

In addition to these direct payments, the Company contributed in wages, salaries, social and health services, and building and administrative expenses which helped considerably not only in the country's direct economic expansion but also in meeting its foreign currency needs. [34]

The main bones of contention between the Company and the Government were the royalties paid to Iraq and the employment of Iraqis in the industry. Unlike the Anglo-Iranian Oil Company, the Iraq Petroleum Company paid its royalties on a per ton production basis. At first the Khanaqin concession in the "transferred territories," which was part of the original Anglo-Iranian concession, paid on a percentage basis, as in the rest of its concession; later, however, this was converted to a per ton basis. This arrangement made the Iraqis constantly demand increased production so that their income would be larger; they argued that the Company was not exploiting all the oil-bearing possibilities and was even pursuing a policy of limited production. In view of changed world conditions following World War II, the Company was willing to increase the per ton royalty and to increase production with the new, added pipelines. When the Arabian American Oil Company introduced the fifty-fifty profit basis arrangement in Saudi Arabia, and the agitation for nationalization in Iran was growing in volume, Iraq also obtained a fifty-fifty profit-sharing arrangement. [35]

THE COMPANY'S EMPLOYMENT PRACTICES

The question of the employment of Iraqis in the industry was next in importance. Because IPC did not build its refineries in Iraq, as did the AIOC in Abadan, the number of Iraqis employed in the oil industry was considerably smaller than Iranians in Iran. At the end of 1947 it employed a total of 14,556 workers, of whom 13,900 were Iraqis. [36] The number was not much greater in 1951, and, as was inevitable, the higher positions, both technical and administrative, were filled by non-Iraqis. [37]

In 1951 the IPC established at Kirkuk an industrial training center with a five-year course for clerical and technical personnel for the Company's needs. The

Table XVI. *Iraq Royalty Payments*

Year	Payments (in ID)	Per Cent of Revenue (from all sources)
1927/28 –		
1933–34 (average)	644,691	–
1934/35	1,019,304	14.1
1935/36	598,202	20.2
1936/37	599,968	9.9
1937/38	730,731	10.5
1938/39	1,977,458	25.2
1939/40	2,014,088	21.9
1940/41	1,575,915	16.0
1941/42	1,463,370	14.4
1942/43	1,463,371	10.6
1943/44	1,794,245	9.9
1944/45	2,132,405	11.3
1945/46	2,315,599	11.4
1946/47	2,326,968	9.3
1947/48	2,346,280	9.3
1948	2,012,000	9.0
1949	4,388,000	12.5
1950	5,285,000	16.6[a]

a. Both the amounts and the percentages are official figures of the Government of Iraq; the years 1927/28–47/48 are from the Government of Iraq, Ministry of Economics, Principal Bureau of Statistics, *Statistical Abstract 1939, 1940, 1941, 1942, 1943, 1944–1945, 1946, 1947,* and *1948;* the years 1948–1950 are from Directorate of Statistics and Reserach, National Bank of Iraq, *The Annual Report of the National Bank of Iraq* (Baghdad, 1951), 52. There are a number of discrepancies between the amounts, particularly in the percentages as given in the different *Abstracts,* and between the *Abstracts* and the *Bank Report;* moreover, there are also differences between those in *The States-man's Year-Book* for the years covered and other sources. However, the general picture both as to the amounts received by the Government and the percentage in the annual income – an average of 13.5 per cent over the entire period – is basically correct and shows graphically the role of oil royalties in the economy of the country. See also UN, *Review 1949–50,* 63; – *Public Finance Iraq,* 21, 41.

training was closely integrated with practical work in the Company's plants. During the course the apprentices were paid a monthly wage which increased progressively from ID 10 on entry to ID 21 during the last year. [38] In the same year the Company undertook to send fifty students to England to continue their studies both in the technical and scientific fields. The maximum number of such students in Britain at any given time was to be two hundred and fifty. Before leaving, the students were obliged to pledge themselves to work either for the Iraq Government or for the Company for twice the length of time they spent on their studies abroad. [39] This training, together with the development of the Government refineries, introduced more Iraqis into the higher positions of the oil industry.

While there could be no doubt that a number of company-trained employees had served in other parts of the industrial sector of Iraq's economy, the overall effect could not possibly have been too great. The physical conditions of the industry and the general historical background of the concession — so different from those in Iran — made Company employment a much greater attraction and carried with it greater permanency, so that local industry could not bid for the trainees. Nevertheless, the Company's great need was for skilled, trained technical and administrative workers, which its own efforts failed to supply. [40]

Although not to the same extent as the Anglo-Iranian Oil Company, IPC had provided its employees with medical services and social amenities. In 1952 the Company had three modern hospitals, eleven clinics and twenty-three dispensaries in the Kirkuk area; one hospital, two clinics and three dispensaries in the Basrah area. [41]

Since the Company was from the very beginning composed of four different national groups and was only a producer of crude, its profits were kept at a minimal level. The prices charged to the partners were arbitrarily low and had no relation whatsoever to prevailing prices. For taxes and other purposes IPC's profits were reduced merely to meet the tax liability to the British Government, and thus afforded those who refined and marketed the crude a maximum of profits. The very nature of the IPC precluded public financial reports, such as were put out by the Anglo-Iranian Oil Company. Nevertheless, the little data which was available about the Company's operations clearly indicated that the partners reaped very considerable profits. In February 1937, Standard of New Jersey, only a one-eighth holder of the concession, estimated that its share in the IPC was worth between $119 and $143 million exclusive of the value of the Basrah, Mosul and Qatar areas. The total investment of Standard in IPC to the end of 1939 was given as $13,940,000.

The profits to the partners from IPC crude were very high. Standard of New Jersey realized a profit of about 52¢ per barrel on its Iraq crude during the years 1934–39, more than twice the approximately 25¢ per barrel paid to the Iraq Government in 1938 for royalty and taxes combined. The total profits of Standard of New Jersey from the sale of Iraqi crude to the end of 1937 was $10,400,000, while total investment, as stated above, was $13,940,000 to the end of 1939. [42]

Multiplying these profits and the increased values of investment by eight, and adding Gulbenkian's five per cent, the venture of the Iraq Petroleum Company becomes not less impressive than that of the Anglo-Iranian Oil Company. Though IPC was limited by its very nature to the pruduction of crude, its manifold activities in the "red line" area were quite imposing. [43]

NOTES

1. Tabulations of the returns of the first census, taken on Oct. 19, 1947, showed a population of 4,799,500. US Dept of Commerce, *International Reference Service*, Dec. 1948, 2.
2. Sir Ernest Dowson, *An Inquiry into Land Tenure and Related Questions. Proposals for the Initiative of Reform* (Letchworth, England, [1932]), 11.
3. Great Britain, Dept. of Overseas Trade, *Iraq Review of Commercial Conditions* (London, 1945), 3. (Henceforth cited Great Britain, *Iraq Review.*)
4. UN, *Review 1949–50*, 45.
5. Great Britain, Overseas Economic Survey, *Iraq* (London, 1949), 18.
6. International Bank, *Iraq Mission Report*, 8. The Dept. of Agriculture supplied the Mission with the following statistics: cultivable area about 48.1 million donums – 32.1 million in the irrigation zone and 16 million in the rainfall zone; under cultivation about 17.5 million donums; and actually under crops not more than 11.1 million: 7.6 in the irrigation zone and 3.5 million in the rainfall zone. *Ibid.*, 137. For an analysis of the agrarian problem of Iraq, especially the most aggravating one – the tenure aspect – see Gabriel Baer, "The Agragrian Problem of Iraq," *Middle Eastern Affairs*, III, 381–391, Dec., 1952.
7. International Bank, *Iraq Mission Report*, 126.
8. *Ibid.*, 49.
9. *Ibid.*, 53; Hashim Al Witry, *Health Services in Iraq* (1944), 8. In 1945 there were, according to a British report, 569 doctors registered in Iraq. Since 308 were in Baghdad, this made approximately one doctor for every 15,400 in the population in the provinces and one doctor to every 2,700 in Baghdad. Great Britain, Overseas Economic Survey, *Iraq*, 30–31.
10. International Bank, *Iraq Mission Report*, 62–63; "The Economic Development of Iraq," *The Times Review of Industry* (London), Sept., 1953, 93.
11. UN, *Review 1949–50*, 12; "The Economic Development of Iraq," *The Times Review of Industry*, Sept., 1953, 93.
12. US, Dept. of Commerce, *International Reference Service*, Dec., 1948, 3.
13. Great Britain, *Iraq*, 16.
14. International Bank, *Iraq Mission Report*, 2, 129. Another estimate gives the total number of industrial workers in 1950 as not more than 72,000, including those employed in the oil industry. "The Economic Development of Iraq," *The Times Review of Industry*, Sept., 1953, 93.
15. United Nations, Dept. of Economic Affairs, *Review of Economic Conditions in the Middle East 1951–52* (New York, 1953), 43. (Henceforth cited UN, *Review 1951–52.*)

16. UN, *Review 1951–52*, 62; – *Summary 1952–53*, 57; Anglo-Iranian Oil Co. Ltd., *Annual Report and Accounts for the Year ending 31st December 1952*, 20; – *Annual 1953*, 19.

17. United Nations, Dept. of Economic Affairs, *Public Finance Information Papers, Iraq* (New York, 1951), 21. (Henceforth cited UN, *Public Finance Iraq.*) For a further breakdown of the composition of income tax payers between 1944 and 1947, see *ibid.* The British report stated that in 1949 there were 24,000 income tax payers. Great Britain, *Iraq Review*, 22.

18. This was done, for instance, in 1931 and 1932. Ireland, *Iraq: A Study in Political Development*, 437; *Report to the Council of the League of Nations, 1932*, 30.

19. For a table of both ordinary and extraordinary budgets for the years 1938–1949, as well as a brief discussion of Iraqi finances, see United Nations, Conciliation Commission for Palestine, *Final Report of the United Nations Economic Survey Mission for the Middle East*, I (New York, 1949), 43–44; for the years 1949–1951, International Bank, *Iraq Mission Report*, 170–176. See also Alfred Michaelis, "Economic Recovery and Development of Iraq," *Middle Eastern Affairs*, III, 101–106, Apr., 1952.

20. In the prewar years when the ordinary budget showed a surplus, oil royalties provided "nearly 30% of the total revenue." Great Britain, *Iraq Review*, 5.

21. UN, *Review 1949–50*, 29; "The Economic Development of Iraq," *The Times Review of Industry*, Sept., 1953, 93.

22. United Nations, *Final Report of the Survey Mission*, I, 36.

23. International Bank, *Iraq Mission Report*, ix. The terms of reference agreed to were: "The Mission will be expected to undertake a general review of Iraq's economic potentialities and to submit recommendations designed to assist the Government of Iraq to formulate a long-term program for the further development of the country's productive resources." *Ibid.*

24. Fahim I. Qubain explained the reason for the reorganization of the Board. "Its budget, under the control of only a few men, exceeded the total regular governmental budget. This caused resentment and even consternation in some political circles." *The Reconstruction of Iraq: 1950–1957* (New York, 1958), 37.

25. In 1952–53, ID 11 million of the Board's allotment remained unspent. *The Manchester Guardian*, Oct. 6, 1953.

26. For a description of the various projects, see UN, *Final Report of the Economic Survey Mission*, I, 71–74; International Bank, *Iraq Mission Report*, 183–457; Norman Burns, "Development Projects in Iraq," *The Middle East Journal*, IV, 362–370, Summer, 1951; Fuad Jameel, "Development Projects and Economic Progress in Iraq," *Al Aswaq al Tijariya*, Apr. 3, 1954; Mohammed Zaki Abdul Kareem, "Beehive of Construction," *Al Aswaq al Tijariya*, June 12, 19, 26, July 3, 1954; *Foreign Commerce Weekly*, Dec. 13, 1954, 3.

27. In 1953 the Government lifted the consumption tax on fresh fruits and vegetables, increased the reductions in the real estate tax and announced substantial cuts in customs and excise taxes. Jameel, "Development Projects," *Al Aswaq al Tijariya*, Apr. 3, 1954.

28. The 1952–53 total Government budget amounted to $134,000,000, of which

about 42 per cent came from direct oil revenue (only 30 per cent of total income, the other 70 per cent being assigned to the Development Board). UN, *Summary 1952–53*, 60.

29. Burns, "Development Projects in Iraq," *The Middle East Journal*, V, 362–366, Summer, 1951.
30. UN, *Summary 1952–53*, 22.
31. *Al Zaman*, July 24, 1953.
32. Qubain stated: "One new hospital in the south, it is reported, remained unopened for two years after completion because no personnel could be found to man it." *Op. cit.*, 230.
33. "The Economic Development of Iraq," *The Times Review of Industry*, Sept., 1953, 93.
34. Economic Cooperation Administration Special Mission to the United Kingdom, *The Sterling Area. An American Analysis* (London, 1951), 408; UN, *Review 1951–52*, 99.
35. The basis of calculating the profit, since all the transactions between the partners in the Iraq Petroleum Co. were in crude, became a source of friction. In the 1952 agreement, in which specific provisions were made for calculating the profits, the Company had decided advantages. Late in 1953 Iraq, seconded by Saudi Arabia, began to demand changes in the arrangements. *The New York Times*, Sept. 13, 1954. Early in 1955 it was reported from Baghdad that an agreement in principle for a new price-fixing formula had been arrived at, and as a result the Government's oil revenue would increase by one-fifth. *The New York Times*, Jan. 12, 1955. The 1952 agreement provided that the profits be calculated by deducting production costs from the base selling price for crude oil in the Middle East. The companies were permitted to discount this base by about 17 per cent. The new formula would reduce this discount to only 2 per cent. On March 24, 1955 a new arrangement was agreed upon between the Government and the IPC which reduced the brokerage rates from 17 to 2 per cent. This increased the royalty for 1954 by ID 10.5 million, making a total of ID 58 million for the year. *Mid East Mirror*, 14, March 26, 1955.
36. International Labor Organization, *Manpower Problems, Vocational Training, and Employment Service* (Geneva, 1951). Regional Conference for the Near and Middle East, Teheran, 1951, 10. The total number employed at the end of Dec. 1948 is given by the UN as 14,241, of whom 13,463 were Iraqis. UN, *Review 1949–50*, 63.
37. The breakdown of locally engaged personnel at the end of 1951 is given as follows:

Company	Clerical and Supervisory	Other	Total
IPC	1,134	5,961	7,095
BPC	438	2,259	2,697
MPC	93	545	638
Total	1,665	8,765	10,430

Iraq Oil in 1951, 11. The figures given in *Iraq Oil in 1952*, 29, apparently include personnel employed by the IPC also in Syria, Lebanon, Jordan and Israel, as well as non-local personnel.

38. "Training Schemes That Will Benefit Iraqis," *Al Assr*, Sept. 18, 25, Oct. 2, 1954.

39. *Iraq Oil in 1953*, 5. The Company also made provision for a limited period — six to twelve months — of special training in Britain for Company employees sent by their respective Company departments. In 1953 twelve trainees were in Britain for such training. *Ibid.*

40. For a description of the benefits and advantages which the Company employees enjoyed, as compared with other industrial workers, see Great Britain, *Iraq Review*, 29–30.

41. *Iraq Oil in 1952*, 18, 24.

42. US, Federal Trade Commission, *International Petroleum Cartel*, 95–96.

43. For a full discussion of these activities, see *ibid.*, 84–89.

Chapter XIII
REVOLUTION AND PROGRESSIVE
NATIONALIZATION

The most outstanding characteristic of Iraq even by comparison with other major oil-producing Middle Eastern countries and which is a key to its entire history and development is its political and general instability. This was true when the country was under British administration; it was accentuated under the Hashimite monarchical regime and worsened under the republican regime, which was ushered in by the July 1958 revolution led by Abd al-Karim Qasim. The relations between the Iraq Petroleum Company and the Iraq Government were best when the country was closely guided by the British; they became fair when Iraq emerged independent at the end of World War II ruled by the Hashimites and rapidly deteriorated after Iraq became a republic.

Failing or refusing to recognize some of the basic weaknessess of their people and of themselves, Iraqi leaders have been frustrated in their inability to raise their country to the level of a modern prosperous state. At the same time they saw amidst their own helplessness foreign companies earning inordinate profits and growing ever more wealthy from a natural resource which really belonged to them. Moreover, having learned, through their training in Western countries, the new social, economic and political theories, they easily and quickly found the answer to their frustrations. The oil companies were the colonialists, the imperialists, the exploiters who, through the practice of international monopolies, were growing fat at the expense of the Iraqi people. The natural suspicion of foreigners, whom one envies and therefore despises, persuaded the masses to accept willingly the attitude of the leadership.

With the acquirement of general knowledge and the terminology of technology, economics, finance and administration, the leaders attacked the foreign companies in scientific and scholarly terms, which impressed first themselves, their followers and to some extent the outside world.

Yet the practical approach was to point up the paucity of the share of Iraq in the profits of the companies. The demands were, therefore, for higher royalties, for a higher percentage in the taxes, for reducing concession areas, for higher posted prices, for participation and ultimately for nationalization. Nationalization had a strong economic appeal as well as providing emotional fulfillment, especially after the companies had made so many concessions and were still making enormous profits.

But nationalization presented serious problems, and no national leader could completely ignore them. One was lack of technical know-how; the Iraqi leaders knew, in spite of all their boasting, that by themselves they could not, from a purely technological point of view, operate the industry, for which, of course, they blamed the foreign companies for not having trained and prepared them for the task. Another was lack of financial resources with which to operate the industry

267

and the necessary heavy capital investments for expanding and increasing production, transportation and storage and loading facilities for additional oil. In spite
of all the revenues which the Iraqis had received they did not possess the financial
means for taking over the industry. Finally, there was the lack of international
markets to dispose of the oil produced. The foreign companies controlled all three
basic factors of the industry. The Iraqis would have liked to see the companies
provide these three requisites, while they, the Iraqis, became the real masters of the
industry and reaped the fantastic profits therefrom.

Some of the stronger advocates of nationalization reasoned that since the
Western consuming countries were falling short of supplies, the Western companies,
in order to meet the demands of their markets, would — almost out of desperation
— supply Iraq with finances and know-how for the production of the crude, for
which they would become mere agents to sell at prices which Iraq would command.
Ultimately, of course, having gained the knowledge, amassed the huge profits, and
established themselves as the indispensable sources of oil for the Western countries,
they could dispense with the Western companies altogether or let them continue to
serve as middlemen between them and the consumers at a minimal commission.

The period under review, from the late 'fifties to the early 'seventies, was one
of constant deterioration in the relations between the Government and the Company, which culminated in June 1972 in the nationalization of the Iraq Petroleum
Company. The crisis which the Government and the Company are now facing,
though in a different form and under different circumstances, is the same which
Iran and the Anglo-Iranian Oil Company faced in 1951 after Iran nationalized its oil
industry. It would be, however, safe to assume that the outcome is not likely to be
similar to that of Iran. The solution to the Iraqi problem may hold the key to the
future of the entire Middle East oil industry.

Iraq has undergone a period of economic planned development, without clearcut objectives or achievements. Based as they were on the revenue from the oil, the
plans suffered from lack of continuity, lack of cooperation of the Iraqi people with
their operation, and inability to execute them.

PRODUCTION AND REVENUE

As shown in Table XVII production was above the minimum guaranteed in the
1952 agreement, but the closing of the Suez Canal by Egypt and the cutting of the
Iraq Petroleum Company's pipeline by Syria seriously reduced output during 1956
and 1957. The production curve rose again in 1958 and moved slowly upward until
1966—67. Again, the closing of the Suez Canal and the closing of the pipeline by
Syria reduced production in 1967 by almost 15% of the 1966 figure. After that,
production continued to rise until 1971, when it was 2½ times that of 1955.
Obviously, the growth was not as spectacular as that of Iran, and in fact fell short
of the production-rate increase of all the major producers in the region.

Iraq's oil revenue, as recorded in Table XVIII, is rather impressive; from $206
million in 1955 to $840 million in 1971, more than fourfold; but a comparison

Table XVII.   *Iraq Oil Production, 1955–1971*

Year	Long Tons	Barrels
1955	31,760,000	243,318,000
1956	30,603,000	233,421,000
1957	21,361,000	162,722,000
1958	34,800,000	270,537,000
1959	40,898,000	305,505,000
1960	46,534,000	355,752,000
1961	48,055,000	367,830,000
1962	48,215,000	366,551,000
1963	55,577,000	421,962,000
1964	60,350,000	456,814,000
1965	63,151,000	480,032,000
1966	66,679,000	506,605,000
1967	58,939,000	447,952,000
1968	72,631,000	549,332,000
1969	73,309,000	552,000,000
1970	76,841,108	564,163,000
1971	82,448,000[a]	618,407,000[b]

a. Iraq Petroleum Company, *Iraq Oil in 1960*, 18; – *1961; – 1962.*
   2; – 1963; – 1964; – 1965; – 1966; – 1967; – 1968, 5; – *1969;*
   – *1970,* 3; – *1971;* Republic of Iraq, *Oil & Minerals in Iraq*
   (Baghdad, 1971), 45–46.
b. *World Oil,* CXLVIII, 108, Feb. 15, 1959; – C L, 95, Feb. 15,
   1960; – CLII, 111, Feb. 15, 1961; – CLIV, 120, Feb. 15, 1962; –
   CLVI, 125, Feb. 15, 1963; – CLVIII, 137, Feb. 15, 1964; –
   CLXI, 169, Aug. 15, 1965; – CLXIII, 170, Aug. 15, 1966; –
   CLXV, 142, Aug. 15, 1967; – CLXVIII, 114, Feb. 15, 1969; –
   CLXXI, 170, Aug. 15, 1970; – CLXXII, 106, Feb. 15, 1971; –
   CLXXIV, 63, Feb. 15, 1972.

with Iran (Table IV) would clearly show that both absolute figures and percentage increase were much smaller, even after allowing for Iran's low basis of growth after nationalization.

RELATIONS WITH THE COMPANY

At first there were no major differences between the Government and the Company. On April 10, 1957, Baghdad reported that the Company had agreed to reduce the discount from two to one per cent.[1] At the time it meant an additional income of 500,000 Iraqi Dinars.

In March 1955, the Company turned over the asphalt refinery at Qaiyarah,

Table XVIII.   *Iraq Oil Revenue, 1955–1971*

Year	£	$
1955	73,824,000	206,000,000
1956	69,165,000	193,000,000
1957	51,523,000	144,000,000
1958	84,604,000	237,000,000
1959	86,819,000	252,000,000
1960	95,538,000	266,000,000
1961	95,094,000	266,000,000
1962	95,124,000	267,000,000
1963	110,045,000	325,000,000
1964	126,200,000	353,000,000
1965	131,346,000	375,000,000
1966	140,800,000	394,000,000
1967	131,700,000	361,000,000
1968	203,300,000	476,000,000
1969	195,600,000	484,000,000
1970	213,600,000	513,000,000
1971	349,700,000[a]	840,000,000[b]

a. Iraq Petroleum Company, *Iraq Oil in 1961; – 1962*, 2; *– 1963; – 1964; – 1965; – 1966; – 1967; – 1968*, 2; *– 1969; – 1970*, 6, 11, 13; *– 1971;* Republic of Iraq, *Oil & Minerals in Iraq*, 44–45.
b. United Nations, *Economic Developments in the Middle East, 1958–1959*, 77; – Economic and Social Office in Beirut, *Studies on Selected Problems in Various Countries in the Middle East, [1967]*, 67; – *Studies, 1971*, 108; *Petroleum Press Service*, XXXVII, 324, Sept. 1970; – XXXVIII, 326, Sept. 1971; – XXXIX, 322, Sept. 1972.

south of Mosul, with an annual capacity of 60,000 tons, to the Govenrment's Oil Refineries Administration.

As stated above, Iraq was seriously affected by the Suez Canal crisis of 1956–1957, for not only was the Canal closed on October 3, 1956 to the movement of Iraq's oil from the southern fields, but the willful cutting of the pipeline by Syria brought to a halt the flow of oil from the northern fields. This experience of Iraq went beyond the mere temporary stoppage of the oil flow. Iraq had anticipated greater revenue from increased oil production; this was dependent on additional transportation facilities. In line with the Government's demand for increased production, IPC announced in May 1955 that it planned to construct a new 24-inch pipeline. After the Suez crisis it became obvious that the Western oil companies would not build additional pipelines in hostile territory. Iraq's oil would henceforth, even at the existing levels of production, be demonstrably exposed to the mercies of the rulers of Egypt and Syria.

Iraq, therefore, apparently not without the assistance of the United States, attempted to join with Saudi Arabia in an oil bloc against Egypt. It was hoped that such a coalition would considerably reduce Gamal Abd an-Nasser's influence and prestige and to that extent lessen the danger to the oil. King Saud realized that Nasser of Egypt represented a serious danger to his oil and closed ranks with King Feisal against Syria, but he soon felt that he was not strong enough to resist the dynamism of Egypt's President and retreated. Before Iraq could fully recover from the Suez crisis, and before it could devise new means of challenging Nasser's hegemony, the latter succeeded in uniting Egypt with Syria, and was openly bidding for supreme command of all the Arabs. Iraq made a counter-move: it united with Jordan and then asked Saudi Arabia to join the new monarchical federation. Saudi Arabia did not respond.

The Arab Union (Iraq and Jordan) did not last very long; it was dissolved by the bloody coup d'état of July 14, 1958, led by Abd al-Karim Qasim. The Western powers believed at first that Qasim was doing Nasser's bidding and they were, therefore, ready to challenge Nasser's intentions in Iraq by landing troops in Lebanon and Jordon.[2]

While active in organizing an Arab united front against the Western companies, the Iraq revolutionary regime made it clear time and again during 1959 that it had no intention of nationalizing the oil industry,[3] and repeatedly assured the people of Iraq, as well as the Iraq Petroleum Company, and the world at large, that it was determined to facilitate the flow of oil to Western world markets. Two issues, however, became the subject of pressing negotiations between the Government and the companies. The Government strove to obtain the maximum under the provisions of the existing agreements in terms of revenue, training of Iraqi personnel, greater Iraqi participation in management, and the release of non-exploited areas. It also sought to increase production of oil both because of the drop in the price of oil and the urgent need for additional revenue for the regular budget and for the development program.

The companies responded positively. The Ministry of Economy announced at the end of April 1959 that IPC had promised to increase production, doubling it in three years.[4] At the end of 1958, the Iraq Government announced that the Basrah Petroleum Company had ceded its concessionary right in Iraqi territorial waters. At the end of 1959, Premier Qasim reported that as a result of the Government's efforts, 157 foreign technicians in the refineries and 50 foreign expert engineers in the distribution field had been liquidated, and he concluded: "As a result of the negotiations, we were able to secure a promise or a pledge from the Company to give up 90,000 square kilometers of its present concession areas. But we did not accept this, we want this Company to give up 60% of the present areas."

At the end of November 1958, the Minister of Economy informed the Iraqi people that in accordance with the terms of the agreement with the British Petroleum Company subsidiary, the Khanaqin Oil Company, the Government had terminated the Naft Khaneh Concession and had taken over the oil fields. On June 30, 1959, the Khanaqin Oil Company transferred to the Government the marketing operations which it had run as managing agent on behalf of the Board of Oil Affairs.[5]

EXPANSION AND CRISIS

In order to facilitate increased output the Company expanded its production, transportation, storage and loading facilities, both in the north and the south.[6]

While taking the initiative in organizing OPEC, and actively participating in the Second Arab Petroleum Congress, Iraq was pressing the Company for greater revenue, and for relinquishment of concession areas. Premier Qasim threatened that, as much as he did not want to, if the Company forced his hand, he would have to resort to nationalization. The immediate cause for the crisis were the cargo and port-loading fees for oil exported from the port of Basrah. The Government maintained that the port duties and export of oil fees from Fao were established late in 1951 at 184 fils per ton (1000 fils to the Iraqi Dinar). When the Basrah Petroleum Company began to export oil from Fao it proposed that the per ton duty be replaced by an annual lump sum of 100,000 Dinars on exportation of up to 8 million tons, and 12.5 fils for every additional ton. The Government made counter-proposals. Finally on March 24, 1955, an agreement was reached on an annual lump sum of 187,000 Dinars for up to 8 million tons, and a pro-rata basis for additional exports.

After the revolution, the Ports Authority ruled that the Company was not entitled to specific exemption from the regular port dues and proposed to raise the duties to the level imposed on other exports. The Government, therefore, announced at the end of June 1960 that the fees were to be 280 fils per ton. Under the lump sum arrangement the rate amounted to 23.4 fils per ton; the new imposition meant a twelvefold increase. The Company maintained that the 1955 agreement was in force; when, therefore, the Ports Authority demanded payment of the increased charge, the Company threatened to reduce production at the Basrah oil fields from twelve to three million tons a year. This was a direct challenge to the Qasim regime, and he warned the Company that if it would reduce production he would take proper action.

In spite of the rising tension, a face-saving compromise was worked out. The Company declared that it did not threaten to cut production, and paid the imposed duties under protest.[7]

While both sides agreed not to pursue, at the time, the port fees issue, the other issues were the cause of a continuing crisis. The Iraqi press, led by the Premier, called for the Iraqization of the Company. The Premier reiterated, however, that he would not use force against the Company. Late in 1960 Qasim himself joined the negotiations, but no progress was made, and they recessed at the end of the year.[8]

The crucial year in the relations between the Government and the Company was 1961; and the end of the year marked the downhill direction of developments which ended in the nationalization of IPC in June 1972. Both sides assumed uncompromising positions thinking that the prevailing conditions favored each side against the other. The Company felt that the glut in the international market and the drop in the price of oil would deter the Iraqis from taking precipitous action which might endanger their income from the oil revenue, especially when their economy was in straitened circumstances. The Iraqis, on their part, felt that the threat of nationalization, for which there was a strong sentiment among the Iraqi nationalists, the reduced influence of the Western nations in the Middle East and the consequent readiness of the Soviet Union to assist in oil and other economic problems would induce the Company to submit to their demands. There were also two different basic approaches to the major issue which determined their ultimate positions. The Company viewed its concessions in Iraq as a commercial under-taking, with legal contractual obligations which both sides were bound to recognize and honor. It was determined to employ the well-tried methods of negotiations based on its legal rights and on its experience as bargainer. Its representatives displayed an extraordinary amount of patience and they never tired of repeating, over and over again, their arguments. When circumstances demanded, they made practical compromises, even what they considered concessions, in order to arrive at agreed solutions.

The Iraqis, on the other hand, approached the issue from a strictly emotional-national point of view. To them, the foreign companies were exploiters of Iraqi resources, thieving imperialists who had managed, through trickery and pressure, and with the cooperation of mercenary traitors, to obtain extraordinary con-cessions and privileges which the revolutionary regime was honor-bound to revoke. The compromises and the concessions offered by the Company representatives were, in their eyes, clear proof of the correctness of their own attitude. The position of the Company seemed to have been consistent and logical, while the position of the Iraqi Government was erratic and heavily overlaid with emotional argumentation.[9]

The issues as outlined by the Oil Ministry on April 10, 1961 were: 1) calcu-lation of the cost of oil production; 2) method used in fixing prices; 3) abolishing the discount which the companies took from the posted prices; 4) appointment of Iraqi directors to the boards of directors of the companies and their active partici-pation on the boards; control of companies' expenditure by the Iraqi Government in a manner which would safeguard Iraq's interests; 5) gradual Iraqization of companies' posts; 6) companies' relinquishment of unexploited areas; 7) companies to relinquish surplus natural gas and not flare it; 8) Iraqi tankers to be employed in transporting the oil; 9) Iraqi participation in the companies' interest to the extent of no less than 20%; 10) increase in royalty rates; 11) royalties to be paid in con-vertible currency; and 12) "elimination of the injustice and harm which befell the Iraqi state as a result of the oppressive agreements and their vague provisions."

Five days prior to this announcement Premier Qasim told the Company to stop prospecting and drilling operations outside the areas which were actually under exploitation until a just agreement was reached between the two sides; he also warned them not to reduce production.

The best example of the differences of approach was that of the major issue of relinquishment of non-exploited territories. As mentioned above, at the end of 1959 the Premier reported that the Company offered to relinquish 90,000 square kilometres, but that the Iraq Government refused the offer and asked for a 60% relinquishment. However, when the Company agreed to the 60% figure, Iraq demanded 75%; when the Company acceded to that percentage, the Iraqi negotiators demanded the entire area with the exception of the sections under actual operation. Meanwhile, early in March the Government promulgated a new oil law which replaced all the old oil laws. It reorganized the Oil and Minerals Ministry, into which were merged the Oil Administration Council, the General Board of Oil Affairs, the Government Oil Refinery Administration, and the Petroleum Products Department. The new Ministry was to lay down and execute oil policy, and was to be responsible for the development of petroleum and natural gas resources and their exploitation, including all negotiations with private enterprise for concession agreements.[10]

After a year of bitter and highly emotional negotiations, the Iraq Government issued, on December 11, Law No. 80. It consisted of seven articles, which defined the areas allocated to each of the petroleum companies operating in Iraq. It ordered the companies to submit, within three months, all geophysical and geological information on the territory covered by the concessions, as well as other engineering data. If they failed to supply such information they would be obliged to compensate the Iraq Government for any damages caused by their delay. The law restricted the Company to only 740 square miles (less than ½ of one per cent of the total original concession areas) and denied them sections where wells had already been drilled and oil found (North Rumaila field).

The Company protested against the law and called for arbitration in accordance with the provisions in its agreements.[11]

The impotence of the old foreign concessionary Company against the new national sovereignty of the oil-producing country was decisive. While Iraq could, if it wanted to, have nationalized the entire industry, it chose to let the companies continue with their operation and derive the revenue therefrom, but the possibility of nationalization loomed larger than ever.

The Company went through the motions of inviting the Iraq Government to arbitration, named a former president of the International Court of Justice as its nominee, and asked Iraq to appoint its arbitrator within thirty days. The British and United States governments sent diplomatic notes to Iraq supporting the Company's request — an act which was resented by Iraq; and the Premier described the two notes as a means of applying pressure "under the threat of military concentration and naval movements." Iraq completely ignored the Company request, while the Company complied with the provisions of Law No. 80 and declared that they were doing so "without prejudice, and reserving their legal rights." The rate of production was increased, and consequently the revenue.

The action of Iraq inevitably had repercussions in other Middle Eastern countries. Relinquishments of concession areas became almost the voluntary pattern for most if not all the concessionary companies.[12]

The relations between the Company and the Government could perhaps be described as that of a married couple who knew that they would be divorced, but until this were to come about life somehow had to go on. Yet the Company always put up a pose as if reconciliation was possible and probable, while the Government more and more played the role of the inevitable divorced member. Moreover, the instability of Iraq produced more changes.

Abd al-Karim Qasim had succeeded in retaining power for almost five years by maintaining a political balance between the communists and the nationalists; with the aid of the oil revenue and economic agreements with the Communist bloc countries, he met, even though precariously, his budgetary needs and held the unity and loyalty of the army. He managed to eliminate all his internal opponents and even resisted successfully the Egyptian President. His maintenance of power was, in Iraqi politics, of very high longevity.

In 1961, however, Qasim's position began to weaken; he could not control the Kurdish revolts, and he launched a campaign to annex Kuwait, claiming that it was an integral part of Iraq which he intended to restore to the fatherland. [13] Internally he could no longer balance the political forces, and, perhaps the most fatal blow, he lost the support of the army. He was doomed. On February 9, 1963, Qasim was assissinated, and his colleague in revolution and later mortal opponent Abd as-Salam Arif replaced him.

Arif was strongly pro-Nasser, and his cabinet was composed mostly of army officers. On July 14, 1964, the sixth anniversary of the revolution, Arif took steps to bring about the union of Iraq and the UAR, and some of the political institutions of the UAR were also decreed for Iraq. In addition, all banks and large businesses were nationalized. Some two years later, when the realities had made union with the UAR impractical, and the pro-Nasserite ministers resigned from the cabinet, Arif was killed or assassinated on April 13, 1966. He was succeeded by his brother Abd al-Rahman Arif. The new president could not cope, like his predecessors, with the political and economic problems of Iraq, and not being a very strong man was soon, after the Arab-Israel 1967 war, replaced in a coup d'état in July 1968 by Ahmad Hasan Bakr.

Right after the enactment of Law 80 the companies assumed the position of the wronged parties who hoped to regain their lost rights, while the Government assumed the position of the victor who regained some of the losses from the companies and was anxious to make the most of the new situation. For the greatest need of the Government was additional revenue, and the expropriated areas were useless unless they were exploited.

The change from Qasim to Arif seemed to have created, even though to a very limited extent, a better atmosphere in Government-Company relations. On April 1, 1963 a temporary agreement, to run for a year, was reached between the Government and the BPC for reducing the dues and port fees at Basrah and for an increase in production. After the agreement, production increased from 179,000 to 259,000 barrels a day. In the following year the agreement was extended for another year, for which the Basrah Petroleum Company committed itself to raise the daily production to 325,000 barrels. [14] However, long and difficult negotiations for royalty expensing which was successfully concluded between the companies and the other governments of the area bore no fruit in Iraq.

FROM CRISIS TO NATIONALIZATION

Ever since the enactment of the 1961 Law No. 80 the Company carried on discussions with the Iraq Government off and on about possible modifications of the law or some provisions thereof, but with little or no success. In the summer of 1965 the companies reported that an agreement had been reached between the Iraq Ministry of Oil and representatives of the Company. It called for the implementation of article 3 of the 1961 law which would have granted the companies additional areas for exploration equal to the areas which they actually exploited. A new company was to be formed in which the companies would have a 2/3 interest and INOC 1/3, for the purpose of exploring and developing the areas chosen by common consent. It also called for the settlement of current disputes about existing agreements. The major disputed issues were: marketing allowances and royalty expensing.

As a token of goodwill, the companies raised the total daily production in all Iraqi fields from 1,225,000 to 1,315,000 barrels. [15] Even though the Iraq Government pledged on August 20, 1965 to make public the oil agreement initiated the previous July, which would have meant at least a delay if not outright rejection, President Arif stated on November 5, publicly, that a law would be issued that would exempt oil and other large companies from the nationalization decree. The Government requested the companies to boost output by 10% annually. But by the end of the year no progress had been made. Early in 1966 a delegation of IPC headed by Geoffrey H. Herridge, Chairman of the Board, arrived in Baghdad for continued negotiations. In the meantime Abd as-Salam Arif was killed and his brother Abd al-Rahman who succeeded him announced on April 29 that the draft oil agreement initialed by the two sides would be publicly debated.

While the Company was awaiting the faintly hoped for modification of the 1961 law, through the ratification of the draft agreement, a new difficulty developed between the Company and the Government of Syria which further intensified the relations between the Company and the Iraq Government. Syria demanded higher transit rates and port-loading fees, and payments for back-dated modifications in the agreement. Negotiations made no progress and the Government of Syria seized IPC's assets. On December 12, Syria stopped the flow of oil, and the export of Iraqi oil was interrupted at the Banias and Tripoli terminals. [16]

This meant reduced revenue for Iraq, and on December 9 the Iraqi Premier Najib Talib told the Company that Syria's seizure of IPC's assets "must not cause interruption of the flow of oil to the Mediterranean." Three days later the Government announced that it would demand oil royalty payments if the oil was interrupted by the Syrian—IPC dispute.

The year 1967 further plagued the relations between the Government and the Company. Agreement with Syria was not reached until March 2, when the flow of oil to the Mediterranean ports was resumed. To meet Iraq's needs, IPC and MPC agreed to advance the Iraq Government £13.5 million as an interest-free loan on future production: £5.9 million to be repaid from the 1967 revenues and the balance of £8 million to be repaid after the end of 1968. [17] Before the Company could resume full operations under the new agreements with Syria and Lebanon the entire oil operation came to a halt on June 6 on the outbreak of hostilities between Israel and Egypt, Jordan and Syria. The flow of oil only for the needs of the Homs refinery in Syria was resumed on June 12, and for the needs of the Tripoli refinery in Lebanon on the 18th. Between June 27 and September 5, Iraq progressively removed its ban on exports, but by the end of the year not all the restrictions imposed by the Government of Syria had been removed. [18]

The Government was apparently more determined than ever that the foreign companies never expand their areas, and on August 6, 1967 enacted Law 97 which assigned to INOC the exclusive right to develop the expropriated IPC areas; it prohibited the granting of any oil concession. It was permitted to operate the oil fields in association with foreign companies, if a law were passed approving such operations. [19]

The closure of the Suez Canal, which made the haul of oil from the Persian Gulf by tanker around the Cape of Good Hope more expensive, had given an advantage to oil shipped from the Mediterranean ports. Libya, Saudi Arabia and Iraq, therefore, demanded a premium for their oil. In 1968 agreement was reached between the Iraq Petroleum Company and the Mosul Petroleum Company and between the Government of Iraq to pay an extra 7 cents a barrel for oil shipped after May 31, 1968 as long as, and to the extent which, the extraordinary circumstances prevailed. In settlement of the shipments from June 1967 to the end of May 1968 the companies paid Iraq a lump sum of £10 million. [20]

One year later, on June 30, 1969, the Iraq Ports Administration concluded an agreement with the Basrah Petroleum Company, which provided for a permanent revised scale of cargo dues to be paid by the Company on crude produced as of April 1, 1969. A lump sum of ID 3.2 million was paid to the Ports Administration in full and final payment of all outstanding claims.

Although there were indications in September that Iraq had agreed to the Company's royalty expensing formula which the Minister of Oil Aziz Al-Wattari calculated would, over a period of years, amount to a total Government share in profits of 56 to 57%, the formula was ultimately rejected. Expensing of royalties was the continuous item of discussion between the Government and the Company. In 1969 the Government claimed a sum of ID 62.5 million in back payments. However, in 1970 the Company refused to discuss the issue of royalty expensing in isolation from all other issues under dispute. [21]

Yet early in 1971 the Company, at least publicly, gave the impression that relations with the Government had improved. The Basrah port-fees issue was settled; the oil exported from the Mediterranean ports was given a special premium, which boosted the Government revenue; and after a lapse of ten years, the Government appointed members to the boards of directors of the companies. [22]

The negotiations carried on in 1970 between OPEC members and the Middle East concessionaire companies produced the Teheran agreement of February 14, 1971, which gave Iraq an increase of about 7 cents a barrel for oil produced in the southern fields and an increase of 5% in the profits share. On June 7, 1971 the Eastern Mediterranean agreement granted Iraq an increase of 6 cents a barrel as well as a 5% increase in the profits share. The Company also reported that the conclusion of the Actual Costs Agreement, "settled finally a number of long outstanding disputes on costs between the Governmene and the companies." Apparently at the request of the Government a 4-year £10 million loan was granted the Government. [23] Government revenue increased from £213.8 million in 1970 to £347.7 in 1971. Yet on December 2, 1970 Dr. Saadoun Hamady, Minister of Oil and Minerals, announced an amendment to Law 80 of 1961 which eliminated article 3 of that law in order to make it absolutely impossible for the Company ever to gain additional areas. [24]

Early in 1972 Iraq participated together with the other Persian Gulf oil-producing countries in negotiations with 16 Western companies on the issue of the devaluation of the dollar, and on January 20 signed an agreement for an immediate increase of 8.49% in posted prices to compensate for the devaluation. This covered both exports from the Persian Gulf and from Mediterranean ports. [25]

In line with the recommendations of OPEC Iraq requested from the Company 20% participation; she also sought mandatory increases in production of 10% annually, 17% from Mediterranean ports; 20% of the seats on the boards of directors for Iraqis; and removal of the Company's head office from London to Baghdad. The Company, on its part, requested arbitration of Iraq's expropriation of the companies' concession areas, including the North Rumaila field. The Government countered by demanding retroactive royalty payment increases amounting to ID 95.6 million. [26]

On May 17, the Government had given the Company a two-week ultimatum, either to meet the Government demands or face legislation. On May 31, Geoffrey Stockwell, Chairman of the Board of the Company, arrived in Baghdad with the final offer; the offer was rejected. These developments represented cause and effect and the resultant crisis. In spite of all the improvements in posted prices, the increase in the profit share, both in the southern and northern fields, and the extra premium for northern oil exports, the Government demands were greater than ever. Moreover, the increase in general world demand for oil, especially in Western Europe, was not as high as anticipated, and as a result there was a substantial drop in tanker rates. This made, according to IPC, the oil from the northern fields non-competitive. As indicated above, the output of IPC dropped while the output of BPC increased to some extent during 1971 and early 1972. While the Company maintained that the reduction of production in the northern fields was dictated by purely economic considerations, the Government was outraged and charged that the action of the Company was an effort to force the Government to return the North Rumaila field. On June 1, President Ahmad Hassan Bakr, in a nationwide radio and television broadcast, announced the decision of the Revolutionary Command Council to nationalize the Iraq Petroleum Company. Neither the Mosul Petroleum Company nor the Basrah Petroleum Company was included in the nationalization act. However, Syria subsequently nationalized all assets of IPC in her territory, covering all the pipelines and terminals; this must have also affected the Mosul Petroleum Company.[27]

On June 9, at an extraordinary conference in Beirut, OPEC passed a resolution supporting Iraq's action and asked members not to allow companies to make up the loss in Iraqi output by increasing production. The Organization of Arab Petroleum Exporting Countries, OAPEC, met in Baghdad and resolved to support Iraq, and to make available £53 million to Iraq and £7.7 million to Syria to tide them over in the face of the loss of revenue.[28]

INOC OPERATIONS

Parallel to the development of the oil industry under the IPC, Iraq experienced — along the lines of Iran — the emergence of a native national oil industrial concern determined ultimately to take over and replace the foreign companies.

In February 1964, the Iraq Government enacted Law No. 11, which provided for the establishment of the Iraq National Oil Company. It was empowered to operate all aspects of the oil industry except refining and distribution within Iraq, both of which were already being performed by Iraq Government agencies. The new Company had an initial capital of ID 25 million, all furnished by the Government. The Company was to pay the Government 50% of its annual profits. The Board of Directors consisted of 9 members, 3 senior officials and 6 full-time executive members. INOC was attached to the Oil Ministry and was obligated to carry out Government policy.[29]

The scope of activities of INOC was limited as compared to that of Iran's NIOC. As mentioned above, Iraq established a Ministry of Oil and Minerals as early as 1959, and in March 1961 the Ministry was reorganized to encompass all oil activities of the country; it remains until this day the outstanding Government agency dealing with Iraq's oil industry. In 1967 and 1968 INOC's scope was expanded to cover the entire territory of the areas expropriated from the IPC companies, including the North Rumaila field. Unlike NIOC, the Iraqi National Oil Company was forbidden to grant concessions, or even to enter joint-venture partnerships.

It took over at first the production operations at Naft Khaneh and at Qaiyarah fields, neither of which were of great magnitude.

At the end of 1967 INOC concluded a contractual-service agreement with the French state company ERAP. The Company was to explore for oil in four different areas where oil had not been proven previously, totaling 8,520 square kilometers on shore and 2,280 square kilometers offshore. The area was to be reduced by 50% at the end of the third year, and by a further 25%[30] at the end of the fifth year, and after the sixth year reduced to the proven area.

ERAP was to finance exploration; if no oil were found Iraq had no obligation to the Company. If oil were discovered, the exploration costs were to be considered an interest-free loan to be repaid by INOC at one-fifteenth of total yearly production or 10 cents a barrel.

Drilling and appraisal costs were to be considered exploration outlay if wells were dry and as part of development expenditure if they were productive. Financing of development, transport and export facilities was to be advanced by ERAP through loans bearing annual interest of not more than 6%, to be repayed within five years from the first shipment of oil. ERAP was to pay a bonus of $2 million on commercial discovery of oil, further bonuses of $2 million each after two, four, six and eight years, and $5 million after 10 years. These bonuses were non-amortizeable. The agreement was for 20 years from the first oil loading. Five years after the beginning of commercial production, provided the development loans had been repaid, INOC was to take over management of all operations, with the cooperation of ERAP.

All oil produced was to belong to INOC, but ERAP was to have the right to purchase 30% of production on certain favorable terms. When daily production reached 75,000 barrels, INOC was to set aside 50% of recoverable oil as national reserves excluded from the development and purchasing arrangements. If requested, ERAP was to help with marketing INOC's 70% share of crude production.

Royalty was to be 13.5% of posted prices and expensed, and ERAP was to be subject to income tax. In marketing INOC oil, ERAP was to be granted a ½ cent commission per barrel for the first 100,000 barrels produced daily and 1.5 cents per barrel for the additional 100,000 barrels of daily production.[31]

NORTH RUMAILA FIELD

The major decision of INOC after the enactment of Law 97 of 1967 was the disposal of the North Rumaila field. It needed development, transportation, terminal and loading facilities, and above all marketing outlets. Should it undertake the operation of the field itself, or should it assign the field to a foreign company?

The French Compagnie Française des Pétroles, though a member of the IPC group with about ¼ interest, submitted to the Iraqi authorities "certain proposals" regarding the development of the field. However, by the end of 1967 the Iraq Government had not yet decided the fate of the field, and in the following year the French Company reported that the Iraq Government had rejected its proposals and turned over the field to INOC.[32]

Not without the encouragement and direct assistance of the Soviet Union, INOC announced early in 1968 that a three-year plan had been worked out to put the field into production of five million tons annually. To obtain the necessary assistance and marketing facilities the Company and the Iraq Government concluded a number of agreements, with what were described by the Iraqis as "friendly countries," to supply Iraq with financial aid, machinery, equipment and technical expertise. The first agreement was with the Soviet Machinoexport for the supply of equipment, materials and technical aid in the amount of $72 million. This was followed by an Iraqi-Soviet Agreement on Economic and Technical Cooperation, signed in Moscow on July 4, 1968. About one year later INOC signed an agreement with the Hungarian Chemokomplex to assist INOC with the building of the petroleum industry.[33]

In the meantime INOC was looking for marketing outlets for its crude oil to be produced at North Rumaila field. In October 1971, Iraq concluded a Technical and Economic Cooperation accord with Rumania which extended a $55 million loan to Iraq at an annual interest rate of 2.5% repayable in crude oil deliveries.[34] In addition a number of barter agreements were signed with East Germany, Czechoslovakia, Italy, Spain and Ceylon. Late in 1971 INOC concluded an agreement with India for the purchase of £5 million worth of crude oil up to 1975.[35]

With the help of the technical and financial assistance of the Communist countries, collecting and pumping stations were erected at the field, a pipeline was laid and the terminal loading facilities at Fao were expanded to accommodate tankers up to 30,000 tons.

On April 7, 1972, the commissioning ceremonies of the operation of the North Rumaila field took place in the presence of the President of the Republic and the Prime Minister of the Soviet Union Alexei Kosygin. Regular deliveries to the Communist bloc countries from Fao terminal began on April 25, 1972, with the loading of two Soviet tankers which left for Rostock in East Germany.[36]

NATURAL GAS

The other major area of INOC's efforts was that of utilization of natural gas. It would seem that the increased utilization of natural gas was one of the characteristics of the coming of age of the oil-producing countries in their relations with the concessionary companies, and with the growth in the native development of the industry.

In Iraq, as in Iran, natural gas exists as associated gas in the crude oil produced, and in gas fields which although already discovered have not as yet been exploited. At first some of the gas was utilized by the companies for the pumping stations, for injecting water in the oil wells to keep up the pressure and for generating power. Gas was utilized, subsequently, for fueling the electric power stations of the Iraq Government; then came the development of liquefaction of the gas for local consumption, the production of sulphur, and the growth of petrochemicals and possible export of dry or liquefied gas.

In 1960 the Basrah Petroleum Company laid a new 65-mile 30/32-inch pipeline from Zubair to Fao. With the completion of the line in September, the old 12/16-inch oil pipeline was reconditioned to carry gas from Zubair to Fao where it was utilized in fueling the Company's pumping stations.[37]

In the following year gas turbines were installed in the pumping stations of the main pipeline. The gas was taken from the Bai Hassan degassing station, dehydrated and, through a 16-inch pipeline, carried to the different stations. Gas turbines were also installed in the Company's Dibbis water injection station. Natural gas was also utilized in the Kirkuk process plant, in the Company power plant, and in the Government Dibbis Power Station.[38]

In 1964 IPC utilized 9 billion cubic feet of natural gas from its oil fields for the Dibbis Power Station and for its own uses. BPC utilized 4 billion cubic feet from Rumaila and Zubair fields for the Government Najibiya Power Station at Muftiyah through a 14-inch 24-kilometer pipeline from Rumaila, and for Company needs.[39]

Three years later, IPC increased its utilization of natural gas to 13.236 billion cubic feet; BPC supplied 1.66 billion cubic feet to the Najibiya Power Station, and utilized for its own needs 2.857 billion; MPC utilized 676 million cubic feet from its Ain Zalah and Butmah fields for Company operations.[40]

In 1971 IPC supplied the Dibbis Power Station and the Government sulphur recovery plant at Kirkuk with 6.954 billion cubic feet of natural gas, and utilized for its own needs 17.984 billion cubic feet. BPC supplied the Najibiyah Power Station 3.403 billion cubic feet, and utilized for its own needs 2.347 billion; MPC utilized 337 million cubic feet of natural gas.[41]

With all this growth in natural gas utilization, the Iraq Government calculated that out of a total of 595.75 million cubic feet produced daily in the seven oil fields — excluding Naft Khaneh — only 73.50 million or 13% was utilized; the rest was flared.

Although, as mentioned above, no estimates for Iraq's natural gas reserves have been established, the Government asserted that gas from the oil fields and gas fields could supply Iraq for the next 60 years at a daily rate of one billion cubic feet.[42]

In 1958 the Daura refinery commenced commercial production of liquefied petroleum gas with an annual 44,000-ton capacity of propane and LPG. The Taji gas-processing plant, completed in 1968 at a cost of ID 3.4 million, had an annual capacity of 136,000 tons of LPG, 100,000 tons of propane and 74,000 tons of natural gasoline.

The distribution of LPG in Iraq was handled by LPG stations of the Oil Products Administration. 125,148 customers consumed 20,000 tons of LPG in 1969. By the end of 1970 the number of customers rose to 130,814. The growth in consumption of LPG in 11 years had been 80-fold, from 250 tons in 1959 to 20,000.[43]

The petrochemical industry in Iraq was in 1972 still in its infancy.

NOTES

1. See footnote 35, Chapter XII.
2. Following a meeting of Secretary of State John Foster Dulles and British Foreign Secretary Selwyn Lloyd with President Eisenhower in Washington, it was reported on July 17 from the American capital: "Intervention will not be extended to Iraq as long as the revolutionary Government in Iraq respects

Western oil interests." *The New York Times,* July 18, 1958.
3. On Apr. 9, the Minister of Economy, Ibrahim Kubbah, declared: "The nationalization of oil has never been discussed between the oil companies and the Iraqi Government. In fact, the Government is still clinging to the oil policy which it declared on the first day of the Iraqi revolution. This policy was based on keeping the oil agreements with the companies working in Iraq, provided that these agreements will not be amended without the acceptance of the aforementioned companies." For the original statement made by Premier Qasim, see *The Power Struggle in Iraq* (New York, 1960), 27n.
4. *The New York Times,* Apr. 17, 1959.
5. *Petroleum Press Service,* XXVI, 278, July, 1959; Radio Baghdad, Nov. 18, 1958; *The New York Times,* Mar. 15, June 30, 1959.
6. IPC, *Iraq Oil in 1961;–1963.* See the next chapter.
7. Having concessions both in the north and the south, production from the northern fields being shipped by pipeline to Mediterranean ports, and from the southern field through the Iraq port on the Persian Gulf, the Company tried to take advantage of the best possible terms of either. When it was costly to ship by way of the Persian Gulf, the Company increased production from the northern fields; when it became too costly to ship via the Mediterranean ports it increased production from the southern fields. Thus, while in 1960 the IPC area produced 33,841,000 tons and the Basrah Petroleum Company produced 11,396,000 tons, after the difficulties with the southern port fees IPC production rose to 36,994,000 tons in 1961 and BPC production dropped to 9,781,000. The reverse was the case in 1971. In 1970 IPC fields produced 56,893,000 tons and BPC produced 17,667,000 tons; in 1971 IPC had to pay an extra surcharge for Iraq oil lifted from the Mediterranean ports because of the advantage over Persian Gulf oil due to the closure of the Suez Canal. Tanker rates dropped and it became cheaper to ship from the Persian Gulf than to pay the extra surcharge. Production from IPC fields therefore dropped to 51,115,000 while the output from BPC fields rose to 30,074,000 tons. IPC, *Iraq Oil in 1960,* 8; *– 1961; – 1970,* 3, 9, 12; *– 1971.*
8. Out of desperation from not being able to make headway with the Company, Iraq made attempts to develop the oil resources without the Company, and invited outsiders to offer bids, but, at the time, no tangible results were achieved in that direction.
9. See above, Chapter V, 116–117, dealing with the same problem in the case of the Iranians and the Anglo-Iranian Oil Company.
10. *International Oilman,* XV, 24, Mar.–Apr., 1961.
11. In its statement of April 10, 1961, the Iraq Oil Ministry announced, to the great embarrassment of the Company representatives, that in spite of their secret nature it would broadcast the minutes of the negotiations-meetings over the radio. They were subsequently broadcast, and served as one of the sources for this period; another major source was IPC, *The Iraq Oil Negotiations* (London, 1962).
12. It was discussed above in the case of Iran with the Consortium, and in the agreements of all the joint-venture and contractual-service agreements. Relinquishments in the other areas will be dealt with in chapters below.

13. "The Kuwait Incident," *Middle Eastern Affairs*, XIII, 2–13, Jan., 43–53, Feb., 1962.
14. Compagnie Française des Pétroles, *Exercice, 1963*, 9; – *Annual Report, 1964*, 9; BP, *Annual Report, 1964*, 23, 25; IPC, *Iraq Oil in 1963*. Production from the Basrah fields in 1962 was 8,842,000 tons; it rose in 1963 to 12,676,000 and in 1963 jumped to 17,347,000 tons. IPC, *Iraq Oil in 1962*, 8; – *1963*; – *1964*.
15. CFP, *Annual Report, 1965*, 5; BP, *Annual Report, 1965*, 23.
16. BP, *Annual Report, 1966*, 25; CFP, *Exercice, 1966*, 2–3; IPC, *Review for 1966*. For details see below, Chapter XXIII.
17. IPC, *Review for 1967*.
18. IPC, *Review for 1967*; CFP, *Exercice, 1966*, 2–3; BP, *Annual Report, 1966*, 25; *World Oil*, CLXV, 142, Aug. 15, 1967. There was an increase in production in BPC and a decrease in IPC and they canceled each other out. But in 1967 there was an increase in BPC production of some 800,000 tons over 1966 and the drop in IPC was 10,000,000 tons from the 1966 level. IPC, *Review for 1965*; – *1966*; – *1967*.
19. *Petroleum Press Service*, XXXIV, 324, Sept., 1967.
20. IPC, *Review for 1968*, 2.
21. *World Petroleum Report*, XVI, 37, 1970; – XVII, 71, 1971.
22. IPC, *Review for 1970*, 1. The Company went as far as to state: "During the last few months of 1970, in discussions between the Government of Iraq and the three companies, we were able to make an encouraging start in resolving the differences that have been outstanding between us for too long."
23. IPC, *Review for 1971; Petroleum Press Service*, XXXVII, 442, Dec., 1970; *Petroleum Times*, LXXV, 6, July 2, 1971.
24. Republic of Iraq, *Oil & Minerals in Iraq*, 149. Article 3 read: "The Government of the Republic of Iraq may at its discretion allocate other lands as reserve to the companies, provided that such lands shall not exceed the extent of the defined area of each company." The full text of the Law is in *Middle Eastern Affairs*, XIII, 268, Nov., 1962.
25. *Petroleum Times*, LXXVI, 37, Feb. 11, 1972.
26. *World Oil*, CLXXIV, 7, May, 1972.
27. IPC, *Review for 1971; New York Times*, June 2, 3, 1972; *Petroleum Press Service*, XXXIX, 238–239, July, 1972.
28. *Petroleum Press Service*, XXXIX, 239, July, 1972; *New York Times*, June 11, 1972.
29. *World Oil*, CLIX, 157, Aug. 15, 1964; *Petroleum Press Service*, XXXI, 86–88, Mar., 1964; Republic of Iraq, *Oil & Minerals in Iraq*, 11.
30. In 1971 it was reported that ERAP returned 2,700 square miles of its original area. *World Oil*, CLXXV, 125, Aug. 15, 1972.
31. *Petroleum Press Service*, XXXIV, 464, Nov., 1967; – XXXV, 104–105, Mar., 1968; *World Oil*, CLXVII, 201–202, Aug. 15, 1968; – CLXIX, 188, Aug. 15, 1969. The agreement with ERAP was legitimized by the enactment of Law No. 5, 1968. The details were presented in order to appreciate, should this agreement bear fruit, the Government's attitude to the IPC conditions of operation, and its determination to convert the old concessions to the terms of the new arrangements.

32. CFP, *Annual Report, 1967, 7; − 1968, 11; Petroleum Press Service,* XXXV, 185, May, 1968. According to *The New York Times* the French were hopeful in 1967 to obtain the North Rumaila field for Mirage jets, for which the Iraqis were then shopping. However, when Iraq bought Russian Migs, the North Rumaila field went to the Russians. Jan. 22, 1970.

33. *World Petroleum Report,* XVII, 71, 1970; INOC, *Press Release* (1972); Republic of Iraq, *Oil & Minerals in Iraq,* 97. Said the Iraq Government report: "The Soviet side is obliged accordingly to prepare and operate North Rumaila field, and to construct a pipeline joining the field and the gas terminal." The Soviet Government offered a loan of 60 million rubles at an annual 2½% interest rate.

34. *Petroleum Press Service,* XXXVIII, 473, Dec., 1971; INOC, *Weekly News Bulletin,* 4−5, June 8, 1972.

35. *Petroleum Press Service,* XXXVIII, 434, Nov., 1971; Republic of Iraq, *Oil & Minerals in Iraq,* 97.

36. INOC, *Weekly News Bulletin,* 13, June 24, 1972.

37. IPC, *Iraq Oil in 1960, 12.*

38. IPC, *Iraq Oil in 1961.*

39. IPC, *Oil in Iraq 1964.*

40. IPC, *Review for 1967.*

41. IPC, *Review, 1971.*

42. Republic of Iraq, *Oil & Minerals in Iraq,* 47.

43. *Ibid.,* 53, 57.

Chapter XIV
DEVELOPMENT AND PROSPECTS

GENERAL DEVELOPMENT

Iraq's increase in "published proved" reserves was very modest; in 1962 it was 26 billion barrels, in 1969 it went up to 28.505 billion, and at the end of 1971 it was given as 35.990 billion barrels. Production increase was the smallest of all the major Middle East producing countries. While the average annual increase in production for the Middle East during the period 1966/1971 was 11.7%, with Abu Dhabi leading with 21% followed by Iran with 16.5%, Iraq was the lowest with an average of 4.1%.[1] Nor was there any substantial increase in the number of oil fields. By 1971 there was a total of 7 oil fields operated by the IPC companies and three operated by INOC; in all three oil was discovered by the companies, and INOC developed only one. IPC had three fields: Kirkuk, Bai Hasan and Jambur; BPC worked Rumaila and Zubair, and MPC produced from Ain Zalah and Butmah; INOC operated Naft Khaneh and Qaiyarah, which were turned over to it by the companies and North Rumaila. The total number of producing wells in 1970 was 284, and the daily total production was 1,565,000 barrels, making an average of some 5,510 barrels per well; this was a lower average than of any of the major Middle East producers.[2]

This abnormal rate of growth, the limited productivity and consequently the low rate of revenue were all the direct result of the difficult relations between the Government and the Company.

REFINING

After production the next step in the oil industry is refining. Unlike Iran, Iraq never had a big refinery which could supply products for export on a large scale. Even at this late date, most of the refineries' output is primarily for local consumption. The Iraq Government estimated in 1968 that only 6.7% of total production was exported as processed oil products.

As mentioned above the Oil Refineries Administration was organized in 1952; it was given the responsibility of refining crude, maintaining all Government refineries, and if necessary building new ones. The first refinery to come under its control was Alwand in Khanaqin. It remained the only supplier of oil products until 1953, when Muftiyah refinery in Basrah was constructed. Two years later it completed the Daura refinery near Baghdad. In March 1955 the Qaiyarah asphalt refinery was turned over to the Oil Refineries Administration.

In 1966 the Government agency took over the Haditha refinery from IPC; two other plants — Sulphur Recovery in Kirkuk and the Taji Petroleum Gasses near

Baghdad — were under its operation. In 1959 Daura refinery was operated entirely by Iraqi technicians.[3]

The crude for all these refineries was supplied by IPC, except Alwand, which was supplied by Naft Khaneh crude. In addition the Company supplied crude to the Syrian Government refinery at Homs, and to its own refinery in Tripoli, Lebanon.

In 1961 it was reported that the Iraq Government signed an agreement with Technoexport of Czechoslovakia to build an oil refinery at Basrah with an annual 1.3 million-ton capacity;[4] yet in 1971 it was reported that the Czechoslovak firm was building a refinery near Basrah with a daily 70,000-barrel throughput at a cost of ID 23 million, of which ID 10 ($28 million) was to be paid in crude oil. It was to be completed in 1972, and most of the processed products were to be exported. It was also stated that some products of the Haditha refinery were exported to Syria.[5]

All the refineries were constantly enlarged. In 1970 the Daura capacity was raised to 67,000 barrels per day, Alwand processing capacity was 12,500 barrels a day, Muftiyah's capacity was 4,280 barrels per day. The Qaiyarah refinery increased its throughput capacity to 23,000 barrels a day.[6]

In October 1957 a 25,000-ton annual capacity lubricating oil refinery was completed in Daura at a cost of ID 4 million.[7]

The quantities of crude delivered by the companies to the different refineries during the period 1962–1971 is presented in Table XIX.

Table XIX.   *Iraq Companies' Deliveries to Refineries, 1962–1971*

			Long Tons		
Year	*Daura*	*Muftiyah*	*Haditha*	*Homs*	*Tripoli*
1962	1,852,688	204,147		773,288	190,573
1963	1,879,884	161,192		886,609	321,805
1964	1,886,556	174,778		998,898	555,849
1965	2,050,000	180,000		1,000,000	740,000
1966	2,215,230	174,588	103,218	1,038,175	810,372
1967	2,386,962	167,407	94,518	1,051,358	889,888
1968	2,597,497	164,373	129,190	1,047,369	928,964
1969	2,809,310	196,348	124,142	1,304,930	1,058,930
1970	2,888,575	205,501	135,784	1,231,121	1,188,900
1971	3,123,941	204,569	194,000	703,301	1,206,853[a]

a. IPC, *Iraq Oil in 1962*, 4; – *1963;* – *1964;* – *1965;* – *1966;* – *1967;* – *1968;* – *1969;* – *1970*, 3, 10, 16; – *1971*. The Company reported that because of the closure of the pipelines by Syria during the first two months of 1967 some 205,988 tons of crude oil were delivered by tankers to the Tripoli refinery. It was not explained from where the tankers obtained the oil.

TRANSPORTATION & TERMINALS

Transportation of oil in Iraq differed basically from that of Iran, and as a result presented problems. For in addition to transporting oil from fields to refineries and terminal and loading stations within the country, Iraq oil had to be hauled from the oil fields in the north to distant ports on the Mediterranean through pipelines which traversed foreign lands.[8]

As the countries through whose territories the pipelines ran became independent and more assertive in their demands, both Iraq and IPC were subject to the danger of having the flow of the oil stopped through demands for higher transit rates and port-loading fees.

With the growth of production, additional pipelines and greater terminal facilities were needed. In 1961, therefore, a new 530-mile, 30/32-inch pipeline was completed from Kirkuk to Tripoli. In the same year a 24-inch, 60-mile branch line was laid to connect to the 30/32-inch pipeline at a point near Homs to increase the export capacity of the Banias terminal. The sea-lines at Tripoli terminal were extended and enlarged to handle larger tankers, and the tank-farm capacity was expanded.[9]

In the south, the BPC completed a new $61.5 million deep-water loading port known as Khor-al-Amaya terminal. It was built on steel legs standing in about 80 feet of water in the open sea of the Persian Gulf about 12 miles out from the mouth of the Shatt-al-Arab. It augmented facilities of the original terminal at Fao; it could accommodate simultaneously two large tankers and had a combined loading rate of 8,000 tons an hour. Oil was pumped from a tank-farm through two 32-inch concrete encased submarine pipelines. In 1971 plans were made for the expansion of the deepwater terminal.[10]

The story of IPC relations with, and payments to, Syria and Lebanon will be presented in Chapter XXIII. Here it should merely be pointed out that despite the considerably increased transit fees granted in the middle of 1971, Syria nationalized all the IPC pipelines in June 1972.[11]

LONG-RANGE DEVELOPMENT

The oil revenues, or a good portion thereof, have made a decisive contribution to the long-range economic development programs. While the increase in oil production, as indicated above, was not spectacular, and its rate was a cause for constant friction between Government and Company, the steadily mounting payments, as was seen in Table XVIII, were impressive. Nevertheless the question of whether the revenue was most effectively utilized for the economic growth of Iraq cannot categorically be answered positively.

Early in 1955, before the original five-year plan ran its course, the Development Board submitted a new six-year plan (April 1955–April 1961) which called for the expenditure of almost twice the amount of the first five-year plan. One year later the Board revised the cost still upward to a total of ID 500,000,000. The

Board divided its projects into major and long-range undertakings, to be under its own direction, and minor, short-range undertakings to be in the charge of the respective Ministries. This was an attempt to meet the pressure of immediate political necessities.[12]

Most of the projects were long-range national investments for the expansion of the agricultural potentialities of the country and were contracted by Western, mostly British, concerns.

The 1958 Qasim revolution brought to a sudden halt all the long-range economic projects which were managed by the Development Board. Most of the contracts with Western companies were canceled, and no new ones were signed. The reasons were: suspicion of the old regime's development program, suspicion of the Western companies that were implementing the projects, and disagreement with the long-range objectives of the development program as a whole. The cessation of the Development Ministry's activities created economic dislocation throughout the country, and the general drop in the price of oil beginning in February 1957 actually reduced the Government's oil revenue. Premier Qasim, intent, for ideological as well as political reasons, to concentrate his efforts on social services — housing, health and education — announced that his Government was devoting only 50% of the oil revenue to economic development, while the other 50% were assigned to the current Government budget.[13]

In 1959 the Development Board was abolished and replaced by an Economic Planning Board attached to the new Ministry of Planning and presided over by the Premier; the members were Ministers, and its functions were reduced to formulation of plans and follow-up; the execution of the projects was transferred to the various Ministries. Thus, the Board completely lost its autonomy and became a political instrumentality of the Government.

In November of that year Qasim was discussing a four-year development plan with an expenditure of ID 400,000,000; it stressed housing, public health and cultural projects, and in line with the spirit of the revolution allotted to industry a much larger share than in any of the previous schemes. The plan was to be financed by the oil receipts, Soviet Union credits, internal loans, private capital and, if necessary, currency reserves.[14]

In December 1959 a new four-year provisional economic development program (1959/60–1962/63) with an expenditure of ID 392.2 million was adopted. Two years later it was replaced by a permanent five-year plan (1961/62–1965/66). The total investment projected was ID 556.3 million. The highest priority was given to industry with an allotment of 29.4% of the total; agriculture was granted 20%, and most of the projects were related to irrigation and drainage, water storage for both irrigation and generation of electric power and financing agrarian reform.[15]

Despite the concentration on industrial development, which was the great hope of the Republican revolutionary regime, and despite the high allotment for it in the development plan, by March 1963 only ID 4.5 million was reported to have been invested out of a total of ID 39.1 million allotted.[16]

Even in the agricultural sector progress was very slow. The most outstanding aspect of social legislation was the Agragrian Reform Law of 1958. Enacted in September it limited land ownership to 1,000 donums (1 donum approximately ¼ acre) of irrigated land and 2,000 donums of rain-fed land. Expropriated land was to be paid for in Government bonds bearing 3% annual interest and redeemable within 20 years. The same terms, in reverse, were to apply to the purchases of the expro-priated land, after a 3-year grace period. Ten million donums of land were to be reformed and 2,253 landowners were subject to this reform. In addition 2 million donums of Government-owned land were to be parcelled out among landless far-mers. By the end of 1959, an area of 2.3 million donums had been appropriated from 256 landowners. It was estimated at the time that there were some 700,000 landless families in Iraq, and of that number only between 250 and 350,000 might be accommodated on expropriated land.[17] By the end of the third year of the enactment of the agrarian law, out of ten million donums expropriated only 3,524,000 donums had been distributed. The slowdown was caused by Government administrative inefficiency but more so by the social, psychological and technical unreadiness on the part of the landowners and the farmers for the new system. The net result was serious deficiencies and far-reaching dislocations in agricultural production.[18]

In the middle of 1963 a total of some 4 million donums was distributed. In 1968 it was reported that because of restrictions of the agrarian reform law of 1958 and the nationalization law of 1964 major changes in economic and social con-ditions had taken place in Iraq.[19]

Despite the agrarian reform law, the per capita income in Iraq did not advance rapidly; in 1965 it was only $210. The United Nations report noting the lack of productivity of Iraq agriculture stated: "The observed stagnation in Iraq's agri-cultural output should not . . . be attributed to the lack of capital formation in this sector. In fact, public investment in dams and irrigation schemes absorbed a con-siderable part of total investment in the public sector. Corresponding progress was not, however, achieved in meeting the human and organizational requirements of greater production."[20]

The political change in 1963 from Qasim to Arif terminated the five-year plan two years ahead of schedule. In July 1965 a new five-year economic plan (1965/66–1969/70) was adopted with a total expenditure of ID 561.2 million.[21] The 1968 report of the United Nations stated that in the last five-year plan, unlike all the previous plans, serious efforts were made to relate investment allocations to an analytical framework of well-defined objectives and criteria for investment allocations, and to gauge the effect of planned projects on domestic growth and foreign exchange needs, taking both the direct and indirect effects into consider-ation.[22]

Before we evaluate the total planning program, a few factors such as education and health conditions should be noted, and how the planning programs affected them.

The Development Board realized the deplorable state of education in the country and began building schools by the hundreds. Between 1950/51 and

1956/57, 776 new elementary schools were added; this amounted, at the time, to 70% of the number of schools opened over the previous 30 years. However, because of the great turnover in attendance, which registered in 1957 only 35% of the school age population, "almost universal illiteracy, despite evident increases in expenditure, schools and enrollment," persisted. The school age children in 1956/57 have been estimated to have been over a million, and only 438,773 were registered; in that year over 2/3 of primary school age children received no instruction.[23]

The last five-year development plan envisaged the need for 10,829 elementary school teachers for the years 1965/66—1969/70, an average of 2,166 teachers per year. The United Nations study of 1968 was optimistic about the attainment of that goal as regarded newly trained teachers, but the training of unqualified or underqualified teachers remained a serious problem. In 1965/66, for instance, out of a total of 44,028 primary and kindergarten teachers, 5,171, or 11.7%, had not received professional training. Thus in spite of long-range economic planning, in spite of extensive budgets over a period of at least some 20 years, Iraq was not in a position to train enough teachers for its schools.[24]

As for health services, in the 1955—56 Iraq Government ordinary budget the percentage for health services, which in Iraq were almost exclusively administered by the Government, was only 7.8%. In 1955 there were 1,014 doctors in the country, 305 pharmacists, 781 nurses, 1,309 male dressers and 767 midwives; of the total 1,014 doctors, 650, or more than half, were concentrated in Baghdad province.[25] Nine years later the number of physicians increased only to 1,436, and the objective of doubling that number during the last five-year development plan was not considered feasible. Even this impossible doubling would have meant a reduction from one doctor to 4,760 to one doctor to 2,623 of the population, and this only statistically; for most of the doctors were still concentrated in Baghdad and vicinity. The 1968 report stated that the medical colleges, even with the projected expansion in Baghdad, Basrah and Mosul, could not graduate physicians at the rate required to meet the projected need. The shortage of nurses was even more serious, for the number of nurses was slightly larger than the number of physicians.[26]

In viewing the total long-range development plans, the picture is not very encouraging. It would seem that the basic difficulty of Iraq — instability — inevitably affected basically all the various plans, their frequent changes, and the results achieved. The 1969 United Nations study of the entire range of Iraq's development plans summarized the situation: "Between 1951 and 1958 four development programs were introduced, the first and the third of which (both five-year programs) did not last for even a year. The situation was better in the post-revolution period: the first four-year provisional program ran for less than two years, but the five-year detailed plan lasted about four years; and the existing five-year economic plan is expected to run its full course."[27]

So much for durability; what about effectiveness? As was noted above, Iraq had the money for development but could not spend it. From 1951/52 to 1966/67 all the development programs and plans spent only 55% of their allotted financial resources. It would seem that those who studied the economic development plans of Iraq arrived at negative conclusions. After analyzing the various plans, Abbas Alnasrawi concluded that there was no planning in Iraq in terms of targets, policies, policy instruments, and coordination among various sectors. The plans might be regarded more as summations of public investment projects with no over-all analytical framework to determine their feasibility and their coordination to the national income.

If all the long-range development plans were designed to make Iraq independent of the oil sector, then they have failed completely, but even the aim of general development did not achieve impressive results. "The failure of development policy is clearly illustrated by Iraq's inability not only to maintain her agricultural exports but also by her inability to be even self-sufficient."[28]

Kathleen M. Langley, studying the economic problems of Iraq, said about the Planning Board that its "industrial policy was rather haphazard." It was preoccupied with the issue of flood control and had completely ignored the private sector.[29]

Concluding a detailed study of all the development plans, the United Nations office in Beirut declared: "Plan implementation, which in essence is execution of projects, has been adversely affected not only by the absence or inadequacy of prior feasibility studies but also by such factors as inadequacies in coordination and follow-up of initial action, decisions of implementative work, by dificiencies in public administration and by failures of contractors; the utilization of foreign aid,

Table XX.   *Iraq.   Home Consumption of Oil Products – Millions of Gallons, 1954/55–1969/70*

54/55	232.02
55/56	296.15
56/57	326.72
57/58	381.04
58/59	381.84
59/60	388.69
60/61	466.35
61/62	503.91
62/63	524.11
63/64	546.09
64/65	614.51
65/66	674.02
66/67	691.67
67/68	726.86
68/69	765.41
69/70	804.73[a]

a. Rebublic of Iraq, *Oil & Minerals in Iraq,* 86.

particularly that made available to Iraq by the USSR, clearly reflects these inadequacies and deficiencies."[30]

As inadequate as the development plans might have been, the country had nevertheless made great strides in major long-range projects affecting the water resources and the cultivable land areas. It also made progress in general economic development as reflected in the home consumption of oil products (Table XX), the refining and distribution of which was performed by the Iraqis themselves; in the consumption of the LPG as stated above; and in the increase of consumption of natural gas by the two Government power stations in Dibbis and Najibiyah from 208 million cubic feet in 1959 to 10.357 billion cubic feet in 1971.[31]

WHAT OF THE FUTURE?

There can be no doubt that since the overthrow of the monarchical regime, in spite of the chaotic conditions resulting from the revolution and from the subsequent changes in administration through assassinations and coup d'états — with all the political international vagaries involved — Iraq has made substantial progress, in production, in refining, in transportation and in marketing; and it is on the threshold of developing the natural gas-petrochemical industries. Both internal and external circumstances would augur well for the progress of these potentialities. However, the nationalization of IPC in June 1972 must be considered the greatest challenge to Iraq's financial, technological and marketing resources.

For in spite of the impressive advances made, the financial, technological and even marketing facilities were actually achieved through foreign efforts, primarily, if not exclusively, of the Communist bloc. And as the Iraqis themselves well understand, the basis is barter: Iraqi crude for materials, equipment, machinery and expert services. So far INOC has obtained very few cash customers. This would mean dependence on the Communist countries for almost all import goods. Moreover, in spite of the many agreements and contracts, the Communist countries could absorb only a small fraction of the total crude oil and products. Unless Iraq finds huge markets, assuming that she could, through her own efforts, and produces on the same scale as IPC, the entire economic structure might collapse. Moreover, in her desperation to sell the oil, Iraq might break the artificially maintained international price structure of crude by cutting prices.

The first move of Iraq in disposing of about ¼ of production of IPC was successful. On July 26, 1972, a permanent agreement was signed with Compagnie Française des Pétroles which provided for the French firm to lift its normal share of the oil for a period of 10 years, on pre-nationalization terms. The first tanker loaded in Banias on July 28, 1972.[32]

In order to operate the nationalized IPC, the Government of Iraq established as part of the Ministry of Oil & Minerals the Iraq Oil Operations Company. Ghanim Abdul Jabil, the director of the new company, stated late in June that the production and the pumping of oil had never been interrupted since his administration took over.[33]

INOC, which was in charge of marketing the output of North Rumaila and IPC, sought new buyers. On July 7, 1972 Brazil signed an agreement with INOC for the purchase of crude in the amount of £5 million. The delivery was to be from North Rumaila, and the Basrah Petroleum Company took legal action against the sale.[34]

Under an agreement between INOC and the Greek Niarchos Concern, a tanker loaded with nationalized oil left Banias on July 16, 1972 for Greece.[35]

An INOC spokesman stated that 4½ million barrels of Kirkuk oil was exported during the two months June 1–July 31, 1972. This was not a very impressive figure; it was not more than four days' production at the 1971 rate (1,060,100 barrels a day), which in turn was below the 1970 average of 1,181,700 barrels a day.[36]

In its efforts to expand its marketing outlets INOC seemed to have given way on two basic issues. It was ready to depart from the established posted prices. The following explanation may be stylistically obscure but the intent is clear. It declared: "In its present and future contracts, the National Oil Company makes precise account of current standard of prices in international markets. It tries to avoid clashes in prices with crude oil sellers wherever they are, either national or other commercial companies. Most important for the Company is the non-decrease of prices in international markets and to stick to regulations set by (OPEC) organization aiming at the obligation of the policy of no-price decreasing in their markets."[37]

The other issue was that of refining. INOC held almost as an article of faith, in line with the basic objective of industrialization, that the oil produced must be refined in Iraq and sold as products; this would give Iraq not only additional employment and technical know-how, but the added value of the oil products would contribute to the national wealth of the country. The test of this almost cardinal principle came first with the export of North Rumaila oil and later with the nationalized IPC oil. The bulk of Iraq's customers were the Eastern European countries, and they apparently refused to buy, even on a barter basis, the products; they were determined to obtain crude and process it themselves, and France would accept, as before, only crude. Of course, the reality was that Iraq did not possess the refining facilities to be able to sell products, but the explanation is instructive. "A new principle to sell crude directly to those refineries in friendly countries was adopted. These countries received Iraqi crude oil from the nationalized company at East Mediterranean ports. The new principle is meant to bear no effect on these countries' interests as a result of nationalization."[38]

The question of the future of the oil industry in Iraq must be seen in terms of the action of the Iraq Government and the reaction of the Company. As was mentioned above, Iraq only nationalized one, although the largest, of the three Iraq Petroleum companies. While the Company protested the nationalization of its concession it did not take any action. Moreover, as was mentioned above, CFP agreed to continue to lift its share of the oil from the Iraq Government company. Indeed, what is even more interesting, late in August it was reported that Dr. Nadhim Pachachi, Secretary-General of OPEC, and Mr. Jean Duroc-Danner, CFP Middle

East Director, had been appointed to mediate between the Government and the Company. It was not made clear by whom they were appointed, but they both must have had the approval of the respective sides. The Iraq Petroleum Company announced that while it considered the Government action a breach of agreement and a violation of international law, and asserted its right to the oil of Kirkuk, Jambur and Bai Hassan fields, it would not take any action to protect its rights until October 12, unless either side should declare that the mediation had failed.[39] The mediators reported progress, and the Company extended the date to the end of December. On February 28, 1973, the President of Iraq, Ahmad Hassan al-Bakr, announced that the efforts of the mediators had resulted in a settlement. The Company was to pay Iraq £141 million for past debts and was to receive 15 million tons of crude oil as compensation for nationalization of its concession. The Company also agreed to relinquish the Mosul Petroleum Company concession without any compensation. This left the Company only with the Basrah Petroleum Company concession.

In a radio and television broadcast in Baghdad, the President declared that the agreement was a victory for the "Arab masses against the Western oil monopolies. It guarantees our sovereignty over our natural resources and gives the companies the compensation they deserved."[40]

The Iraqis may be gambling, and they said so openly and clearly, on the demand in Western and other consuming countries, which was constantly rising, while native supplies were steadily dwindling.

The solution would, therefore, have to lie in some sort of arrangement that the Company continue to serve as the marketing agent of Iraq, and continue to receive the oil on favorable terms, which would maintain the price structure for the Company to make reasonable profits.

NOTES

1. BP, *Statistical Review, 1963*, 5; − *1971*, 6; − Arabian American Oil Company, *Facts and Figures 1971*.
2. BP, *Statistical Review, 1970*, 6; *World Oil*, CLXXIV, 114, Aug. 15, 1971.
3. Republic of Iraq, *Oil & Minerals in Iraq*, 63−64.
4. *Foreign Commerce Weekly*, Sept. 4, 1961.
5. *Petroleum Press Service*, XXXVII, 443, Dec., 1970; Republic of Iraq, *Oil & Minerals in Iraq*, 76.
6. *World Petroleum Report*, XIII, 68, 1967; − XVII, 71, 1971.
7. *World Petroleum Report*, XXX, 122, Jan. 15, 1957; IPC, *Iraq Oil 1957*, 29.
8. The development of pipelines up to 1955 was described above, Chapter XI, 240−243.
9. IPC, *Iraq Oil in 1961*.
10. IPC, *Iraq Oil in 1962*, 2, 8; − *1963;* − *1971; Petroleum Press Service*, XXXI, 75, Feb., 1964.
11. The helplessness of the companies in being pressured between Iraq and Syria in early 1967, as mentioned above, was reflected in the sadly plaintive comment

of the Chairman of the Board of the British Petroleum Company. "Despite the events which led to the closing of the Iraq pipeline from Kirkuk to the Mediterranean for a period of about 2½ months, our relations with the governments of the producing countries generally remained good. With the increasing use of long-distance pipelines, there are now a number of countries which are dependent, either as exporters or importers, on transit facilities through neighboring countries. It seem unfortunate that there should not be any machinery at international level for impartial arbitration to settle any differences affecting transit payments which often, as in this case, inflict greater hardship on those who are not the actual parties to the dispute." BP, *Annual Report and Accounts 1966*, 5—6. The first part of the statement was not quite accurate; not only did Iraq threaten IPC, but the Kuwait Government, in the oil company of which BP had a 50% interest, issued a communiqué on Dec. 25, 1966, declaring: "Kuwait supports the struggle of the Iraqi and Syrian peoples to obtain their just demands and denounces the machinations of IPC."

12. Government of Iraq, The Development Board and Ministry of Development, *Second Development Week* (n.p.n.d.); *Iraq Times*, Mar. 27, 1956; "Marketing in Iraq," *World Trade Information Service*, Pt. I, No. 56—73, Sept., 1956.

13. For further discussion of the economic aspects of the 1958 revolution see *The Power Struggle in Iraq*, (New York, 1960), Chapter VI. On Apr. 29, 1959, the Minister of Finance explained: "As for the increase in the ordinary budget's share of oil royalties to cover part of the deficit, this is an interim measure necessitated by the transitory circumstances and the drive of the Republican Government to exploit a large part of the country's revenue to secure social and educational services of which the people have been denied for long, and to reinforce the army, the Republic's shield."

14. Under the revolutionary regime Iraq signed a series of financial and economic agreements with the Eastern European Communist countries. Loans from the Soviet Union in March 1959 and Aug. 1960 for credits amounted to ID 65 million and an agreement with Czechoslovakia in Oct. 1960 granted Iraq ID 12 million credit. UN, *Studies 1969*, 4.

15. UN, *Economic Development 1961—1963*, 20.

16. *Ibid.*, 37. Merchandise exports as percentage of the GNP, including the oil sector, for the years 1956—1962 was 45, but when oil was excluded it was reduced to 3. *Ibid.*, 70. In 1968 crude oil remained the most important single export item, accounting for about 95% of the total commodity exports. UN, *Studies, 1968*, 2.

17. UN, *Economic Development, 1958—1959*, 3—4.

18. Oded Remba, "The Middle East in 1961 — Economic Survey," *Middle Eastern Affairs*, XIII, 73—74, Mar., 1962.

19. UN, *Economic Development, 1961—1963*, 22; — *Studies*, 1968, 1.

20. UN, *Studies, 1968*, 2.

21. *Ibid.*; — *Studies, 1969*, 5.

22. UN, *Studies, 1968*, 2.

23. Qubain, *op. cit.*, 209, 223.

24. UN, *Studies, 1968*, 37. Of the five Arab countries Jordan, Kuwait, Syria, Bahrain, and Iraq the illiteracy rate was the highest in the last-named, 85.4%; — *Studies, 1969*, 61. For a generally sympathetic yet realistic description of the

educational situation in Iraq see Victor Clark, *Compulsary Education in Iraq,* Number IV of the Studies in Compulsary Education series published by UNESCO (n.d., n.p.).

25. Qubain, *op. cit.,* 230.
26. UN, *Studies, 1968,* 38. Qubain noted the reason for the lack of nurses. "The Jewish girls who used to be the main source of supply for nurses have gone with the general exodus of Jews in 1950." *Op. cit.,* 230.
27. UN, *Studies, 1969,* 5.
28. Abbas Alnasrawi, *Financing Economic Development in Iraq. The Role of Oil in the Middle East Economy,* (New York, 1967), 87–88, 158–160.
29. Kathleen M. Langley, *Industrialization of Iraq* (Cambridge, 1961), 235.
30. UN, *Studies, 1969,* 16. For a detailed study and analysis of all the plans see this study of the United Nations, 1–17.
31. Republic of Iraq, *Oil & Minerals in Iraq,* 87; IPC, *Review for 1970,* 3, 9; – *1971.*
32. *Petroleum Press Service,* XXXIX, 342, Sept., 1972; INOC, *Weekly News Bulletin,* 2, June 24, 1972; – 7, Aug. 19, 1972. This latter source added that France undertook to assist Iraq technically and financially in developing oil projects. The full text of the agreement is produced, and one of the articles read: "The Iraqi side expresses its concern for the continued presence of ERAP in Iraq. Hence, it has been agreed that ERAP should strive to establish co-operation with the INOC for carrying out projects of common interest by forming a mixed company (comprising ERAP and INOC) charged with carrying out all oil processes in accordance with the detailed agreements to be held between the two companies." – 10, July 1, 1972. This no doubt is a clumsy rendering in English from the Arabic which was a translation of the original French. This new, close Franco-Iraqi friendship might be compared, ironically, with the report of *Al Ahram* of Cairo on Jan. 1, 1961, that Iraq's Foreign Minister Hashim Jawad was to advise an Arab Foreign Ministers' Conference that his country decided to nationalize the share of French companies in Iraq oil.
33. INOC, *Weekly News Bulletin,* 13, June 17, 1972; – 11, June 24, 1972.
34. *Petroleum Press Service,* XXXIX, 342, Sept., 1972; INOC, *Weekly News Bulletin,* 3, July 10, 1972.
35. INOC, *Weekly News Bulletin,* 9, July 22, 1972.
36. *Petroleum Press Service,* XXXIX, 342, Sept., 1972; IPC, *Review for 1970,* 3; – *1971.* During Dec. 1972 INOC exported a total of 2,759,221 tons from North Rumaila and IPC fields, which was about half of what was exported from IPC fields alone in Dec. 1969. INOC, *Weekly Bulletin,* 4, Feb., 1973; Republic of Iraq, *Oil & Minerals in Iraq,* 37.
37. Republic of Iraq, *Oil & Minerals in Iraq,* 104. On June 3, it was reported from Beirut that Iraq had offered her nationalized oil for sale at Meditarranean terminals at "reduced and competitive" prices, *The New York Times,* June 4, 1972.
38. INOC, *Weekly News Bulletin,* 13, June 17, 1972.
39. *Petroleum Press Service,* XXXIX, 301, Aug., 420, Nov., 1972.
40. *The New York Times,* Mar. 1, 1973.

Section Four

SAUDI ARABIA

Chapter XV
FROM CONCESSIONS TO OPERATIONS

The most dramatic manifestation of the revolutionary impact of the oil industry on a primitive desert country appeared in Saudi Arabia. Nowhere in the Middle East, up to that time, had oil development produced such striking contrasts between he indigenous nomadic mode of life and the Western technological innovations introduced by the regimes of the oil companies. In none of the Middle Eastern countries had the change been so radical and so artificially superimposed as in Saudi Arabia, and nowhere did it seem to be so rootless.

The history of the oil question in Arabia goes back to the time when there was no Saudi Arabia, before Ibn Saud, the hero of united Arabia, came into the picture. It goes back to the outbreak of World War I when the Sultan of Najd was struggling to maintain his semi-independent position and to guarantee himself against his neighbor-enemy, Sherif Hussein of Mecca. Following the procedure of the other shaikhs of the Persian Gulf coast, Ibn Saud signed a treaty with Great Britain on December 26, 1915 which established a protectorate relationship. Britain recognized the independence and territorial integrity of Najd, and granted its Sultan a monthly subsidy of £5,000 on condition that he not attack Hussein. Ibn Saud recognized British control of the foreign relations of Najd, and agreed not to grant any concessions without the approval of his new protectors. Article IV of the treaty stated: "Ibn Saud hereby undertakes that he will absolutely not cede, sell, mortgage, lease, or otherwise dispose of the above territories or any part of them, or grant concessions within those territories to any Foreign Power, or to the subjects of any Foreign Power, without the consent of the British Government."[1]

THE HOLMES CONCESSION

That provision of the treaty, combined with the self-denial clause of the "Foreign Office" agreement of the Turkish Petroleum Company, brought no rush of concession seekers to Najd. In the early 'twenties, a British company, Eastern and General Syndicate Ltd., through a British major, Frank Holmes, attempted to obtain oil concessions in the Persian Gulf area with the purpose of selling rather than operating them. These attempts annoyed not only the dominant British oil concern in the area, the Anglo-Persian Oil Company, but also the British Government representatives in the Persian Gulf.

Holmes tried for a long time to contact Ibn Saud with a view to obtaining an oil concession in the al-Hasa province, but Baghdad authorities denied him traveling facilities. When the serious clashes between the Wahhabis of Najd and the Aneiza tribes of Iraq brought Sir Percy Cox, High Commissioner for Iraq, and Ibn Saud to Uqair for a peace conference in November 1922, Holmes also appeared. While Cox did not altogether discourage Holmes, he was reported to have told him not to press for the concession, that the time was not ripe, and that the British Government could not afford to extend any protection to Holmes' company. At that time the Anglo-Persian Oil Company was interested in the concession, and Sir Percy was strongly supporting Anglo-Persian. Sir Arnold Wilson, then representing Anglo-Persian in the area, wrote that he was coming to see the Sultan and hoped to strike a deal with him on oil.

Ibn Saud indicated his willingness to leave the matter to Sir Percy and to accept whatever he and Holmes agreed upon; Sir Percy subsequently advised the Sultan to write Holmes that he could not give his decision until he had consulted with the British. Ibn Saud wrote the letter and Holmes returned to Baghdad to await the decision. In August 1923 he was informed that Ibn Saud had granted the Eastern and General Syndicate the oil concession of al-Hasa.[2]

The two most important provisions of the concession were that the Company pay in advance an annual rental of £2,000 for prospecting rights, with a lien on a concession to be negotiated when oil was discovered; and that the Company take immediate steps to explore the area. Should the Company not pay the rental on the date due, or discontinue exploration work for a period of eighteen months, the Government would have the right to cancel the concession. For the first two years the rental was paid, and a Belgian geologist explored the area for the Company, but without results. In the third year the Company defaulted on its payment and the geologist left the area. In 1928 Ibn Saud warned the Company that unless the arrears, £6,000, were paid immediately and exploration work resumed, he would cancel the concession. The Company never acted on this warning.[3]

In the meantime, momentous events were taking place in the life of Ibn Saud and in the history of Arabia. After the British realized that their backing of Sherif Hussein had not resulted in what they had hoped for, and that his proclamation of himself as Caliph had caused serious disturbances in Egypt and India as well as in Arabia, they stopped their subsidy to Ibn Saud.[4] This meant that the King no longer needed to restrain his Wahhabis, and a full-scale war on the Sherif broke out. It resulted in the capture of the Hejaz and the triumphant entry of Ibn Saud into Mecca; in January 1926 he was named King of the Hejaz. The relationship between Great Britain and the new conqueror of Arabia could no longer be that of a protector and protected; the Treaty of Jidda of 1927 between Ibn Saud and Great Britain removed the restrictive provisions of the old treaty and recognized Ibn Saud as an independent ruler.[5]

Ibn Saud faced the double task of consolidating his position within his own Wahhabi-Ikhwan movement, and obtaining means for administering his new kingdom as well as providing "gifts" for the various tribes in his territories to keep them loyal. The traditional income from the pilgrimage to Mecca was considerably reduced as a result of the war between Ibn Saud and Hussein and because of the mistrust and suspicion by the Muslims outside Arabia of the Wahhabis, and Ibn Saud was frantic in his efforts to obtain financial assistance. Because of his desperate situation, although he was basically opposed to the coming of foreigners to the Holy Land of Islam, defiling it by their ungodly ways, he was willing to permit the exploitation of the natural resources, if there were any, of his new kingdom. During a visit to Jidda in February 1931, Charles Crane, who actively participated in the Paris Peace Conference and who had made himself a patron of things Arabic, discussed with Ibn Saud the various economic possibilities for the development of Arabia. The outcome was that Crane undertook to provide the Saudi Arabian Government with the services of a mining engineer to examine the mineral resources of Arabia and other development possibilities. About two months later Karl S. Twitchell appeared in Jidda.[6] While Twitchell's report on water resources, the primary objective of the King, was pessimistic, it was encouraging as regards mineral possibilities, such as gold, and especially oil.

It was at this juncture that the Bahrain oil story came into prominence. Through a complex of circumstances (details of which are discussed in Chapter XX), the oil concession obtained in December 1925 by Major Frank Holmes for the Eastern and General Syndicate from the Shaikh of Bahrain, covering some 100,000 acres, came in December 1928 into the ultimate possession of the Standard Oil Company of California. Oil was discovered in Bahrain in 1932;[7] this caused the Iraq Petroleum Company to seek a concession in Saudi Arabia, if for no other reason that to prevent any other company from entering the area of the Red Line Agreement. This caused Ibn Saud to look to oil development for his financial needs, since al-Hasa was of the same geological structure as Bahrain.

Philby reports that during the summer of 1932 he was approached by Messrs. M. E. Lombardi and Loomis of the Standard Oil Company of California on the possibility of obtaining an oil concession from the Saudi Arabian Government. In October he officially inquired about terms and wired them to the Standard Oil Company representatives as a basis for negotiations. Subsequently Lloyd Hamilton, accompanied by Twitchell as technical adviser, arrived in Jidda to negotiate for Standard of California.[8]

It would seem that Twitchell and the King were operating on different planes. Twitchell's encouragements led Ibn Saud, who understood very little of the technical and financial complications of oil exploitation, to believe that Saudi Arabia could develop the minerals of the country, with capital obtained in the United States. At least that was the impression he gave Twitchell, and he believed that Twitchell could supply not only the technical needs, but also the financial means, and the oil resources would be developed by Saudi Arabia. Twitchell on the other hand thought that if the King offered American companies concessions on attractive terms, the possibilities for development by them would be good.

Twitchell thus reports that at the end of May 1932 he discussed the question of mineral development with the King's advisers, Shaikh Abdullah Sulaiman and Shaikh Yusif Yassin. The latter declared, in the name of the King, that a drop in the pilgrimage income had made it impossible to carry out the development plans, and Twitchell was requested to find the necessary capital for the scheme agreed upon. Twitchell protested that he was an engineer and not a financial operator; however, he would agree to become the promotion man provided he had proper authorization from the King for offering an oil concession to the American oil companies.

Such authorization reached him in New York in July 1932, and he began to peddle the Saudi Arabian oil possibilities to the American oil concerns. He saw James Terry Duce of the Texas Oil Company, but Duce suggested that the Near East Development Corporation and Standard of California be approached. He saw the representative of Near East Development and was told that IPC, having missed the boat in Bahrain, would not be caught napping again in Saudi Arabia. Twitchell then went to Gulf Oil Company, but was told that the Company's association with IPC precluded its seeking a concession within the "red line" area. He finally turned to Standard of California. He met with Mr. Loomis and later with M. E. Lombardi, the Company's production manager, and they decided to seek the concession. Lombardi gave Twitchell power of attorney to act on behalf of Standard.[9] When Standard's representatives arrived in Jidda they found there Stephen Hemsley Longrigg,[10] representing IPC, and Frank Holmes, representing the Eastern and General Syndicate, competing for the concession. The latter, according to Philby, soon left, after it was intimated to him that before he could be considered he would be expected to pay up the arrears in rent, amounting to about £6,000 when his concession was cancelled. This reduced the competition to two. After prolonged negotiations with the representatives of the groups, which took up the first four months of 1933, the only issue outstanding was the down payment of a lump sum of £100,000 by the prospective concessionaire on account of future royalties. The IPC, whose geologists advised negatively, apparently felt that Ibn Saud's terms were too high; it was not ready to consider a lump sum of more than £10,000. On May 5, 1933 the directors of IPC decided not to apply for an oil concession in al-Hasa. The field was thus left completely free to Standard of California, which obtained the concession on May 29, 1933.[11]

On July 7 the King confirmed by decree the concession agreement and it was published in *Umm al-Qura* of Mecca, the Government journal, on the 14th, when it officially went into effect. Its major provisions were:

1) The concession was granted for a period of sixty years (solar) beginning with the date exploration started.[12] The Company was given the absolute right to explore, drill, produce, transport, handle and export oil and oil products and all other carbonic substances (Arts. 1 and 30).

2) The Company was to advance the Government an initial loan and an annual rent until oil was discovered and produced in commercial quantities (Arts. 3 and 4).[13]

3) The Company was to start operation of the concession not later than September 1933; if it did not start for three years from the end of September 1933, the Government had the right to cancel the concession (Arts. 6 and 7).

4) The Company was to pay a royalty of 4 gold shillings per ton of crude produced. Oil needed for the Company's operations in Saudi Arabia, as well as the gasoline and kerosene to be supplied to the Government free of charge, were not subject to the royalty fee.

The payment was to be made either in 4 gold shillings or, at the choice of the Company, in one US dollar plus the difference in market value of 4 gold shillings less 10 cents. "For example, if the rate of exchange is $1.14 in United States currency for every 4 gold shillings, then the value of the fee for one ton net of crude oil will be $1.04 in United States currency." (Art. 11).

5) Royalties were to be paid at the end of every half year beginning with the discovery of oil in commercial quantities. In case of disagreement about the sum due, the Company was to pay the sum claimed by the Government, and the dispute was afterwards to be settled through regular arbitration machinery as provided in Article 28 (Art. 13).

6) After discovery of oil in commercial quantities, the Company was to erect a plant in Saudi Arabia for producing sufficient quantities of gasoline and kerosene for the regular needs of the Government, but not including resale of oil products within the country or abroad. Following completion of the plant, the Company was to give the Government each year, free of charge, 200,000 gallons of gasoline and 100,000 gallons of kerosene (Art. 16).

7) The Company and its operations were to be exempt from all direct and indirect taxation, and from both import and export custom duties (Art. 18).

8) The Company was granted rights to construct roads, camps and barracks, to install machines and conduct all other activities necessary for the production and export of oil, including the construction of harbors and quays and lines of sea transportation. "The Company likewise is entitled to take and use whatever is necessary for its operations detailed in the present agreement of other natural resources belonging to the Government such as soil, wood and stone, gypsum and so forth." (Art. 19).

9) The Company was to employ as many Saudi Arabian subjects as feasible, and as long as it could find suitable staff among the Saudi Arabians it was not to engage subjects of any other Government (Art. 20).

10) The Company was to submit to the Government copies of all topographical maps and geological reports, in their approved and final form, which were the results of exploration and exploitation of the concession area, as well as an annual report of its operations, "on condition that the Government should treat these maps and reports as secret" (Art. 23).

11) Disputes between the Company and the Government over the execution of the provisions of the concession, or their interpretation, were to be settled by mutual agreement; failing this, they were to be submitted to two arbitrators, one selected by each party. If the arbitrators failed to reach an agreement both parties were to appoint a single arbiter; if no agreement were reached on the appointment of the arbiter, they were to request the President of the Permanent Court of International Justice to appoint him. The arbiter's decision was to be final (Art. 28).

12) The Company was not to transfer its rights and obligations under the agreement without the consent of the Government. It did have the right to transfer the concession to a company organized by itself, after notification to the Government, and it had the right to organize companies for the purpose of exploiting its concession. Should such companies offer shares for sale to the general public, Saudi Arabian subjects were to be given an opportunity to buy at least 20 per cent of the shares offered (Art. 29).

13) Article 31 provided: "The date on which this agreement will go into effect shall be the date it shall be published in Saudi Arabian territory following the ratification of the agreement on the part of the Company." [14]

The agreement was signed on May 29, 1933 by Finance Minister Abdullah as-Sulaiman al-Hamdan, on behalf of the Government, and by L. N. Hamilton on behalf of the Standard Oil Company of California. [15]

In spite of the official publication of the concession agreement by King Ibn Saud, and despite Lebkicher's assertion that the concession was "a simple document which sets forth the rights and responsibilities of the Company and the returns to the Government," [16] and a statement by Twitchell that it was an "able and simple form of an agreement," [17] Aramco persistently refused to make the concession public in the United States. The Company even presented affidavits to the United States Federal District Court of Southern New York that the concession had never been made public and that its contents could not be divulged, assertions which are baffling.

James A. Moffett brought suit in the Federal District Court of Southern New York on January 23, 1947 against Aramco for $6,000,000, for services rendered in 1941. [18] During the preliminary hearings the Aramco representative refused to reveal the terms of the concession, and consequently the Court issued a subpoena, at the request of the plaintiff, to the defendant to produce the original concession. On July 5 James Terry Duce, vice-president of Aramco, and on July 7 Charles Evans Hughes, Jr., counsel for the Company, requested the Court to quash the subpoena. Hughes declared that this was required to "protect the defendant, the Government of Saudi Arabia and the Government of the United States from annoyance and embarrassment." In his affidavit, Duce stated: "I am informed and believe that the full text of said agreements have not been made public by the Saudi Arabian Government in Saudi Arabia or elsewhere. Defendant has not made public the full text of said agreements within the United States or elsewhere. It has been and is the understanding of defendant that the Government of Saudi Arabia does not desire such publication." He further stated: "The question of the production of said agreements, moreover, has a significance transcending the private interests of either of the parties herein. The Saudi Arabian Government is a sovereign state. It has oil resources upon which the United States is increasingly dependent. It is an area in which the international situation at the moment is delicate. Its views with respect to agreements which constitute a major source of its income are entitled to respect, and the Government of the United States itself has an interest in whether any action should be taken which would be contrary to the wishes of the Saudi Arabian Government. Accordingly, I suggest that, since as I am advised the relevance of the text of said agreements is at best dubious, the Court should not resolve the doubt in favor of requiring their production; or at least should not so resolve without seeking the advice and recommendation of the Department of State of the United States, which might well wish to consult the Saudi Arabian Government to ascertain the views of said Government." [19]

The Court granted the request of the defendants on July 22, 1947. In view of the publication of the concession agreement in the official Saudi Arabian Government journal, and in view of the terms of the agreement itself which required publication, it is difficult to understand how Duce could have declared that the agreement had never been made public by the Saudi Arabian Government, or how he could have asserted that it was the desire of that Government that it not be made public. Nor is it possible to find in the articles of the concession itself any provisions that could have embarrassed either Aramco or the Saudi Arabian Government, or for that matter, the United States Government.[20]

On August 25, 1933 the Company made the loan payment in Jidda. At the end of the year the concession was assigned to the California-Arabian Standard Oil Company, which was organized under the laws of Delaware with a capitalization of $700,000,[21] and wholly owned by the Standard Oil Company of California.

Exploration work was begun almost immediately and although the results were not very impressive, they were apparently promising enough to induce the Texas Oil Company to become, in 1936, a half-partner in the California-Arabian Oil Company.[22] In July 1936, in return for a half interest in the Bahrain Petroleum Company, Texas gave its marketing facilities east of Suez, which Standard of California lacked, to be turned over to a subsidiary, California Texas Oil Company (Caltex).[23] In December, for a half interest in the Saudi Arabian concession, it paid $3 million in cash and undertook to pay $18 million out of the oil produced in Saudi Arabia.[24]

OPERATIONS UP TO OUTBREAK OF WAR

The first commercial field was discovered in Dammam in March 1938, and in September of the same year a six-inch pipeline from Dammam to al-Khobar was laid; from there barges took the oil to Bahrain. In the same year Ras Tanura was selected as the terminal site of a 39-mile pipeline where tankers were to be loaded; the first tanker called on May 1, 1939.

The success of the Dammam field brought a flock of concession hunters to Saudi Arabia, among them Germans, Italians, Japanese, and the IPC; and of course, California-Arabian Standard was also among the bidders for the additional territory. On May 31, 1939, W. J. Lenahan, general representative of Standard of California, succeeded in signing the supplemental agreement, which extended the original area by 80,000 square miles, to make a total of 440,000 square miles. The new area covered the neutral zones between Saudi Arabia and Kuwait and between Saudi Arabia and Iraq.[25]

Other important provisions of the agreement were: With the entry into effect of the supplemental agreement, the Company was to pay the Government a lump sum of £140,000 in gold or equivalent; in addition, it was to pay an annual rental of £20,000 until oil in commercial quantities was discovered in the additional areas, or until the Company withdrew completely from those areas; and a lump sum of £100,000 in gold or equivalent at the time of discovery of oil in commercial quantities.[26] The per ton royalty was the same as in the original concession. With the discovery of oil in commercial quantities, the Company was to increase the allotment of free gasoline to the Government to 2,300,000 gallons. The supplemental agreement was ratified by the Company, and after King Ibn Saud issued a decree of confirmation it was officially published on July 24, 1939, as required by one of its provisions.[27]

During 1940, before the full impact of the war was felt in Saudi Arabia, the development of oil progressed satisfactorily. In the autumn of that year a small 3,000-barrel-capacity refinery was completed to meet local needs. Daily production at Dammam reached the 30,000-barrel mark, and in February of that year the number of Company employees in Saudi Arabia reached its peak, up to that day, of 371 Americans and 3,300 non-Americans and Arabs. Soon, however, Italy entered the war and the effects reached Saudi Arabia and the oil enterprise. The Ras Tanura refinery was shut down in June; in October Dhahran was bombed by the Italians; and by the end of the year the number of Americans in Saudi Arabia was reduced to 226, while Ras Tanura was completely closed down.[28] At the same time the other major source of the Saudi Government's income, the pilgrimage, was brought to a complete stoppage. The King began to press the Company for advances on future royalties.[29] His insistent demands, combined with the advances he was getting from the British and the resultant consequences to American interests, prompted the Company representatives to approach the American Government with a suggestion which would not only remove the onus of supplying the King with his financial needs, but also strengthen its position in Saudi Arabia and its concession vis-à-vis both the King and the British Government.

SUMMARY

During the early 'twenties there was no special interest in the oil potentialities of Saudi Arabia; the concession option obtained by Eastern and General Syndicate through Major Frank Holmes was offered to British concerns but not one was willing to pick it up. Exploration work carried on by the Syndicate geologist for two years was far from encouraging, and the Syndicate was not willing to pay the annual rental of £2,000 a year to preserve its concession rights. Since Saudi Arabia was included in the "red line" area, the IPC displayed no interest in developing the potential oil resources of that country, perhaps because it would only have added more oil to the international market which, during the middle 'twenties, was oversupplied, and perhaps also because of the unfavorable opinion on the geology of the area.[30] After the conquest of the Hejaz by Ibn Saud and the internal consolidation of his kingdom, the mineral explorations of his territories became a decisive factor for his very existence. However, not until oil was commercially worked in Bahrain in 1932 did the various oil companies display any interest in obtaining concessions in Saudi Arabia.

Standard of California was willing to explore the possibilities even before that date, but the King was not willing to permit geological exploration without a concession. After the discovery of oil in Bahrain, Standard tried again, through Philby, to obtain exploration rights with an option for a concession; when the King still insisted on a concession before exploration, the Company this time agreed to meet the King's terms.

Actually the Iraq Petroleum Company was also a bidder for the concession and so far as Ibn Saud was concerned he might have granted it to IPC had the latter met his monetary demands. The directors of that Company, considering the terms too high, decided on May 5, 1933 to withdraw, and the field was left open to Standard of California.

As the exploration work of the American company progressed with promising results and the world market demand increased, Anglo-Iranian displayed an interest and apparently was willing to buy out Standard.[31] In 1936 California sold half an interest in its concession to the Texas Oil Company for about $21 million — $3 million in cash and $18 million to be paid out twenty-five cents from each barrel of oil produced.

After the discovery in 1938 of oil in commercial quantities in the Dammam field, development was rapid, pipelines were laid from the field to al-Khobar and later to Ras Tanura. The success of Dammam brought a number of concession seekers for the rest of the Saudi Arabian potential oil-bearing area, among whom were IPC, Italians, Germans and Japanese. In May of 1939 Ibn Saud granted the concession in the form of a supplemental agreement to the American company. The King has been represented as preferring Americans to all the others in spite of the pressure exerted on him by the governments of the other foreign concerns, and as actually accepting lower terms from the Americans. In view of Ibn Saud's conduct during the 1933 negotiations and his conditions for extending the area of the concession, one is inclined to think that his only consideration in making his choice was monetary.

The terms of both the 1933 and 1939 concessions were on the same general lines as the Iraqi concessions: four shillings gold per metric ton of crude produced plus some fixed rental charges until oil was produced in commercial quantities. The Company had exclusive rights to search, produce, refine and transport oil and oil products; it was free from all taxation and imposition; it was to supply free of charge limited quantities of gasoline and kerosene to the Government for ordinary needs, which was defined as not to be sold by the Government either to individuals in Saudi Arabia or abroad. Between 1938 and 1940 the work of the Company progressed, the number of American employees as well as the number of non-Americans and Saudis increased, and production at Dammam reached the 30,000-barrel per day mark; a small refinery for local needs was erected in 1940. With the entry of Italy into the war, its work was seriously curtailed, subsequently creating acute problems for Ibn Saud and for the Company. These made the Company turn to the United States Government for assistance.

NOTES

1. Charles T. Wilson, *Loyalties, Mesopotamia 1914–1917* (London, 1930), 314–315; Arnold Toynbee, *Survey of International Affairs 1925*, I. *The Islamic World*, 282. (Henceforth cited *Survey, 1925*, I.)
2. Ameen Rihani, *Maker of Modern Arabia* (Boston, 1928), 79–88; H. St. J. B. Philby, *Arabian Jubilee* (London, 1952), 68–69.
3. Philby, *Arabian Jubilee*, 68–69.
4. *PDC*, 131, col. 2346, July 14, 1920; *PDC*, 167, col. 512, July 25, 1923; Toynbee, *Survey, 1925*, I, 284.
5. Great Britain, Foreign Off., *Treaty Between His Majesty and the King of Hejaz and of Nejd and Its Dependencies, May 20, 1927* (London, 1927), Cmd. 2951.
6. Philby, *Arabian Jubilee*, 164–165, 176–177; K. S. Twitchell, *Saudi Arabia* (Princeton, 1947), 139–141.
7. United States Senate, *The International Petroleum Cartel. Staff Report to the Federal Trade Commission submitted to the Subcommittee on Monopoly of the Select Committee on Small Business* (Washington, 1952), 71–73 et passim. (Henceforth cited U.S. Senate, *International Petroleum Cartel*.)
8. Philby, *Arabian Jubilee*, 177.
9. Twitchell, *Saudi Arabia*, 147–150. Curiously, Twitchell does not mention Philby's role in helping Standard of California obtain the concession. On the other hand, Roy Lebkicher, in his book which was prepared as a handbook for the Company's American employees in Saudi Arabia, states that Standard of California became interested in Saudi Arabia as far back as the spring of 1930. At that time, realizing that Ibn Saud was not aware of the Company's standing, it was willing to let Major Holmes, an old acquaintance of the King's, approach him on its behalf. Holmes, however, constantly postponed his visit to Ibn Saud. A direct approach to the King by the Company, made in a telegram through the Saudi Arabian Minister in London in the autumn of 1930, asking permission to send a geologist to make a preliminary examination of al-Hasa, was rebuffed. In June 1932, oil was discovered in Bahrain and, according to Lebkicher, in early November 1932 the Company telegraphed Philby in Jidda and asked him to propose to the Saudi Arabian Government that Standard be permitted to make a preliminary geological examination of al-Hasa with a view to negotiating a concession if conditions justified further exploration. The Government wanted to negotiate a concession before any geological work was begun. Lebkicher thus confirms Philby's role, but minimizes Twitchell's. Roy Lebkicher, *Aramco and World Oil* (New York, 1952), 25–26.
10. US Senate, *International Petroleum Cartel*, 74. "The IPC directors were slow and cautious in their offers and would speak only of rupees when gold was demanded. Their negotiator [the writer], so handicapped, could do little." Longrigg, *Oil in the Middle East*, 107.
11. Philby's statement and the records of the IPC decisively demolish the legend built up by the American oil companies that Ibn Saud's decision in favor of the American company was motivated by his suspicion of Great Britain and his preference for American policies and ideology. Cordell Hull, *Memoirs* (New York, 1948), II, 1511. The only motivation for Ibn Saud's decision was monetary. IPC's withdrawal because of the King's demands left Standard as the only

bidder willing to go beyond the £10,000 offered by IPC. It would therefore seem that the report in *The New York Times* of July 15, 1933 that fear of British political control was a factor in influencing the King to award the concession cannot be substantiated from the facts. The general manager of the IPC reported in 1935 that he had told his groups that the al-Hasa concession "would be theirs within 48 hours if they plunked down £50,000 (gold); they replied I was far too hasty − 'Try Ibn Saud with £30,000 sterling'; within 48 hours California deposited £50,000 (gold) and walked away with it." US Senate, *International Petroleum Cartel,* 114.

In 1964 the Middle East Institute in Washington issued a new book by Philby dealing with his role in the oil concessions in Arabia. In it he gives a rather minute detailed day by day, if not hour by hour, report of developments in the negotiations. The information supplied does not add to the then available knowledge, except that it stresses the role of the writer in the decision of the King to grant the concession. The book also reveals some of the practices and ethical standards of Philby. On April 30, or May 1, Longrigg paid him a visit, and he, Philby, gave him "an interesting bit of information. If you are prepared to offer the sum of £100,000 originally asked for by the Government, you will still be in the running." To which Longrigg exclaimed incredulously, "One hundred thousand pounds. Why no one would think of that. Do you know what I am authorized to offer? The maximum I am able to offer is £5,000 gold."

Perhaps the most graphic episode presented is the final decision to sign the concession. "On May 8th the agreed final draft of the concession was ready for the expression of the King's pleasure; and that night in Mecca I attended the privy council meeting, convened to hear and consider the details of the document. It was a long and dreary screed that Abdullah Sulaiman [Finance Minister] read out, clause by clause, interrupted by an occasional remark or question from the royal lips, and the appropriate responses by the lieges. As the proceedings lengthened, the King grew drowsy and dozed off at intervals until he could hear no more. The meeting stood adjourned till the morrow, when we re-assembled in Maabida palace to hear the remainder of the text. Abdullah droned on, clause by clause as before; and this time the King went fast asleep, only to wake with a start when the reading ended. 'Must have been asleep,' he said. 'Well what do you all think about it?' Of course, he already knew all the essential points of the agreement, especially the details of the payments to ensue; and everybody knew that the concession had his blessing. So everybody expressed his satisfaction at the arrangement arrived at. 'And what do you think, Philby?' he asked, fixing me with his good eye. I expressed my pleasure at the successful termination of negotiations, which seemed to spell great future prosperity for his people. 'Very well!' said the King, turning to Abdullah Sulaiman, 'Put your trust in God, and sign.' "

He further emphasized his own role by the following episode which took place some twenty years later in his presence. "Amir Abdullah, the old King's brother, was telling the story to a group of Syrian and Lebanese visitors, who had just dined with him. 'Even at the last stages of the discussion about the concession,' he said, 'the King was anxious to give it to the British company, as he had been pressed to do by the British Minister; but most of us were for accepting the best terms we could get. And in the end it was this man's advice' (pointing to me) 'that persuaded the King to change his mind, and grant the concession to the Americans.' " H. St. J. B. Philby, *Arabian Oil Ventures* (Washington, 1964), 119, 124, 126–127.

12. US Senate, *International Petroleum Cartel*, 114. The area limitation was subject to a special agreement and hence there are different estimates. Longrigg estimated it to be 56,000 square miles, *ibid.*, 74; others gave it as 249,000 square miles. E. Willard Miller, "The Role of Petroleum in the Middle East." *The Scientific Monthly*, LVII, 243–244, Sept., 1943; see also James Terry Duce, *Middle East Oil Developments* (n.p., 1952), 13 (henceforth cited Duce, *MEOD)*, who gives a 1952 figure of 440,000.

13. The specific amounts, which were subject to separate understandings and were not specified in this agreement, were an initial loan of £30,000 gold or its equivalent to be made 15 days after the agreement went into effect and, after 18 months, unless the agreement was terminated, a second loan of £20,000. Upon discovery of oil in commercial quantities, the Company was to advance £50,000 and a similar amount one year later. These loans were not recoverable but were deductible from future royalties. The annual rent was £5,000.

14. Since the King confirmed the agreement, by decree, on July 7, and it was published in Mecca on the 14th, the conclusion must be drawn that it was ratified by the Company, after which it was published by Saudi Arabia. Actually, Art. 33 provided that if, within 15 days after the Company received the agreement at its headquarters in San Francisco, it was not ratified, it was to be cancelled. The same Article provided that upon ratification by the Company, the agreement must be published in Saudi Arabia. Lebkicher, who no doubt speaks with authority for the Company, states: "It became effective on July 14 by publication in the official Government journal." *Aramco and World Oil*, 26.

15. *Umm al-Qura*, July 14, 21, 1933. See also Philby, *Arabian Jubilee*, 177–179; U.S. Senate, *International Petroleum Cartel*, 79, 114; Twitchell, *Saudi Arabia*, 150–151.

16. Lebkicher, *Aramco and World Oil*, 28.

17. Twitchell, *Saudi Arabia*, 151.

18. Moffett's efforts on behalf of the California-Arabian Oil Co. are presented in Chapter XVI.

19. United States District Court of Southern New York, File Civ. 39–779.

20. From the supplemental agreement of 1939, which was also made public by the Saudi Arabian Government, it would appear that there was a "Second Main Agreement," addressed by L. N. Hamilton on behalf of the Company to Shaikh Abdullah as-Sulaiman al-Hamdan, and confirmed by him, bearing the same date as the original concession, and which was not made public. This "Second Main Agreement" apparently merely stated the specific amounts to be advanced as loans as well as the exact area of the concession, items not spelled out in the concession agreement, and accordingly it added a number of articles to the concession. *Umm al-Qura*, July 24, 1939.

21. Lebkicher, *Aramco and World Oil*, 26–28; Twitchell, *Saudi Arabia*, 152; *Fortune*, XXXV, 179, May, 1947; US Senate, *International Petroleum Cartel*, 114.

22. It was reported that Anglo-Iranian had become interested in the Saudi Arabian

concession in 1935 and had offered Standard of California £6 million for it. Anton Zischka, *Die Auferstehung Arabiens* (Leipzig, 1942), 232; see also US Senate, *International Petroleum Cartel*, 74.

23. Organized under the laws of the Bahamas, BWI, with a capitalization of $1,000,000.

24. US Senate, *International Petroleum Cartel*, 74, 115 *et passim*. For the marketing problems of Standard of California as part of the international oil market, see *ibid.*, 75 *et passim;* Lebkicher, *Aramco and World Oil*, 27; United States Senate, *Hearings Before a Special Committee Investigating the National Defense Program. Part 41, Petroleum Arrangements with Saudi Arabia* (Washington, 1948), 24708–24709, 24803, 24902–24905, 25016, 25029 (henceforth cited US Senate, *Hearings Saudi Arabia*).

25. US Senate, *International Petroleum Cartel*, 117; Lebkicher, *Aramco and World Oil*, 30; Duce, *MEOD*, 13; Twitchell, *Saudi Arabia*, 153–154. In a memorandum prepared by the Company for Secretary of the Interior Harold L. Ickes, dated Feb. 8, 1943, the Company stated that the area of the concession was 250,000 square miles. US Senate, *Hearings Saudi Arabia*, 25385; *Congressional Record – Senate*, 94, part 4, 4944, Apr. 28, 1948. (Henceforth cited *CR.*)

26. *The New International Yearbook 1939*, 34, gives the figures, in dollars: down payment $1.5 million; rental $750,000. The supplemental agreement was for a period of sixty years from 1939; Duce, *MEOD*, 13, gives the period as sixty-six years but states that it expires 1999.

27. *Umm al-Qura*, July 24, 1939. See also US Senate, *International Petroleum Cartel*, 117. The American companies represented – and this was subsequently repeated by all who advocated aid to Saudi Arabia in one form or another – that Ibn Saud was under great pressure by the governments of the other nationals who were seeking the concession, but that he withstood all the pressures in his determination to award the additional areas to the American companies; moreover, that he accepted much lower terms from the Americans than those offered by the other bidders. Twitchell, *Saudi Arabia*, 153–154; Raymond F. Mikesell and Hollis B. Chenery, *Arabian Oil: America's Stake in the Middle East* (Chapel Hill, N. C., 1949), 53; United States Senate, *American Petroleum Interests in Foreign Countries. Hearings before the Special Committee Investigating Petroleum Resources* (Washington, 1946), 319 (henceforth cited US Senate, *American Petroleum Interests);* Hull, *Memoirs*, II, 1511.

In a memorandum to Secretary Ickes on Feb. 8, 1943, the representatives of the American companies which held the concession in Saudi Arabia stated: "This concession was granted by King Ibn Saud to American interests after keen competition and pressure from the nationals and governments of the major powers of Europe – the King at that time having clearly expressed his preference for dealing with Americans." US Senate, *Hearings Saudi Arabia*, 25385. Nevertheless, it would seem that since the King's decision in 1933 was determined primarily by monetary considerations and since in 1933 he had granted the American company preferential consideration in the adjoining territory on condition that it meet the terms others might offer, he would have followed the methods he did in 1933. It is very unlikely that he would have

preferred Aramco, if it had offered less, just because of ideological considerations.

28. According to Sanger, by the end of 1942 the American staff along the oil coast
had dropped to 92, most of them maintenance workers. Richard R. Sanger,
The Arabian Peninsula (Ithaca, 1954), 104.

29. Philby, *Arabian Jubilee,* 227; Lebkicher, *Aramco and World Oil,* 38–42.

30. This may perhaps explain why, after first insisting that Ibn Saud, in accordance
with the 1915 treaty, should not grant the concession to Holmes, the British
subsequently permitted him to make the grant.

31. On July 9, 1936 King Ibn Saud signed an agreement with Stephen Hemsley
Longrigg, representing the IPC's subsidiary, Petroleum Concessions Limited,
granting the company an oil exploitation concession along the Saudi Arabian
coasts on the Red Sea, *Umm al-Qura,* July 31, 1936. In 1941 the concession
was abandoned. Longrigg, *Oil in the Middle East,* 138.

Chapter XVI

FROM LEND-LEASE TO PIPELINE

The year 1940 saw a curtailment in the activities of the California-Arabian Standard Oil Company and a complete stoppage of the pilgrimage to Mecca, the two major sources of income of the Saudi Arabian Government. Ibn Saud turned for assistance to the American company and to the British Government. During the year 1940 the British Government advanced Ibn Saud £100,000 and California-Arabian advanced him $2,980,988 over and above what was due for royalties and rentals; this was in addition to $1,711,692 advanced in the previous year.[1] The King's demands for further advances and the increasing advances of the British, combined with the general uncertainty of the war, created a very difficult situation for the Company. Could it risk pouring more millions of dollars into Saudi Arabia without a reasonable guarantee of recovering it? Would the British, with their influence increased as a consequence of their financial advances to Ibn Saud, help the Company recover its money and guarantee its concession, or would they attempt to get hold of at least part of the concession for themselves, or, through their influence, make the American position untenable? Perhaps the American Government might help Ibn Saud; this would not only relieve the Company from advancing him money but would also strengthen its position with both the King and the British.

Late in that year, Frederick A. Davies, president of the California-Arabian Standard Oil Company, accompanied by Lloyd Hamilton, went to Saudi Arabia to discuss the financial question with the King. Ibn Saud demanded $6,000,000 annually from the Company; his annual budget was about $10,000,000 and the balance he expected to obtain from the British. After prolonged discussions and negotiations, a financial arrangement was established in an exchange of letters, dated January 18 and 19, 1941, between Davies and Abdullah Sulaiman, Minister of Finance of the Saudi Arabian Government. Davies undertook to advance to Saudi Arabia during 1941 a loan of $3,000,000, and to endeavor to increase this loan to $6,000,000. The Saudi Finance Minister stated that as a result of this promise his Government was counting on the $6,000,000 as part of its budget for the year. On its part, Saudi Arabia agreed to extend the term of the concession by two years, and during the following two years the Company would be allowed to deduct ten cents from the royalty on every ton of crude produced. In view of war conditions, Saudi Arabia also agreed to the curtailment of the Company's activities in the concession area provided that such curtailment did not affect "the Government's revenue from the export of oil."[2]

These arrangements were purely and simply between the Saudi Government and the California-Arabian Standard Oil Company. According to Davies, he was authorized by the Company to promise the $3,000,000 loan outright; the additional $3,000,000 he would endeavor to obtain for the Saudi Government, also from the Company. There was no indication either orally, or by implication in the exchange of letters, that the additional $3,000,000 loan, or the entire $6,000,000, was to come from the United States Government. The Company, however, was not only unwilling to advance the additional $3,000,000 which Davies had promised to obtain, but even the original $3,000,000 authorized; it determined to obtain the entire $6,000,000 from the American Government. By April 1941 the Company had decided, as a result of lend-lease legislation, to seek assistance for Saudi Arabia from the President. It was important that the Company's approach to the United States Government be supported by a request from the King outlining the economic situation of his country and stressing the need for financial assistance, and Davies so instructed the Company's representative in Saudi Arabia, William J. Lenahan.[3]

As a matter of fact, President Roosevelt had already been approached, but Lenahan was advised on April 11, 1941 that there was little hope of achieving the objective without a formal request from the King. Ibn Saud refused to go along; he maintained that the Company had broken its word in spite of all the concessions he had made to it in return for the promised loan. Davies promised to persuade the Company itself to increase the loan. He declared that he knew that he could borrow the money from the United States Government if he so desired, and he could do so directly and not through the Company. Although the King later retreated somewhat from his negative position, the Company proceeded to try to obtain assistance for Saudi Arabia from the United States Government.

THE MOFFETT MISSION

James Andrew Moffett, a man with long experience in both the domestic and international oil business, had been retained by the Standard Oil Company of California in 1933 to handle its foreign concessions. In 1936 Moffett became chairman of the Bahrain Petroleum Company, as well as of Caltex, Bahrain's subsidiary and distributor of its products. Early in April of 1941, after their return from Saudi Arabia, Davies and Hamilton approached Moffett, who was a personal friend of President Roosevelt, and asked him to discuss the situation with Secretary of State Cordell Hull or Jesse Jones, president of the Reconstruction Finance Corporation, or with the President himself on behalf of the California-Arabian Oil Company. They explained that Ibn Saud's budget was about $10 million annually; the Company had been obliged to advance him between $10 and $12 million, but he was asking for a further advance of $6,000,000 annually for five years. The Company could not possibly advance him $30 million when royalties from production were running only to about $1.5 million a year. The Lend-Lease Act was promulgated on March 11, 1941, and Moffett saw President Roosevelt on April 9. After telling the President of California-Arabian's difficulties, Moffett suggested a loan from lend-lease funds, with the royalties from the oil as collateral. The President explained that there was no legislation that would permit making such a loan from lend-lease funds, but he evinced great interest in the problem. Moffett then consulted with Davies and Hamilton, and on April 16 he sent the President a letter in which the request of California-Arabian was officially set forth. A memorandum attached to the letter dealt with the situation in Arabia, the importance of the concession, and a plan for meeting Ibn Saud's financial needs. It warned: "We believe that unless this is done, and soon, this independent kingdom, and perhaps with it the entire Arab world, will be thrown into chaos."[4] The emphasis was not on the possibility of the King's cancelling the concession should he not receive the amount he asked, or that the British might get the concession, but on the internal stability of Saudi Arabia. The memorandum pointed out that the oil company had already spent $27.5 million on the development of the Saudi oil resources and had advanced $6,800,000 to Ibn Saud against future royalties; the King had told the Company's representative that unless the necessary financial assistance were immediately forthcoming, the stability of his country would be in grave danger.

The Company proposed that instead of advancing money for oil in the ground, as suggested in the first conversation between the President and Moffett, the United States Government purchase from the Saudi Arabian Government finished petroleum products — gasoline, diesel and fuel oils — at very low prices, to the value of $6,000,000 annually for a period of six years. The Company would contract with Ibn Saud to produce, manufacture, and load such products for his account at a Persian Gulf port, while the King on his part would waive the royalty on an amount of crude oil corresponding, in current royalty rates, to $6,000,000 annually.[5]

That the Company was disturbed by the growing influence of the British in Saudi Arabia is proved by the fact that while it recommended that the State Department urge the British Government to increase its advances to the King, it demanded that "any British advances should be on a political and military basis and should not involve their getting any oil from this concession, the British at the present time being well supplied from Iran, Iraq and Bahrain, et cetera."[6] The President sent the Moffett memorandum to the State and Navy Departments where, after it had been carefully examined and discussed,[7] it was decided that the United States could not do anything to assist Ibn Saud directly. It was felt that since Saudi Arabia was politically in the British sphere of influence, it was up to the British to assist the King. The RFC was granting Britain a loan of $425 million and the President indicated that the British should take care of the King's financial requirements.

At the direction of President Roosevelt, Moffett saw Jones, and on October 9, 1941 Jones wrote Moffett that while the American Government could not advance any funds to Ibn Saud under the Lend-Lease Act or from any other Government agency, he had suggested to the British Ambassador, at the instance of the President and the Secretary of State, that his Government might provide the King with such funds as in its opinion were necessary to meet Ibn Saud's requirements. The oil companies might, Jones wrote, if they wished to do so, work out some arrangement between the British Government and the King.[8]

The Moffett mission must be considered partly successful for while it did not obtain direct assistance for Ibn Saud from the United States Government — which would have strengthened the Company's position with the King — it did succeed in relieving the Company from advancing large sums to him, sums which he received in increasing amounts from the British during the years 1941–1942.[9]

LEND-LEASE

Though California-Arabian was successful in obtaining financial support for Ibn Saud through the British, it was not successful in achieving one of its major objectives in approaching the United States Government: to lessen British influence and increase American power in the area. In their testimony before the Special Committee Investigating the National Defense Program, the Company representative maintained that through their financial advances to the King with American funds, the British had increased their influence in Saudi Arabia at American expense. The Company became alarmed at the plans which the British were developing for completely controlling the Saudi Arabian monetary system. Said Frederick A. Davies, president of California-Arabian: "We felt all through here that the money that the Saudi Arabian Government was receiving from the British was in effect money that we have passed over to the British; and we, as a company, felt that the United States should be getting the credit for it rather that the British." [10]

Referring to a plan to arrange with the British what Moffett had proposed to the American Government in his memorandum to the President, to which the Company objected, Davies explained: "That was one of the most obvious steps that had been made along the route that we have been afraid of all the time over there, the encroachment of the British into the oil picture in Saudi Arabia. I never feared the loss of the concession, but I was mightily afraid that we might, through some ingenious means, such as this, lose a portion of our control." [11] W. S. S. Rodgers, chairman of the Board of the Texas Company, stated that the companies felt that, for political considerations, help to Saudi Arabia should come from the United States through lend-lease. As regards the British he declared: "We felt that if it were left to Britain to make all the necessary advances to the king, Britain would be able to do this because of the aid she would be receiving from the United States through lend-lease; but if this happened Britain's prestige and influence in Saudi Arabia would be so enhanced that ultimately that country might be drawn into the so-called sterling area and we might not be able to maintain the American character of our enterprise. We felt that the United States should recieve the credit for these advances, which would actually come from American resources, and that it should use this opportunity of cementing friendship between Saudi Arabia and the United States by going directly to the financial assistance of the King." [12]

Although the representatives of the Company kept on protesting that they did not fear that Ibn Saud might cancel the concession, the long correspondence between the Company president and its representative in Saudi Arabia during 1941–1942 clearly indicates that the Company was gravely concerned over its future relations with Ibn Saud, and that it hoped to improve relations and guarantee its concession by demonstrating to the King that it was through the Company's efforts that he was receiving his financial assistance from the British. The King maintained, however, that the Company had not lived up to its agreement of January 18–19, 1941.[13] Early in 1943 the Company determined to try to persuade the United States Government to grant Ibn Saud direct lend-lease aid.

On December 2, 1942, Harold L. Ickes, Secretary of the Interior, was made Petroleum Administrator for War. About two months later Harry D. Collier, president of the Standard Oil Company of California, and W. S. S. Rodgers, the chairman of the Texas Oil Company, parent companies of the California-Arabian Company, saw Ickes and told him that Saudi Arabia had probably one of the largest and richest oil pools in the world and they were afraid that the British were trying to edge in on the concession, having advanced Ibn Saud some $20 million. The companies had tried to obtain direct United States assistance to Saudi Arabia but the State Department had blocked their efforts, and they feared that the American oil concession would be cancelled and given over to British interests. They impressed Ickes with the urgent need for fast action to save the situation. At his request, Ickes testified, the two representatives supplied him immediately with the pertinent facts so that he could present the case to the President.[14]

On February 8, 1943, Rodgers sent Ickes a letter, in the name of the companies, accompanied by a memorandum outlining the situation in Saudi Arabia and urging that that country be granted lend-lease. During the next few days Rodgers and Collier saw the Secretaries of the Navy, Army and State and other high officials and impressed upon them the need to grant lend-lease aid to Saudi Arabia. On the 16th Ickes[15] met with the President. After a week of hectic activity on the part of the Company representatives, Secretary Ickes and others, President Roosevelt wrote the following letter on February 18, 1943:

> My dear Mr. Stettinius:
>   For the purpose of implementing the authority conferred upon you as Lend-Lease Administrator, by Executive Order No. 8926, dated October 28, 1941, and in order to enable you to arrange for lend-lease aid to the Government of Saudi Arabia, I hereby find that the defense of Saudi Arabia is vital to the defense of the United States.[16]

This was the crowning achievement of almost three years of effort on the part of the Texas Oil Company and the Standard Oil Company of California to obtain direct aid for Saudi Arabia from the United States Government. It relieved them of the burden of financing the King; it enhanced the American company's prestige and importance; it committed the United States to the protection of the American concession in Saudi Arabia and removed the possibility of British penetration.

What, according to Company spokesmen, was the reason the Company had rejected the previous arrangement — that Saudi Arabia receive aid through the British — which had worked well during the previous two years? W. S. S. Rodgers testified that in order to counteract British influence, the Company itself had made advances to the King in 1941–1942, but in 1943 the situation had become so bad that it was no longer possible for the Company alone to meet it. He declared that American military men were even more disturbed by British activity than was the

Company; they suspected that the British were trying to take the concession away from the Americans. A British expedition, ostensibly to combat locusts, arrived in Saudi Arabia with strong military detachments, and there was a suspicion that there were many geologists in the group. What most seriously disturbed the Company, according to its spokesmen, was a British attempt to open a bank in Jidda, at the end of 1942, to issue paper money. This would have put Saudi Arabia in the sterling area and would have made it impossible for the Company to carry on its activities as an American concern. The time for action had therefore arrived.[17]

GOVERNMENT OWNERSHIP

Ickes, who as Petroleum Administrator for War had the immediate responsibility for oil supplies, was impressed by the arguments of the Company representatives as to the importance of Saudi Arabian oil to the national welfare of the United States. Moreover, his general liberal point of view disposed him to accept the charges which the Company made, that the British were attempting to take over the American concession in Saudi Arabia, especially as the British Government was itself an owner of oil rights. He was therefore instrumental in helping California and Texas obtain lend-lease for Ibn Saud. And this very understanding led Ickes to the next logical step. If the oil were so plentiful and so important for the national welfare, if there were danger to the American concession both from Ibn Saud and from the British, while at the same time American home reserves were being rapidly exhausted, why should not the American Government take over the Saudi Arabian concession and thus guarantee itself the necessary petroleum reserve?

In a letter to the President dated June 10, 1943, Ickes declared that by the end of 1944 there would be a shortage of crude oil for meeting the requirements of the armed services and the essential industrial and civilian economy. To meet this dangerous situation it was imperative for the United States to take "immediate action to acquire a proprietary and managerial interest in foreign petroleum reserves." He recommended that the President instruct the Secretary of Commerce, who headed the Reconstruction Finance Corporation, to organize a Petroleum Reserves Corporation which would be authorized to acquire and participate in the development of foreign oil reserves. He emphasized that in view of past history any realistic appraisal of the problem "compels the conclusion that American participation must be of a sovereign character compatible with the strength of the competitive forces encountered in any such undertaking." After outlining the functions and operations of the proposed corporation, he added: "It is suggested that the first order of business of the Corporation should be the acquisition of a participating and managerial interest in the crude oil concessions now held in Saudi Arabia by an American Company. The potential crude oil reserves underlying this concession have been estimated at approximately 20,000,000,000 barrels − as large as the total current known crude oil reserves of the entire United States.

"Their acquisition will serve to meet an immediate demand by the Army and Navy for large volumes of petroleum products in or near Arabia and will also serve to counteract certain known activities of a foreign power which presently are jeopardizing American interests in Arabian oil reserves." [18]

Ickes fully accepted the arguments of the companies. The major contention was that British oil interests were the interests of the British Government, since it owned about 56 per cent in Anglo-Iranian, and Anglo-Iranian owned about 25 per cent of the Iraq Petroleum Company and a half interest in Kuwait, and that private American companies could not compete with the British Government. The only way to guarantee the American concession against cancellation by Ibn Saud was direct United States Government influence, hence direct lend-lease aid to the King. Saudi Arabia's huge oil reserves were offered to the United States Government in various proposals for military and naval needs, the main argument for the Saudi Arabian concession being the depletion of American home reserves. Accepting all the arguments, Ickes came to the inevitable conclusion — Government ownership. It should be noted that while the Petroleum Reserves Corporation, as conceived by Ickes and his colleagues, was of a very broad nature and theoretically covered every possible area in the world, the specific objective for which it was proposed was the Saudi Arabian concession.

But it was not Ickes alone who arrived at this conclusion and advocated it to the President. The two outstanding facts for all concerned with the problem of oil supply were: 1) the American reserves were being consumed very rapidly, and 2) there was, at least potentially, an enormous supply of oil in Saudi Arabia, and production in the entire British-controlled Middle East area was far below capacity. When these facts were combined, the only possible conclusion was that the United States must guarantee its absolute control over the Arabian resources by overcoming the two obstacles: the British and Ibn Saud; and the method: a United States Government agency to obtain the concession. A series of events led to the President's final act in creating the Petroleum Reserves Corporation.

On June 8, 1943 the Joint Chiefs of Staff submitted a report to the President describing the dangerous depletion of American resources, and pressing for adoption of measures to guarantee supplies outside the United States. On the 14th the State Department sent a memorandum to him attacking the same problem from the diplomatic angle. Two days earlier, on June 12, at a meeting in the White House presided over by James F. Byrnes, Director of the Office of War Mobilization and attended by Secretary of War Henry L. Stimson, Secretary of the Navy Frank Knox, Secretary of the Interior Harold L. Ickes, Dr. Herbert Feis of the State Department and General Boykin Wright of the War Department, it had been agreed that the United States Government must acquire foreign petroleum reserves at the earliest practicable moment. All concurred specifically that a corporation for this purpose should be formed prior to July 1, 1943,[19] and "immediately to initiate steps looking to the acquisition of an interest in the highly important Saudi Arabian fields."

On June 17, 19, 21 and 24, an interdepartmental committee representing the Departments of State, War, Navy and Interior met to consider the petroleum reserve question and arrived at unanimous recommendations. Among these were that the RFC be requested to organize a corporation, to be known as the Petroleum Reserves Corporation, with broad powers to acquire interest or ownership in petroleum reserves outside the continental United States, and to finance, retail, develop, export, or lease such reserves; that the Board of Directors of the Corporation be the Secretaries of State, War, Navy and Interior. As regards Saudi Arabia, the committee recommended:[20]

> The interest to be acquired by our Government in the Saudi Arabian oil reserve shall be the ownership of 100 per cent of the stock of the corporation now owning the oil concessions. The corporation which now owns the oil concessions is the California-Arabian Standard Oil Co., a Delaware corporation, and its stock is understood to be owned 50 percent by the Standard Oil Co. of California and 50 percent by the Texas Corp.
>
> In view of the inherent uncertainty as to the quantity of oil which can be obtained from those or any similar oil fields, it is proposed that payment for the stock of the California-Arabian Standard Oil Co. shall be made by providing that (a) the present owners shall receive a proportion or percentage of the oil to be produced to be paid in kind or at the option of our Government in the United States currency, and (b) that such owners shall receive a payment, either in money or in oil, computed upon the basis of reimbursing them for the net expenditures made by them to date in connection with the concession. The percentage or proportion of oil to be received by the present owners should be the minimum amount which under the circumstances it is fair to accord them.
>
> In the event the Petroleum Reserves Corporation shall determine that it is advisable to enter into an operating and management contract covering all or part of the Saudi Arabian oil fields, the two American corporations now owning the concessions may be afforded an opportunity to operate and manage the oil fields in question pursuant to a contract containing such terms and conditions as shall be stipulated by the Petroleum Reserves Corporation (or by its subsidiary the California-Arabian Standard Oil Co.) including appropriate provisions placing in the Petroleum Reserves Corporation, or the California-Arabian Standard Oil Co., the right to exercise control over the rate of production, the development of oil structures, and the sale or other disposition of all oil produced from such fields.[21]

On June 26, 1943, Secretaries Hull, Stimson and Ickes and Acting Navy Secretary Forrestal reported to the President that having met to consider the oil question and the imperative need to assure adequate foreign reserves, and having given particular attention to the situation in Saudi Arabia, they had all agreed on a signed report which they submitted to him; this report contained the recommendations of the interdepartmental committee.[22]

On June 30, 1943, with the President's authorization, the Reconstruction Finance Corporation organized the Petroleum Reserves Corporation. It was given authority "to buy or otherwise acquire reserves of proved petroleum from sources outside the United States, including the purchase or acquisition of stock in cor-

porations owning such reserves or interests therein, and to store, transport, produce, process, manufacture, sell, market and otherwise dispose of such crude petroleum and the products derived therefrom." [23]

According to Ickes, President Roosevelt summoned to his office on July 14,
1943 the Secretaries of State, War, Navy and Interior, the Director of the Foreign
Economic Administration, and Dr. Herbert Feis, Economic Adviser of the State
Department, and discussed future foreign oil policies with them; he indicated as
members of the Board of Directors of the newly organized Corporation all those
present except Dr. Feis. Ickes was designated president. [24]

Ickes immediately began negotiations with Collier and Rodgers, the top
officers of California and Texas, owners of the California-Arabian Standard Oil
Company, for the purchase of the entire Saudi Arabian concession by the United
States Government. The representatives of both these companies expressed shock
and outrage at the proposal; they feared that their attempts to involve the United
States Government in their Saudi Arabian concession might boomerang and deprive
them of the huge wealth-producing resources which they had acquired in the
deserts of Arabia. Finding them reluctant to part with 100 per cent of their ownership, Ickes went down first to 70 per cent and then to 51 per cent. Finally, the
companies' representatives indicated that they might be willing to give up 33 1/3
per cent of their holding.

Negotiations were carried on from August to October 1943 and then, according to Ickes, the attitude of the companies changed. He attributed this to the
military fortunes of the Allies. When it looked as if Rommel might break through in
North Africa and overrun the Middle East, the companies were eager to have the
Government in their Middle East concession, but after Rommel was chased out of
North Africa and the concession was secure, they were no longer disposed to permit
the Government to purchase even a small part of their Saudi Arabian holdings. The
negotiations broke off. [25]

MOTIVATION

The American companies in the Middle East had never felt completely secure.
They were aware of the hostility of the British, not only because the Americans had
obtained concessions which the British thought should have been theirs, but also
because the Americans were intruding in an area which was regarded as an exclusive
sphere of British influence. Moreover, the British oil companies operating in the
Middle East were at least partly government-owned, and the relations between the
companies and the local authorities were those of two governments, while the
American companies were private concerns and subject to the sovereignty of the
governments which had granted the concessions.

When war broke out and Ibn Saud began to make greater and greater demands
on the companies for advances, they were not quite sure where these demands
might lead. Would they be in a position, if the advances became extraordinarily
large, even to deduct them from future royalties? They knew that the British

Government was advancing to the local shaikhs and to Ibn Saud subsidies which did not come from the oil companies, and the advances did not have to be repaid. Ibn Saud might accept these British subsidies, which carried no obligation for repayment, and in return he might bestow on his British friends the only favor which they would be anxious to receive — oil concessions. To be sure, he would not cancel the entire American concession — at least the companies hoped so — but there was always a danger that he might reduce their concession and transfer part of it to the British.

The only alternative was for the American Government to come to the companies' assistance. Not only was monetary consideration included, but also protection and influence; perhaps a rectification of the unequal position of the American companies vis-à-vis the British. With the passage of the Lend-Lease Act, the companies saw an avenue through which the United States Government could help them. The scheme worked out by the companies would have met all the conditions, and the amounts advanced to Ibn Saud would not have to be repaid; they would be for oil sold to the United States Government. The relations would be between the Saudi Arabian and American Governments, the companies merely acting for the account of the King, and the British would not dare to interfere with the United States Government.

President Roosevelt was fully cognizant of the oil resources of Saudi Arabia and their importance to the war effort as well as to the peacetime economy, and he was willing to help Ibn Saud in his difficulties, but he felt that at that early stage he could not stretch the Lend-Lease Act to include Arabia without incurring domestic criticism; nor would he at that stage of the war antagonize the British, in their fully recognized exclusive sphere of influence, the Middle East. He therefore arranged that Ibn Saud should get his financial assistance, but through the British.[26]

The companies were not too happy, but half a loaf was better than none. The promises to the King to make available to him $6 million over and above the royalty and rentals due him did not have to come entirely from their pockets; at least $3 million was to come from the British, and to that extent they were relieved. They were irked, however, by the fact that the assistance Ibn Saud got from the British was actually American and yet they could not get the credit for it. Though they made numerous attempts to convince him that it was in effect through their efforts that the British were granting him the additional advances, the fact remained that the British were making those advances.[27] After two years, when on the one hand the war demands were taxing the American oil resources heavily, and on the other hand public opinion was more amenable to Saudi Arabia's being included in the lend-lease class, and when President Roosevelt was no longer so concerned with British sensibilities, the companies tried again. Early in 1943, with Ickes the Petroleum Administrator for War sympathetic to their pleas, as were the Secretaries of War and Navy,[28] success was achieved. Saudi Arabia was declared by the President eligible for lend-lease.

But the very arguments which the companies had advanced for the granting of lend-lease to Saudi Arabia — British attempts to take over at least part of the concession, the insecurity of their position with Ibn Saud, the importance of the Saudi Arabian oil to the national welfare in view of American depletion — caused Ickes and his colleagues to propose that the Government purchase the entire Saudi Arabian concession or at least a good part of it. It was for this purpose that the Petroleum Reserves Corporation was organized, and the companies were requested to sell their concession. They had not counted on such a possibility; after having protected their concession from all the conceivable dangers they were not ready to sell it. The traditional cry against the Government's entering into business was raised, even though oil was considered vital to national defense, and the Government's attempt to purchase the concession had to be abandoned.

NOTES

1. US Senate, *Hearing Saudi Arabia,* 25381. According to the testimony of Frederick A. Davies, president of the California-Arabian Standard Oil Company, the 1940 advance above royalties and rent was $3,509,835. *Ibid.,* 25051.
2. *Ibid.,* 25052–25057, 25409–25411.
3. *Ibid.,* 25392.
4. *Ibid.,* 25379 *et passim.*
5. According to Sen. Claude Pepper, the purpose behind this arrangement between the Government of the US and Saudi Arabia was to involve the US Government in the Saudi Arabian concession instead of having the US buy directly from the Company; the King would then not be constantly threatening the Company. *Ibid.,* 24725–24726.

   During the discussions between Moffett and various US Government officials in 1941, Jesse Jones, Administrator of the Federal Loan Agency, suggested that the desired objective might more easily be accomplished through some arrangement between the US Government and the California-Arabian Oil Co. rather than through the proposed arrangement with the King. To this Moffett replied on July 22, 1941: "The proposal outlined in that letter was based on the thought that the plan to be worked out shall take into consideration primarily the political and military aspects of the situation, but to facilitate the transaction, we suggested an arrangement on our part which might provide security necessary for the financial aid to the King. After further consideration, we still think that the political and military aspects should receive primary consideration. We further think that direct and immediate aid by the United States Government to the King at this critical time will bolster the British position amongst the Arabs and the Moslems the world over." US District Court of Southern New York, File Civ. 39–779.
6. US Senate, *Hearings Saudi Arabia,* 24728 *et passim.*
7. On June 14 1941, Harry L. Hopkins wrote the following letter from the White House to Jesse Jones:

Personal & Confidential.

The Honorable Jesse Jones,

    Administrator, Federal Loan Agency.

Dear Jesse:

    The President is anxious to find a way to do something about this matter. I am enclosing confidential correspondence from the White House so you can see what goes on. Will you return it as soon as you have read it?

    I am not sure what techniques there are to use. It occurred to me that some of it might be done in the shipment of food direct under the Lend-Lease Bill, although just how we could call that outfit a "democracy" I don't know. Perhaps instead of using his royalties on oil as collateral we could use his royalties on the tips he will get in the future on the pilgrims to Mecca.

    The RFC has done some funny things since that man from Houston took charge of it.

Cordially yours,

Ibid., 25415.

    In a cable dated July 11, 1941 to William J. Lenahan, the Company's representative in Jidda, Frederick A. Davies stated: "For your information President has approved in principle financial assistance by US Government and officials in Washington are trying to develop feasible method of accomplishment under present legislation." *Ibid.*, 25424. On July 22, 1941 Jones replied to Hopkins:

Dear Harry:

    You wrote me June 14th, hoping that I could find some way to assist King Ibn Saud. There appears no legal way that we can help the King, so, with the approval of the President, I suggested to Lord Halifax and Sir Frederick Phillips, also Mr. Neville Butler, that they arrange to continue taking care of the King.

Sincerely,
Jesse H. Jones
Administrator, Federal Loan Agency

US District Court of Southern New York, File Civ. 39–779.

8. US Senate, *Hearings Saudi Arabia*, 24745. On Aug. 6, 1941 Jones wrote to Secretary of State Cordell Hull that he had spoken twice about assistance to Saudi Arabia to the British Ambassador, Lord Halifax, before the conclusion of the $425 million loan, and that he had shown Sir Frederick Phillips the note the President wrote to Jones on July 18, 1941:

Jess—

    Will you tell the British I hope they can take care of the King of Saudi Arabia. This is a little far afield for us!

F.D.R.

US District Court of Southern New York, File Civ. 39–779.

9. During 1941 the British advanced Ibn Saud £850,000 and 10 million riyals (equivalent to $5,285,500) and during 1942 £3 million (equivalent to $12,090,000). During those same years the Company advanced him, above royalties and rentals due, $2,433,222 and $2,307,023, respectively, actually

less in each year than the 1940 advance. US Senate, *Hearings Saudi Arabia,* 25381.

10. *Ibid.,* 25091.
11. *Ibid.,* 25094—25095.
12. *Ibid.,* 24805. In the original proposal prepared for Mr. Moffett on April 8, 1941, there was the following provision: "The United States Government should procure a commitment from the British that neither the Government or any company of British nationality, or a British National will during the life of the present concession, either directly or indirectly, attempt to acquire any interest therein, or take any other action in derogation of such concession." *Ibid.,* 25380.
13. Cables between Davies and Lenahan and other Company representatives in Saudi Arabia, *ibid.,* 25417—25435. In fact, Moffett, Davies and Rodgers admitted under questioning that they feared at least partial cancellation of the concession by Ibn Saud.
14. *Ibid.,* 25232.
15. *CR—Senate,* 94, part 4, 4948, Aprl 28, 1948.
16. US Senåte, *Hearings Saudi Arabia,* 24861; *CR—Senate,* 94, part 4, 4948, Apr. 28, 1948.
17. US Senate, *Hearings Saudi Arabia,* 24829—24830. In the memorandum attached to his letter of Feb. 8, 1943 to Ickes, Rodgers said: "In lieu of continuing these cash advances, the British Government is now insisting that Saudi Arabian financial requirements be met by an internal note issue backed by a British bank and controlled by the British Currency Control Board — the practice followed in Britain or British-controlled territory." *Ibid.,* 25386; *CR—Senate,* 94, part 4, 4948, Apr. 28, 1948.
18. US Senate, *Hearings Saudi Arabia,* 25237—25238.
19. Before the expiration of the powers of the RFC to organize corporations.
20. *CR—Senate,* 94, part 4, 4961—2, Apr. 28, 1948.
21. The following recommendation of the committee is interesting as a sidelight on the relations between Congress and the Executive and the methods employed to coordinate the two. "After definite determination has been made as to the program to be pursued, it is suggested that Mr. Byrnes on a confidential basis should inform certain Members of the Congress of this program, of this endeavor, to obtain their informal approval in advance of the initiation of the negotiations with the two American companies now owning concession." *Ibid.,* 4962.
22, *Ibid.*
23. *Federal Register,* VIII, 9044, July 2, 1943.
24. Harold Ickes, "Oil and Peace," *Collier's,* 55, Dec. 2, 1944.
25. US Senate, *Hearings Saudi Arabia,* 25240—25241. Rodgers reported that the two sides had actually not been far apart in the negotiations. He personally felt that one morning when he was talking to Secretary Ickes they were "about of a meeting of minds." However, that afternoon, without any previous warning, Ickes said that the deal was all off. "I have done a lot of trading in my life, but I have never had anything like that happen before." *Ibid.,* 24868. Ickes recollected that the negotiations had reached a point where he thought: "They were

doing two things: they were reaching into certain Members of the Senate and House to oppose any deal of any sort, and at the same time they were just pollyfoxing us and stringing us along, and I thought that the time had come when I wanted to call Mr. Rodgers' bluff, and I did call it.

"I think he was somewhat less than ingenuous when he said that he came fully prepared to accept that, or any other proposition. I do not think that they were ready to accept any proposition." *Ibid.*, 25241–25242.

26. Hull, *Memoirs*, II, 1511–1512.
27. For a Company version of its efforts to help Ibn Saud, its efforts and failures with the American Government, and its fears of the British, see Roy Lebkicher *et al., The Arabia of Ibn Saud* (New York, 1952), 65–68.
28. Said Secretary Hull: "Both the War and Navy Departments were interested in securing oil reserves in Saudi Arabian ground." *Memoirs*, II, 1512.

Chapter XVII
PROPOSED PIPELINE AND THE ANGLO-AMERICAN AGREEMENT

Though the negotiations for direct Government acquisition of the stock of the California-Arabian Standard Oil Company from its two parent companies (Standard Oil of California and the Texas Oil Company) broke up abruptly, neither side was willing to give up the effort to involve the United States Government, through the instrumentality of the Petroleum Reserves Corporation, in the development of the oil resources of Saudi Arabia. The Government was still anxious to assure the petroleum reserves of that area for the United States through some governmental action; the companies were willing to benefit from that same action, thereby strengthening their position and increasing the value of the concession.

The question of the refining capacity in Saudi Arabia for immediate military needs was being seriously discussed by the interested authorities even before the organization of the Petroleum Reserves Corporation. In fact, in the recommendations of the interdepartmental committee, mentioned in the previous chapter, Paragraph 5 stated: "Matters connected with the construction of an oil refinery for the Saudi Arabian fields are primarily questions of military policy and as such should be determined by the Army and Navy Petroleum Board. In no event should a consideration of problems as to the acquisition of these oil reserves or the formation of the proposed corporation be permitted to cause delay in carrying out any program for oil refinery construction in Saudi Arabia and elsewhere determined to be advisable by the military services. Instead it is suggested that such programs of refinery construction go forward without delay and that questions as to the ultimate financing and ownership be reserved for future determination."[1]

When it became obvious that the Government could not acquire direct ownership of the Saudi concession, it was proposed that it build a refinery of 100–150,000 barrels' capacity; at that time the oil companies declared the proposal impractical. After long-drawn-out negotiations involving a multitude of proposals, the companies informed the Government that they had decided to build a 50,000-barrel-capacity refinery in Saudi Arabia with their own means.[2]

In the meantime, the Petroleum Reserves Corporation had sent out, in November 1943, a mission headed by Everette Lee DeGolyer, the well-known oil geologist, and including Dr. W. E. Wrather, Director of the United States Geological Survey, and C. S. Snodgrass, Director of the Foreign Refining Division of the Petroleum Administration for War, to inspect the reserve potentialities of Saudi Arabia and other countries in the area. The mission's preliminary report, made public early in January 1944, stated that the center of gravity of world oil production was shifting from the Gulf-Caribbean to the Middle East-Persian Gulf area and was likely to continue to shift. It asserted that estimates of reserves proved by developed fields and indicated by fields discovered but not yet fully explored were

329

about nine billion barrels in Kuwait, six–seven billion barrels in Iran, five billion in Saudi Arabia, and one billion in Qatar.[3]

These at the time very optimistic estimates encouraged the Petroleum Reserves Corporation to seek new means of securing petroleum reserves in the Middle East for the United States through some direct governmental undertaking with or for the oil companies. Admiral Andrew Carter, Petroleum Administrator of the Navy, after visiting Saudi Arabia and conferring with representatives of the companies and local governments, came up with a proposal that the United States Government build a pipeline from the Saudi oil fields to a point of outlet on the eastern Mediterranean. Ickes immediately took up this suggestion; he felt that if such a pipeline were set up as a common carrier, built, paid for and administered by the United States Government, it would be almost as good as a direct interest in the concession. He began at once to negotiate with the representatives of the Arabian-American Oil Company[4] and the Gulf Exploration Company.[5]

THE PROPOSED PIPELINE

On February 6, 1944, Secretary Ickes announced that the Petroleum Reserves Corporation, of which he was president, had been authorized by its Board of Directors and with the approval of the President and the State Department, and the recommendation of the War and Navy Departments and the Joint Chiefs of Staff and the Army and Navy Petroleum Board, to enter into an agreement in principle with the Arabian-American Oil Company and the Gulf Exploration Company, under which the United States Government would construct a pipeline for the transportation of petroleum products from the Persian Gulf area to a point on the eastern Mediterranean. At the same time he made public the "Outline of Principles of Proposed Agreement."[6] The preamble of the proposed agreement declared that it had been concluded "in appreciation of the practical importance of reserve of petroleum in war and peace and of the necessity of assuring to the military forces of the Nation and to the people of the United States adequate petroleum supplies." The major provisions of the proposed agreement were that the Government would construct, own, and maintain a main trunk pipeline system with requisite facilities for the transportation of crude petroleum from a point near the oil fields in Saudi Arabia and Kuwait to a port at the eastern end of the Mediterranean. Charges for the pipeline service would be such as to cover current maintenance and operating costs and to amortize the entire investment within a period of twenty-five years.

The companies agreed to maintain a crude oil reserve, to be available to the military forces, of one billion barrels, or 20 per cent of the recoverable oil content of their reserves, should they be less than five billion barrels. The Government would have the option of purchasing from the reserve oil, at a discount of 25 per cent below the market price in the Persian Gulf region or the average of market prices in the United States, whichever was the lower. The companies were to notify the State Department of any negotiations with foreign governments on the sale of petroleum or petroleum products from their concessions in Saudi Arabia and

Kuwait, and if the State Department objected to such sales the companies were to abide by the Department's decision. The agreement was to be approved by the respective governments of Saudi Arabia and Kuwait. Paragraph 11 emphasized that the intention of the signatories to the agreement was not only to promote the development of petroleum in the areas covered, but also to promote the interest of the governments of those areas, and to respect their sovereignty and protect their rights. "It is the desire of the United States that American nationals that enjoy privileges with respect to petroleum in countries under foreign governments shall have an active concern for the peace and prosperity of such countries and shall exercise their rights with due regard to the rights, including that of political integrity, of the governments of such countries."

The last paragraph stated that the proposed agreement was subject to approval by the respective Boards of Directors of the Petroleum Reserves Corporation and the oil companies. Harold L. Ickes signed for the Petroleum Reserves Corporation; for the Arabian-American Oil Company, Frederick A. Davies; and for the Gulf Exploration Company, J. F. Drake. For the parent companies of the Arabian-American Oil Company, H. D. Collier, president of the Standard Oil Company of California, and W. S. S. Rodgers, president of the Texas Oil Company, also signed. Drake of Gulf added the reservation that "nothing herein shall require action in violation of existing contracts with the British Government or with any corporation in which the British Government has an ownership interest."[7]

The announcement by Secretary Ickes of the agreement created in the United States an extremely controversial issue, with the insinuations and accusations that discussions on oil, especially foreign oil, traditionally evoked. In evaluating the controversy, the personalities, business concerns and organizations which attacked the proposal should be kept in mind. Of all the oil companies, including the major ones, only three stood to benefit from the pipeline — Texas, California and Gulf. All the others feared the advantages which these companies might derive from the pipeline and hence they used every conceivable argument against the proposal. The companies directly involved were mostly satisfied to let Government spokesmen advocate the project.

The arguments in favor of the pipeline, as advocated by its two major protagonists, Secretaries Ickes and Knox, were that American oil resources were being rapidly depleted; that the pipeline was necessary for the war effort; that it would be a protection against any attempt to take the concession away from the Americans; and that it would be a very profitable undertaking. The counter-arguments were that there was no danger of domestic oil depletion if home companies were given enough drilling equipment to open new fields; the pipeline could not be considered a war measure since it would take at least two years to build; it would be indefensible in time of war; it would involve the United States in the politically turbulent Middle East; that protecting the pipeline would involve military commitments and hence increase the danger to American security and to world peace; that it was not economically sound because when the war was over there would be an oversupply of tankers and it would be cheaper to transport the oil by tanker; that it was economically expensive and would increase the cost of oil in the United States; that it was going to decrease the price of oil in New York and cause a depression in the American home oil industry; that it was fascist, that it was communist, that it was against the Atlantic Charter provision for equal access to raw materials; that it would antagonize Great Britain and thus endanger world peace. But the loudest and perhaps the most decisive argument against it was that it was an attempt on the part of the Government to enter into the oil business and to interfere with private enterprise.

The first major attack came from Senator E. H. Moore of Oklahoma, an independent oilman and anti-Administration Senator. Assailing the Petroleum Reserves Corporation,[8] he and Senator Owen Brewster of Maine, leading Republican opponent of the Administration, introduced a resolution in the Senate on February 9 calling for a committee of nine, representing the Senate committees on Foreign Affairs, Commerce and Interstate Commerce, to investigate the national oil policy and make recommendations.[9] The Senate subsequently approved the resolution[10] and a special eleven-man committee was appointed.[11]

On March 1 James A. Moffett attacked both the plan and Ickes and demanded the latter's immediate removal as Government Petroleum Administrator.[12] On the 22nd the Petroleum War Industry Council, which consisted of fifty-five American companies, adopted a resolution demanding that the Government stay out of the oil industry and instead pursue an "open door" policy for private enterprise in formulating a long-range oil plan. All but the three companies directly affected voted for the resolution.[13]

On March 16 the issue was debated over the radio on the Town Meeting of the Air by Secretary Ickes and Senator Moore. While Ickes repeated his reasons for the pipeline and constantly emphasized that there was no attempt at Government entering the oil business, Senator Moore attacked the pipeline, declaring it "needless and useless and impractical," a reckless venture in imperialism "fraught with international complications, dangers and hazards." Ralph T. Zook, president of the Independent Oil-Producing Association, also attacked the pipeline; he declared that it had all the potentials of running the American companies out of business — it would increase the supply from the Middle East and close the European and African markets to American concerns; it would seriously compete with American oil in the United States; the cry that the United States sources were dwindling was not based on facts; nor would it be militarily feasible to protect the line in time of war.[14]

A more systematic attack on the project was made by Herbert Feis, who had sat in as a representative of the State Department at the birth of the Petroleum Reserves Corporation. He pleaded the case of American private enterprise and pointed up the dangers involved should the Government build the pipeline. Eugene Holman, vice-president of the Standard Oil Company of New Jersey, which did not then own a share in Aramco, attacked the pipeline scheme indirectly on April 6, 1944, by refuting the assertion advanced by its protagonists that the United States was running out of oil. He declared that "conservative estimates place the ultimate amount of oil which can be derived from these sources as sufficient to supply our needs at the present rate for more than a thousand years to come." [15]

On April 3 *Time* published an interview with DeGolyer, who was in favor of the project. To the argument that it would involve the United States Government in the turbulent Middle East and would necessitate military protection and commitments, DeGolyer declared, according to *Time*: "It seems fair to assume that the Government is under no greater obligation to go to war to protect its own investment in what is essentially a commercial enterprise than it is to protect the investment of its nationals in similar enterprises. To leave the American companies in a weak position is to invite disaster to the American position in oil." DeGolyer felt keenly that the American companies needed strong Government backing and that the traditional free enterprise in foreign countries did not apply in the case of oil. "It is difficult for our people to realize the degree to which the chancellories of great European nations are willing to interfere particularly in support of the business interests of their nationals or the degree of economic vassalage accepted by the smaller states of the Persian Gulf in the treaties by which they are allied to Britain. Able as American business may be, it cannot support itself against such unequal competition."

DeGolyer realized that the motives of both the opponents and proponents of the pipeline were quite selfish, but he believed that a much higher prize was at stake. "Building the Arabian pipeline is to the advantage of one group and they are for it. It threatens the markets of another group and they are against it. Actually the problem is not one to be settled by the oil industry. I submit that whether or not objections to the present enterprise, its initiation or method of handling are valid, it was conceived, as one editor put it, 'in the interest of national security and for no other purpose.' " [16] Nevertheless with a few exceptions, the general expression of opinion in the press and other media of public information was against the Government's building the pipeline, and the intensity of the attack increased.

THE BRITISH POSITION

Since one of the major objectives of the companies and the Petroleum Reserves Corporation was to guarantee the American concessions against the British, Secretary Ickes suggested to President Roosevelt that an oil agreement be negotiated with the British. According to Ickes, the President thought well of the idea and appointed a committee (with Secretary of State Cordell Hull as chairman and Ickes as vice-chairman) and instructed Hull to open negotiations through diplomatic channels. The British, however, held back and would not appoint an appropriate cabinet committee to meet with the American body; they were apparently reluctant to enter into high level negotiations on the international aspects of oil. When Admiral Carter suggested the pipeline project, Ickes reported: "Well, I thought that, and he understood from the start, that the primary purpose in that suggestion was to alert the British to the idea that we really meant business in the Middle East on oil." [17]

The reaction to the pipeline project both in the British press and in Parliament was bitter, and there was particular resentment against the motivation behind the proposal. Foreign Secretary Anthony Eden, evading the issue so far as British public questioning was concerned, stated in the House of Commons that the project was in the very early stages and that before it was developed it would no doubt be reasonable for the United States to approach the other governments concerned. In reply to a question whether the British Government should have been consulted, even though the project was only in the preliminary stage, Eden said: "Yes, Sir, I am expecting a reply from His Majesty's Ambassador in Washington on the subject." [18] But he had not more information about it even at the end of April when he was asked again about the pipeline proposal. [19]

The agitation in the press continued. [20] Summarizing the situation at the end of March, *The New York Times* correspondent stated that there were three distinct reactions: first, irritation that the plan should have been developed without consultation between the allied governments; second, anxiety over the strategic implications of the project, especially over the possibility of an American naval base being erected in the Mediterranean; third, a belief that Washington was guilty of underestimating the grave political dangers inherent in the plan. [21] The British were also bitterly resentful of the proposal from another aspect. At that time they were clamoring for American materials to build new pipelines from Iraq in order to increase the 90,000-barrel daily capacity to 300,000 barrels. The American Government had refused the materials, yet here it was planning to build a pipeline more than twice as long; the production of American oil in the Middle East would be increased at the expense of the oil of the Iraq Petroleum Company.

But what appeared on the surface was nothing compared to what was going on behind the scenes. The President, eager to tie the hands of the British, demanded a high level conference on the petroleum question which would force the British to recognize the American position in the Middle East. In this he was wholeheartedly supported by Ickes, who had never trusted the British and who was motivated in his entire oil policy by the determination to prevent them from gaining control over Middle Eastern oil. The British stubbornly refused to commit themselves. According to Hull, they were seriously disturbed by Ickes' moves and by the President's insistence that the discussions be on cabinet level. On February 20, 1944 Prime Minister Winston Churchill cabled the President that certain British quarters were apprehensive that the United States wished to deprive the British of their Near Eastern oil interests and that any announcement of a conference on Near East oil, with the American delegation led by the Secretary of State, was sure to raise questions in Parliament which he, the Prime Minister, would be unable to answer. President Roosevelt replied that he was concerned about a rumored British desire to "horn in" on the oil reserves of Saudi Arabia, that it was necessary to clear up all the rumors and fears through a conference and agreement, and that the negotiators must be of cabinet rank.

After a preliminary exchange of assurances with respect to British oil interests in Iran and Iraq and American interests in Saudi Arabia, a compromise was worked out that an announcement would be made of forthcoming discussions, first on a technical and then on a higher or cabinet level. [22] This announcement was made on March 7 [23] and early in April the British sent a technical group, headed by Sir William Brown, to the United States. The American group was headed by Charles Rayner, Petroleum Adviser to the State Department. The meetings began April 13 and ended on May 3. [24]

The content of the discussions, as officially announced at the conclusion of the meetings, was the exploration by the two groups of the full range of both countries' interests in petroleum, on the basis of broad principles looking to a long-range development of abundant oil supplies. It was stated that after a thorough discussion of the broad principles, the two groups had reviewed various specific matters of mutual interest relating to production, distribution and transportation of oil. These included oil production "particularly in the Middle East; the proposed Trans-Arabian pipeline; and the Iraq Petroleum Company's project for an additional pipeline from Kirkuk, Iraq, to Haifa." [25]

In spite of the statement of the conferees that the "groups shared views that the peacetime inter-governmental aspects of such matters should be resolved as between the two governments, within the framework of the broad principles which had been discussed," the British did not follow up with the appointment of a cabinet-rank mission. [26] According to Ickes, the British failure to appoint their delegation caused anxiety in Washington that "some serious hitch had occurred which would break off abruptly what had started apparently as a most harmonious undertaking." [27]

In the meantime the pipeline proposal met tough going in Congress as well as in the press and on the radio. After several executive sessions, the chairman of the special Senate committee of eleven, Senator Francis Maloney, announced that he had been assured by the Executive branch of the Government that the Congress would be fully consulted in advance of final commitments. [28] Members of the committee held discussions with the Interior and State Departments, after which the latter instructed the American Legation in Jidda, on June 27, that there was an informal undertaking within the United States Government that no future action would be taken on the pipeline as a Government project until the British and American cabinet-level discussions, which were to take place shortly, were ended. [29]

It would seem, therefore, that the British reluctance to proceed with high level discussions was based on their objection to the Government pipeline project. As soon as they were assured that the proposal would be deferred, they announced (on July 12) the appointment of a cabinet delegation headed by Lord Beaverbrook, Lord Privy Seal. The American group was headed by Secretary Hull as chairman, and Ickes as vice-chairman. [30] On August 8 an agreement between the two delegations was signed, [31] and on the 24th President Roosevelt sent it to the Senate for approval. [32]

Actually, the agreement said nothing specifically about the Middle East or about the pipeline; its overall objective was the orderly development of oil reserves with a definite eye to the Middle East. The American delegation apparently felt that it had obtained British consent to future American expansion in the Middle East. The public reaction in the United States was, however, almost identical to the reaction to the pipeline agreement. George A. Hill, president of the Houston Oil Company, attacked it violently;[33] Howard Pew, president of Sun Oil Company, opposed it;[34] and they were followed by all the others who had previously attacked the pipeline.[35] The American oil industry attacked it as envisioning a "cartel" system, and above all, because it had not been consulted before the Government concluded the agreement. So vociferous was the opposition that on December 2 Senator Connally, chairman of the Senate Foreign Relations Committee, predicted that the treaty would never be ratified.[36]

On January 10, 1945, President Roosevelt asked the Senate to return the agreement, because of the protests both in the United States and England about some of the provisions, with a view to redrafting it so as to remove any grounds for misunderstanding.[37] On the 15th the Senate returned it to the President.[38] A revised agreement, meeting the objections raised by the American petroleum industry, was informally worked out between the Senate Foreign Relations Committee, representatives of the oil industry, and the State Department. It was ready at the end of February, but it was not until September 7 that Harold Ickes, Petroleum Administrator, announced that renegotiation of the Anglo-American oil agreement would begin in London in the latter part of the month, with himself heading the American delegation.[39]

On September 17, in London, Secretary Ickes indicated that the talks would be mainly concerned with obtaining a more equitable share of Middle East oil for the United States rather than with any basic revision of the proposed Anglo-American oil agreement. As regards the agreement, he said that it was a good one except that it needed drafting changes, but on the question of Middle East oil reserves, something more than a redrafting job was ahead for the American delegation.[40]

The new Anglo-American oil agreement was signed on September 24, 1945, by Ickes for the United States and by Emanuel Shinwell, Minister for Fuel and Power, who headed the British delegation.[41] On November 1 President Truman sent it to the Senate for approval.[42] Although hearings were scheduled to begin on January 20, 1946, they did not commence until June 1947.[43] There was opposition from independent oil producers[44] who argued that the new agreement was the first step toward placing the American oil industry under the control of an international body, and that if it were approved, the United States would be flooded with Middle East oil and the entire domestic oil industry would be ruined.[45] Although some Senators also expressed grave doubts about the new agreement, the Senate Foreign Relations Committee approved it on July 1, 1947 and sent it to the Senate where the matter lapsed.[46] This was the end of the effort to get some commitment from Great Britain for the development of American oil reserves in the Middle East.

To return to the pipeline proper. Writing in *Collier's* on December 2, 1944, Ickes stated that many American commentators and newspapers had "done to death without benefit of clergy" the pipeline plan which he proposed and championed. Nevertheless, he was still determined that the project should be built by the

Government. While negotiations with the British were taking place and the special Senate committee was investigating the oil policies of the Government, the attacks on the pipeline project as well as on the Anglo-American agreement were so powerful that the companies realized that the only way for them to assure increased oil production from their concessions in the Middle East and quick delivery to world markets would be to build a pipeline themselves. By the end of February 1945 it was reported that the companies would do so.[47]

Indeed, on March 14, W. S. S. Rodgers, chairman, and Harry T. Klein, president of the Texas Oil Company, informed their stockholders, in a statement accompanying the annual report, that the pipeline would be built, but that Government participation was doubtful and for the time in abeyance; that the Arabian-American Oil Company intended to obtain the rights of way for the line as soon as possible; and that it had already made preliminary surveys.[48]

When the companies finally gave up hope that the Government would build the pipeline and decided that they would build it themselves, they faced two major problems: obtaining the necessary steel for the pipeline; and acquiring rights of way from the various countries and territories through which the line passed and in which the terminus was to be located.

PIPELINE CONSTRUCTED

The *sine qua non* for the full development of the Saudi Arabian concession was rapid and cheap transportation of the oil to world markets — a pipeline to the eastern Mediterranean. In July 1945 the companies organized the Trans-Arabian Pipeline Company (Tapline) with Burton E. Hull as president, on the same basis as their ownership in Aramco. In December 1946 the Board of Directors of the Company approved the pipeline project.[49] The most immediate question was to obtain wayleaves from the different states and territories through which the pipeline would pass, as well as to determine the location of the terminus. There were several alternatives for the site of the terminus and the construction of the last section of the line. One was to have the line end at Haifa after crossing Transjordan and Palestine. From the point of view of facilities and general terms, Palestine was the most ideal location. However, because of the tension between the Arabs and the Jews, the Company preferred the possibility of having the line cross Syria and Lebanon and terminate on the Levant coast.

First the Company proceeded with Palestine. On January 7, 1946 Sir Alan G. Cunningham, High Commissioner for Palestine, signed an agreement in Jerusalem with W. J. Lenahan, the Middle Eastern representative of the Trans-Arabian Pipeline Company, granting the Company a concession to construct the pipeline across the country.[50] Some four months later George Hall, the British Colonial Secretary, explained in the House of Commons that the concession was practically identical with that granted the Iraq Petroleum Company in 1931, since the Anglo-American Convention on Palestine of December 3, 1924[51] precluded discrimination against United States nationals or companies in the granting of concessions in Palestine.[52]

The concession to IPC having been granted free of charge, the concession to Trans-Arabian was on the same basis, and Hall stated that Palestine would "receive no direct financial benefit from the pipe-line, but its construction and maintenance" would provide increased employment and indirect revenue.[53]

At the time negotiations with the Palestine High Commissioner were begun, Transjordan was still a British mandate, but when the Company began to negotiate with Amman, it encountered obstacles, for in the meantime Transjordan had been granted independence by Great Britain.[54] The discussions between the Company and the Transjordanian Government resulted in an agreement, signed August 8, 1946, which granted Tapline the right to construct and maintain an oil pipeline in return for an annual payment of about $250,000.[55]

Dealings with Lebanon and Syria proved more difficult. Jealous of their recently acquired sovereign status, these two states made heavy demands on the Company; moreover, the relations between themselves, since they were united economically, complicated the situation and created an impasse. Lebanon, anxious to have the terminus on her territory, signed an agreement with the Company granting it the right to construct the line for an annual payment of $180,000. Syria, however, balked; so high were the terms she demanded that the Company threatened to build the terminus in Palestine. This brought about a diplomatic crisis not only between Syria and Lebanon, but also between the United States and Syria. The American diplomatic representatives in the area were apparently determined to keep the pipeline out of Palestine and divert it to the Levantine coast. A compromise was finally worked out after a two-week conference in Beirut attended by the American Ambassador to Iraq, James Wadsworth, the Minister to Lebanon, Lowell Pinkerton, and the Company's representative, W. J. Lenahan, and on September 1 a convention was signed between Syrian Premier Jamil Mardam and Lenahan.[56] However, the convention was never submitted to the Syrian Parliament for ratification. Meanwhile, tension in the area mounted, finally culminating in the war of the neighboring Arab states against the United Nations' newly created State of Israel. This, combined with internal instability in Syria, brought to a halt the construction of the line from the western end.

Immediately after the armistice between Israel and Egypt was agreed upon, on January 6, 1949, Lebanon and Syria took steps to make possible the construction of the pipeline through their territories. With the armistice agreements with Israel in sight, they perhaps feared that Tapline might revert to Palestine. On January 28 they signed an agreement providing for the division of the income from the Trans-Arabian pipeline between themselves, and on May 16 the Syrian government of dictator Colonel Husni Zaim (who had seized power on March 30 and had been quickly recognized by the United States Government) signed an agreement in Damascus with the representatives of Tapline.[57] The Company was granted the right to construct and maintain a pipeline for a period of seventy years, at the end of which time all its immovable property and those fixtures which were part of the project were to be given to the Syrian Government free of charge. The Company was to pay £.0015 per ton of oil moving across Syria, with an annual minimum of

£20,000, which fee was to be in lieu of import, export and all other taxes on petroleum and petroleum products flowing through the pipeline. Materials and equipment necessary for the construction and operation of the pipeline were to be imported duty-free. The parent companies were to make available to the Syrian Government a maximum of 200,000 tons of crude oil annually at current world prices for use in Syria. The Company and its employees were to be free from all Syrian taxes. The Government was to take all reasonable measures to protect the property and employees of the Company; in return, the latter was to pay a minimum of £40,000 to cover the costs.[58]

While the Company was encountering obstacles in the Middle East, it also faced the problem at home of obtaining the necessary steel for the construction of the projected 30/31-inch diameter, 1040-mile pipeline. The question of supplying the Company with the scarce steel was bound to raise serious objections from the independent American oil companies who were clamoring for steel for developing new oil resources to increase local production; they maintained that if Tapline got the steel it was asking for, they would be at a real disadvantage vis-à-vis Aramco: Middle East oil production would be increased at the expense of home production.

On September 26, 1947, right after the Trans-Arabian agreement with Syria was signed, the Department of Commerce approved an export license for 20,000 tons of steel on the ground that Arabian oil would serve "American strategic, political and economic interests." About one month later the Senate Small Business Committee began a battle against the shipment of the steel. At the very opening of the Committee's hearings in October, Government representatives declared that the decision to allot the steel, at the height of the domestic shortage, had been made at cabinet level in the public interest. But Senator Kenneth S. Wherry, reporting for the Committee, declared that hearings over the period of a year[59] had failed to reveal what the public interest was.[60]

On June 8, 1948 the Committee succeeded in getting a commitment from the Department of Commerce that shipment of steel piping for the main-line construction would be held up until the Department had investigated further. The Secretaries of State, National Defense and Commerce all agreed that no further shipments of main-line pipe would be made between that time and January 1, 1949 (the expiration date of the Committee) without prior advice to the Senate Small Business Committee.

On February 24, 1949, when it looked as if Syria were about to approve the agreement, the Secretary of Commerce announced that export licensing of steel pipe to Trans-Arabian would be resumed at the recommendation of the Departments of State and Interior, the National Military Establishment, and the Economic Cooperation Administration. Aware of the fact that the companies would derive considerable profits from the new pipeline at a time of national shortage, the Secretary of Commerce declared that the saving to the Arabian-American Oil Company which would result from the construction of the pipeline would be shared with the National Military Establishment. "Negotiations between the Company and the National Military Establishment on this question were begun in late November and concluded in February 1949. The Company has now agreed to transport for the National Military Establishment, at cost, substantial quantities of oil from the Persian Gulf to the Mediterranean, for a period of ten years after completion of the pipeline." [61]

The Defense Department was apparently not thoroughly convinced of the desirability of using scarce steel material for the pipeline, because of its vulnerability, but favored the construction of tankers. Two days later, therefore, it announced that it had not participated in the decision to renew the export license of steel piping to Trans-Arabian, and that justification for such licenses must be based primarily on considerations other than military. "The economic and political considerations involved" were matters upon which other departments were more qualified to comment than was the National Military Establishment.

The Independent Petroleum Association of America issued a blistering statement on March 1 against the Government's decision to issue export licenses for steel to Tapline and attacked Secretary of Commerce Charles Sawyer's announcement. It declared: "It is impossible to view this latest development as anything except one more step in a decision made long ago. Through all the evasions, the dodgings and the squirming that has occurred over a period of eight years, there has run one consistent thread. The Government of the United States was committed by certain officials to a course of securing the position of the oil companies which hold the Arabian concession." [62] At this stage of development the public and press displayed no such interest in the issue, as it had in the question of the pipeline being built by the Government; after Tapline signed its agreement with Syria and the Department of National Defense ruled in favor of the pipeline, the project progressed rapidly. [63]

Construction of the western section of the line was pushed vigorously and the final weld connecting the section from the Levantine coast with the eastern section from Saudi Arabia was made on September 25, 1950. The first oil reached the terminus at Sidon on the 10th of November, and the first tanker was loaded on the 2nd of December.

The pipeline, which had a daily capacity of about 310,000 barrels, ran for a distance of 1,040 miles from the Saudi Arabian oil fields to the coast of Lebanon. It cost some $200 million to build. [64] The five-berth tanker-loading terminal at Sidon and the pipeline from there to Qaisumah, a distance of 753 miles, was owned and operated by Tapline, while the tank farm at Qaisumah and the pipeline from there to the oil fields, a distance of some 270 miles, was owned and operated by Aramco.

AGREEMENT REVISED

While Tapline was working feverishly to complete the line, the Saudi Arabian Government began to press Aramco for larger royalties. There were two major factors contributing to this demand: the profit-sharing arrangement between foreign oil companies and the Venezuelan Government, and the large royalties granted to Saudi Arabia by the Western Pacific Oil Corporation in the neutral zone between Saudi Arabia and Kuwait.[65] During July the Saudi Government started a press campaign calling for a revision of the existing agreement which continued unabated until late in November. Among the demands were increased royalty payments, Government sharing of the Company's profits, payments in hard currency, and improved working conditions.[66] One constant source of irritation between Aramco and Saudi Arabia throughout the entire period was the gold evaluation, the former maintaining that it should be based on the official international market rate, and the latter demanding that it be based on the free local market rate, which was considerably higher.

After long negotiations a supplemental agreement was signed at Jidda on December 30, 1950, retroactive to January 1 of that year. It provided that the Government was to have a total participation of up to one-half of the Company's net operating revenue; this share was to be inclusive of the fixed per ton royalty which the Company was required to pay regardless of earnings and was to be a ceiling covering all payments — royalty, miscellaneous levies, and income taxes. All monetary transactions between the Company and the Government were to be based on official internationally accepted exchange rates. The Company was to make its payments in any currency which it received and in the same proportion that such currency bore to its total receipts. The Company's gross income was to be subject to four classes of deductions: 1) operating expenses; 2) exploration and development expenses; 3) depreciation; and 4) foreign government taxes, including United States income taxes. In October of the same year, for the purpose of this revised agreement, the Saudi Government introduced, for the first time, an income tax, to which the Company agreed.[67]

Although the tempo of production and general development increased considerably as soon as the danger of war was removed from the area — from 645,860 tons in 1943, to 1,034,603 tons in 1944, and to 26,196,852 tons in 1950 — the greatest and most phenomenal single jump was in 1951 after completion of the pipeline when production rose to 36,608,585 tons.[68] Of this, 14,221,589 tons were shipped via the pipeline.[69]

CONCLUSION

Standard Oil of California and the Texas Oil Company, the companies directly involved, were desirous of strengthening their position in the Middle East not only vis-à-vis the British, but above all, vis-à-vis King Ibn Saud; to do this they tried to involve the United States Government directly in the area, but without in any way permitting the Government to become a part owner in their concessions. The building of a pipeline from the Saudi Arabian oil fields to the eastern Mediterranean would not only have given them an advantage over other oil companies, but would have reduced their costs. Moreover, it would have relieved them of a most difficult and delicate task — obtaining rights of way in the complex and turbulent Middle East. The Government on its part, as personified by Secretaries Ickes and Knox, felt that it would be rendering a great service to the nation by strengthening the position of American nationals in an area where Great Britain was predominant and where the low rate of production from British concessions appeared to be an attempt to exploit United States' oil while preserving for themselves all the fabulous oil of the Middle East. Moreover, the proposed agreement on the pipeline would have placed a billion-barrel reserve at the Government's disposal to draw upon in any future war emergency, and at a considerably reduced price.

The American oil companies not directly involved were determined not to let Texas, California and Gulf get any advantages, not only from such powerful Government protection as direct ownership of the pipeline, but also from the economic benefits of a government-built pipeline. Their arguments against the proposal — Government interference and penetration into the oil business — were successful in stopping the Government from pursuing the scheme further. It would appear that the battle was actually fought between two powerful economic interests, without the public at large feeling itself in any way involved.

The Government was at best on the defensive, for the proposal in itself was a radical departure from traditional American government policy, and since one of the underlying motives was to gain the upper hand over an ally — England — the campaign could not be waged too vigorously. Moreover, the more vigorous the campaign the stronger would be the accusation that the Government was planning to gain control over the oil industry. Nor were the companies concerned in a position to fight too strenuously, for the advantages they stood to derive were so obvious that a campaign to have the Government spend between $100 and $200 million for their benefit would have condemned them not only in the eyes of their opponents, but very likely in the eyes of the public at large as well. Their policy, therefore, was to let the Government do their fighting. At best they could say that the issue should be decided by "those in authority who are entirely neutral and have only the welfare of the American people at heart."

The opposition, on the other hand, able to act without any restraints, had as its greatest weapon the traditional free enterprise suspicion of any Government interference with or penetration into business. The press and radio were overwhelmingly with the opposition. The result was that the Government permitted the agreement in principle to die and only tried to obtain from the British, through the negotiations for an oil agreement, first in 1944 and then in 1945, their consent to the expansion of American oil production in the Middle East. The companies decided to build the pipeline themselves.

From the point of view of availability of facilities and payments for rights, Palestine was ideal for a terminus of the line, but because of political considerations — the Arab-Jewish conflict — the Company sought rights from the governments of Syria and Lebanon. Lebanon soon agreed, but because of difficulties between that country and Syria, and Syria's exorbitant demands, Tapline threatened to build the terminus in Palestine. Through the efforts of the American diplomatic representatives in the area, a compromise was worked out and in September 1947 an agreement was signed with the Government of Syria; it was not submitted to the Syrian Parliament for ratification.

The pipeline, which was to run some 1,040 miles, was partly owned by Aramco (some 270 miles) and the rest by Tapline, a subsidiary of the parent companies of Aramco. Construction was to have been pushed from the western end, on the Mediterranean coast, and from the eastern end, from Saudi Arabia. The eastern end proceeded according to schedule, but the western end — because of the war of the Arab countries against Israel and internal instability in Syria — was seriously interrupted. As soon as the armistice between Israel and Egypt was accepted early in January 1949, Syria and Lebanon came to an agreement between themselves on the location of the terminus and the distribution of the earnings therefrom, and subsequently the Syrian regime under dictator Colonel Husni Zaim signed an agreement with Tapline. From then on construction of the pipeline from the western end proceeded rapidly.

In the meantime, the Company had difficulties at home. Steel piping was in scarce supply and the independent oil companies were clamoring for steel equipment in order to increase their production. The Senate Small Business Committee under Senator Kenneth S. Wherry, investigating the Department of Commerce's licensing for steel to Tapline, clashed with the Executive branch of the Government and for a while succeeded in halting shipments of steel for the Saudi Arabian line, but after the Committee expired on January 1, 1949 the Commerce Department renewed its export licensing of steel to the Company.

The pipeline enabled Aramco to increase production of Saudi Arabian crude oil. Although the first flow through the pipe did not reach Sidon, the terminus, until November 10, during the remainder of the year 1,234,672 tons of oil were delivered through the pipeline. In the next full year of operations the flow was 14,221,589 tons; the increase in total production was about 10 million tons.

At the very time Tapline was rushing the completion of the pipeline, the Saudi Arabian Government was demanding revisions in royalty payments. The contributing factors for these demands were the profit-sharing arrangements between foreign oil companies and the Venezuelan Government, the high royalties granted King Ibn Saud by the Pacific Western Oil Corporation which operated in the neutral zone between Saudi Arabia and Kuwait, and the general agitation in Iraq and Iran for higher royalties. The upshot was a supplemental agreement establishing the income of the Government on a fifty per cent profit-sharing basis. After deductions for expenses, exploration, development and production, as well as depreciation and taxes paid to foreign governments, especially United States income tax, the Saudi Arabian Government was to receive a maximum of 50 per cent of the profits. It was, however, guaranteed, regardless what the profits were, a minimum income equivalent to the royalty and other levies paid up to that time.

This profit-sharing arrangement, the first in the Middle Eastern region and at the time very advantageous to the Saudi Arabian Government, together with the completion of the pipeline, made possible an accelerated production of oil and the establishment of good relations between Ibn Saud and Aramco. The newly expanded facilities were ready for the emergency which was soon to come with the shutdown of Iranian production.

NOTES

1. *CR—Senate*, 94, part 4, 4962, Apr. 28, 1948.
2. US Senate, *Hearings Saudi Arabia*, 24863—24865 *et passim;* Herbert Feis, *Petroleum in American Foreign Policy* (Stanford, 1944), 38—39; Ickes, "Oil and Peace," *Collier's*, 55, Dec. 2, 1944.
3. *CR—Senate*, 94, part 4, 4962, Apr. 28, 1948; Lebkicher, *Aramco and World Oil*, 45.
4. At the end of 1943 the name of California-Arabian Standard Oil was changed to Arabian-American Oil Company, Aramco for short. *CR—Senate*, 94, part 4, 4945, Apr. 28, 1948.
5. US Senate, *Hearings Saudi Arabia*, 25243; see also memorandum of Jan. 25, 1944 from James Francis Byrnes, Director, Office of War Mobilization, "Middle East Oil Proposal," to Pres. Roosevelt. *Ibid.*, 25387—25388.
6. *CR—Senate*, 90, part 2, 1466—1471, Feb. 9, 1944.
7. *Ibid.;* US Senate, *Additional Report of the Special Committee Investigating the National Defense Program. Report of Subcommittee Concerning Investigations Overseas, Section 1 — Petroleum Matters* (Washington, 1944), 77—79.
8. *CR Senate*, 90, part 1, 1135—1138, Feb. 3, 1944.
9. *CR Senate*, 90, part 2, 1466, 1468—1471, Feb. 9, 1944.
10. *Ibid.*, 2489—2490, Mar. 13, 1944.
11. *Ibid.*, 2559—2560, Mar. 14, 1944.
12. *The New York Times*, Mar. 2, 1944.
13. *Ibid.*, Mar. 23, 1944. On Mar. 6 George A. Hill, an oilman from Houston, in a study prepared for the Independent Oil-Producing Association, called the plan fascist. *United States Oil Policy and Petroleum Reserves Corporation. An Analysis of the Effect of the Proposed Saudi Arabian Pipeline*, Petroleum Industry War Council, Washington [1941].
14. *The Proposed Arabian Pipeline. A Threat to our National Security.* Address by Ralph T. Zook, president of the Independent Oil-Producing Association of America, Apr. 28, 1944 [n.d., n.p.].
15. *The New York Times*, Apr. 7, 1944.
16 *Time*, 78, Apr. 3, 1944. It should be noted that W. S. S. Rodgers, retiring president of the Texas Co., in the annual report to the stockholders, declared that the pipeline issue should not be settled by the oil industry, but "by those in authority who are entirely neutral and have only the welfare of the American people at heart. In my mind this question transcends not only the interests of the companies involved, but also those who feel their interests might be adversely affected. Furthermore, I feel very strongly that the decision should not be unduly influenced by publicity emanating from interested companies." *The New York Times*, Apr. 26, 1944.
17. US Senate, *Hearings Saudi Arabia*, 25243—25244.
18. *PDC*, 396, col. 1744, Feb. 9, 1944.

19. *PDC*, 399, col. 786, Apr. 26, 1944.
20. *The Economist* (London) called the attempt to build the pipeline an American incursion into Middle Eastern politics out of keeping with the line of close cooperation with the British, and declared that it was a definite indication of permanent American intervention in the affairs of the region. It further declared that the manner in which the pipeline scheme had been made public showed that there had been no prior consultation with the British, nor any attempt to relate the new scheme to the interests of the Middle East as a whole. "America in the Middle East." *The Economist*, CXLVI, 328–329, Mar. 11, 1944.
21. *The New York Times*, Mar. 23, 1944.
22. Hull, *Memoirs*, II, 1522–1523.
23. *Department of State Bulletin*, X, 238, Mar. 11, 1944.
24. *Ibid.*, 411, May 6, 1944.
25. *Ibid.*
26. At this very time tension between the US and Great Britain over Saudi Arabia was mounting. Hull reported that contrary to his official diplomatic assurances to the people of the US, the State Dept. had received reports in March and April 1944 of increasing British activity in Saudi Arabia prejudicial to American interests. The British Minister in Jidda, S. R. Jordan, was working in numerous ways to supplant American interests in Saudi Arabia by British interests. On March 31, 1944 the American Minister to Jidda, James S. Moose, reported that Jordan had persuaded Ibn Saud to remove certain key officials friendly to the US and to appoint a British economic adviser and possibly a British petroleum adviser as well. Hull, *Memoirs*, II, 1513–1514.
27. Ickes, "Oil and Peace," *Collier's*, 55, Dec. 2, 1944.
28. *CR–Senate*, 90, part 3, 3615–3616, Apr. 21, 1944.
29. Hull, *Memoirs*, II, 1524.
30. For an account of the clash of personalities and polities of Secretary Hull and Secretary Ickes and their different interpretations of the acts of President Roosevelt, see Hull, *Memoirs*, II, 1517–1527; US Senate, *Hearings Saudi Arabia*, 25243–25244 *et passim*
31. *Department of State Bulletin*, XI, 153–156, Aug. 13, 1944.
32. *CR–Senate*, 90, part 6, 7304–7305, Aug. 25, 1944. Attached was a letter of the same date from Secretary of State Hull to the President explaining the agreement.
33. *CR–Senate*, 90, part 11, A4065–A4072, Sept. 15, 1944.
34. *The New York Times*, Aug. 21, 1944.
35. It is interesting to note that W. S. S. Rodgers, chairman of the Texas Co., said on Aug. 9, 1944 that in his opinion the agreement on petroleum production and distribution was constructive and that it would be helpful to the petroleum industry of the United States and foreign countries. *The New York Times*, Aug. 10, 1944.
36. *The New York Times*, Dec. 3, 1944. Senator Connally stated: "It is my view that the treaty is unfair to the American oil industry and is not necessary for the general welfare. I have been opposed to ratification since it was first submitted to the committee."

37. *CR—Senate*, 91, part 1, 179—180, Jan. 10, 1945; *Department of State Bulletin*, XII, 63, Jan. 14, 1945.
38. *CR—Senate*, 91, part 1, 259, Jan. 15, 1945.
39. *Department of State Bulletin*, XIII, 385—386, Sept. 9, 1945.
40. *The New York Times*, Sept. 18, 1945.
41. *Department of State Bulletin*, XIII, 481—483, Sept. 30, 1945.
42. The full text in *CR—Senate*, 91, part 8, 10323—10324, Nov. 2, 1945.
43. US Senate, *Petroleum Agreement with Great Britain and Northern Ireland. Hearings before the Committee on Foreign Relations*, Eightieth Congress, First Session, June 2—25, 1947.
44. In the meantime, two of the major American oil companies, Socony-Vacuum and Standard of New Jersey, acquired shares in Aramco. Discussion in Chapter XIV.
45. *The Christian Science Monitor*, Jan. 4, 1946.
46. US Senate, *Anglo-American Oil Agreement. Report from Committee on Foreign Relations to Accompany Executive H.* Seventy-Ninth Congress, First Session, July 1, 1947.
47. *The New York World-Telegram*, Feb. 27, 1945.
48. *The New York Herald Tribune*, Mar. 15, 1945.
49. Burton E. Hull, "Tapline Presents Great Organization Problem," *The Oil Forum*, II, 450, Nov., 1948.
50. Palestine, *Official Gazette of the Government of Palestine*, Supplement No. 3, to No. 1469 of 24th Jan., 1946, 3—13.
51. United States, Department of State, *Mandate for Palestine*, Near Eastern Series No. 1 (Washington, 1927), 107—115.
52. *PDC*, 422, col. 1860, May 15, 1946.
53. *Ibid.*, col. 87, May 8, 1946; see also *Haboker*, Feb. 19, 1946.
54. Great Britain, Foreign Office, *Treaty of Allinace Between His Majesty in Respect of the United Kingdom and His Highness the Amir of Trans-Jordan*, Cmd. 6779, 1946.
55. *The New York Times*, Aug. 9, 1946.
56. *Ibid.*, Sept. 2, 1947.
57. For an interpretation of Syria's policies with regard to the pipeline, see *Haaretz*, July 7, 1949.
58. "Synopsis of Tapline Convention with the Government of the Republic of Syria," July 21, 1949, supplied by Arabian-American Oil Co.
59. United States Senate, *Problems of American Small Business. Hearings before the Special Committee to Study Problems of American Small Business.* Parts 21—24, 25—28, 33—34, 36, 38, and 40—44 dealt with oil supply and distribution problems as well as steel supplies. The hearings began Oct. 15, 1947 and ended July 10, 1948.
60. *CR—Senate*, 95, part 2, 2222, Mar. 11, 1949.
61. *Ibid.*, 2223.
62. *Ibid.*, 2225.
63. *The New York Herald Tribune*, May 19, 1949.
64. US Senate, *International Petroleum Cartel*, 125; Lebkicher, *Aramco and World Oil*, 50, 55; "Tapline Operations," *World Petroleum*, XXIII, 84—89, Sept., 1952. For a description of the technical aspects of building the line, see Hull,

"Tapline Presents Great Organization Problem," *The Oil Forum*, II, 449–454, Nov., 1948.
65. See below, Chapter XXI.
66. US, Dept. of Commerce, *Foreign Commerce Weekly*, XXXVII, 18, Nov. 14, 1949.
67. United Nations, Department of Economic Affairs, *Summary of Recent Economic Developments in the Middle East 1950–51*, 33–35 (henceforth cited UN, *Summary*, and year); US Senate, *International Petroleum Cartel*, 128–129; Lebkicher, *Aramco and World Oil*, 28; Arabian-American Oil Co., *Report of Operations, 1950*, 3. (Henceforth cited *Report*, with year.)
68. *Report, 1951*, 1; *Report, 1952*, 13; *Report, 1953*, 14.
69. *Report, 1952*, 15.

Chapter XVIII
ARAMCO AND SAUDI ARABIA: 1938–1955

From a relatively modest beginning in 1933, Aramco's Saudi Arabia venture developed into one of the greatest oil undertakings in the world. As was noted above, the first commercial oil field was discovered at Dammam in March 1938; by the end of 1951 it was producing 90,000 barrels a day. In March 1940 the Abu Hadriya field was discovered. Abqaiq was discovered in November of that year and by the end of 1951 was producing 590,000 barrels a day; Qatif and Ain Dar came in in June 1945, with a daily production of 20,000 and 150,000 barrels respectively. Other fields were discovered in January 1949 and April 1951, which were shut in, and the off-shore well, Safaniya, in May 1951. By the end of 1953, 137 wells were producing about 845,000 barrels daily.

Table XXI. *Saudi Arabian Oil Production, 1938–1954*

Year	In Tons	In Barrels	Daily Average (in barrels)
1938	65,618	580,000[b]	
1939	521,214	3,934,000	11,000
1940	672,154	5,075,000	50,000
1941	570,046	4,310,000	12,000
1942	600,351	4,530,000	12,000
1943	645,860	4,868,000	13,000
1944	1,034,603	7,794,000	21,000
1945	2,825,990	21,311,000	59,000
1946	7,899,675	59,944,000	165,000
1947	11,813,668	89,852,000	246,000
1948	18,751,270	142,853,000	390,000
1949	22,820,783	174,009,000	477,000
1950	26,196,852	199,547,000	547,000
1951	36,608,585	277,963,000	770,000
1952	39,870,805	301,861,000	825,000
1953	40,887,754	308,294,000	845,000
1954	46,174,073[a]	347,800,000[c]	953,000[d]

a. *Report, 1953*, 14; *Report, 1954*, 10–11.
b. Up to 1939.
c. Arabian-American Oil Company, *Arabian Oil and World Oil Needs* (1948), 13; Arthur N. Young, "Saudi Arabia Currency and Finance," *The Middle East Journal*, VII, 533, Autumn, 1953; *Report, 1953*, v; *International Financial News Survey*, Feb. 4, 1955; *Report, 1954*, 10.
d. US Senate, *International Petroleum Cartel*, 113; Duce, *Middle East Oil Developments*, 15; *Report, 1953*, v; *Report, 1954*, vi.

The Company built a network of some 330 miles of pipeline ranging from 10 to 30 inches in diameter and from 18 to 45 miles in length, in addition to the main pipeline to the Mediterranean, to bring the oil from the fields to the refineries, to the main pipeline, and to tanker loading stations.[1] Table XXI gives a statistical picture of the enormous production achieved in the comparatively short period of fifteen years.

By the end of 1945 the refinery at Ras Tanura had been expanded to a 50,000-barrel daily capacity, and after further expansion the run to the refinery in 1953 amounted to 74,559,673 barrels.[2]

The Saudi Arabian Government received from Aramco in direct oil revenues:

Table XXII. *Saudi Arabian Government Direct Oil Revenues*

Year	US Dollars	Year	US Dollars
1939	166,890	1948	31,860,000
1940	1,523,649	1949	66,000,000
1941	1,070,550	1950	112,000,000
1942	1,107,302	1951	155,000,000
1944	1,832,000	1952	212,000,000
1945	4,820,000	1953	166,000,000
1946	13,500,000	1954	260,000,000[a]
1947	20,380,000		

a. For the years 1939–1942, US Senate, *Hearings Saudi Arabia,* 25381 *et passim;* for 1946, United Nations, *Review 1951–1952,* 59; for 1949–1951, United Nations, *Summary 1952–1953,* 64; for 1952, United Nations, Secretariat Economic Commission for Europe, *The Price of Oil in Western Europe* (Geneva, 1955), 15; and for 1953–1954, "Oil and Social Change in the Middle East," *The Economist,* 3, July 2, 1955.

The number of persons employed by the Company increased constantly, except for the first few years of the war, from 3,085 in 1938 to 22,345 at the end of 1953.[3] Both because of the great expense involved in bringing American workers to Saudi Arabia (transportation to and from, vacations in the United States, high scale of wages and salaries) and because of the desire to maintain good relations with the Saudi Government and people, the Company pursued a policy of employing as many Saudi Arabs as feasible. The achievement of such an objective was not easy, for not only was the necessary labor not available, but some of the very basic notions of steady, regularized and disciplined methods of work were strange and incomprehensible to the nomad population of the country. It was inevitable that at first the percentage of natives in the total number of workers would be small; it was hoped that after a period of training the number would slowly increase. But even after considerable time had elapsed, very few of the Saudi Arabs attained higher positions. The reverse was the case with the American staff; as more and more Saudis were trained for the lower positions and thus replaced the Americans, the greater was the percentage increase of Americans in the higher positions, and the greater their percentage decrease in the lower staff. By 1953 Arabs constituted 60 per cent of the total staff, Americans 18 per cent, and others, both Middle Easterners and Europeans, composed the balance.[4]

Because of the general stage of the country's development, the Company limited its training to its immediate operational needs. In 1947, 171 men received limited training; the number was increased two years later to 3,556, or about 24 per cent of the total Saudi Arabs employed by Aramco.[5] Not until September 1949 did the Company initiate its five-year training program. The Production Training Center developed three major types of training activities: the vestibule; one-eighth time; and advanced trade training. Vestibule training, which was began in December 1950, was actually a screening for employment on a probationary basis. One-eighth time training was an arrangement which released the employee for one-eighth of his time for training; it aimed to equip the Saudi Arabs with manipulative skill and to give them some understanding of the related information needed for the performance of their jobs. The great bulk of the enrolled trainees were in this division. Finally, the advanced training unit in Dhahran, inaugurated in 1950 with an experimental group of thirty-one, was on a full-time intensive basis for as long a period as necessary up to two years to "develop qualified uni-skilled Saudi Arab employees into highly trained multi-skilled craftsmen."[6]

By the end of 1950, out of total of some 12,000 Saudi employees, 2,496 were enrolled for the three different types of training. In 1951, 35.6 per cent of all Saudi employees were enrolled in all training categories, with the following classifications: pre-job training, job training, voluntary training, supervisory training, and advanced trade and professional training. In 1953, the total number of Saudi Arabian employees in the different training divisions was 4,875, out of a total of 13,555.[7] In addition to the advanced training course at Dhahran, the Company opened special training indoctrination centers in Long Island, New York, and at the American University of Beirut, Lebanon. The center in Long Island was later closed; in 1953 the Company reported that 25 Saudi employees were enrolled at the American University of Beirut, at Aleppo College, and at the Kennedy Memorial Hospital in Tripoli, Lebanon.[8]

In spite of this extensive training program, the number of Saudi Arabians who attained higher positions was very small, if not insignificant, as indicated by the Table XXIII.

Because of the very primitive nature of the country and especially of the eastern coast where the oil fields were located, Aramco had to supply its staff — Americans and Europeans, of course, first and natives afterwards — with housing and other accommodations. The Company's building activity involved investment of many millions of riyals and introduced standards of housing and living, particularly for its Western employees, unheard of or undreamt of by the natives. Air-conditioned bungalows and mess halls, swimming pools and electric lighting were

Table XXIII. *Distribution of Saudi Employees by Grade, 1949–1953*

Grade	1949	1950	1951	1952	1953
Senior Staff		3	4	6	11
Intermediate					
10					4
9	7	6	4	2	10
8	4	6	13	22	24
7	19	25	58	85	119
6	54	128	143	351	574
General					
5	417	555	802	1,018	1,373
4	1,372	1,752	2,189	2,684	2,897
3	1,943	2,088	2,424	3,034	3,424
2	2,313	2,265	3,540	4,050	3,199
1	3,897	3,939	4,609	3,567	1,920
Total	10,026	10,767	13,786	14,819	13,555[a]

a. *Report, 1953,* 45; in 1954 it increased to 14,182. *Report, 1954,* 42.

some of the features. In fact, practically Western-type towns sprang up on the eastern coast of Saudi Arabia which were the direct result of the Company's activities above and beyond the direct construction needs of oil production.[9] The Company worked out schemes by which it helped the native employees get better housing; and with housing went medical care. The Company built hospitals, clinics and health centers in the oil production and refinery locations along the eastern coast. In 1948 the medical staff of the Company consisted of 26 doctors, 97 nurses and 276 other medical employees. During 1950, 1,254 Americans, 4,571 Arabs, and 996 Italians, a total of 8,621 persons, were in-patients in the Company's hospitals. In the same year its clinics treated 310,197 out-patients, of whom 58,624 were Americans and 237,214 Arabs. During 1951, 402,761 out-patient treatments were provided by the clinics, and 7,078 patients were hospitalized. The medical staff was increased that year to 651, of whom 42 were doctors and 142 nurses. [10] During 1952 the Company supplied 532,168 out-patient treatments — to 424,748 Aramco employees or members of their families, 52,911 employees of Saudi contractors operating for the Company, and 54,509 others. Hospitalization was provided for 8,424 persons. The medical staff was increased to 770; there were now 52 doctors, 158 nurses, and 104 technicians. In addition to curative medicine, the Company cooperated with the Government in many preventive measures. [11]

In addition to its oil production activities, Aramco, perhaps unlike the Anglo-Iranian Oil Company in Iran and the Iraq Petroleum Company in Iraq, entered into contracts with the Saudi Arabian Government for various economic undertakings.

It built a railroad for the King from Dammam to Riyad at a cost of some $160 million; managed the al-Kharj agricultural experiment for Ibn Saud; built roads and ports, and made borings for water; and it evolved a policy of contracting with local Saudi merchants and contractors for the development of business projects.

With all the expenses connected with its operations and the various services rendered to the American staff and the Saudi and other staff, has the venture been an economic success The answer can only be: fabulous.

Standard Oil of California's original investment in Saudi Arabia was $50,000, and when the California-Arabian Standard Oil Company was organized for the concession, its capitalization was $700,000. In 1936, California sold to Texas a half interest in its Saudi concession, as mentioned above, for about $21 million. The vice-president of Aramco, Frederick A. Davies, declared that by the end of 1945 the parent companies had invested in Aramco $100,293,540; this was broken down as capital stock $700,000, earned surplus $4,700,000, capital surplus $3,150,000, and loans from the parent companies $91 million.[12] In 1946 the assets of the Company were estimated at $150 million,[13] which was apparently a very conservative estimate.

At about this time Standard of New Jersey and Socony-Vacuum, the partners of the Near East Development Corporation (the American group in IPC) were seeking additional crude supplies from the Middle East to meet the increased demands of their markets. They began negotiations with AIOC and the Kuwait Petroleum Company for the purchase of large quantities of crude and for the construction of a pipeline from the Persian Gulf area to the eastern Mediterranean through which to transport the new quantities of oil to be acquired. At the same time they were also negotiating with Aramco for the purchase of crude oil, and even for the purchase of shares in that Company. Objections were raised by members of IPC on the ground that the actions of Standard of New Jersey and Socony-Vacuum were in violation of the Red Line Agreement, for Saudi Arabia was within the area covered by that agreement, and the case eventually came into the British courts.[14] Yet the negotiations between Aramco and the two companies were continued and in December 1946 an agreement in principle was concluded. It provided for Standard of New Jersey to purchase 30 per cent of the Aramco stock and Socony-Vacuum to purchase 10 per cent, and the shares of Texas and California to be reduced to 30 per cent each. The purchase price was to be $102 million – $76.5 million for Standard's share and $25.5 million for Socony-Vacuum's share. However, until the deal could be concluded, pending the outcome of the litigation between those two companies and the IPC members, a bank loan of $102 million was to be arranged and Standard of New Jersey and Socony-Vacuum wer' to guarantee it according to their shares. When acquisition of Aramco stock had been completed, the bank loan was to be retired.[15]

While the litigation was still going on in the British courts, Aramco signed a series of agreements with Standard of New Jersey and Socony-Vacuum on March 12, 1947. In addition to the main agreement outlined above, the two companies now became the same percentage-share owners in the Trans-Arabian Pipeline

Company and guaranteed their respective shares in a $125 million loan for co
struction of the pipeline.

With the receipts of the guaranteed loan of $102 million, Aramco paid o
$79.8 million to its parent companies and declared a dividend of about $22
million, which also meant a return to each of the parent companies of $11
million. But this was not all the parent companies received for the shares they so
to Standard and Socony.

The new companies agreed to grant Standard of California and the Tex
Company a prior claim to Aramco's earnings. Standard of New Jersey's a
Socony's shares in Aramco were excluded from participation in dividends until t
aggregate amount of dividends paid on Standard of California and Texas sha
equalled $37,234,758 in 1947, $15 million in 1948, $15 million in 1949, a
beginning in 1950, a sum equal to ten cents per barrel for each barrel of cru
produced by Aramco until dividends based on the total production of three billi
barrels ($300 million) had been paid. All the dividend payments were to be cun
lative, that is, if they were not paid in the year due they were to be carried over
the following year. Thus Standard of New Jersey and Socony-Vacuum paid
their 40 per cent share in Aramco about half a billion dollars.[16]

It would seem that Aramco's profits were such that this tremendous sum p
by the two new companies for their shares was justified. Under questioni
Howard Herron, chairman of the Board of Directors of the Bahrain Petrole
Company, declared before a Senatorial committee that the cost of producing cr
oil at Bahrain was approximately twenty-five cents per barrel, including a royalt
approximately 15 cents.[17] Since the royalty to Ibn Saud was approximately
cents, that would make the cost of production per barrel about 32 cents, w
Aramco sold crude oil to the United States Government under contract at $1
and $1.13.[18]

It had been estimated that the pipeline from the oil fields to the Mediterran
cost between $135 and $200 million. The capacity of the line was about 1
million tons per year. United Nations experts stated that the price differentia
Sidon, terminal of the line, and in the Persian Gulf was 66 cents per barrel (ab
$4.90 a ton); that amounted to a saving of $75 million a year, though to be sur
did not take into account amortization and operating costs,[19] but it was ne
theless a great saving which further increased the Company's profits. But the n
interesting fact about the immensity of the profits was brought out by the se
tariat of the United Nations Economic Commission for Europe. Total paymen
respect of taxes and royalties by Aramco to the Saudi Arabian Government in 1
amounted to $212 million. This sum, under the terms of the agreement, re
sented 50 per cent of the net profits realized only from the production of crude
and after the deduction of exploration and development expenses, and Ur
States income tax, in addition to operating expenses and depreciation. By addir
this Aramco's share, the amount realized on the 300 million barrels produced
year would be $424 million. Since crude oil was then selling at $1.75 per barrel
net profit was about $1.40 per barrel.[20]

These profits represented only the production aspects and did not take into account the profits which the parent companies derived from the crude which they transported, refined, exported and marketed. No wonder that Standard of New Jersey and Socony-Vacuum were willing to pay approximately a half billion dollars for their 40 per cent share in Aramco. The question then arises why Standard of California and the Texas Company had been willing to sell part of their concession. It is possible that these two companies, realizing the immense size of their concession, were anxious to have as partners the two other major American oil companies which had concessions abroad and possessed international markets. It should be remembered that both of these companies had vigorously opposed the construction of the pipeline by the Government and that Standard of California and Texas might have wished to have them as partners in their venture. It is also possible that taking into account the great new supply of oil which would result from their expanded operations, they were anxious to obtain additional international markets which these two companies could readily supply.

In a memorandum of a meeting between Harry T. Klein, president of the Texas Company, and representatives of Standard of New Jersey and Socony-Vacuum on September 4, 1946 at which the proposed purchase of an interest by these companies in Aramco was discussed, Klein explained "that Aramco owned a large concession in Saudi Arabia; that in order to keep King Ibn Saud satisfied with the operation of the concession, it is important that production be increased substantially so that the King would receive greater royalties; that added production and the increase in royalties which would flow therefrom would tend to add stability to the concession;[21] that Standard Oil Co. of California and the Texas Co., owners of Aramco, did not have outlets for the volume of crude that should be produced."[22] It would seem that the next to the last reason given was perhaps the chief determinant for the Company's action. In spite of all protestations to the contrary, the sense of insecurity of the parent companies in the concession they were holding from King Ibn Saud was the underlying motivation for all their actions – from their efforts in 1941 through Moffett to obtain lend-lease assistance for the King, down to the selling of shares in the concession to Standard of New Jersey and Socony-Vacuum. As the years advanced the value of the concession went up; the estimates of the Saudi Arabian reserves were constantly revised upwards. In January 1944 the DeGolyer mission estimated the proved reserves as two billion barrels, but taking into account developed fields and fields discovered but not yet fully explored the reserves were estimated as between four and five billion barrels.[23] In 1948 the president of the Trans-Arabian Pipeline Company estimated the proved reserves as seven billion barrels.[24] Testifying before a Senate committee in the same year, W.S.S. Rodgers, then president of the Texas Oil Company, declared that a conservative estimate was between four and five billion barrels; a non-conservative estimate was 20 billion barrels. Secretary of Defense James Forrestal declared that it was testified before the Armed Services Committee that Arabia had a reserve of 30 billion barrels.[25] In January 1949 the estimate was given as nine billion barrels.[26] In 1949 United Nations' experts estimated Saudi

Arabian reserves as 1,213,400,000 tons;[27] in 1951 they were estimated at 1,482,900,000 tons,[28] and in 1953 at 2,426,500,000 tons.[29] The estimate of reserves continued to go up and the value of the concession increased accordingly.

RECAPITULATION

Except perhaps for Kuwait, the Saudi Arabian concession covers the greatest oil-producing field in the entire Middle East. The modest beginnings of the Standard Oil Company of California in 1933 emerged by 1954 into a multi-billion dollar venture. Forty per cent of Aramco was sold in 1947 for about a half billion dollars, and since then Saudi Arabia's reserves have been estimated as three times greater than at the time of the sale.

As a matter of necessity as well as policy, Aramco introduced a system of training for Saudi Arabian employees and provided a standard of living for them and their families hitherto undreamed of by the desert-roaming Bedouins. It built towns and villages on Western models, and for its American employees provided all the Western amenities possible in the desert. It built hospitals and clinics for its employees and helped some Saudi Arabians develop individual economic undertakings; the Company itself became the contractor for many Government projects. It dug wells for water, managed the al-Kharj agricultural venture, constructed ports and roads, built and ran a railroad from Dammam to Riyad, and became the sole distributing agent in the country for petroleum and petroleum products. But despite all these activities, its great investments and the good relations with the King, the parent companies of Aramco — Standard Oil of California and the Texas Oil Company — were in constant fear that Ibn Saud might cancel the concession or modify it extensively. Throughout the period, from 1941 on, they therefore sought to strengthen their position and guarantee themselves against King Ibn Saud and his successor. It is in this fear that one should seek the reason for the Company's efforts to get lend-lease aid for Ibn Saud from the United States Government in 1941; it was this fear which prompted them to seek lend-lease aid for the King again in 1943, when they actually obtained it. It was this fear that made the companies consider the possibility of selling the United States Government a one-third interest in Aramco; that made them discuss the possibility of the Government's building a refinery in Saudi Arabia; and that made them anxious that the Government build the pipeline from Saudi Arabia to the eastern Mediterranean. It would seem that this fear was a determining factor in their decision to sell 40 per cent of their shares in Aramco to Standard of New Jersey and Socony-Vacuum.

Aramco must be considered one of the most fabulous commercial undertakings of modern times. In the year 1952, from production of oil alone in Saudi Arabia, after deducting operating, exploring, development and depreciation expenses as well as United States income tax, the Company's profits amounted to $424 million. This does not take account of the profits derived from the transportation of the oil through the pipeline, and the profits the parent companies made on the crude obtained from Saudi Arabia. The development of the oil resources of Saudi Arabia was a tremendous boon to the American companies. What has it done to Saudi Arabia?

OIL AND SAUDI ARABIA

The population of Saudi Arabia has been variously estimated as between two-and-a-half and five million; no census has ever been taken, but the experts are of the opinion that the lower estimate is closer to reality.[30]

The area of the country is given as 1,546,000 square kilometers, but since most of it is desert, size is of no decisive importance. Before the inroads of the oil industry, the primitive nomadic pattern of existence dominated the country. Sedentary life, even of the type introduced by the Ikhwan movement to develop agriculture, did not achieve the success its advocates hoped for. The geographic and climatic conditions of the desert developed over the centuries a mode of living which austere Wahhabi puritanism only pointed up. The desert, with its swirling, blinding sandstorms, could not afford the roaming tribes of the area more than a bare subsistence, made possible by the scattered oases, and the fatalistic acceptance of the existing order, based on strong religious precepts, made society practically static. The innovations of modern Western techniques were considered abominations. Even such means of communication as the telephone and radio were condemned by the ulemas as works of the devil. It took ingenuity on the part of King Ibn Saud — reading passeges of the Koran over the radio — to convince the learned doctors of the law that these instruments could not be the handiwork of the devil.

Not only were the tribesmen poverty-stricken and struggling hard for their daily existence, but the ruler and the Government had very limited means at their command to support themselves against the many hostile tribes and to run the limited Governmental services. The main and only source of income was the pilgrimage tax levied on the devout Muslims of the world who came to the holy city of Mecca to perform the Hajj. The population was overwhelmingly illiterate, very few indeed even of the royal family could read and write, and fewer were acquainted with and were aware of the world around them. Practically the entire population was ignorant and superstitious. But even this primitive society could not subsist on its own means, and when the King faced difficulties in consolidating his newly enlarged kingdom, and the demands for "gifts" from the tribes for their continued loyalty became pressing, he was willing, though reluctantly, to let the infidels come and exploit the mineral resources, if there were any, of his lands.

To this primitive society, then, the American oil companies came with their twentieth century technology. Some of the Saudi Arabians saw a Western man for the first time in their lives when the oil men arrived, and the paraphernalia of derricks, monster trucks and airplanes could not but bewilder them. When the Company began to recruit labor from among the natives it took patience and perseverance to impress them with the desirability of steady, uninterrupted, disciplined work. The labor turnover at the beginning was overwhelming, for it was a real revolution in the life of a nomad Arab to settle down in one place and work at one job for any considerable length of time.

But soon the Arabs began to enjoy the rewards of labor and the advantages of Western technology. The benefits of good housing, compared to primitive tents, good food and clothing and the other amenities which the camp life of the oil company afforded, proved very attractive to the natives, and they began to show remarkable adaptability. Though they still looked with distrust and suspicion on the social behavior of the infidels, they began to imitate those very infidels in their mode of life. The first thing to crack was their fatalism, their acceptance of the existing order. The education and training which they received in the Company schools or outside the schools, limited as it was, created a sense of discontent, and instead of being grateful to the Company for the benefits they had received, they were inclined to blame it for the differences between themselves and the American employees, and in spite of a Government injunction, they actually proclaimed a strike[31] in 1953 demanding higher wages and additional employment benefits.[32]

To be sure, the total number of Saudis employed by the Company was only a very small percentage of the total population (as noted above, they numbered only 13,555 in 1953), but many more thousands were working for the native contractors who operated for the Company and for others who catered to the needs of the American staff and to the other activities of the Company on the eastern coast and along the pipeline which stretched across the country.

The same quick accommodation to the Western way of living, without under-standing or appreciation of the forces behind the achievements which made such a way of life possible, affected the rulers and their servants. While in 1938, for instance, Saudi Arabia's entire government budget was only about $7 million, or 25 million riyals, and the King met all his obligations, in 1952—53 the budget went up to 758 million riyals, equivalent to $205 million, and the King could not meet all his obligations.[33] This huge increase in the government income was almost exclusively from oil revenues. As against one million riyals in 1938 from oil in the total government revenue of 25 million, it was estimated that at least three-fourths of the income in the 1952—53 budget came from that source.[34] What did King Ibn Saud do, and what has his son Saud done, with these great incomes from the oil resources of their country?

It had been estimated that life expectancy in Saudi Arabia was thirty-three years; tuberculosis was prevalent, 70 per cent of the population had trachoma, and at least 40 per cent suffered from syphilis.[35] In 1950 it was estimated that the per capita annual income was only $45.[36] Even if one accepts these generous estimates on health conditions and per capita income, the situation must be considered deplorable. Since the oil — no matter what estimates of reserves be adopted — is exhaustible, it would be only reasonable to expect the great revenues from oil to be used for investment in human resources and in economic wealth-producing projects. A careful examination of the government budget of 1952/3 revealed that the oil income had not been fully used for constructive purposes.

Ibn Saud fully understood neither the structure nor the purpose of budgets, and he appreciated less the necessity of publicizing them; he published no budgets until the year 1947–48 when a SR215 million budget (about $55 million) was made public. After that only two budgets were issued, the ones for 1951–52 and for 1952–53. Let us carefully examine the budget for the year 1952–53 (ending March 16, 1953).

Table XXIV. *Saudi Arabian Budget, 1952–1953*

A. *Income*

Item	Per Cent	Riyals	US Dollar Equivalent
Taxes (mostly income)	48.7	356,000,000	96,400,000
Royalties	30.1	220,800,000	59,700,000
Foreign trade	10.3	75,200,000	20,300,000
Pilgrims and traders	2.3	16,800,000	4,500,000
Government enterprises	3.0	22,300,000	6,000,000
Other revenue	1.5	10,900,000	2,900,000
Issue of new coins	4.1	30,000,000	8,100,000
Total	100.0	732,000,000	197,900,000[a]

a. Young, "Saudi Arabian Currency and Finance," *MEJ*, VII, 553, Autumn, 1953; UN, *Summary 1952–1953*, 62–63.

B. *Expenditures*

Item	Per Cent	Riyals	US Dollar Equivalent
General development, including agriculture	21.5	163,100,000	44,000,000
Defense	17.8	135,200,000	36,500,000
"Riyadh Affairs"	13.6	103,100,000	27,900,000
Internal security and subsidy to tribes	10.4	79,100,000	21,400,000
Ministry departments	9.9	75,000,000	20,300,000
Health, education, social and religious services	5.3	39,800,000	10,700,000
Government enterprises	3.3	25,200,000	6,800,000
Debts	13.2	100,000,000	27,000,000
Purchase of new coins	4.0	30,000,000	8,100,000
Reserve and adjustment	1.0	7,500,000	2,000,000
Total	100.0	758,000,000	204,700,000[a]

a. Young, *loc. cit.*, 555; UN, *Summary 1952–1953*, 64.

The item Taxes was primarily from oil, thus percentage-wise the direct income from oil was about 80 per cent. The 10.3 per cent from Foreign Trade must also be attributed indirectly to oil, for it was through income from the oil that imports increased, enabling the Government to collect such a large sum from custom duties.

The item "Riyadh Affairs," which means the royal household, was allotted $27,900,000, almost four times the country's total budget for 1938. The $21,400,000 allotment for Internal Security and Subsidy to Tribes, three times the entire 1938 budget, was to maintain the Saudi dynasty in power. On the other hand, Health, Education, Social and Religious Services all combined, were allotted only $10,700,000, or 5.3 per cent of the budget, and that in a theocratic state. Payments for debts, $27,000,000, 13.2 per cent of the budget, is a strange item in a country which derived more than $200,000,000 a year in oil revenues. The $36,500,000 for Defense, 17.8 per cent of the total, seems quite out of proportion in a budget which gives all social, educational and health services only 5.3 per cent. Nor is it easy to understand why a country surrounded by sister Arab states pro-testing unity and organized into one League should need such a large budget for defense. It is even less understandable, in view of the general praise heaped by Company representatives on King Ibn Saud for the amazing internal security he established, that it should be necessary to allot 10.4 per cent of the total budget for such security.[37] General Development, the largest single item in the budget, amounting to 21.5 per cent, gives no clue as to whether it was for wealth-producing projects or some capricious projects of the King's. An excellent example of Ibn Saud's spending, which very likely came under the heading of General Develop-ment, was the railroad from Dammam to Riyad.

In the winter of 1945, when Ibn Saud talked with President Roosevelt on the cruiser *Quincy*, the President asked the King how many miles of railroad there were in Saudi Arabi. The King, who had never ridden on a railroad or even seen one, was reported to have replied that although there were several hundred miles of railroad none of it was being worked at that time. He was probably referring to the former Hejaz railroad which ran from Damascus to Medina, which had been built by the Turks early in the century and practically destroyed in World War I. The question worked on the King; he was further stimulated when during his visit to Egypt in 1946 Ibn Saud actually rode on a train. It was supposedly then and there that he made up his mind to have a railroad in Saudi Arabia, come what may. The fol-lowing year he instructed his Minister in Washington to inquire about the possi-bilities of an American loan to finance a railroad from the Persian Gulf to his capital. The United States Government's reply was that the line which the King proposed was more a political matter than an economically sound project, and that the United States Government would prefer to lend Saudi Arabia money for such improvements as harbors, roads, public utilities and hospitals, but not for the railroad.[38] Ibn Saud was not, however, to be deterred from his objective. Although the experts consulted advised a motor road, he still wanted a railroad and Aramco undertook to build it for him. The Company began to build the line in the middle of 1947 and completed it in October 1951, at an approximate cost of $160 million which Aramco advanced to the King and which was to be repaid out of royalties before 1960.[39] One of the greatest admirers of Ibn Saud and of Saudi Arabia, St. John Philby, declared that the oil made it possible for Arabia to indulge in extravagances out of its own revenues, and it was done literally on a princely scale, "leading off with a despatch of a dozen princes to the New World to inaugurate the new era of the United Nations, and to ransack America for motor-cars and other aids to the enjoyment of life. Other such expeditions followed, one led by the Crown Prince and another by 'Abdullah Sulaiman himself [Finance Minister]: each bringing back to Arabia substantial mementos of its invasion of the richest country in the world." Philby reported that he had sat down to *alfresco* dinners in the Crown Prince's garden estate at Riyad at which every item on the menu had come from America on refrigerator planes.[40] He stated that even the warmest admirers of Saudi Arabia could not claim that all the income of the Government had been used wisely "or even legitimately. There has unquestionably been a great deal of leakage of a far from creditable nature, and there has been gross extravagance, while many of the undoubtedly useful development works that have been undertaken have been contracted for on an unnecessarily expensive basis."[41]

Because of the absolutist character of its government and the utter backwardness of the people, Saudi Arabia did not have the benefit of the services of experts to present a general plan of development, as was done in Iran and Iraq. All decisions depended on the King, but his wisdom and knowledge of development were woefully lacking, as illustrated by the railroad. And even when he decided on worthwhile projects, he lacked the necessary men of vision and the determination to execute the projects. Corruption, most pronounced where no sense of national consciousness and public responsibility exist, also plagued Saudi Arabia and played its part in the disposal of the wealth derived from oil. Philby declared that with Italy's entry into the war in 1940 and the resultant halt of the pilgrimage, the King's economic position seemed hopeless. "Then the British and American Governments came forward with generous measures of help, to which the Arabian response was a further orgy of extravagance and mismanagement accompanied by the growth of corruption on a large scale in the highest quarters."[42]

A very revealing insight into King Ibn Saud's attitude on the general Arab question is afforded by Philby's remark about the Saudi Arabian Government's action in connection with the Arab's anti-American agitation because of United States' support of the Jews in the Palestine question. "The only really effective action which the Saudi Arabian Government could have taken, namely the cancellation or suspension of the American oil concession as a policy against American championship of the Jewish cause, was never considered seriously, or considered only to be rejected out of hand as prejudicial to the economic interests of the country, in spite of a spate of propaganda sponsored by unfriendly elements in Iraq and Trans-Jordan."[43] Knowing that it was dependent on the King's good will for the continued fully successful operation of its concession, Aramco followed a policy of catering to his whims.[44]

More so than in Iran and even in Iraq, the specially trained Saudi Arabian employees of the Company could not find employment anywhere else in the country, because nowhere in Saudi Arabia could their training and skill be utilized, and having broken with their previous mode of living they became more and more dependent upon the Company. The industrial stage of the country is perhaps best illustrated by the following table of the internal consumption of petroleum and petroleum products.

Table XXV. *Saudi Arabian Internal Petroleum Consumption*

(In Barrels)

1948	387,353
1949	392,755
1950	554,658
1951	784,284
1952	1,113,815
1953	1,466,627
1954	1,750,655[a]

a. *Report, 1950,* 17; *Report, 1953,* 23; For consumption in tons, see UN, *Review 1951–1952,* 62; UN, *Summary 1952–1953,* 57; *Report, 1954,* 14.

Keeping in mind that the original concession called for a free supply of petroleum products in considerable quantities to the Saudi Government, that motor vehicles were very widely used, and that the railroad was diesel-powered, the above amounts are small indeed; it would seem that most of the petroleum consumed was for transportation rather than for industrial purposes.

Nevertheless, the impact of the oil industry in Saudi Arabia has affected practically the entire population. With all the shortcomings of the Government's efforts, the income from oil redounded to the benefit of the whole country. Well-boring projects have resulted in an increased supply of water, making an expansion in agriculture possible; new roads have been laid out, ports constructed and improved; the railroad from Dammam to Riyad built; a number of towns and villages electrified; and the airplane has become a common feature of the desert sky. The efforts of the Company and the Government have brought about a transformation of the former placid desert country.[45] at least in some parts, into a teeming community. With this transformation have come deep social stirrings. The Arabs — both those directly employed by the oil company and those whose employment is indirectly connected with the Company's activities — have learned at least some of the ways of the West and to some extent have broken with their former nomadic way of life. These elements have become a leavening force for discontent as they have abandoned their former fatalistic attitude toward life and toward the existing regime.

Moreover, the substitution for the former austerity of a life of oriental extravagance in the ruling circles, made possible by the easy and fabulous oil income, had become a disturbing force and a model for imitation, and the tribes had been making greater and greater demands.[46] It is, of course, only natural that the abysmal general ignorance and illiteracy should have slowed down the general process of progress and the concomitant dissatisfaction with the status quo. But the deep revolutionary changes which the oil industry had brought to this primitive desert country were of such magnitude that they could not but seriously upset the social and economic patterns of centuries-old traditions. As long as the world demand for oil was great and prices were high and income constantly mounted, no serious disturbances were to be expected. However, should production drop or prices go down, the King could no longer meet the ever-expanding demands of the royal house and the subsidized tribes, and the workers in the oil industry would be thrown out of work. The desert kingdom might then be rocked by serious social and political convulsions.[47]

NOTES

1. *Report, 1953*, 12, 41; Lebkicher, *Aramco and World Oil*, 55.
2. *Report, 1953*, v, 41. The spectacular expansion of the refinery in a period of less than ten years is shown by the following table of throughput (in barrels);

1945	2,953,623
1946	29,297,101
1947	39,065,060
1948	45,086,139
1949	46,269,619
1950	38,364,333
1951	58,107,534
1952	62,204,161
1953	74,559,673
1954	79,800,000

 Ibid., 19; *International Financial News Service*, Feb. 4, 1955; *Report, 1954*, 12.
3. UN, *Summary 1952–1953*, 63–64; Lebkicher, *Aramco and World Oil*, 55; *Report, 1953*, 39.
4. *Report, 1949*, 12–13; *Report, 1950*, 18; *Report, 1951*, 35; *Report, 1952*, 39.
5. *Report, 1948*, 23; *Report, 1949*, 26.
6. *Report, 1950*, 28.
7. *Report, 1950*, 28; *Report, 1951*, 23–29; *Report, 1952*, 44–46; *Report, 1953*, 45–46.
8. *Report, 1950*, 28; *Report, 1953*, 46.
9. Richard A. Sanger, *The Arabian Peninsula* (Ithaca, 1954), 109 *et passim*; *Report, 1952*, 33–37.

10. *Report, 1948,* 20; *Report, 1950,* 47; *Report, 1951,* 31—33.
11. *Report, 1953,* 41—43.
12. US Senate, *Hearings Saudi Arabia,* 24756, 25101—25122 *et passim.*
13. *Fortune,* May, 1947, 179.
14. See Chap. XI above.
15. In an editorial reacting to this arrangement *World Petroleum* states: "Viewed in perspective, the recent agreements among Middle East companies appear as ordinary commercial undertakings inspired by commercial motives. At the same time the oil companies are to be credited with having introduced an element of stability into the Middle East oil situation which the diplomats failed to accomplish." XVIII, 27, February, 1947.
16. US Senate, *International Petroleum Cartel,* 120—125; *CR-Senate,* 94, part 4, 4941, 4955, Apr. 28, 1948; *The New York Times,* Dec. 27, 1946, Mar. 13, 1947.
17. US Senate, *Hearings Saudi Arabia,* 25022.
18. *Ibid.,* 25105. James Moffett calculated that the Company's investment, including advances to Ibn Saud, and the building of the refinery, which amounted to $82 million, was recovered from the sale of oil to the US Navy in 1944 and 1945, *Ibid.,* 24732.
19. UN, *Summary 1950—51,* 19.
20. United Nations, *The Price of Oil in Western Europe,* 15.
21. This theme was presented as early as 1944 when Ickes explained to the American public why the Government wanted to build the proposed pipeline. He stated that Ibn Saud wanted royalties and in considerable quantities, and unless the American company supplied him with such royalties, and quickly, it might find itself without a concession. Ickes, "Oil and Peace," *Collier's,* 55, Dec. 2, 1944.
22. US Senate, *International Petroleum Cartel,* 121.
23. E. DeGolyer, "The Oil Fields of the Middle East," *Problems of the Middle East* (New York, 1947), 83; US Senate, *Additional Report of the Special Committee Investigating the National Defense Program. Report of the Subcommittee Concerning Investigations Overseas. Section 1 — Petroleum Matters.* 78th Congress, Report No. 10, Part 15 (Washington, 1948), 8.
24. Hull, "Tapline Presents Great Organization Problem," *The Oil Forum,* II, 449, Nov., 1948.
25. US Senate, *Hearing Saudi Arabia,* 24839, 25159, 25288.
26. US Senate, *International Petroleum Cartel,* 6.
27. UN *Review 1949—1950,* 60.
28. UN, *Review 1951—1952,* 53.
29. UN, *Summary 1952—1953,* 39—40.
30. United States, Department of Commerce, *Business Information World Trade Series,* Dec., 1952, 2. Lebkicher, *Aramco and World Oil,* 95, gives an estimate of 3.5 million, while the United Nations gives the population in 1947 as 6 million. UN, *Review 1949—1950,* 43.
31. So serious was the strike that it led to the imposition of martial law and to the arrest of one hundred strike leaders. *The New York Times* reported that the progressive breakdown of many of the traditional and religious loyalties had

made the oil workers increasingly receptive to the appeals of native agitators and the Moscow radio. The most outstanding aspect of the strike was that while it was originally against Aramco for better working conditions, it turned against the Saudi Government for arresting some of the strike leaders. *The New York Times,* Dec. 20; *The New York Herald Tribune,* Oct. 20; *The Christian Science Monitor,* Nov. 12, 1953.

32. UN, *Summary 1952–1953,* 63–64. A. C. Sedgwick reported from Cairo on the impact of Western standards on Saudi Arabian oil workers: "Arab workers in Saudi Arabia, already used to higher living standards, had requested drinking water. When they received water that had been warmed by the sun, but which they and their forefathers used to drink, they spat it out and demanded 'American' water, meaning ice water carried in vacuum jugs." "And," he added, "it is expected that workers who are not able to achieve the standard set by the Americans will clamor for it nonetheless." *The New York Times,* Feb. 10, 1944.

33. Young, "Saudi Arabian Currency and Finance," *MEJ,* VII, 553, Autumn, 1953.

34. *Ibid.;* Sanger in *The Arabian Peninsula,* 111, states that 90 per cent of the King's total income came from oil revenues. The UN states that in 1953, 80 per cent was derived from oil. UN, *Summary 1952–1953,* 64.

35. Sanger, *The Arabian Peninsula,* 108–109.

36. UN, *Review 1949–1950,* 12. This is based on the income from oil.

37. Speaking of the financial factors of Saudi Arabia, Philby states: "During the past decade for instance, the year's budget has only thrice been published for general information; and in none of the three cases have the published figures (rising from a total estimated revenue of £25 million in 1947 to £100 million last year) given the public any real idea of how the income of the country has been laid out for its benefit." H. St. J. B. Philby, "The New Reign in Saudi Arabia," *Foreign Affairs,* XXXII, 453, Apr. 1954.

38. On Mar. 3, 1947 the Department of State revealed that the Saudi Arabian Legation had submitted a report to the Department, prepared by an American engineering firm, which declared that the proposed railroad was "not only economically justifiable but also economically feasible." *Department of State Bulletin,* XVI, 506, Mar. 16, 1947.

39. Sanger, *The Arabian Peninsula,* 118. "Against considerable opposition the king insisted on the building of a railroad across the Dahna sands from the Persian Gulf to his capital." *Ibid.,* 39. "Ibn Saud's idea of a major public works project is an unnecessary railroad from his inland political capital to the Persian Gulf." Leigh White, "Allah's Oil: World's Richest Prize," *Saturday Evening Post,* Nov. 27, 1948, 31; Young, *loc. cit.*

40. Philby, *Arabian Jubilee,* 232.

41. *Ibid.,* 182–183.

42. *Ibid.,* 227.

43. *Ibid.,* 218–219. On Feb. 27, 1948 the *New York Herald Tribune* correspondent reported from Dhahran that despite anti-American sentiment in the Middle East because of the American stand on the partition of Palestine, King Ibn Saud had assured the Arabian-American Oil Co. that it would not lose its

concession in his land. In return for his friendship Aramco redoubled its lobbying in Washington to convince the Government that support of Palestine might force abandonment of the potentially greatest American investment in any foreign country. The Company announced that William A. Eddy, former Middle East expert in the State Dept. and former Minister to Saudi Arabia, had been retained to present its case, and that its public relations budget had been increased. *The New York Herald Tribune,* Feb. 27, 1948. Discussing the progress of production in the Middle East, the London *Petroleum Press Service* observed in April 1948: "Notwithstanding the tragic disturbances in Palestine and the political repercussions caused by them, the development of the rich oil resources of the Middle East has made further considerable headway." XV, 80, April, 1948.

44. Reporting from Riyad early in December 1953, the correspondent of *The New York Times,* described the financial conditions of the country and the exorbitant amounts absorbed by the spending of the royal family and its entourage and by the subsidies to the tribes. He noted that some observers were of the opinion that King Saud, more progressive and more international-minded than his father, would introduce financial reforms by reorganizing the governmental machinery. Others, however, asserted that "King Saud has abetted the practice of royal profligacy by his luxurious life in his new palace at Nasiriya, near here [Riyad] with its swimming pool, scores of fountains and floodlit gardens of imported plants." *The New York Times,* Dec. 11, 1950.

45. It would not require too vivid an imagination to realize what the activities of the oil company have done to the otherwise quiet desert, as illustrated by the following statistics of movement connected only with the building of a pipeline across the country to the Mediterranean. In addition to a vast amount of construction equipment such as graders, caterpillar tractors, traveling cranes and ditching machines, as well as other equipment, the Company used an enormous automotive fleet consisting of 150 fifty-ton truck-tractors, 120 tenton truck-tractors, 135 six-ton trucks, 500 four- and eight-wheel tractors; 4 sixty-passenger buses, 10 sixty-passenger trailers, 80 refrigerator trucks for perishable foods, 12 lunch trailers, 60 fuel and water tank trailers, and some 400 jeeps, pickups, station wagons and smaller vehicles. Hull, "Tapline Presents Great Organization Problem," *The Oil Forum,* II, 454, Nov., 1948.

46. The methods of supporting the royal household and subsidizing the tribes are described by the outstanding orientalist, J. Heyworth-Dunne, "Report from Saudi Arabia," *Jewish Observer and Middle East Review,* IV, 13–14, Jan. 14, 1955. Analyzing the social changes which the oil brought about in the Middle East, the *Economist* states: "Saudia Arabia publishes estimates of revenue and expenditure, but no figures for actual expenditure which — despite the astronomic growth of its income — has in recent years regularly and considerably exceeded revenue. Judging by appearances, one reason for this deficit is that a large proportion of the revenue provides a cushioned existence, and palatial private investment in real property abroad, to princes, ministers, rivals for power and other palace connections." "Oil and Social Change in the Middle East," *The Economist,* CLXXVI, 2, July 2, 1955. Special insert.

47. An enthusiastic description of the impact of oil on the country is presented by

Dr. Kheirallah in the chapters, "A Land Transformed," "A People Reawakened," and "Bloom in the Desert," the last of which concludes thus: "The lot of the people has been hard in this desiccated land, but Allah, in his compassion, stored underground a compensation bounty of oil and water; and now we have the stimulating sight of a King with the wisdom and determination to tap both and make the desert bloom." George Kheirallah, *Arabia Reborn* (Albuquerque, 1952), 185–228. See also Lebkicher, *Aramco and World Oil*, 68–71, 99–100; *Reports, 1948–1953*; United States Senate, *American Petroleum Interests in Foreign Countries. Hearings before a Special Committee Investigating Petroleum Resources.* 79th Congress, First Session, June 27 and 28, 1945 (Washington, 1946), 286–288.

Chapter XIX
PROGRESS AND PROSPECTS 1955–1971

PRODUCTION AND REVENUE

Of the countries which have been discussed in the previous sections, Saudi Arabia was and is the most conservative and underdeveloped. The nature of the regime, the general level of development of the population, and the desert-like physical characteristics of the territory, in spite of the impact of the revolutionary innovations caused by the oil industry and all its ramifications, have worked for relative stability and have determined the oil industry pattern of development during the period under review.

As Table XXVI shows, production of oil has been steadily and constantly growing, from 356.6 million barrels in 1955 to 1,704.8 million in 1971, almost a fivefold increase; and neither the 1956 Suez Canal crisis nor the 1967 Arab-Israel war and the consequent Syrian interruptions of the flow of oil through the pipeline has seriously curtailed output. During the same period, production from the Saudi

TABLE XXVI.
Saudi Arabia. Oil Production, 1955–1971, by Companies (Million Barrels)

Year	Aramco	Getty Oil	Arabian Oil	Total
1955	352.2	4.4		356.6
1956	360.9	5.8		366.7
1957	362.1	11.6		373.7
1958	370.5	14.7		385.2
1959	398.8	21.2		421.0
1960	456.4	24.9		481.3
1961	508.3	28.7	3.7	540.7
1962	555.0	33.7	11.0	599.7
1963	594.6	33.1	24.1	651.8
1964	628.1	34.4	31.8	694.3
1965	739.1	32.6	33.1	804.8
1966	873.3	30.2	46.5	950.0
1967	948.1	25.1	50.6	1,023.8
1968	1,035.8	23.2	55.1	1,114.1
1969	1,092.3	22.7	58.8	1,173.8
1970	1,386.7	28.7	62.7	1,386.7
1971	1,641.6	33.7	65.5	1,740.8[a]

a. Saudi Arabian Monetary Agency, *Annual Report 1389–90 A.H. (1970)*, 87; – *Statistical Summary*, 47, Sept., 1972; Aramco, *Facts and Figures 1971*.

Arabian share of the Getty Oil Company in the Neutral Zone rose from 4.4 million barrels in 1955 to 33.7 million in 1971, and the half share in the Arabian Oil Company (Japanese) went up from 3.7 million barrels in 1961 to 65.5 million in 1971.

Similarly the oil revenue rose, as shown in Table XXVII, from $340.8 million in 1955 to $1,149.7 million in 1970, and then jumped to $1,944.9 in 1971, making more than a fivefold increase. The Saudi Arabian half share in the Getty Oil Company in the Neutral Zone went up from $2.6 million in 1955 to $20.6 million in 1971, and the half share in the Arabian Oil Company rose from $2.5 million in 1959 to $44.2 million in 1971. In the last year the Government collected from new companies exploring for oil $13.7 million. A very impressive accomplishment.

Table XXVII. *Saudi Arabia. Oil Revenue, 1955–1971, by Companies (Million $)*

Year	Aramco	Getty Oil	Arabian Oil	Other Oil Companies	Total
1955	338.2	2.6			340.8
1956	286.8	3.4			290.2
1957	286.5	9.8			296.3
1958	287.4	10.2			297.6
1959	295.3	15.3	2.5		313.1
1960	312.8	18.4	2.5		333.7
1961	352.2	22.9	2.5		377.6
1962	381.7	25.0	3.0		409.7
1963	571.1	23.0	13.6		607.7
1964	482.1	27.7	17.4		523.2
1965	618.4	23.7	20.4		662.6
1966	745.5	20.6	22.3	1.3	789.7
1967	859.4	17.8	31.8	0.1	909.1
1968	872.0	13.6	34.3	6.9	926.8
1969	895.2	15.2	37.1	1.5	949.0
1970	1,088.4	17.2	40.3	3.8	1,149.7
1971	1,866.4	20.6	44.2	13.7	1,944.9[a]

a. Saudi Arabian Monetary Agency, *Annual Report 1389–90 A.H. (1970)*, 88; – *Statistical Summary*, 44, Sept., 1971; – 48; Sept., 1972; UN, *Economic Development 1958–1959*, 77; *Petroleum Press Service*, XXXIII, 226, Sept., 1966; – XXXVII, 324, Sept., 1970; – XXXVIII, 326, Sept., 1971.

Relations with the oil companies, especially with Aramco, the major and by far the most decisive both in scope of operations and contribution of revenue, was good and at times even friendly. The Government demands from the Company were mostly met. Aramco agreed to eliminate gradually the marketing allowances; it worked out a formula, after recognizing the principle, for royalty expensing, which ultimately expensed royalty in totality; it practically always met the request for increased production, which resulted in expansion of refining capacity, pipeline extension and additional storage tanks, terminal and loading facilities. In line with the other companies in the region, Aramco agreed in 1971 to increase the posted prices, raise the Government share of the profits by 5%, and granted a premium for the oil shipped by pipe through the Mediterranean, and in 1972 agreed to adjust payments due to the devaluation of the dollar. Finally Aramco agreed in principle to Saudi Arabian Government participation of up to 25% in the Company. The Company systematically relinquished sections from both its exclusive and preferential concession areas.

ARAMCO DEVELOPMENT

In June 1956, a new producing field, Khursaniyah, about 8 miles northeast of Dhahran, was discovered. In April 1957 the offshore field of Safaniya was producing at the rate of 50,000 barrels a day.[1] By the end of the year Aramco had a network of approximately 600 miles of pipelines, ranging in diameter from 10 to 32 inches, and from 18 to 142 miles in length. Along the Tapline the Company installed five fully automatic pumping units, increasing its capacity by 440,000 barrels a day, or by about 25%.[2]

In 1959 Aramco expanded the North Pier at Ras Tanura's terminal by 800 feet, which gave it a total of ten tanker berths.[3] In spite of a worldwide surplus of oil production and the Soviet Union's increased oil exports to world markets, Aramco reported in 1960 an increase of 13.9% over 1959.[4] In the following year the Company discovered a new offshore oilfield, Manifa, and added 29 miles of 30/32-inch pipeline onshore and 30 miles of new offshore submarine pipeline, and new pumping stations.[5]

Five years later the Company expanded oilfield development and raised the daily crude production by 390,000 barrels in the Safaniya offshore field and by 165,000 barrels in the Ghawar field. It added 165 miles of trunk pipeline to the crude transport network, and it completed the last 28-mile section of the third 130-mile pipeline between Safaniya field and Ras Tanura. In order to expand delivery capacity and accommodate the new giant oil tankers, Aramco built a two-berth deep-water sea-island two miles off Ras Tanura. In the same year the Company completed building three 500,000-barrel crude oil tanks and one 200,000-barrel butane tank at the Ras Tanura tank-farm. Three additional 500,000-barrel oil tanks and one 400,000-barrel propane tank were under construction. The offshore fields of Abu Safah and Manifa were brought into production.[6]

In 1967, the Company expanded the sea-island from two to four berths, and a fifth was nearing completion; it accommodated tankers of 200,000-ton capacity. Work was started on the construction of two 630,000-barrel crude oil tanks. Facilities at Ras Tanura were extended to accommodate tankers of up to 400,000 tons.[7] One year later the Company built an additional 146 miles of pipelines, bringing the total system, excluding flowlines, to a total of 1,879 miles.[8] In 1969, the Company added 135 miles of pipeline; a sixth berth of the sea-island was completed.[9]

The Company continued to expand its facilities in 1970. A million-barrel crude oil storage tank was completed at Ras Tanura terminal, and a second one was nearing completion. 26.2 miles of pipeline were laid, and an additional 36 miles of pipeline were under construction.[10] In 1971, Aramco took delivery of a 230,000-ton tanker for use as a floating storage vessel. It was stationed 40 miles off Safaniya and moored to a buoy in water about 130 feet deep. It could accommodate the largest tanker afloat.[11] As is shown in Table XXVIII, the capacity of the Ras Tanura refinery was constantly expanded, from a throughput of 9,858,907 tons in 1955 to 25,139,080 tons in 1971.[12]

Table XXVIII. *Saudi Arabia. Aramco Disposition of Crude — Tons*

Year	Total Production	To Tapline	To Bahrain	To Ras Tanura Refinery	To Tankers	Others
1955	46,784,693	15,674,363	8,352,131	8,858,907	12,858,203	41,089
1956	47,935,041	15,992,909	7,676,236	9,683,050	14,484,915	97,131
1957	48,229,690	16,860,873	5,930,547	9,460,410	15,724,129	253,731
1958	49,339,006	18,010,623	6,465,085	8,093,168	16,609,491	160,639
1959	53,307 390	16,524,325	6,732,275	8,469,115	21,502,139	79,536
1960	61.087,931	12,244,940	7,769,905	10,988,743	29,934,148	150,095
1961	68.138,424	15,328,857	8,321,077	12,135,887	32,305,672	46,931
1962	74,554,207	16,833,963	9,202,502	12,200,829	36,052,623	264,290
1963	79,768,933	19,028,060	8,753,810	13,009,348	38,690,209	287,506
1964	84,442,667	21,305,334	7,517,627	13,443,703	42,009,989	166,014
1965	89,436,852	21,951,439	6,465,630	14,737,069	55,795,940	486,774
1966	117,568 598	22,975,595	6,550,126	14,869,700	72,934,233	238,944
1967	127,262,215	16,578,999	8,301,758	15,870,390	85,848,509	662,559
1968	138,776,535	23,527,249	7,415,214	18,548,146	88,561,415	724,511
1969	146,494,684	16,768,062	7,727,986	19,441,523	101,794,233	762,880
1970	174,056,992	8,585,380	8,440,060	26,332,213	130,044,765	654,574
1971	219.983,830	17,146,689	8,776,023	25,139,080	167,673,761	1,248,277[a]

a. Aramco, *Report of Operations 1959*, 2, 5; — *Aramco 1963*, 7; — *1967*; — *1970*; — *Facts and Figures 1971*.

ARAMCO RELATIONS WITH THE GOVERNMENT

After the Saudi Arabian Government nominated Abdullah at-Tariqi and Hafiz Wahaba, they were elected on May 21, 1959 to the Board of Directors of the Company, which consisted of 15 members, four representing the four parent companies, two Saudi Arabia and nine the management of the Company.[13]

The basis of calculating the profits of oil delivered in Sidon was an issue between Aramco and the Government. From October 6, 1953 to December 31, 1963, the basis was the Ras Tanura posted prices; the Government demanded that the basis be the Sidon posted prices minus Tapline's transport charges including transit and other payments to the Governments of Saudi Arabia, Jordan, Syria and Lebanon. Aramco agreed to pay the Government income tax on the excess of the profits so calculated over the profits calculated hitherto. Beginning with January 1, 1964 Aramco's profits from deliveries at Sidon were to be calculated on the basis of Sidon posted prices, minus any marketing allowances and discounts granted by the Company.

In the same year the Saudi Government reached agreement on Tapline transit payments effective as of January 1, 1963. In settlement of all claims up to that time, the Company was to make a payment of $2,178,989.[14]

At the end of 1964, the Government accepted the Company's formula for royalty expensing effective as of January 1, 1964. The increase in Government revenue from this formula amounted to $32 million in 1965.[15]

On July 1, 1966, Aramco sold Petromin (about which later) its marketing facilities in Saudi Arabia; it included bulk plants, airport fueling units and a storage depot.[16] In the middle of June 1969, the Company signed an agreement with Petromin to supply the latter with LPG for local distribution, and to supply natural gas to local industries, including Saudi Arabian Fertilizers Company, Saudi Cement Company and the Dhahran Power Company.[17]

Aramco was one of the signatories of the Teheran agreement of February 14, 1971, which increased the Government profit share by 5% and posted prices by 33 cents per barrel plus 2 cents in settlement of freight differentials. All marketing allowances were eliminated, and an increase of 2.5% was made in posted prices of crude in addition to an annual increase of 5 cents per barrel starting from June 1, 1971 to compensate for inflation. In June 1971, Aramco signed an agreement with the Government to raise posted prices of oil exported from Sidon from $2.37 a barrel to $3.18 as of March 20, 1971, and of course a 5% increase in Government profits share.[18]

On July 2, 1960 the Company relinquished 33,700 square miles of its concession area, which was reduced to 306,000 square miles; this was the fifth relinquishment under the terms of the agreement, making a total of 140,413 square miles in the exclusive and 135,200 square miles in the preferential areas, respectively.[19]

Under the impact of the Iraqi act at the end of 1961, discussed above, the other Middle Eastern oil-producing countries began pressing their companies for relinquishment. Saudi Arabia entered into negotiations with Aramco for the same purpose. On March 26, 1963, an agreement was reached and approved by King Saud. It provided for the immediate relinquishment of all but 125,000 square miles of the concession areas; this was about 75% of the original concession. It also provided for six additional relinquishments at five-year intervals, so that by the end of the concession period in 1993, the area retained by the Company was to be about 20,000 square miles, or 3% of the original grant.[20] Consequently, in line with the provisions of this agreement, five years later Aramco relinquished 20,000 square miles from its exclusive area.[21]

Aramco followed a policy of steadily, even though slowly, increasing the number of Saudi Arabians in its employ in Saudi Arabia. In 1959, for instance, Saudi Arabians comprised about 72% of all persons employed by the Company, Americans about 15% and others about 13%. In 1970, while the total number of employees went down, the percentage of Saudi Arabians went up to 82.4; the American percentage went down to 9.2, and that of the other nationals to 8.4. The average annual salary of the Saudi Arabian employees went up steadily from $2,187 in 1963 to $3,956 in 1971. Moreover, as a measure of economy demanded by the Saudi Arabian Government, the Company reduced its staff in New York headquarters in 1960 from 530 to 372.[22] The Company provided its employees and their dependents with medical care and health programs, which cost the Company in 1970, for instance, $18,760,000.[23]

Aramco also purchased local materials and services, which added to the country's economic wealth. It estimated that these purchases, salaries and wages of employees, as well as personal taxes, etc., contributed, for instance, in 1961, in addition to direct payments, $72 million, and this increased steadily and reached over $127 million in 1971.[24]

OTHER GOVERNMENT OIL ACTIVITIES

Early in December 1957, the Government granted a concession to the Japanese Petroleum Company to exploit Saudi Arabia's undivided share of the offshore area of the Neutral Zone, between Saudi Arabia and Kuwait. The Government was to receive 56% of the profits; it was also entitled to acquire, at a minimal price, 10 to 20% of the Company's shares. A basic departure from all previous concessions was the provision that the Government was to share with the producing company in the profits made outside as well as inside Saudi Arabia. The Japanese concessionaire was to set up a company which was to produce, refine, transport and market oil. The Company was to pay the Government a 20% expensed royalty in kind or in cash, and was to make available 20% of produced oil at a price five per cent below that charged other customers. The Government was to name one of the members of the Board of Directors, and the Company renounced its "right of diplomatic recourse." The Company's management and books were to be located in Saudi Arabia, and 70% of the total labor force was to be Saudi Arabian. The exploitation term was for 40 years after the discovery of oil in commercial quantities. The Company was to pay, until discovery, a rental fee of $1,500,000, and on discovery a bonus of $1 million; it was to erect a refinery. On February 5, 1958, the Japanese Arabian Oil Company was formed in Tokyo.[25]

The major oil spokesman of the Saudi Arabian Government was Abdullah at-Tariqi, the Director of Mines and Petroleum Directorate. He stated the new Government policy: "We believe that each petroleum concessionaire within our borders should create and maintain an integral petroleum operation based on its Saudi Arabian production, including refining, transportation and marketing, in addition to production. We believe that this integrated operation of each concessionaire should not be limited to the geographic boundaries of this nation but should extend to and take place in other parts of the world where an integral activity is legally and practically possible."[26]

In line with the innovations of new concession forms in Iran, Saudi Arabia entered in 1963 into an agreement with AUXIRAP, a subsidiary of the French state firm RAP (Regie Autonome du Pétrole), for exploring for oil for a period of 2 years in the Red Sea area, during which time the French group was to spend $5 million; drilling was to begin in the second year. A 30-year exploitation concession was to be granted upon commercial discovery of oil. The agreement called for the relinquishment of 20% of the unexploited area within 3 years and an additional 20% every succeeding five-year period. AUXIRAP was to pay a bonus of $500,000 on signature; other bonuses were to be $1 million upon discovery of commercial quantities of oil and $4 million when daily production reached 70,000 barrels for 90 consecutive days. Rentals were to range from $5 per square kilometer in the first year to $500 after the 25th year. Royalties, fully expensed, ranged from 12½ to 20% of posted prices, depending on the magnitude of production, and could be taken either in cash or in kind.

In the joint-company to be set up Petromin could take up to a 40% share of ownership (but 50% of voting rights) in the production phase, and from 30 to 50% in a fully integrated company. This company was to be financed from 30% of net profits on production. AUXIRAP was to market half of its crude; and oil and products not marketed were to be sold 60% by AUXIRAP and 40% by Petromin. The French company was to be subject to Saudi Arabian income tax. In 1968 AUXIRAP assigned, with the consent of the Government, 1/3 of its exploration rights to the American Tenneco.[27]

PETROMIN

In line with the other major oil-producing countries in the Persian Gulf in establishing national oil companies — Iran, Kuwait and Iraq — a royal decree issued in Mecca on November 22, 1962 set up the General Petroleum and Minerals Organization; later it became known as Petromin. Its immediate task was to supervise all oil affairs in the country, and it was empowered to enter into any phase of oil and mineral activity. It was granted Rls. 1 million as an operating budget for the fiscal year 1962/1963; it had no capitalization. A nine-member board of directors, under the chairmanship of the Minister of Petroleum and Mineral Resources, was to be responsible to the Government; it was attached to the Ministry of Petroleum.[28]

In 1969 a Saudi Arabian official source defined the aim of Petromin as "participating in different industrial and commercial aspects of oil and minerals so as to establish and develop, in cooperation with private enterprise, oil and mineral industries and industries closely related to this sector."[29] The stress on, at least partial, participation of private enterprise in the various undertakings, whatever its motivation, was a factor in the financial structures and operations of Petromin.

In April 1964, Petromin organized the Arabian Drilling Company with a capital of Rls. 10 million. Petromin held 51% interest and 49% was held by two French firms. Early in 1966 Petromin established the Arabian Geophysical and Survey Company, in which it had a 51% share with 49% owned by a French Company.[30]

In March 1964, Petromin acquired Aramco's bulk plant at Jidda, and in July 1966, as was mentioned above, Aramco sold all distribution facilities in a number of towns and the storage facilities in Dhahran to the national company, which subsequently set up the Petromin Marketing Department. In 1968 a total of 2.5 million gallons of products were sold.[31]

Petromin entered, in 1964, into an agreement with Occidental Petroleum Company of the United States for the supervision of the construction of a fertilizer plant in Dammam and for managing and operating it. International Ore and Fertilizer Company (Interore), an Occidental affiliate, was to purchase, at prevailing world prices, the entire production of the plant and market it throughout the world for a period of 17 years, at a commission of 5% of the purchase price. One year later Petromin organized the Saudi Arabian Fertilizers Company (SAFCO) with a capitalization of Rls. 100 million. Petromin owned 51%, and 49% was offered for public subscription which, it was reported, was immediately taken up.[32]

In 1967, Petromin reached an accord with two affiliates of Occidental for building a plant to extract and produce sulphur out of natural gas available in the eastern region of the country, and on its financing, administration and marketing. The plant was to have been built in Abqaiq with a daily 500-ton production capacity. The capitalization of the Company, which was called Petromin Sulphur Company, was to be Rls. 120 million, half to have been raised through international loans, and of the remaining half, Petromin together with the Saudi private sector were to have contributed two-thirds and the two companies one-third.[33]

In the fifth year of its organization Petromin entered the field of oil exploration. On April 10, 1968, it signed an agreement with the Ministry of Petroleum for a three-year oil exploration concession covering two areas: 76,475 square kilometers in the Rub-al-Khali section and 9,107 square kilometers in the al-Hasa section. Petromin, in turn, while retaining its legal title, assigned its rights and obligations under the concession agreement to ENI's subsidiary AGIP under the following terms: Exploration period, three years, with a possible extension for an additional three years. Exploitation was to be for a period of 30 years. Within three years of the granting of the exploitation concession, following commercial discovery, 20% of the unexploited area was to be relinquished, and a further 20% every five years. Annual rents ranged from $10 per square kilometer for the first five years to $625 from the 26th year onwards. A bonus of $2 million was to be paid on signature; additional bonuses were: $3 million on granting the exploitation rights and $8 million when daily production reached 300,000 barrels. Royalty, fully expensed, ranging from 12½ to 14% of posted prices depending on rate of production, was to be taken in cash or in kind. Tax was to be 50% of profits, or any higher rate which might be imposed.

In the Saudi Arabian Oil Company to be set up by AGIP to operate the concession, Petromin could participate up to 30% and was to pay 30% of exploration expenditure. It had the option of raising its participation to 40% when daily production reached 300,000 barrels, and 50% when daily production went up to 600,000 barrels; in each case the national company was to contribute the respective part of the investment. Should no oil be found, AGIP was to bear all exploration costs. From the very beginning of exploitation a long-range program of integrated operations was to be planned. When daily production reached 200,000 barrels for a period of 90 days, Petromin was to build a 30,000-barrel daily capacity refinery within three years. At least 30% of both partners' net profits from crude oil production were to be made available for investmment in integrated operations. With the consent of the Government AGIP transferred 50% of its interests to Phillips Saudi Arabia, a subsidiary of the American Phillips Oil Company. In the same year Petromin obtained a 3-year concession to explore in two areas along the Red Sea coast. It then entered into an agreement with two United States firms, Sinclair and Natomas Oil companies, and Pakistan on terms similar to those of AGIP.[34]

REFINING

Like all the other Middle Eastern countries, Saudi Arabia aspired to develop refining facilities of its own, even though it had on its territory a very large refinery at Ras Tanura. The lack of refining facilities was particularly felt in the western region of the country, which was far from the oilfields. On May 30, 1960, therefore, King Saud granted his son Sab a concession to erect a 20,000-barrel daily capacity refinery at Jidda. It was to be for a period of forty years and renewable, and it granted the prince exclusive rights in the western region of the country for all operations connected with transportation, refining, importing and exporting of oil and oil products.

Subsequently the Saudi Arabian Refining Company was organized, owned 75% by Petromin and 25% by Saudi private investors. In November 1968, Petromin and SARCO organized the Jidda Oil Refinery with a capital of Rls. 70 million, 75% owned by Petromin and 25% by SARCO. Late in 1968, the refinery, which was contracted in 1966 by Petromin with a Japanese chemical and engineering firm at a cost of $7 million, went on stream with a daily capacity of 12,000 barrels. Plans were worked out for the extension of the Jidda refinery and for the construction of a refinery at Riyadh.[35]

NATURAL GAS AND PETROCHEMICALS

The question of the utilization of natural gas in Saudi Arabia is in some respects similar to that of Iran and Iraq and in some respects quite different. As in the other countries, gas is found as associated with crude oil and in gas fields, and as in the others it was mostly flared. As late as 1969, the Saudi Arabian Government maintained that 60% of all the natural gas produced had been flared. However, Aramco made many efforts to increase the utilization of the gas. Four major uses were made of the natural gas: generating electric power, fueling industrial plants, injecting in oil wells to keep up the pressure and liquefying for domestic use and for export.

As early as 1959, Aramco launched a program to promote industrial use of natural gas; and it attempted, through the availability of the gas, to attract new industries to the eastern province. In the same year a gas injection plant in the Ain Dar area, the largest such installation in the Middle East, began operations. By the end of the year the plant was injecting high-pressure gas at a daily rate of 160 million cubic feet. The Company also added a new section at the Ras Tanura refinery, at a cost of $8 million, which produced 4,000 barrels a day of refrigerated gas, LPG, for sale abroad. It built a special plant, at a cost of $6.5 million, for the purpose of exporting LPG. During 1960 Aramco began to export LPG, and added another LPG plant, costing $7 million, at Abqaiq. Two years later the capacity to produce refrigerated liquefied petroleum gas was increased to 12,000 barrels a day. Exports in 1963 totalled 1,674,322 barrels.[36]

Three years later Aramco shipped daily 15,356 barrels of refrigerated liquefied petroleum gas, injected daily 337 million cubic feet of gas, and completed a second pipeline for refrigerated liquefied petroleum gas from the refinery to the terminal to load simultaneously propane and butane.[37]

Natural gas is the primary material for the development of a petrochemical industry. Two Petromin projects which depend on natural gas were mentioned above. Here it should be added that in 1967 Petromin concluded an agreement with another ENI subsidiary, ANIC, to set up a petrochemical company in which Petromin was to have a 50% share.[38]

Following the pattern of Iran and Iraq, Saudi Arabia entered into a barter agreement with Rumania for the supply of 9–12 million tons of crude during the 1968–1971 period for Rumanian industrial goods.[39]

In order to achieve its major objective, Petromin decided that it must develop its own transportation facilities and ordered in 1972 two 40,000-ton tankers from Yugoslavia in a $53.4 million barter deal; Saudi Arabia was to supply Yugoslavia with 1.8 million metric tons of oil during the 1972–1974 period for sale outside the country.[40]

Petromin has gained experience and knowledge in the fields of refining, drilling, distribution and in the petrochemical industry. However, the great bulk of the industry and indeed the overwhelming source of Saudi Arabia's existence depended on the operations and revenue of the major oil company, Aramco, and even the manifold activities of Petromin depended in great measure on the technical resources and financial means of foreign companies.

DEVELOPMENT

The strict personal character of the Saudi Arabian regime and the general backwardness of the country made the revenue from the oil a matter of personal disposal, first by King Ibn Saud and after his death by his son Saud. The general governmental machinery was autocratic and primitive. The last act of King Ibn Saud before his death to expand the basis of governmental authority took place in October 1953 when he created the Council of Ministers to act as a cabinet and consultative body. However, the decisions of this Council had to be approved by the monarch. With the ascendance of Saud in 1953, the financial situation of the country deteriorated very rapidly despite the steady increase in the oil revenue. In order to meet his obligations the King had to ask for advances from the Company. As a result the royal powers of Saud were taken over twice — 1958–1960, 1962 — by his brother, crown-prince Feisal; and in the end Saud was forced, by his brother and the religious leaders, to abdicate in favor of Feisal. Saud was sent into exile, and on February 23, 1969, died in Athens.

His brother Feisal was an experienced and skilled administrator, and practiced more enlightened economic and financial policies. He succeeded in rehabilitating the financial structure of the country, and introduced many fiscal and institutional reforms.

It was realized in Saudi Arabia, as it was in Iran and Iraq, even though perhaps not so strongly and not so early, that the oil revenue must be utilized to create other economic resources so that the country would become less dependent on the inevitably ultimately exhaustible resource. In 1960, an Economic Development Committee was set up, and the International Bank for Reconstruction and Development was invited to send a mission to report on the possibilities for economic development. A mission came and subsequently submitted its report. The result was the establishment, in January 1961, of a High Planning Council, "to deal with problems of economic development in an expeditious manner."[41]

The Council was by no means comparable to either Iran's Plan Organization or Iraq's Development Board or Planning Board; it had neither the scope nor the organization of a full economic planning and executing program. What should be noted in the subsequent published Government budgets were the ever larger shares assigned to economic development and social services departments. In 1969 a Five-Year Plan (1970–1975) was adopted. It was prepared by the Central Plan Organization and endorsed by the Council of Ministers on September 7, 1970. The plan projected an outlay of Rls. 41.3 billion, and it stressed primarily health, education and social affairs. This was described by a Saudi official as "an outstanding landmark in the future growth of Saudi Arabia and would undoubtedly help in a more rational and constructive utilization of the country's oil income. The Plan aims at raising the rate of growth of the gross national product, diversification of the economy to reduce its dependence on oil, development of human resources, and in general, laying the foundation for sustained economic growth."[42]

It would be instructive to study some of the indicators of general development. We might begin with health and education. There has been considerable progress in the statistical picture of health services between the years 1961 and 1970, as indicated in the following table:

Health Statistics, 1961–1970

Year	1961	1970
Hospitals	41	80
Beds	4,080	6,974
Dispensaries	59	206
Health Centers	84	325
Physicians	318	818
Male Dressers (1963)	484	1,510
Nurses and Midwives (1963)	424	772
Pharmacists and Assistants (1963)	150	548[a]

a. SAMA, *Annual Report 1383–84 A.H. (1964)*, 14; – *1387–88 A.H. (1968)*, 56; – *1389–90 A.H. (1970)*, 63.

A similar impressive increase had taken place in educational statistics. A comparison between 1959 and 1969 reveals that the education budget for 1959 was Rls. 114.6 million, while ten years later it rose to Rls. 389 million; the number of primary and special schools was 601 in 1959 and went up to 1,318 in 1969; the number of students in these schools rose from 96,060 in 1959 to 253,339 in 1969; the number of intermediate and secondary schools was 40 in 1959 and reached 174 in 1969; the number of students in these schools went from 4,280 to 39,501 in 1969. Perhaps the greatest statistical advance both in the number of schools and the number of students was in the anti-illiteracy schools for adults; in 1959 there were 66 schools attended by 7,168 students and in 1969 the numbers jumped to 598 schools attended by 35,231 students. Progress was also made in the area of higher education.[43]

An indicator of general economic growth is perhaps the increase in local consumption of oil, and oil and natural gas products. It rose from 2,069,450 barrels in 1955 to 14,401,751 in 1969, practically a sevenfold increase in fourteen years.

Indeed, the Saudi Government reported a growth in GNP from Rls. 6,887 million in 1963 to 9,891 million in 1967.[45] However, it would seem that in spite of all the efforts of King Feisal and his loyal advisors, while the country made general economic progress and increased its health and educational facilities, it was dependent more than ever, almost exclusively, on oil revenue for its annual governmental income. In 1961 it was estimated that oil revenue supplied 86% of the budget; in 1967 it went up 90%; and it reached 92.3% in the budget for August 1971–August 1972.[46]

This complete dependence on the oil revenue was demonstrably illustrated in July 1967 after the flow of oil was interrupted by the closing of the Suez Canal when the Saudi Arabian Minister of Oil and Mineral Resources declared: "We will be compelled to suspend some projects to reduce some expenditures, and we might also be compelled to impose a temporary tax to meet the loss."

PROSPECTS

During the last 20 years the relations between Aramco and the Saudi Arabian Government have fundamentally changed; and both sides are facing momentous decisions. In spite of all its efforts and some accomplishments in attempting to master the complex oil industry, and in spite of the intensified agitation of Abdullah at-Tariqi, who was first the head of the Directorate of Petroleum and Mineral Affairs and later promoted to Minister when King Saud reorganized his cabinet, for the integrated company, and his vociferous demands, after he was relieved of his position, for nationalization, Saudi Arabia, it would seem, never was and still is not ready to nationalize the oil industry.

Indeed in 1960, when world oil prices slumped as a result of a glutted international oil market, Aramco unequivocally told the Saudi Arabian Government to be prepared for a cut in prices. It reported: "The world-wide surplus of oil-producing capacity continued in 1959 and it appears unlikely that surplus petroleum will be absorbed by mounting demand within the next several years. At the same time, the international search for additional sources has not slackened. The prospect is that substantial quantities of oil from North Africa soon will enter markets now served with oil from the Middle East. These economic circumstances in the oil industry require that every effort be made by the company and His Majesty's Government to improve the competitive position of Aramco's production." The Company clearly invited the Government to take a cut in profits in order to make Saudi Arabian oil competitive, and the Government did not resist.[47]

Early in 1962, Tariqi was relieved of his office as Minister, and was replaced by Ahmad Zaki Yamani, who has been at the head of the Saudi Arabian oil ship of state ever since. Early in his career as Minister he tried to explain in Beirut, on November 8, 1963, that his Government's economic policy was based on private enterprise, and it was, therefore, opposed to the principle of nationalization. However, this, in an autocratic country like Saudi Arabia, could hardly have been a decisive deterrent against a highly emotional drive to nationalize the oil and oust the foreign oppressive companies.

The answer must be sought in another direction. The general conditions of the producing countries have been tremendously improved. The exporting countries have obtained many and far-reaching concessions, in all areas, through OPEC but primarily because of the increase in demand for oil, and the great, almost panicky fear of the Western countries that they are running out of oil supplies. The producing countries could practically dictate their own terms to the foreign companies.

Nevertheless, of all the major oil-producing countries in the region, Saudi Arabia would seem the least ready for nationalization, and Ahmad Zaki Yamani has been pleading with the companies against nationalization. Saudi Arabia does not possess the technological knowledge, it lacks the investment capital necessary to maintain the operation and development of the industry, and it lacks the international markets. At the same time it must have been obvious to all concerned, in spite of the improved terms of the Aramco concession, that things could not continue as they were. Yamani, therefore, proposed, after OPEC's demand for Government participation and the acceptance by the Aramco in principle of a 20% Government participation, that the Company accept the Government goal of ultimate 51% participation. Such an arrangement would benefit both sides. It would give the Government control of the industry and would satisfy national ambitions; it would give the Government a sufficient increase in revenue to compensate for participation and accumulate investment capital for both home oil developments and downstream operations — investments abroad. In the process of gradual increase of participation the Government agencies would master completely all the operations of the industry, including marketing. Moreover, perhaps most important, the price structure would be maintained for the benefit of Saudi Arabia and the Company, and the international markets would be secure. Aramco on its part would be guaranteed with supplies for the consuming countries. The alternative would be nationalization, under which, Yamani claimed, the Company would be the loser. What Yamani did not reveal was his fear of the inevitability of price cutting in open competition by the various producers for the world markets once all have nationalized.[48]

Aramco was faced with a very serious problem. In the face of Iraq's nationalization and the pressure from OPEC as well as the fear of loss of supplies, it had two alternatives: either to accept Yamani's offer, with an attempt to bargain for the best possible terms for participation payments and supply guarantees, or to force the issue and let the Government nationalize the entire industry, obtain compensation and become a buyer of oil from the Government in a free competitive market. The Saudi Government left to its own resources could not possibly compensate for nationalization except with crude oil to be sold at high posted prices, and the level of compensation would be determined by the Government. The Government would not have the necessary investment resources to continue with the development of the industry, which would affect supplies and reserves.[49]

The conclusion, therefore, must be that in the struggle between Saudi Arabia and Aramco, although it would attempt, through hard bargaining, to obtain the best possible terms under the circumstances, the Company would submit to the Yamani Plan.

NOTES

1. Aramco, *Report 1957*, 9, 14, 15.
2. *Ibid.*, 16; *International Oilman*, XII, 260, Sept., 1958.
3. Aramco, *Report 1959*, 11.
4. *Aramco 1960*, 1.
5. *Aramco 1961*, 6.
6. *Aramco 1966*, 2, 4, 8.
7. *Aramco 1967*, 6; *World Petroleum*, XXXIX, 26., 1968.
8. *Aramco 1968*, 6; SAMA, *Annual Report 1387−88 A.H. (1968)*, 11.
9. *Aramco 1969*, 6; *World Petroleum*, XL, 22, Aug., 1969.
10. *Aramco 1970*, 4.

11. *Petroleum Press Service*, XXXVIII, 227, June, 1971. The magnitude of the loading capacity increase may be gauged from the shipment of both crude and products at Ras Tanura; in 1961 it was 319,710,146 barrels; in 1971 it reached 1,250,151,014 barrels, almost a fourfold increase in ten years. *Aramco 1961*, 11; – *Facts and Figures 1971*.

12. The same table illustrates the flexibility of the flow through the pipeline; when tanker rates were high the flow through the pipeline was the greatest (1968) and when tanker rates were low the flow through the pipeline was the smallest (1960).

13. Aramco, *Report 1959*, ii.

14. *Petroleum Press Service*, XXX, 151, Apr., 174–5, May, 1963; *World Petroleum Report*, X, 80, Mar. 15, 1964; *World Petroleum*, XXXIV, 31, July, 1963. For the Tapline issue see below, Chapter XXIII.

15. *World Petroleum Report*, XII, 82, Mar. 15, 1966. The details of the formula were given in Chapter VII dealing with Iran.

16. *Aramco 1967*, 6.

17. *Aramco 1969*, 6.

18. SAMA, *Annual Report 1389–90 A.H. (1970)*, 18–19. For the same settlements in Iran and Iraq see above.

19. *Aramco 1960*, 15.

20. SAMA, *Annual Report 1381–82 A.H. (1962)*, 16–17; *Aramco 1963, World Oil*, CLVII, 180–185, Aug. 15, 1963.

21. *Aramco 1968*, 4; *World Oil*, CLXIX, 196, Aug. 15, 1969.

22. Aramco, *Annual Report 1959*, 21; – *Aramco 1960*, 19; – *Aramco 1963; Aramco 1971*, 4.

23. *Aramco 1970*, 16.

24. *Aramco 1961*, 28–29; *Aramco 1971*, 6.

25. *Petroleum Press Service*, XXV, 46–48, Feb., 1958; "New Oil Agreements in the Middle East," *The World Today*, XIV, 141–142, Apr., 1958. For further details of developments in the Neutral Zone, see below, Chapter XXI.

26. "Saudi Arabian Demands," *International Oilman*, XII, 371, Nov., 1958.

27. SAMA, *Annual Report 1383–84 A.H. (1964)*, 12; *World Petroleum*, XXVI, 28, July, 1965; *World Oil*, CLXI, 184, Aug. 15, 1965; – CLXXI, 168, Aug. 15, 1970; *World Petroleum Report*, XV, 89, 1970; General Petroleum and Mineral Organization, *Annual Report 1969*, 14 (henceforth cited Petromin, *Annual 1969*).

28. *World Petroleum Report*, IX, 97, Feb. 15, 1963; *Petroleum Press Service*, XXX, 31, Jan., 1963.

29. SAMA, *Statistical Summary*, 10, Dec., 1968–Jan., 1969.

30. *Ibid.*, 10–11; Petromin, *Annual 1969*, 16–19, 20–23; *World Oil*, CLIX, 175, Aug. 15, 1964.

31. SAMA, *Statistical Summary*, 11, Dec., 1968–Jan., 1969; Petromin, *Annual 1969*, 29–34; *World Petroleum Report*, XV, 89, 1969.

32. SAMA, *Statistical Summary*, 12, Dec., 1968–Jan., 1969; Petromin, *Annual 1969*, 45–48. The Saudi Arabian Embassy in Washington reported that the American Import-Export Bank authorized a $12 million credit to SAFCO to help finance the plant construction. *Saudi Arabia Today*, VI, 3, July–Aug., 1968.

33. SAMA, *Statistical Summary*, 12, Dec., 1968–Jan., 1969; Petromin, *Annual 1969*, 48–49.
34. SAMA, *Statistical Summary*, 13, Dec., 1968–Jan., 1969; – Petromin, *Annual 1969*, 14; *Petroleum Press Service*, XXXV, 4–6, Jan., 1968; *World Oil*, CLXVII, 212, Aug. 15, 1968.
35. SAMA, *Statistical Summary*, 9, Dec., 1968–Jan., 1969; Petromin, *Annual 1969*, 24–29; *World Petroleum Report*, VII, 231, Feb. 15, 1961; – XV, 89, 1969; – XVI, 84, 1970; *Foreign Commerce Weekly*, July 5, 1960. Around 1968, the national concern formed the Petromin Lubricating Oils Company. It had a capitalization of Rls. 10 million. Petromin held 71%, 29% was held by Mobil Oil Investments. The Company was to produce lubricating oils and distribute them both in Saudi Arabia and abroad.
36. Aramco, *Report 1959*, 3, 11, 16; *Aramco 1961*, 11–12; – *1963*, 18; UN, *Economic Development 1959–1961*, 60–61. The Company estimated that the natural gas available in 24 hours, in million cubic feet, were: 1962 – 194; 1963 – 204; 1964 – 206. Aramco, *Natural Gas for Sale in Saudi Arabia*, 3.
37. *Aramco 1966*, 2, 6.
38. SAMA, *Statistical Summary*, 13, Dec., 1968–Jan., 1969; *Petroleum Press Service*, XXXV, 4–6, Jan., 1963.
39. *World Petroleum Report*, XV, 90, 1969.
40. *Petroleum Press Service*, XXXIX, 310, Aug., 1972.
41. SAMA, *Annual Report 1380 A.H.*, 13.
42. *Ibid.*, – *Annual Report 1389–90 A.H. (1970)*, 40. The allocations of the plan were given as 22.5% for education, health and social affairs, 18.1% for transport and communications, 11.1% for public utilities and urban development, 12.3% for Petromin projects, and the balance was distributed for industry, commerce, agriculture and water resources. The total investment for industry was given as Rls. 235.9 million. *Ibid.*, 41.
43. SAMA, *Statistical Summary*, 12–14, Sept., 1969. The total number of students in the entire educational system went up from 2,300 in 1939 to 400,000 three decades later. *World Petroleum*, XL, 39, Jan., 1969.
44. Aramco, *Report 1959*, 14; – *1969*, 22.
45. SAMA, *Annual Report 1387–88 A.H. (1968)*, 6.
46. SAMA, *Statistical Summary*, 5, 44–45, Sept., 1972. Even the other source of government income, customs duties, which averaged about 4%, must be attributed, in great measure, to the oil industry.
47. Aramco, *Report 1959*, I.
48. *Petroleum Press Service*, XXXIX, 382, Oct., 1972. It was subsequently reported that negotiations on participation of up to 51% between the Governments and the Companies were agreed upon.
49. Early in the autumn of 1972, speaking at the annual Middle East Institute Conference in Washington, Ahmad Zaki Yamani proposed that the United States permit the import of Saudi Arabian oil free of import duty, and in return the Saudi Government would invest in the downstream United States domestic oil industry operations. James E. Akins, Director, Office of Fuels and Energy in the United States Department of State, responded that the offer deserved careful study. He added that the idea might be distasteful to some in

the petroleum industry; however, few alternatives existed. The industry would need $500 billion in capital over the next 10 years but would not be able to borrow all the money required. Therefore, he concluded, there was no reason that the Saudi Arabian Government should not participate in the United States companies. He added that the United States would welcome the investments to offset payments for oil imports, and he suggested that the plan would help solidify a new interdependency between the Middle East oil-producing nations and the United States. *World Oil,* CLXXV, 19, Nov., 1972. For a full presentation of the Saudi Arabian proposals and Mr. Akins' presentation, see *World Energy Demands and the Middle East* (Washington, 1972), I., 75–104.

Section Five

PERSIAN GULF

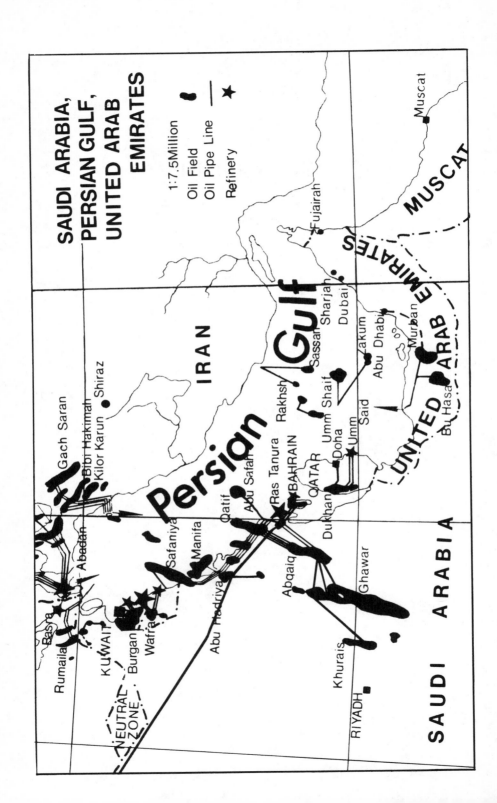

SAUDI ARABIA,
PERSIAN GULF,
UNITED ARAB
EMIRATES

1:7.5Million

Oil Field

Oil Pipe Line

★ Refinery

Muscat

MUSCAT

Fujairah

EMIRATES

Persian Gulf

Sharjah
Dubai
Sassan

Akum
Abu Dhabi

Murban
UNITED ARAB

Bu Hasa

IRAN

Gach Saran

Bibi Hakimah
Kilor Karun
Shiraz

Rakhsh

Umm Shaif
Umm
Said

Ras Tanura
BAHRAIN
Doha
QATAR
Dukhan

Abu Safah

Qatif

Manifa

Safaniya

Abqaiq

Ghawar

Abadan

Basra

Rumaila

KUWAIT
Burgan
Wafra

Abu Hedriya

Khurais

RIYADH

SAUDI   ARABIA

NEUTRAL
ZONE

Chapter XX
BAHRAIN

In discussing Iran, Iraq and Saudi Arabia, two basic differences were noted between the two former and the latter. The regimes of Iran and Iraq were not as completely autocratic as that of Saudi Arabia; the income from the oil, while very important and constantly increasing, was only one of many sources of income for the Government. In both countries the political leadership realized the importance of the oil revenue to long-range economic development and appreciated their own inability to envision and project plans for development, and they therefore invited foreign experts to prepare large-scale as well as long-range plans for the realization of the economic potentialities of their countries. The basic difficulty in both Iran and Iraq was implementation; both being sovereign and deeply suspicious of foreigners, especially Westerners, they would not voluntarily surrender their powers to foreign specialists to exercise supreme authority in the utilization of the oil revenue for development. They devised means within political-legislative limits and set up authorities to implement the plans prepared by the foreign experts. The limitations of the political leadership in control of these authorities — the Seven Year Plan Organization in Iran, and the Development Board in Iraq — hampered the successful achievement of the desired objectives.

Saudi Arabia differed from Iran and Iraq in two important aspects, and the problem of development became much more acute. Because the country was in a very primitive stage of development, it became almost exclusively (about 90 per cent) dependent on the income from oil. The absolutist King saw no need for foreign experts to work out a long-range program of development; whatever projects were undertaken were decided upon by the ruler, who could not by any stretch of the imagination be considered an economic expert with knowledge, training and experience in development.

The British protectorates in the Persian Gulf presented another situation, and because of that the results of the development of these territories were different from those in Iraq and Iran as well as in Saudi Arabia. While in theory the shaikhdoms of Bahrain, Kuwait and Qatar were nominally independent, they were far from sovereign and were subject in different degrees and by different methods to British guidance and authority. While all three had become increasingly dependent on oil revenue for almost their entire governmental budget, the disposal of the revenue was subject to British control, exercised sometimes openly, sometimes subtly, and in consequence a good portion of the income had been used for the economic development of the territories. It should be underlined that this had been primarily due to the efforts and determination of outsiders rather than to the wisdom and foresight of the native rulers.

BAHRAIN

The shaikhdom of Bahrain is composed of a group of islands in the Persian Gulf, situated about fifteen miles off the coast of Saudi Arabia between the province of al-Hasa and the Qatar peninsula. The capital, Manama, is on the main island; the second largest island is Muharraq. The islands, with a population estimated in 1954 at 120,000, were ruled by the al-Khalifah family. The traditional economic sources of income were pearling at Muharraq and trading for the Arabian mainland at the port of Manama.

Since December 2, 1880 the shaikhdom had been a British protectorate; on that date the Shaikh of Bahrain signed an agreement in which he declared: "I Isa bin Ali al-Khalifah, Chief of Bahrain, hereby bind myself and successors in the Government of Bahrain to the British Government to abstain from entering into negotiations or making treaties of any sort with any State or Government other than the British without the consent of the said British Government, and to refuse permission to any other Government than the British to establish diplomatic or consular agencies or coaling depots in our territory, unless with the consent of the British Government."[1]

In 1910 geologists of the British Indian Survey discovered seepages of oil in Bahrain. Coming as it did after the successful discovery of oil in Persia and the organization of the Anglo-Persian Oil Company, there was apparently considerable discussion of oil possibilities in Bahrain. The British pressed the Shaikh for a definite commitment not to permit the exploitation of oil by non-British concerns. On May 14, 1914 the Shaikh wrote the British Political Agent in Bahrain that he would not himself embark on the exploitation of oil in his territory, nor would he entertain overtures regarding such exploitation from any quarters without consulting the British Political Agent and without the approval of the British Government.[2]

Oil, however, was not discovered. In the early 'twenties Major Frank Holmes, representing the Eastern and General Syndicate, was seeking oil concessions in the Persian Gulf area, and as noted above, succeeded in 1924 in obtaining a concession from Ibn Saud in the al-Hasa province. On December 2, 1925 Holmes acquired from the Shaikh of Bahrain for his British syndicate an exclusive oil exploitation concession over an area of about 100,000 acres. It was provided that if the concession rights were not exercised within four years, it would lapse. The Syndicate tried to interest British concerns in the concession, but without success.[3] At the end of November 1927 the Eastern Gulf Oil Company, a subsidiary of the American Gulf Oil Corporation, entered into an option contract to purchase the Syndicate's rights in Bahrain, and was given until January 1, 1929 to exercise its option. Gulf, at the time a member of the American group in the Turkish Petroleum Company, had in the meanwhile signed the Red Line Agreement which covered the Bahrain islands, and the Board of Directors of the Turkish Petroleum Company ruled that for Gulf to exercise its option would be a violation of Article 10 of the Agreement.[4] On December 21, 1928 Gulf, with the consent of the British

Syndicate, transferred its rights to the Standard Oil Company of California for $50,000.[5]

Under the option contract, the Syndicate was to have secured from the British Colonial Office a one-year renewal of the concession. When the Colonial Office was approached, it made the renewal contingent on the insertion in the concession agreement of the British Nationality Clause, whice provided that the managing director and a majority of the other directors of the company must be British subjects, that the concessionaire company must be British-registered, and that none of the rights and privileges which the Shaikh had granted in the concession must be controlled directly or indirectly by foreigners. To this the United States Government took serious objection, for such a provision would have meant the absolute exclusion of Americans from holding or operating the concession, which was contrary to the American Open Door policy. On March 28, 1929 the Secretary of State instructed the American Chargé d'Affaires in London to discuss the Bahrain concession question informally with the proper authorities. The legal difficulty involved, so far as the American claims were concerned, was the status of the concession; the legal owner was the British Syndicate and the negotiations between the Colonial Office and the Syndicate were purely British affairs, and the position of the American Government was not legally clear. The British Foreign Office replied on May 30 that it had no objection in principle to American participation in the concession, subject to the British Government being satisfied as to the conditions under which American capital would participate, and particularly as to the nationality of the operating company, its chairman and director and the personnel employed on the islands. The British Government suggested that since the Eastern and General Syndicate was still the existing concessionaire, the Syndicate should negotiate with the Colonial Office.[6]

Negotiations between the Colonial Office and the Syndicate were carried on for about a year and a half. On June 12, 1930 the Syndicate signed an agreement with the Shaikh, and transferred it to the Bahrain Petroleum Company, a Canadian subsidiary of the Standard Oil Company of California. The major provisions of the agreement were: The Bahrain Petroleum Company must remain a British company; one of the directors must be a British subject and *persona grata* to the British Government; the Company's chief local representative must be approved by the British Government and all communications with the Shaikh must be through him and the Political Agent; as many of the employees as was consistent with efficiency must be subjects of Britain and Bahrain.[7]

Standard of California immediately sent out geologists and drillers. On May 31, 1932 the first commercially producing well was brought on. Drilling was continued, tanks were built, and in December 1934 the first tanker load of Bahrain oil entered the world market. By 1935 there were sixteen producing wells. Standard of California, searching for markets for its increased production of crude, and with the anticipated output from Saudi Arabia, sold one-half of its Bahrain concession to the Texas Oil Company for a half share in the latter's numerous marketing facilities east of Suez. A new company registered in the Bahamas, the California-Texas Company

(Caltex), was organized for the purpose of conducting Bahrain's marketing operations.[8]

In June 1940 the Bahrain Petroleum Company (Bapco) received an extension of its concession covering all the lands and other areas under the jurisdiction of the Shaikh, an area of some 1,644,000 acres,[9] for a period of fifty-five years from June 19, 1944.

PRODUCTION

From 1936 production progressed rapidly, reaching the 20,000 barrel per day mark in 1939. During the years 1942–1944 some wells were shut in as a war measure, and production was reduced by some 2,000 barrels per day. By 1951 there were 67 producing wells, with an average per well output of 430 barrels per day. The following table gives production in tons and barrels:

TABLE XXIX. *Bahrain Oil Production, 1933–1954*

Year	In Tons	In Barrels
1933	4,000	31,000
1934	39,008	285,000
1935	173,072	1,265,000
1936	635,555	4,645,000
1937	1,058,511	7,762,000
1938	1,138,000	8,298,000
1939	1,041,000	7,589,000
1940	971,000	7,074,000
1941	932,000	6,794,000
1942	856,000	6,241,000
1943	902,000	6,572,000
1944	921,000	6,714,000
1945	1,003,000	7,309,000
1946	1,099,000	8,010,000
1947	1,291,000	9,410,710
1948	1,496,000	10,915,000
1949	1,508,000	10,985,484
1950	1,512,000	11,016,000
1951	1,508,000	11,008,000
1951	1,510,000	11,004,000
1953	1,506,000	10,978,380
1954[a]	1,570,000	10,994,000[b]

a. UN, *Review 1949–1950*, 60; – *Review 1951–1952*, 55; – *Summary 1952–1953*, 47; Economic Cooperation Administration, Special Mission to the United Kingdom, *The Sterling Area. An American Analysis* (London, 1951), 418; *World Petroleum*, X, 43, July, 1939.

b. United States Senate, *Petroleum Requirements – Postwar. Hearings Before A Special Committee Investigating Petroleum Resources* (Washington, 1946), 40; UN, *Summary 1950–1951*, 22; George P. Stevens, "Saudi Arabia's Petroleum Resources," *Economic Geography*, XXV, 221, July, 1949; Faroughy, *The Bahrain Islands*, 43; *Le Commerce du Levant*, Oct. 15, 1954; DeGolyer and MacNaughton, *Twentieth Century Petroleum Statistics 1954*, 9; *World Oil*, CXL, 198, Feb. 15, 1955.

ROYALTIES

The original concession called for 3 1/2 rupees per ton of crude oil produced. This was equivalent to 14 cents per barrel, or $1.05 per ton, before the devaluation of the rupee in September 1949.[10] After the devaluation, and because of the general demands for higher royalties throughout the Persian Gulf region, Bapco increased the royalty to the Bahrain Government from 3 1/2 rupees to 10 rupees, approximately 29 cents per barrel, or about $2.10 per ton, effective as of January 1, 1950. In 1951 another supplemental agreement increased payments by an annual lump sum, and the following year, in line with the general tendency in the area, a 50/50 profit-sharing of net income from production and sale of oil was agreed on, effective as of January 1, 1952.[11]

The following table gives an approximate picture of the steady increase in royalties.

Table **XXX**. *Royalty Payments to Bahrain (selected years)*

Year	In Dollars
1933	8,800
1938	305,000
1940	1,000,000
1945	780,000
1946	1,200,000
1948	1,508,876
1949	1,500,000
1950	3,300,000
1951	3,800,000
1952	6,300,000
1953	6,500,000
1954	8,100,000[a]

a. J.H.D.B., "Oil and Bahrain," *The World Today,* VII, 77, Feb., 1951; UN, *Summary 1950–51,* 37–38; – *Review 1951–52,* 58–59; "Oil and Social Change," *The Economist,* 3, July 2, 1955.

The nature of the Bahrain islands as the commercial port of the Saudi Arabian coast and the Qatar peninsula made it an ideal location for a refinery, and as soon as oil was discovered in commercial quantities in Bahrain and the prospects for Saudi Arabia looked promising, Bapco built a refinery which went into operation in July 1935. Increased production and the demand for oil for war needs made it necessary to expand the refinery and by 1945 the Company had spent about $27.5 million. By 1951 the refinery had a capacity of 155,000 barrels daily. Most of the crude came from the Saudi Arabian mainland as the maximum of local production was only 30,000 barrels per day. At first the oil was brought by tanker, later it came by way of a 12-inch 34-mile-long pipeline, two-thirds of which was submarine. During 1947 another 12-inch pipeline was laid, bringing the total capacity of the pipeline to 126,000 barrels per day. [12] By 1949 about 85 per cent of the refinery's throughput came from Saudi Arabia. The output in 1950 was 56,059,000 barrels; in 1951, 8,040,000 tons; in 1952, 8,621,000 tons; and in 1953, 10,408,000 tons. Up to 1955 the parent companies had invested about $75 million in the Bahrain undertaking. [13]

RESERVES

Of all the major oil-producing areas of the Middle East, Bahrain was the least promising as regards reserves. From the very beginning the reserve estimates were generally static. In January 1949 they were estimated as 170 million barrels; two years later Faroughy gave them as 200 million barrels; at the end of 1954 they were estimated at 235 million barrels. The United Nations experts estimated the reserves as 40,900,000 tons in 1951 and the same in 1953. Compared to Saudi Arabia and Kuwait, the Bahrain reserves were considered of little significance. [14]

COMPANY POLICIES AND ACTIVITIES

In accordance with the conditions of the concession, laid down by the British Colonial Office, Bapco employed a considerable number of Bahrainis in its undertakings, almost exclusively in the lower grades. The total number of employees in 1948 was 6,078, and of these 4,650 were local employees. While this was only 4.23 per cent of the total population, then estimated at about 110,000, [15] it was 76.5 per cent of the total employed by the company. The staff was increased in 1951 to 7,749, and of this number 4,937 were Bahrain nationals. [16] This drop in the percentage of native employees, from 76.5 to 64, was not the result of Company policy but was due to the greater demand for workers in the expanded oil-producing areas of the Gulf — Saudi Arabia, Kuwait and Qatar. Many Bahrainis, finding the wages offered in other areas more attractive, left their home islands, which made it necessary for Bapco to employ more foreigners such as Indians, Pakistanis, Persians and others. This created a peculiar phenomenon: educated and semi-educated Bahrainis sought work abroad, while destitute foreigners flocked to the islands.

In addition to direct payments in revenue and indirect payments in terms of wages and purchases of local materials for the operation of the concession, the Company contributed directly to many social and economic projects aimed at bettering the general conditions; these included housing, electric lighting, water supply, and sanitary facilities. It operated a hospital and clinics, and also subsidized the American Mission Hospital. In the field of education it assisted the Government's technical school for native boys; upon graduation students from that school were employed by the Company, which provided further mechanical training.[17] It also cooperated in many agricultural and irrigation projects.[18]

However, the impact of the oil industry on the people of Bahrain came primarily from the utilization of the revenue from the oil. The oil payments were divided into three parts: one went to the Shaikh for the maintenance of his household and family; one was for current Government expenses and development projects; and one for investments abroad as a reserve for future contingencies. With the increase in income — directly from oil revenue and indirectly from customs duties — as a result of the Company's activities, Government services increased and a number of development projects were undertaken. By 1951 there were three hospitals on the islands — the American Mission Hospital and the Company hospital mentioned above, a Government hospital for men and women operated by a British staff, including a section for the mentally ill — as well as a number of municipal clinics. General health conditions were not good: malaria and other fevers and dysentery were prevalent; bilharzia, trachoma and hookworm took a heavy toll; and above all, a large proportion of the population was infected with venereal diseases. Medical services were supplied to every Bahraini free of charge.

Though there had been some educational facilities before the development of the oil industry,[19] it was not until 1934 that the Government engaged a number of Iraqi teachers and enlarged the facilities. By 1951 there were nine elementary schools, one secondary school, founded in 1940, and one school for girls opened in 1939, all tuition-free. The schools were staffed with 258 teachers, who instructed some 6,000 boys and girls. The Government, in conjunction with the Company, also provided scholarships to graduates of the secondary schools for study abroad.

The Government built roads, bridges, harbors and a large airport; water was supplied from artesian wells instead of being brought in by boat; and efforts were made to improve housing conditions. Though without a single newspaper in 1951, Bahrain had a broadcasting station, built in 1940 for Allied war purposes, a dial telephone system, forty miles of roads and 2,000 automobiles. Most of the streets of the main towns were electrically lighted, and a sewerage system had been installed. The net result was a higher standard of living for the Bahrainis, not only in comparison with their own pre-oil standard, but also with the people of the neighboring states.[20]

There have also been other implications. The Shaikh, the Government and the people had become increasingly, in fact almost exclusively, dependent on the oil. In 1932 the entire government budget was about $143,600; in 1951 the total government revenue was $5,271,000, of which approximately 61 per cent came from oil and 34.9 per cent from customs duties,[21] which must be considered an indirect result of the oil industry. In 1953 the Government revenue was derived 68 percent from oil, 24 per cent from customs duties, and 5 per cent from interest on the reserve fund.[22] The oil boom also created inflationary pressures which caused many Bahrainis to seek employment in other oil-producing areas in the Gulf. Because of its central location, the IPC subsidiary, Petroleum Concessions Ltd., which held concessions in Qatar and on the Trucial coast, had made its Gulf headquarters in Bahrain and built large offices and houses for its staff. Bahrain had become a resort and shopping center for the oil employees, especially Americans and Europeans, in Saudi Arabia and Qatar. These not only added to the inflationary pressures, but so greatly increased the social vices of gambling, drinking, prostitution and their concomitants as to cause grave concern to those responsible for the health and welfare of the population.[23]

The Bahrain islands are limited in area and in resources; oil, in a more definitely limited quantity than that of any of the other areas in the Gulf, was the only resource supporting the Shaikh, the Government and its extended services, and also to a considerable degree the people at large. Though perhaps originally a little more advanced than some of their neighbors, the people of Bahrain, prior to the discovery of oil, had a low standard of living and consequently a very low standard of governmental and social services. When the pearling industry suffered a serious setback as a result of the world depression and the development of cultured pearls, Bahrain was in a desperate situation. With the arrival of the Bahrain Petroleum Company, the islands entered a new era. Governmental services, education, health, housing, sanitation and other services gave the Bahrainis one of the best living standards of the area. This was the result not only of the income derived from the Company — although the Company contributed directly to the raising of the standards and the direct revenue from its operations made these improvements possible — but was primarily due to the Government's determination to utilize the income for constructive purposes. This achievement must be ascribed first to the British Government which (in spite of all disclaimers) controlled the government of the Shaikh, and, secondly, to the Shaikh's adviser, J.H.W. Belgrave. It was no doubt the British Government's decision to divide the revenue from the oil into three parts, allowing the Shaikh only one-third, earmarking another third for current expenses and general development, and leaving the rest to be invested. This was perfectly illustrated by the Government's expenditure budget for 1953.[24]

But with the improvements also came drawbacks; a serious breakdown in social and religious loyalties, and an alarming rise in the social vices. Education deteriorated to an emphasis on those subjects which guaranteed employment in the petroleum industry. It had been observed that some Bahraini students knew more English, even though only superficially, than Arabic; the eagerness for employment caused many parents to take their children out of school long before their graduation even from primary school. Yet even with this limited education and the experience of employment with Bapco, some of the workers began to show symptoms of discontent. They demanded that the Company allow trade unions in its operations, and that Bahrainis have more control over the Company.[25] All in all, considering the conditions of the islands and the general stage of the people's development, one must conclude that the discovery of oil had brought Bahrain prosperity and a higher standard of living, and that the oil revenue there had been better employed than in other areas. The credit for that must go to the British Government and the British adviser to the Shaikh.[26]

What of the Bahrain Petroleum Company? As mentioned above, Standard of California paid $50,000 to the Gulf Corporation for the concession. Bahrain's capitalization was $100,000. At the end of 1947, H.M. Herron, chairman of the Board of Directors, testifying before the Special Senate Committee Investigating the National Defense Program, stated that the Company's earned surplus was $55,472,255. In addition, it had paid-in capital surplus of $28,191,709 and surplus reserves of $8,705,779. From its organization until 1947 it had paid no dividends.[27]

The reason for these enormous profits was given by Mr. Herron when he reluctantly revealed to the Committee that the cost of producing a barrel of oil in Bahrain, including royalties paid to the Shaikh, was about 25 cents.[28] If crude was selling in the Persian Gulf at the time at $1.05–$1.15 a barrel, that would have produced a profit on each barrel of 80–95 cents.

Caltex, Bapco's subsidiary marketing company, was organized in the Bahamas in 1936 with a capitalization of $1,000,000. As of December 31, 1946, ten years later, it had an earned surplus of $17,192,529, and a paid capital surplus of $8,195,144. No dividends were paid.[29]

Thus the profits of Bapco and Caltex – on an ititial limited capitalization of $100,000 and $1,000,000 respectively – though not as large as those of Saudi Arabia – were exceptionally high. In spite of the various measures undertaken by the Company to assist the Bahrainis to improve their standard of living, the latter actually received only a very small portion of the income from the oil, compared with the Company's profits. Only after the 1952 profit-sharing arrangement became effective did Bahrain begin to get a little larger share of the profits from the production and sale of crude. But since all the oil produced by Bapco was sold to its subsidiary, Caltex, for marketing, the price of Bahrain oil was not that of the open market, but a price arbitrarily established by the parent company and the marketing subsidiary. The basis for the calculation of the profit was therefore narrow.[30]

PROGRESS AND CHANGES, 1955–1971

In spite of the number of major events which occurred during the period under review, the basic role of oil in the economy of Bahrain, the relations between the Government and Bapco, and the function of the large company refinery at Awali remained relatively the same. Table XXXI notes the rate of production of crude oil. To be sure, it more than doubled during this period, but the total amount was not of significant importance in the total picture of Middle East oil output growth. The increase in reserves was very limited indeed, from 243,000,000 barrels in 1955 to 631,000,000 barrels in 1971; while in itself it was almost a threefold increment, it was of small importance in the fabulous reserves increases of the Middle East.[31]

The growth of the refinery was similar. In 1958 it processed 70,415,036 barrels, or a daily average of 192,918 barrels, to which Bahrain contributed only some

Table XXXI. *Bahrain Oil Production, 1955–1971*

Year	Tons	Barrels
1955	1,502,000	10,951,000
1956	1,537,000	11,014,000
1957	1,670,000	11,690,000
1958	2,102,000	14,718,000
1959	2,250,000	16,473,378
1960	2,249,540	16,500,424
1961	2,257,000	16,444,492
1962	2,250,000	16,445,932
1963	2,249,580	16,502,868
1964	2,438,967	17,999,821
1965	2,793,670	20,787,617
1966	3,076,620	22,520,831
1967	3,465,854	25,370,000
1968	3,506,538	27,597,728
1969	3,800,000	27,735,000
1970	3,813,648	28,349,000
1971	3,750,000[a]	27,346,498[b]

a. UN, *Economic Development 1955–1956*, 47; *Petroleum Press Service*, XXVII, 5, January 5, 1960; – XXIX, 4, January, 1962; – XXX, 4, January, 1963; – XXXVII, 6, January, 1970; – XXXIX, 10, January, 1972; *Petroleum Times*, LXVIII, 325, January 26, 1964; – LXIX, 315, June 11, 1965; – LXXI, 978, July 7, 1967; – LXXII, 991; July 5, 1968; – LXXV, 22, July 30, 1971.
b. Bapco, *1968; – 1969; – 1970*, 2; – *1971*, 4; *World Oil*, CXLII, 186, February 15, 1956; – CXLVI, 196, February, 1958.

15 million barrels, and the rest came mostly from Saudi Arabia; in 1971, the throughput of the refinery was only 94,089,342 barrels, or a daily average of about 257,779 barrels, to which Bahrain contributed some 27 million barrels, which was less than a seven per cent increase.[32]

The number of producing wells did not increase spectacularly, nor were new oil fields discovered. In 1959, for instance, there were 148 wells producing oil and five producing natural gas; some 12 years later the number of oil wells went up to 215 and the gas wells rose to eight.

But the growth of the refinery meant an increase in the number of refined products, from fuel oil to aviation fuel as well as LPG, for local and international markets. Bahrain was primarily, if not exclusively, an oil-products exporter. Bapco followed a policy of cooperation with Government, whether it was under British guidance or when Bahrain became independent. It tried to find new oil fields and to improve the productivity of existing wells, and it succeeded not only in maintaining the level of output but even in expanding it. The average annual output per well went up by some 9,000 barrels.

In 1959 Bapco undertook the extension of the oil-loading facilities at Sitra, and the building of a new two-berth island wharf with modern equipment for tankers of up to 45,000 tons; they were completed in 1961. In the same year the storage capacity at the Sitra tank-farm was expanded to 200,000 barrels. Three years later the number of loading berths was increased, and the storage capacity was expanded by 220,000 barrels. In 1971 the Company began erecting two 350,000-barrel products storage-tanks at Sitra. [33]

An outstanding phase of company cooperation was the utilization of natural gas. Bapco used natural gas for fueling the operations of its refinery and supplied gas for firing the generators of the Government electric power station at Jufair, which supplied the major part of Bahrain's electric needs. In 1959 the station's capacity was increased to 28,400 kilowatts. Two years later the Company sold LPG to local consumers through a Bahrain dealer. In 1963 local sales reached 175,781 gallons. Four years later it jumped to 853,440 gallons, and in 1968 rose to 1,340,850 gallons.

Before natural gas could be used it had to be dehydrated, and with the increased demands for gas the dehydration facilities were greatly extended in 1963, and in 1970 a third unit was added. In December 1968, Bapco signed an agreement to supply daily 60 million cubic feet of natural gas to Aluminum Bahrain (Alba), which was to build an aluminum smelter close to the Bapco refinery, and was to commence operations early in 1971. This prompted the Company to drill two gas wells in the Khuff zone underlying the oil reservoir. The wells were completed in 1969, and until the gas was needed for the Alba plant it was injected in the oil wells. [34]

Production of natural gas in 1970 reached about 25.4 billion cubic feet, of which 6.8 billion were used by the refinery, 5.5 billion were delivered to the Government power stations, and the balance was reinjected. In March 1971, deliveries of gas from the Khuff zone wells to the Alba plant commenced, and by the end of the year the daily average was 41 million cubic feet. Production that year amounted to 18.4 billion cubic feet, of which the refinery consumed 6.2 billion, deliveries to the Jufair station were 5.6 billion cubic feet, and the remainder was reinjected. [35]

The Company pursued a policy of constantly increasing the percentage of Bahrainis in the number of its employees. In 1959, it was 70.4%, in 1965 it increased to 79.7%, and in 1971 it reached 89.1%. It may be assumed, and the Company reported some promotions from time to time, that in the increases of the percentage there were a number of supervisory and professional positions, but whether this reflected the percentage increase was not clear. [36]

Bapco continued with its policies of providing its employees and their families with medical care, both in hospitals and outpatient clinics, pension schemes and house building assistance; it reduced the work week to 40 hours without reduction in pay. [37]

Bapco distributed the oil products in Bahrain, but the dispensing stations, for all products, were owned by Bahrainis. Both the number of products and the magnitude of consumption rose with the general development of the shaikhdom. In 1960 the total of all products locally consumed, including aviation fuel at the Muharraq airport, was 9,118,510 gallons; five years later it rose to 19,486,782 gallons and in 1967 it jumped to 24,550,890 gallons.[38]

In addition to direct payments of royalty and income tax the Company contributed to Bahrain's economy by purchasing materials and services locally, a policy instituted in 1957. At first the calculations were in Indian Rupees, the currency of Bahrain. In 1957 it amounted roughly to Rs. 5 million and it went up in 1964 to Rs. 29 million. With the introduction of the Bahrain Dinar in 1965, the purchases amounted to BD 2.4 million. The Company calculated that the total local purchases for the period 1957–1967 amounted to BD 18.7 million.[39]

As will be recalled the marketing outlet for Bahrain's oil products was Caltex, the agency of the parent companies of Bapco. In 1969 it was announced that a new marketing company, Caltex-Al Khalij, was formed, with field headquarters in Dubai, to market Bapco products and Caltex lube oils in the Emirates of the Trucial States; and it was also to handle Caltex business in Kuwait, Muscat, Oman, Qatar and Saudi Arabia.[40]

EVENTS AND DEVELOPMENTS IN BAHRAIN

With the rapid development of offshore oil production in the Persian Gulf the question of demarcation lines resulted in conflicts between Bahrain and Saudi Arabia and between Bahrain and Iran. In 1958 Shaikh Sulman bin Isa of Bahrain paid a visit to King Ibn Saud in Riyad and concluded an agreement establishing the median line between the two countries; the agreement also provided for sharing profits from reserves located in the median area. Aramco's Abu Safah offshore field was located in the median area. When in 1966 Aramco began production in that field Bahrain shared in the profits.[41]

In 1961 Shaikh Sulman died and he was succeeded by his son Shaikh Isa bin Sulman al-Khalifa; the Company cooperated with him as it cooperated with his father. Under his administration important and far-reaching developments took place. He attempted to increase the oil revenues by granting new concessions. In 1965 Continental Oil Company Bahrain, a newly organized subsidiary of Continental Oil Company of the United States, obtained an offshore concession covering an area of 2,500 square kilometers for a period of 45 years. The terms were: royalty and tax rates as prevailed, at the time, in the Persian Gulf area. However, Continental, which was in the meantime joined by Pure Oil Company, relinquished the concession in 1968. In December 1970 the same concession was granted to Superior Oil Company for a period of 35 years. Royalties were to be 12½% of posted prices and were to rise with increased production. Percentage of profit taxes was to be in accordance with the rate prevailing in the Gulf subject to amendment when it changed. In the same year Bahrain reached agreement with Iran on a demarcation median in the offshore area between the two countries.[42]

In line with the practice of other countries in the area, the Finance Department organized in 1967 a Development Bureau. Its members considered many projects proposed by economic experts and recommended 35. The Bureau set up the aluminum smelter at Askar. The cost of the smelter was to have been $161.3 million, and it was to have an annual production capacity of 120,000 tons.[43] In 1970 the Government launched a BD 10 million airport expansion program aimed at enlarging the terminal facilities and extending the runways to accommodate jumbo jets.[44]

The population of Bahrain was growing steadily: in 1951 it was 120,000; the 1959 census counted 143,000. In 1965 the number went up to 182,203, of which about 79% were Bahrainis; in 1972 the population was estimated at 200,000.[45]

Bahrain faced during the period under discussion two major issues; its relation with the British and the Iranian claim that the Bahrain islands were part of Iran. It was obvious ever since the British withdrew in 1947 from India that their role in the Persian Gulf would have to be reduced if not eliminated. In spite of the importance of oil to the British economy and position, first the Laborites and later the Conservatives (for a while hesitating and attempting to reverse the Labor decision), determined to withdraw from the Persian Gulf. When this became inevitable Bahrain was faced with a decision. Should it join the other Arab Emirates of the Trucial Coast and become a member of the projected Arab federation or should it declare its independence. However, before any formal steps could be taken the Iranian claim had to be dealt with.

Early in 1970 Iran laid official claim to Bahrain, and the case reached the United Nations. The British formally objected to the claim and the matter was taken over by the Secretary General to use his good offices to settle the issue. He sent a personal representative to Bahrain, Vittorio Winespeare Guiccicardi, to ascertain the wishes of the Bahrainis regarding the future of their status. On April 30, 1970, he reported his findings to the Secretary General. The Secretary General in turn submitted the report, with his own introduction, to the Security Council, and in doing so declared that he had completed his good-offices undertaking. Both the Iranian and British representatives informed the Secretary General that they would accept the recommendations of the United Nations mission. On May 11, 1970, the Security Council unanimously adopted a resolution welcoming the conclusions and findings of the mission and in particular the statement that "the overwhelming majority of the people of Bahrain wish to gain recognition of their identity in a fully independent and sovereign state free to decide for itself its relations with others."[46]

On August 15, 1971, the British signed an agreement terminating the protectorship, and Bahrain declared its independence. It applied for membership in the Arab League. On the same date as it proclaimed its independence Bahrain applied for membership in the United Nations. The Security Council considered the application, and after the Committee of Admissions approved the application, the Security Council recommended, on August 18, to the General Assembly to accept Bahrain. On September 21, 1971, the General Assembly approved officially the application, and Bahrain became the 128th member.[47]

Iran's renunciation of its claim was part of the general Western solution to the Gulf question. With the total withdrawal of the British the resultant vacuum in the area had to be filled. It was apparently decided, not without some risk and consequent reprisals, to permit and encourage the Iranians to build up their forces as the strongest power in the Gulf. The Iranians therefore occupied the islands of Abu Musa, and Lesser and Greater Tunbs. This, of course, increased the existing tension between Iran and Iraq, and caused Libya to nationalize a British Petroleum Concession in that country. But the price which Iran had to pay was renunciation of its claim to Bahrain. It is this understanding between Iran and the Western powers which would explain the readiness, nay eagerness, with which Iran welcomed the Security Council's decision. For the Iranian representative declared: "With the Council's decision, a major obstacle has been removed from the path of fruitful cooperation in the Persian Gulf."[48]

INTERNAL DEVELOPMENTS

The internal political regime was and to a very large extent remains autocratic and paternalistic. A tradition going back to 1831 established advisory committees in governmental departments; the first one named was that of education. But these committees were at best advisory in nature, and in most cases nonoperative. In January 1970, the Amir created a Council of State composed of governmental department heads, who were as a rule members of the Amir's family. With independence the Council of State became the Cabinet. The 12-member Cabinet was headed by a Prime Minister who was the Amir's brother. An advertisement in the *New York Times* declared that the policy of Government was: "growth and improvement of public services — health, education and social welfare — and a massive investment in public utilities — electricity, water and housing." Its policy was to direct a major portion of the oil revenue to public social services.[49]

Bahrain made considerable progress in the field of education. In 1969 the country celebrated the 50th anniversary of modern education on the islands. In 1971 there were 52,619 boys and girls in state schools, with 2,930 teachers; and the education percentage in the total government budget was 22.[50]

As to health, the first government doctor was appointed in 1925. Two hospitals were then on the islands: the American Mission and the Government of India hospitals. In 1938 the first Bahrain hospital was established. After the arrival of Bapco, hospital and medical facilities were greatly augmented. Tuberculosis was still a serious menace to the country. But on the whole medical services had been greatly improved and all Bahrainis were guaranteed health protection.

In the field of water the Government resorted more to artesian wells. With the help of Bapco's techniques in drilling for oil new water wells were found, and the Government launched an agricultural program based on irrigation.

There could be no doubt that the country was well supplied with electricity for home and industrial uses, generated by the Government Jufair power station.

The two latest economic projects undertaken by the Government successfully were the expansion of the Muharraq airport and the construction of the aluminum smelter at Askar.

NOTES

1. Abbas Faroughy, *The Bahrain Islands (750–1951)* (New York, 1951), 23. This was confirmed by a subsequent agreement on Mar. 13, 1892. *Ibid.*, 123–124.
2. *Ibid.*, 124.
3. The following comments from two different directions are interesting: British — "American interests were first engaged almost fortuitously when oil was struck on the Manama Island of Bahrain, after British prospects had failed. The successful company is the one which has subsequently extended its activities to the great Saudi Arabian oil fields." "The Persian Gulf — A Romance," *The Round Table*, XXXIX, 134, Mar., 1949. American — "It was our good luck that the British failed to discover oil which we later found in great abundance." Harold L. Ickes, "We're Running Out of Oil," *The American*, 84, Jan., 1944. For personal testimony about the oil concessions question in this area see: Thomas Edward Ward, *Negotiations for Oil Concessions in Bahrain, El Hasa (Saudi Arabia), the Neutral Zone, Qatar and Kuwait* (New York, 1965).
4. See above, Chap. XV.
5. US Senate, *International Petroleum Cartel*, 71–72; *Foreign Relations*, 1929, III, 80; Duce, *Middle East Oil Development*, 8, 10; Faroughy, *The Bahrain Islands*, 103–104; United States Senate, *American Petroleum Interests in Foreign Countries*, 318. In Mar., 1926 the Anglo-Persian Oil Co. was ready to take over the concession from Holmes, but in April it advised that its geologists in the area had reported unfavorably and that it was therefore no longer interested. E. DeGolyer, "The Oil Fields of the Middle East," in *Problems of the Middle East*, 77; Sanger, *The Arabian Peninsula*, 100.
6. *Foreign Relations 1929*, III, 80–81.
7. DeGolyer, *loc. cit.*, 77–79; *Foreign Relations 1932*, II, 5. The British backed down from some of their previous demands but insisted on some of the others. The State Department apparently considered this achievement a victory for American interests and claimed credit for it. In its memorandum to the Senate Special Committee Investigating Petroleum Resources, the Department declared: "Here again the prompt and positive action of the State Department has secured results favorable to an American-owned company. By securing the entry of American oil interests into Bahrain, the way was paved for some American interests to obtain concessions in nearby Arabia." US Senate, *American Petroleum Interests in Foreign Countries*, 23–24.
8. US Senate, *International Petroleum Cartel*, 79; Chap. XV above.
9. Faroughy, *The Bahrain Islands*, 111; Sanger, *The Arabian Peninsula*, 147. DeGolyer was of the opinion that the Shaikh delayed granting the extended concession because of the efforts of Holmes in 1937, when he reappeared in Bahrain and attempted to get a concession for the additional area. For suddenly, in the spring of 1937, the Shaikh halted all negotiations with Bapco for one year. "He finally granted it to the Bahrain Petroleum Company on June 19, 1940, but the British officials distilling the last drop of advantage that

could be obtained hooked on a political agreement of June 29, 1940." De-Golyer, *loc. cit.,* 79. "The 'Bahrain unallotted area,' being the difference between the 100,000 acres and the whole extent of the Shaikh's dominions, was from 1934 to 1940 the subject of lengthy negotiations between the Shaikh, Bapco, and another candidate, the Iraq Petroleum Company." Longrigg, *Oil in the Middle East,* 104.

10. UN, *Summary 1950–51,* 37; – *Review 1951–52,* 61; Faroughy stated that royalty was 3 rupees and 8 annas plus one shilling per ton, *The Bahrain Islands,* 45. Adamiyat, who follows Faroughy closely, also gives the royalty as 3½ rupees and one shilling. Fereydoun Adamiyat, *Bahrain Islands. A Legal and Diplomatic Study of the British-Iranian Controversy* (New York, 1955), 200.

11. UN, *Summary 1950–51,* 37–38; – *Review 1951–52,* 61; Adamiyat, *Bahrain Islands,* 200; Economic Cooperation Administration, Special Mission to the United Kingdom, *The Sterling Area. An American Analysis,* 409.

12. For a Company description of the pipeline and the refinery, see "Submarine Pipeline," *Oil Lifestream of Progress,* III, 20, Oct., 1953.

13 UN, *Summary 1950–51,* 24; "Oil and Bahrain," *The World Today,* VII, 78, Feb., 1951; Sanger, *The Arabian Peninsula,* 145; Faroughy, *The Bahrain Islands,* 44–45; *Le Commerce du Levant,* Oct. 15, 1954.

14. US Senate, *International Petroleum Cartel,* 6; UN, *Review 1951–52,* 53; – *Summary 1952–53,* 40; Faroughy, *The Bahrain Islands,* III; *World Oil,* CXL, 205, Feb. 15, 1955.

15. Max Steineke and M.P. Yackel estimated the population at about 125,000, "Saudi Arabia and Bahrain," in *World Geography* (Princeton, 1950), 226.

16. UN, *Review 1949–50,* 60; – *Summary 1952–53,* 58.

17. The Bapco training program is described by its vice-president, Russel M. Brown, "Schooling in Skills," *Oil Lifestream of Progress,* IV, 6–8, Oct., 1954.

18. US Senate, *American Petroleum Interests in Foreign Countries,* 288–289.

19. In 1926 there was one school for boys only, and it was poorly attended. James Bell, "He Said Forward! to the Backward," *Life,* Nov. 17, 1952.

20. UN, *Summary 1952–53,* 58–59; Sanger, *The Arabian Peninsula,* 148–149; Faroughy, *The Bahrain Islands,* 20 *et passim;* "Oil and Bahrain," *The World Today,* VII, 77–79, Feb., 1951; Bell, *loc. cit.*

21. Bahrain Petroleum Company, Limited, *Bahrain 1952.* Henceforth cited Bapco, and year

22. Angus Sinclair, "The Resources and Their Potentialities. III, Petroleum and Minerals," in *Middle East Resources Problems and Prospects* (Washington, 1954), 46. The 68 per cent of the revenue directly from oil did not include the one-third which went to the Shaikh.

23. "Oil and Bahrain," *The World Today,* VII, 79, 81–83, Feb., 1951; Faroughy, *The Bahrain Islands,* 20 *et passim.*

24. Government operating expenses including administration, courts and public protection, $707,000 (9 per cent); health and education, $1,464,000 (18 per cent); public works, $1,453,000 (18 per cent); two capital works projects – gas supply for the city of Manama and a deep water terminal, $1,260,000 (15 per cent); surplus for investment, $2,846,000 (36 per cent). Sinclair, *loc. cit.,* 46.

25. In July 1954 it was reported that a general strike had been proclaimed in Bahrain, directed against the ruler, Sir Salman bin Hamad al-Khalifah, for

failure to reform the despotic regime. The strikers demanded that a commission be set up to examine reforms, including improvements of public health and education facilities. *Daily Telegraph* (London), July 12, 1954.

26. For Belgrave's contributions to the achievements in Bahrain, see Bell, *loc. cit.,* and the many articles and books by Belgrave himself. A more ideal picture of the adjustment of the Bahrainis to the new conditions resulting from the oil development is presented in Bapco, *Bahrain 1952.*

27. US Senate, *Hearings Saudi Arabia,* 25030–25031. Senator Owen Brewster, chairman of the Committee, intimated that had dividends been paid to the parent companies, the latter would have had to pay US taxes. Faroughy states that in 1945 the Bahrain Petroleum Co. estimated that its consolidated net profit for the year, after taxes and all other costs had been deducted, was approximately $10.5 million. In 1946 profits were estimated at $23 million. The total investment, according to Faroughy, was not more than $65 million. *The Bahrain Islands,* 45.

28. US Senate, *Hearings Saudi Arabia,* 25022.

29. US Senate, *Hearings Saudi Arabia,* 25029–25030. The earnings for the years 1942–1946 were: 1942, $1,903,520; 1943, $686,673; 1944, $1,457,509; 1945, $2,416,362; 1946, $1,469,245. While the Bahrain Petroleum Company was a Canadian corporation registered in the British Empire, Mr. Herron admitted that Caltex could have been an American corporation. The inference drawn by Senator Brewster was that it was registered in the Bahamas only to escape United States taxation, and that for the same reasons no dividends were distributed. The Government of the Bahamas did not tax corporations, and the Canadian Government taxed only declared dividends. *Ibid.,* 25016–25018.

30. Iran claimed sovereignty over Bahrain and continuously challenged the British rights to the islands. The Iranian Government tried to bring the case to the League of Nations; it protested against the granting of the concession to Standard of California, and during the oil nationalization period agitated for the return of the islands. A voluminous literature has grown up on the subject. Two complete studies which present the Iranian case are those mentioned above: *The Bahrain Islands,* by Faroughy, and *Bahrain Islands,* by Adamiyat. The Iranian Government's note to the League of Nations claiming sovereignty over the islands, which was circulated to all the members of the League, and the British reply, are in League of Nations, *Official Journal,* 9th year, 1928, 1360–1363; – *Official Journal,* 10th year, 1929, 351, 790–793. See also PDC, 213, col. 843, Feb. 15, 1928; PDL, 81, col. 418, June 25, 1931; PDC, 300, col. 776, Apr. 8, 1935. For the protest note to the United States on May 22, 1934, see Arnold J. Toynbee, *Survey of International Affairs 1934* (London, 1935), 221–224. For a more recent study, see Majid Khadduri, "Iran's Claim to the Sovereignty of Bahrain," *American Journal of International Law,* XLV, 631–647, Oct., 1951.

31. UN, *Economic Development 1961–1963,* 103; Aramco, *Facts and Figures 1971.*

32. Bapco, *1959,* 6–7; – *1971,* 7.

33. Bapco, *1959,* 18; – *1961,* 3, 9, 12; – *1962,* 12; *1964,* 13 – *1971,* 9.

34. Bapco, *1959,* 18; – *1961,* 14; – *1963,* 21; – *1967;* – *1968;* – *1969.*

35. Bapco, *1970,* 2; – *1971,* 4.

36. Bapco, *1959*, 24; – *1962*, 12; – *1965*, 23; – *1968;* – *1971*, 18.
37. Bapco, *1964*, 24; *1966*, 23.
38. Bapco, *1960*, 12; – *1965*, 27; – *1967*.
39. Bapco, *1959*, 14; – *1960*, 14–15; – *1964*, 21; – *1967*. The Bahrain Dinar was valued in 1967 at $2.10.
40. Bapco, *1969*.
41. *World Petroleum Report*, XII, 74, Mar. 15, 1966; – XV, 78, 1969; – XVI, 70, 1970; – XVII, 65, 1971.
42. *World Oil*, CLXIII, 180, Aug. 15, 1966; *World Petroleum Report*, XII, 74, Mar. 15, 1966; – XVII, 65, 1971; – XVIII, 53, 1972; *Petroleum Press Service*, XXXII, 394, Oct., 1965; – XXXVIII, 67–69, Feb., 1971.
43. Bapco, *1967; New York Times*, Jan. 21, 1972.
44. Bapco, *1970*, 10.
45. Bapco, *1959*, 18; UN, Security Council, *Official Records Twenty-fifth Year Supplement for April, May and June 1970.* S/9772, 6, Apr. 28, 1970.
46. UN, Security Council, *Official Records Twenty-Fifth Year Resolutions and Decisions 1970*, 7–8.
47. UN, Security Council, *Official Records Twenty-Sixth Year Resolutions and Decisions 1971*, 12; – General Assembly, *Official Records Twenty-sixth Session Supplement No. 29*, 2; Bapco, *1971*, 2. Bahrain could not become a member of OPEC, for to qualify for membership a country had to be a net exporter of crude oil. However, it became a member of OAPEC (Organization of Arab Petroleum Exporting Countries) in 1970. *World Petroleum Report*, XVII, 65, 1971.
48. UN, Security Council, S/PV. 1536, 16, May 11, 1970.
49. *New York Times*, Jan 21, 1971.
50. *Ibid.*

Chapter XXI
KUWAIT AND THE NEUTRAL ZONE

Perhaps the largest oil field in the Middle East, if not in the world, was located in the Shaikhdom of Kuwait. Situated on the northwestern shore of the Persian Gulf, it covers an area of some 6,000 square miles and borders on the oil-bearing states of Iraq and Saudi Arabia, as well as the neutral zone. The relations between this shaikhdom and Great Britain were based on an agreement signed in 1899 which made the territory a British protectorate. In that agreement the Shaikh pledged himself to secure the approval of Great Britain before entering into any engagement with any foreign government other than the British, and not to dispose of any part of his territory without the approval of the Government of Great Britain. The importance of Kuwait to the British increased early in the century when the southern terminal of the projected Berlin-Baghdad railway was to have been located in the shaikhdom. As regards oil, the Shaikh wrote, in a letter dated October 27, 1913, to the British Political Resident in the Persian Gulf at Bushire: "We are agreeable to everything which you regard as advantageous and if the Admiral honors our company we will associate with him one of our sons to be in his service, to show the place of bitumen in Burgan and elsewhere and if in their view there seems hope of obtaining oil therefrom we shall never give a concession in this matter to anyone except a person appointed from the British Government."[1]

The possibility that there might be oil in Kuwait and its special importance to Great Britain were recognized as early as 1914; in the Foreign Office oil agreement of March 19 of that year Paragraph 10 specifically eliminated Kuwait from the restrictions of what subsequently became known as the "red line" area.[2] The British, however, took no measures to guarantee for themselves whatever oil resources might be in the territory.[3] The story of oil exploitation began in the early 1920's when Major Frank Holmes, representing the Eastern and General Syndicate, obtained a concession from the Shaikh and attempted to sell it to the Anglo-Persian Oil Company. The latter declined, and the Syndicate offered it to an American concern, the Gulf Oil Corporation.[4] The story of Bahrain was repeated, but with different results.

Since Kuwait was not included in the Red Line Agreement, the Syndicate negotiated with the Shaikh to transfer the concession to the Gulf Oil Corporation.[5] The Shaikh apparently was willing to grant the concession on terms acceptable to the American company, but the British Colonial Office intervened and insisted upon the inclusion in the concession of the British Nationality Clause.

On December 3, 1931, after Gulf Oil had complained to the State Department about its difficulties, Secretary of State Henry L. Stimson wrote to the American Chargé d'Affaires in London that unless the British modified their position on the Nationality Clause, as they had done in the case of Bahrain, the American company would be barred from proceeding with the development of the Kuwait concession;

he was requested to intercede with the proper authorities so that action could be taken on the American company's request through the Syndicate. The British were not ready to retreat, as they did in the case of Bahrain. At the end of February 1932 the American Chargé informed the Secretary of State that he had informally protested to the Foreign Office the procrastination in acting on Gulf's request and had pointed out that, according to reports reaching the American Embassy, the Anglo-Persian Oil Company's geologists had arrived in Kuwait with drilling equipment to carry out exploration work. This could only mean that the British company was seeking or had already obtained a concession from the Shaikh, with the approval of the Colonial Office, while the Foreign Office was negotiating with the United States Government about the concession owned by the American company.

Later, Stimson informed the Chargé that the American Legation in Baghdad reported that it had learned from a confidential source in Kuwait that the Shaikh had secretly signed a document granting Anglo-Persian a concession covering the entire shaikhdom.[6] The State Department was convinced that the stalling tactics of the British were aimed at not allowing a recurrence of what had happened in Bahrain. The Americans had succeeded in establishing themselves in two areas in the Persian Gulf — Saudi Arabia and Bahrain — and the British were determined to keep them out of Kuwait. Under the cloak of the Nationality Clause they would eliminate the Americans, and during the prolonged negotiations between the Syndicate, a British concern, and the Colonial Office on behalf of the Gulf Oil Company, the Anglo-Persian Oil Company would acquire a new concession. This suspicion was confirmed on March 14, 1932 when the British Assistant Under Foreign Secretary told the American Chargé in London, informally, that the APOC had made a formal application for a concession long before the Eastern and General Syndicate even appeared on the scene, and that it was carrying on geological exploration with the sanction of the Shaikh. Despairing of the delaying tactics of the British Foreign Office, the Department of State instructed the Chargé on March 26 1932 to discard the informal approach and proceed formally.

A very difficult and delicate situation developed. The British were by no means ready to oppose the Americans openly; they declared they would be willing to relax the Nationality Clause except for the insistence of the Shaikh; the latter, on his part, declared that he would not insist on the Nationality Clause if the British agreed to relax it. This discrepancy between the statements of the British and the Shaikh gave the Americans a trump card in the negotiations. A strong note was delivered to the Foreign Office by the American Chargé. After it had been discussed by the British Cabinet, Sir John Simon, the Foreign Secretary, wrote to the American Ambassador on September 9 reiterating the basic British position in Kuwait: the predecessor of the present Shaikh had given an undertaking.to the British Government that he would not grant an oil concession in his territories without the consent of the British Government; the Shaikh had informed the British Government that he was averse to granting an oil concession to any company except one that was entirely British. Having thus saved face, the British were ready for a compromise. The Foreign Secretary then stated: "On a balance of all the conflicting concessions, His Majesty's Government are, however, now prepared, for their part, not to insist in this case that any concession must contain a clause confining it to British interests, if the Shaikh for his part is willing to grant a concession without such a clause."

But the British were not ready to give up the APOC's claim completely. Sir John suggested that the Eastern and General Syndicate submit an application for the concession without the British Nationality Clause, which application would subsequently be transferred to Gulf, to be considered by the Shaikh together with the other application. Consequently both the Eastern and General Syndicate and APOC submitted applications; Eastern's encountered clearance difficulties in the Colonial Office as the latter found that British interests were not satisfactorily safeguarded. The American Embassy again began to prod the British authorities to take action on the Syndicate's application and thus make it possible for the Shaikh to grant the concession. On November 2, 1932 the American Ambassador, Andrew Mellon, told the British Foreign Office that "the delay in reaching a settlement in the matter of the Kuwait oil concession was becoming exasperating."[7] Delays, however, continued, and although the Syndicate's application was finally approved and submitted to the Shaikh, the British Government's comments on both applications made them unacceptable to him.[8]

The stalemate was finally broken in 1933 when Sir John Cadman of Anglo-Persian, the old troubleshooter, visited the United States, and Gulf's representatives visited London. On December 14, 1933, Gulf and APOC entered into an agreement which provided that they should exercise Eastern Gulf Oil Corporation's option on any concession which the Eastern and General Syndicate might obtain in Kuwait; that the agencies and facilities at the disposal of each of the two companies should be used to obtain the concession on terms not more onerous to the concessionaires than those of a draft concession, which was attached to the agreement; that each party should share equally the expenses subsequently incurred by either one in obtaining the concession, including £36,000 which would be due Eastern and General Syndicate if Eastern Gulf should take up its option from the Syndicate; that an operating company – Kuwait Oil Company Ltd., financed and owned equally by APOC and the Gulf Exploration Company – should be organized; that production was to be shared equally by the parent companies, and that ownership of one party was not to be sold or transferred except with the consent of the other. The capitalization of the Kuwait Oil Company Ltd., which had to be a British concern, was 200,000 shares of £1 each, held equally by APOC and Gulf. About a year later the Kuwait Oil Company Ltd. obtained a 75-year concession covering the entire shaikhdom, about 3,900,000 acres.[9]

OPERATIONS, 1946–1954

Drilling began near al-Bahara in 1936 but without results. Two years later drilling was begun in the Burgan area (the same area mentioned in the Shaikh's

letter of 1913) with considerable success, and by July 1942 nine producing wells
had been drilled. On July 13, 1942 operations were suspended by the British Army
as a war measure and all the wells were shut in.[10] Operations were resumed in
August 1946; from then on production was rapid, and after the shutdown at
Abadan, became spectacular. The following table shows the production of crude
during the period 1946–1954.

Table XXXII. *Kuwait Oil Production, 1946–1954*

Year	Tons	Barrels
1946	800,000	5,927,979
1947	2,185,308	16,227,906
1948	6,400,000	46,546,795
1949	12,378,000	89,930,444
1950	17,280,000	125,722,396
1951	28,327,000	204,909,662
1952	37,631,000	273,432,895
1953	42,654,000	314,592,000
1954	46,969,415[a]	347,955,000[b]

a. AIOC, *Report, 1947*, 8; UN, *Review 1949–50*, 61; – *Summary 1952–53*,
    47; *Le Commerce du Levant-Beyrouth Express*, Mar. 16, 1955.
b. *World Petroleum*, XXIV, 30, Aug., 1953; *Petroleum Press Service*, XVII,
    209, Aug., 1950; DeGolyer and MacNaughton, *Twentieth Century Pet-
    roleum Statistics 1954*, 9; *World Oil*, CXL, 198, Feb. 15, 1955; Kuwait Oil
    Company, Limited, *The Story of Kuwait* (London, 1955), 20.

REFINERY

In November 1949 the Company completed a refinery at Mina al-Ahmadi, with
a capacity of 1,000,000 tons annually, or 25,000 barrels per day; by the end of that
year it had refined 118,000 tons. The following year it refined 1,101,000 tons, and
in 1952 1,326,000 tons. It provided gasoline and kerosene for the local demands of
the shaikhdom and the Company's operational needs, as well as furnace and marine
diesel oil for tankers loading crude at Mina al-Ahmadi.[11]

The constant upward revision of the estimate of Kuwait's reserves was most
interesting. In 1944 the reserves were estimated at 4 billion barrels;[12] DeGolyer
reported that the indicated reserves were about 9 billion barrels.[13] As of January 6,
1949 the estimates were raised to 10,950,000,000 barrels,[14] 13.98 per cent of the
total world reserves and surpassing Saudi Arabian reserves by about 2 billion
barrels. By 1951 they were increased to 16 per cent of the total world reserves,
compared to Saudi Arabia's 10.8 per cent;[15] by 1952 the reserves were estimated
at 16 billion barrels; and at the end of 1954 at 28 billion barrels.[16] Thus Kuwait
had concentrated in a very small area, practically in two fields – Burgan since 1946
and Magwa since 1953 – one of the greatest single oil pools in the world.

ROYALTIES

The original concession provided for a royalty of 3 rupees per ton of crude oil produced, plus four annas in lieu of taxes. At the time this was the lowest payment in the entire Persian Gulf area. When the rupee was devalued in September 1949 the royalty amounted to the equivalent of about 9 cents per barrel. Negotiations between the Shaikh and the Company for revised royalty payments took place during 1951, and on December 3 it was announced in London that a 50/50 profit-sharing arrangement had been worked out, effective as of December 1, 1951. The new arrangement provided that in addition to the royalty previously paid, the Company's earnings were to be subject to a corporate income tax so adjusted that the royalty payments and the income tax would give the Shaikh one-half the Company's production profits before deduction of foreign taxes. This brought up the royalty per barrel to approximately 52 cents. In return, the Shaikh extended the concession for an additional seventeen years, making its term ninety-two years from 1934.[17] The increase in income is dramatically illustrated in the following table.

Table **XXXIII.** *Direct Oil Payments to Kuwait Government*

Year	Dollars
1946	800,000
1948	13,700,000
1949	11,800,000
1950	12,400,000
1951	30,000,000
1952	139,000,000
1953	158,900,000
1954	217,300,000[a]

a. KOC, *Oil in Kuwait*, 24; AIOC, *Report, 1951,*
21–22; UN, *Review 1951–52,* 59, 61; Elizabeth Monroe, "The Shaikhdom of Kuwait,"
International Affairs, XXX, 274, July, 1954;
"Oil and Social Change," *The Economist,* 3,
July 2, 1955.

COMPANY POLICIES AND ACTIVITIES

The operations of the Company were located twenty-five miles south of the town of Kuwait. A forty-seven foot pier into the Persian Gulf was built for loading oil and unloading incoming supplies,[18] and not far from this pier a new city, Ahmadi, grew up. The Company attempted from the very beginning to employ as many Kuwaitis as feasible, but there were obstacles. The actual field of operation was far from the city where more than 80 per cent of the population was concentrated, and the city held strong attractions for the Kuwaitis; furthermore, they not only lacked the necessary skills and knowledge, but also the readiness to work with sustained effort at continuous employment. In 1946, for instance, the labor turnover was 66 per cent. In order to provide itself with workers, the Company organized a training center at Magwa which trained Kuwaitis in the various trades needed by the industry. The trainees were not requested to accept Company employment in return for their training, yet in 1954 no Kuwaitis were employed on the higher staff, even the clerical and technical grades were filled by Indians, Pakistanis and Bahrainis, and only one-half of the labor corps consisted of Kuwaitis.[19]

As inducements, and to assist its staff, the Company built some six hundred permanent houses of different types between 1946 and 1951; it laid a number of miles of road, installed telephone and radio communications; built a drainage system and laid water and gas pipes for domestic distribution; erected a water distillation unit; and provided its employees with medical care. Until 1947 the American Missionary Hospital at Kuwait cared for the Company's employees; then the Company built its own air-conditioned hospital at Magwa, with a 200-bed capacity and the most modern equipment.[20]

It had been calculated that annual local expenditure of the Kuwait Oil Company over and above the direct payments to the Government had been about $15 million.

KUWAIT AND OIL

The Shaikhdom of Kuwait is limited in territory; its 1954 estimated population of 170,000 was primarily concentrated in the city of Kuwait; and natural resources, prior to the discovery of oil, were non-existent. It had no fresh water and drinking water was brought in from Shatt-al-Arab by boat; this ruled out any possibility of agricultural development through irrigation. Kuwait assumed a political importance out of all proportion to its size and population at the end of the last century because of international rivalry: first the British feared that Russia might establish a coaling station, and later, when Germany planned to establish the southern terminal of the proposed Berlin-Baghdad railway in the shaikhdom.

The traditional sources of income were pearling, shipbuilding, fishing, and smuggling into Iraq. Kuwait was very seriously hit during the world depression years and it was estimated that during the early 'thirties the Government's annual revenue was less than $200,000. Governmental and social services were consequently either non-existent or of a very low standard. Until 1936 there were no Government schools, nor did the Government make any medical services available to the people.

The introduction of the oil industry, with its direct royalty payments to the Government and indirect Company expenditure, changed the entire economic structure of the shaikhdom and to some extent also the social structure. In 1936 a Department of Education was set up and four schools, housed in converted homes, began giving instruction to 600 students; one year later a school for girls was opened. As the income from oil increased, the school system expanded; in the school year of 1952–1953 there were twenty-four boys' and fifteen girls' schools, attended by 10,738 students who were instructed by 570 teachers, some of them Kuwaitis. The education budget was equivalent to $7.5 million, making up 12 per cent of the total budget. [21]

Education was only one of the innovations. As a result of the oil revenue, the following governmental departments were established: Justice, Health, Development, Public Works, and Finance, all headed by ministers who were members of the Shaikh's family. Each ministry had as adviser a foreign technical expert who was either British, Syrian or Palestinian. Relations between the Kuwaiti ministers and their foreign advisers were not too cordial, but by the winter of 1952, when the oil revenue had become fabulous indeed (about $800 for each man, woman and child), a six-year development plan was worked out which met with the approval of all concerned. It called for a capital development expenditure during the six years of $252 million, and an equivalent amount for current needs. The plan not only envisaged the expenditure of a gigantic amount of money, but it was of a revolutionary character. The school system was allotted some $39 million, which included the construction of a technical high school and the distribution of meals for the entire school population. Government buildings were allotted $42 million; $28 million was allotted for electric power projects; $36 million for a plant for the distillation of sea water and the distribution of fresh and salt water; $22 million for a harbor; $14 million for an airport; and $9 million for a luxury hotel. The city of Kuwait was to be rebuilt in accordance with the entire new development scheme. [22] A building boom was on in Kuwait, and the Company had difficulties in keeping its employees. By 1954 some progress had been made towards modernizing Kuwait. The city of Kuwait underwent a complete change, some old mud houses were demolished, streets were widened, new modern buildings erected, water towers rising to heights of 100 feet installed throughout the city, an electric power station installed, water storage tanks and sea water distribution plants built. The number of school buildings increased to twenty-six for boys and thirteen for girls, accommodating a total of 12,830 students. About 120 boys were studying abroad on Government scholarships. A state hospital with a 120-bed capacity was completed and outfitted with the latest modern equipment, and a sanitorium for 200 tuberculosis patients was added to the hospital. Two polyclinics were also opened. In March 1953 the first section of the sea water distillation plant with a million-gallon daily capacity was completed. [23]

The Shaikh agreed to the establishment of an investment board in London, to which the six-year plan assigned $152 million. He also depostied in a New York bank his dead-rent money from Aminoil. Behind all these plans and achievements stood the British Political Agent for Kuwait, C.J. Pelly, who served not only in a personal capacity but also as a representative of the protecting power, Great Britain. [24]

Subsequent developments were even more ambitious. The water distillation plant was completed and provided 5 million gallons of water daily. The new electric power station had four units of 750 kilowatts each. Plans had been completed for the building of either a pipe or a canal to bring water from the Shatt-al-Arab, thus making irrigation possible,[25] and plans had been completed for a new 250-bed eye hospital and a new general hospital with a 750-bed capacity.

There could be no doubt that Kuwait, as a result of the oil industry, had established higher standards of health, education, sanitation and general welfare than any of the other Persian Gulf states and territories. There could also be no doubt that this had been in great measure due to the advice which the Shaikh had accepted from the British Political Agent.

One thing was certain: from a very primitive, backward, waterless, poor territory, by 1955 Kuwait emerged as the leading state in the Persian Gulf in standards of health, literacy and general welfare. A hundred and seventy thousand Kuwaitis had abandoned their old methods of living and had experienced an economic and social revolution of great magnitude. This had serious repercussions not only in their own territory, but also in the neighboring states.

KUWAIT OIL COMPANY DEVELOPMENT, 1955–1971

All aspects of the oil industry were steadily advancing; and in line with the times and developments in neighboring oil-producing countries, Kuwait established its own national company, built its own refinery, acquired the internal marketing facilities, increased the utilization of natural gas, entered the petrochemical field, had the KOC relinquish huge chunks of its original concession areas, saw huge increases in oil production, and reaped the consequent growth of the oil revenue. It granted new concessions, and entered into foreign oil undertakings.

The magnitude of the phenomenal increase of production is illustrated by Table XXXIV; and the jump in revenue brought about by production and increment in posted prices and percentage profits is evidenced in Table XXXV. Kuwait had also undergone, during the period under review, internal and external political changes; it continued with the social welfare program, intensified its economic development program, and reached a stage of possibly deliberately limiting oil production.

The Kuwait Oil Company kept production rates rising in spite of the two serious interruptions in the 1956 and 1967 closings of the Suez Canal, the major artery for oil movement to Europe. Production between 1955 and 1971 almost tripled; and payments during the same period more than quadrupled.

In 1955 a new field at Raudhatian was discovered with estimated reserves of 10 billion barrels. In 1970 another new field, Sabiriyah in northern Kuwait, was discovered and began producing. KOC had systematically worked towards developing all the oil potential of Kuwait. In March 1958 the refinery at Ahmadi was expanded, and in the following year its daily capacity was raised to 190,000 barrels at a cost of $39 million. In 1970 the daily throughput was raised to 250,000 barrels. In 1959 the second, north pier was completed; it accommodated at the

Table XXXIV. *Kuwait Oil Production, 1955–1971*

Year	Tons	Barrels
1955	53,894,068	398,593,597
1956	54,117,349	399,874,491
1957	56,375,946	416,045,187
1958	69,117,138	509,382,593
1959	68,437,498	504,855,244
1960	80,537,627	594,278,196
1961	90,721,636	600,226,114
1962	92,050,170	669,284,000
1963	95,666,630	705,471,000
1964	105,033,846	774,815,000
1965	107,322,529	791,903,000
1966	112,548,188	830,537,000
1967	113,355,644	836,719,000
1968	120,162,473	886,125,143
1969	127,502,203	940,040,972
1970	135,228,278	998,109,707
1971	144,468,129a	1,067,795,049b

a. KOC, *The Story of Kuwait* (1959), 24; – *Annual Review 1959*, 14; – *1965*, 19; – *1966*, 10; – *1967*, 16; – *1968*, 7; – *1969*, 8; – *1970*, 6; – *1971*, 9; CFP, *Exercice 1961*, 11; – *1962*, 7; – *1963*, 7; – *Annual Report 1964*, 6; *Petroleum Times*, LXVII, 304–306, June 14, 1963.

b. BP, *Annual Report 1956*, 27; – *Annual Report 1957*, 30; KOC, *Annual Review 1959*, 12; – *1961*, 15; – *1968*, 11; – *1969*, 8; – *1970*, 6; – *1971*, 9; *Petroleum Press Service*, XXXVI, 40, January, 1969.

time five medium-size tankers simultaneously.[26] Storage and pipeline facilities were built at the north pier; a 63-mile 30-inch pipeline was laid between the Raudhatian field and the north tank-farm in Ahmadi, and 38/40-inch gravity lines connected the north tank-farm with the north pier. The storage capacity at the north tank-farm was raised to 3,010,000 barrels. Two years later the daily export capacity at the south pier reached 950,000 barrels and that at the north pier 765,000 barrels. By 1970 the capacity of the two tank-farms at Ahmadi reached 12,500,000 barrels.[27]

In 1967 the Company began building a nine-mile offshore deepwater crude-oil loading terminal with a capacity to accommodate the largest tankers under construction. By the following September the two-berth terminal was completed; it was 1,620 feet long and was capable of accommodating 312,000-ton tankers. A submarine 51,113-foot-long 48-inch pipeline was laid to bring the crude to the terminal.

Indeed, to coordinate the many projects for the expansion of production facilities, an organization unit called COPE (Crude Oil Production Expansion) was set up by the company.[28]

Table XXXV. *Kuwait. Direct Payments to Government, 1955–1971*

Year	$ Million
1955	282
1956	310
1957	338
1958	425
1959	405
1960	465
1961	464
1962	526
1963	557
1964	655
1965	671
1966	707
1967	718
1968	766
1969	812
1970	897
1971	1,395a

a. UN, *Economic Development 1958–1959*, 77; – *Studies 1968*, 69; – *Studies 1971*, 114; *Petroleum Press Service*, XXXVII, 324, September, 1970; – XXXVIII, 326, September, 1971; – XXXIX, 322, September, 1972.

The Company expanded production by discovering new oil fields and by developing more wells. In 1955 there were 6 oil fields with 185 wells which produced more than 398 million barrels; by 1971 there were 8 oil fields with 693 wells which produced some 1,067 million barrels of oil.[29]

The utilization of natural gas was a well-established practice in Kuwait. Since the late '40's gas had been used to fuel the generators of the power stations and the sea-water distillation plant. In 1959 the Company instituted the gas injection program to keep up the pressure in the wells. In 1961 the Company reported: "One of the highlights of our year's work was the advance we made in the utilization of natural gas. We began the injection of approximately 100 million cubic feet of gas a day in the Burgan Field; we started to sell regular quantities of gas to the American Independent Oil Company; and we increased the useful consumption of gas for fuel and for power production." A gas injection plant costing £1.5 million and a £3.5 million LPG plant were completed that year. In the following year the Company began to export LPG.[30]

Ever since 1953, and continuously in increasing quantities, the Company supplied the state with natural gas, which was used for the water distillation plants at Shuaibah, Kuwait City, and the state's Shuaibah Industrial Development Complex, and for the electric generation stations.[31]

In 1965 the Company erected another gas injection plant in the north Kuwait oil field. It supplied a total of 12.5 billion cubic feet of natural gas free of charge to the Government for industrial use and another 1.2 billion only for cost of delivery at Shuaibah. Some 4,184 tons of LPG were supplied to the Kuwait Oil Tanker Company for bottling and local marketing. Total daily consumption reached the 44 million cubic feet mark. [32]

In the following year the Company installed a gas injection plant at the Minagish oilfield. Consumption for the year totalled over 98 billion cubic feet, of which the Government power and other stations and the Shuaibah complex used 18.8 billion cubic feet, and the Company utilized 77.4 billion cubic feet for power generation, fuel and injection in the oil reservoirs. [33]

In 1967 the daily consumption of natural gas for all purposes went up to 327 million cubic feet. The Company exported 1,015,050 tons of propane, butane and natural gasoline. Local consumption of LPG was 10,588 tons. In the following year the daily consumption of natural gas rose to 459 million cubic feet; the increase over the previous year was primarily due to the needs of the Shuaibah complex, the Kuwait National Petroleum Company and the Kuwait Chemical Fertilizer Company. [34]

In 1970 a joint-venture of the Government and the KOC parent companies — 50% for the Government and 25% for each of the companies — began building a $30 million natural gas plant to produce annually 1,000,000 tons of LPG from the Burgan field associated gas to be exported primarily to Japan. Daily consumption of natural gas in 1971 jumped to 474 million cubic feet. [35]

In 1963 the Government signed an agreement with KOC for the latter to deliver, at cost, all the surplus natural gas produced. The disposal of this gas was to be limited to Kuwait and not to be exported. Eight years later, another agreement granted the Government all rights without any restrictions over natural gas not required by the Company in its basic operations and existing LPG plant. [36]

KOC–GOVERNMENT RELATIONS

Generally speaking the relations between the Company and the Government were good. They became a little ruffled when the younger and more sophisticated elements became articulate in the newly established and power-gaining National Assembly. But at no time did the relations reach a critical situation.

Following the first major forced relinquishment of concession areas by Iraq at the end of 1961, Kuwait undertook negotiations on April 21, 1962 with the Company, and in less than a month concluded what the Company constantly referred to as a "voluntary" relinquishment agreement. It agreed to give up some 9,262 square kilometers of its onshore area immediately and some 1,000 square kilometers of the offshore area within five years. Both amounted to about 50% of the original concession. [37]

In 1967 the Company relinquished 1,012 square kilometers and in 1971 an additional 1,152 square kilometers, reducing the concession area to about 1/3 of the original area.

With the establishment of the Kuwait National Petroleum Company, KOC sold its local marketing facilities to the new concern.[38]

After three years of prolonged negotiations the Government signed on November 28, 1966, an agreement on the expensed royalty payments with the Company; the agreement was subsequently ratified by the Kuwait National Assembly.[39]

In line with the Teheran agreement of February 14, 1971, Kuwait signed an agreement with KOC effective as of November 1970 which raised the Government's share of profits to 55% and raised the posted prices along the lines of the other Persian Gulf producers. Similarly, in 1972 Kuwait received the increases in posted prices due to the devaluation of the dollar.

KUWAIT OIL ACTIVITY

Kuwait National Company

As mentioned above, Kuwait followed the practice of the other Middle Eastern oil-producing countries and organized in 1961 its own Kuwait National Petroleum Company (KNPC). There was, however, a pronounced difference; while both the Iran and Iraq national companies were entirely government owned, and Saudi Arabia Petromin was not even capitalized, the Kuwait national company was owned 60% by the Government and 40% by Kuwaiti private investors. The powers of KNPC encompassed all oil industry operations at home and abroad; but its first undertaking was the taking over from KOC of the distribution of all oil products in the country. At first KOC turned over products to the national company; however, when the national company built its $150 million refinery at Shuaibah, KOC supplied crude oil. The refinery was completed and began operations on November 18, 1968, and in that year KOC supplied 1.4 million tons of crude oil. The ultimate refinery daily capacity was to have been 95,000 barrels, the 1972 daily throughput rose to 104,000 barrels.[40]

In 1965 KNPC obtained 49% participation in the Spanish group Hispanoil. which had the right to supply up to 25% of the crude oil needs of Spain during the period 1970–1980. In the following year it was reported that the national company had established a subsidiary in Denmark and was planning to build a 46,000-ton bulk storage plant near Copenhagen, and that it had been ready to retail its oil in Europe. In the same year it obtained a 35-year-old concession covering about 3,860 square miles mostly relinquished by KOC. To work the concession a new concern was to be organized jointly with Hispanoil: the Kuwait Spanish Petroleum Company, in which KNPC was to hold 51% interest and Hispanoil 49%. The Spanish group was to spend $3.6 million over a two-year period and drill a minimum of 80,000 feet over an eight-year period.[41]

Petrochemicals

The Kuwait Petrochemical Industries Company (KNPIC) was established in 1967; it was owned 88% by the Government, 5% by KNPC, 3% by Kuwait National Industries and 4% by Kuwaiti investors. This company organized the Kuwait Chemical Fertilizer Company, which was owned 60% by itself and 20% each by BP and Gulf Oil. The fertilizer plant built at Shuaibah was completed in 1971, and had an annual capacity of 710,000 tons of various products. [42]

New Concessions

Late in December 1960 the Kuwait Government signed an agreement with the Shell International Petroleum Company, a subsidiary of Royal Dutch-Shell, for a concession of approximately 1,500 square miles offshore. The agreement provided that a new company, Kuwait Shell Petroleum, was to be organized by the concessionaire; the period of the concession was to be 45 years. Annual rental was to be $2.8 million for the first two years and $5.6 million for each of the following years until oil exports had begun. The basic payment arrangements were to be 50% of profits and an additional $84 million in bonuses, some on signature of concession and some at designated times depending on the rate of oil output. Shell had the right to surrender the concession at the end of three years on payment of three years' rental, plus the original bonus at signature which was to have been $19.6 million, or a total of $30.8 million. The Government had been granted the right to purchase up to 20% participation when oil was discovered. In 1963 it was reported that Shell suspended operations in its offshore concession. Even as late as 1971 it was reported that the three-sided boundary dispute in the offshore waters of Iran, Kuwait and the Neutral Zone prevented Shell from developing its concession. [43]

In 1964 the Kuwait Government signed an agreement with Aminoil granting an oil concession of the offshore islands of Knur, Qara and Umm al-Maradin, on the same terms as the company's onshore concession in the Neutral Zone. [44]

Transportation

Kuwait also developed its own transportation facilities. In 1959 the Kuwait Oil Tanker Company, a privately owned concern, was organized. In November 1966 it ordered two 207,000-ton tankers from a Japanese shipbuilding firm. In 1970 the tanker company took delivery of its third 207,000-ton tanker. This increased its fleet to six vessels: three 207,000-ton, two 60,000-ton, and one 49,000-ton. The Company ordered a 324,000-ton super tanker from a Spanish concern.

KOTC and the Kuwait Government agreed to establish a company for shipping natural gas to be known as the Kuwait Gas Tanker Company; 51% ownership was to be held by the state and 49% by KOTC.

In 1970 the Kuwait Global Tanker Company was established with a capital-ization of KD 360,000, of which 51% was to be owned by Kuwaiti private interests and 49% by foreign investors.[45]

KUWAIT DEVELOPMENT AND PROSPECTS

Of all the Middle Eastern oil-producing countries Kuwait was the first to achieve the saturation point of oil revenue; and this because of the small size of the population and the enormous concentration of the oil reserves in a comparatively very tiny area. Kuwait has been and still is the most prosperous state in the area, organized on the broadest base of social welfare, part real and part statistical. The progress in the fields of education, health, housing and water supply services ad-vanced even further during 1955–1971 than during the previous period. The Kuwaiti economy, although the Government was still heavily dependent on oil revenue for up to 90–95% of its budget, had greatly expanded, and the Shuaibah Industrial Development Complex had enlarged the industrial sector of the country. A five-year development plan was adopted for the years 1966–1971 and a second ten-year plan was adopted for the years 1972–1982. But the industrial possibilities were inevitably limited, for the shaikhdom has no natural resources other than oil. The Kuwaiti penetration into some of the oil industry branches from refining to transportation, natural gas and petrochemical phases, and even oil ventures abroad have given Kuwait a further economic boost. The statistical game of computing per capita income by dividing the enormous oil revenue by the total population of the country had become the most outstanding feature in discussing the paradise-like existence of Kuwaiti life.

One writer on Middle Eastern development pointed out that while in 1938–1939 the total government income was about $290,000 and the gross per capita income was $40, the Government income from oil in 1965 was $671 million and the gross per capita income was $3,360.[46] One year later the per capita income jumped to $3,410; and statistically it must have risen, in spite of the huge annual increase in population, even higher in the subsequent years as the oil revenue curve would indicate in Table XXXV. Nevertheless, it should be kept in mind that these are only statistical figures. To be sure, education – from kindergarten through high school, academic or vocational, and even at college level – was free, and in some cases parents were given bonuses for sending their children to school; health services were free and no taxes were imposed; housing was made reasonably available through various schemes. But the great bulk of the wealth was amassed by the upper classes: the ruling family, the merchant class and the rich social elite. More-over, the wealth did not stay in Kuwait, for there was nothing to be done with it, and as the above-mentioned writer put it: "Kuwaitis rapidly became the landlords of Cairo, Beirut and Lausanne apartment houses."[47]

For years it was the practice, established by the British advisors, to invest some of the oil revenue abroad, especially in Britain, which induced Dr. Polk to assert that Kuwaiti deposits in London became the important factor in the stability of the

British pound.[48] Under pressure from Arab nationalists, especially in Egypt, Kuwait established in 1961 the Kuwait Fund for Economic Development of Arab Countries with the first allocation of $140 million from the regular Government budget. A Kuwaiti mission visited Arab countries and submitted recommendations for credits to finance various projects. Two years later the assets of the Fund were doubled, and were being augmented ever since. During 1962 the Fund granted loans of about $20 million each to Jordan and Lebanon. By 1967 the Fund's assets were KD (Kuwaiti Dinar) 200 million; ten Arab countries were granted long-term loans to the extent of some KD 60 million.[49]

DEFENSE PROBLEM

A tiny country hemmed in by ill-defined boundaries between two powerful neighbors — Saudi Arabia and Iraq — though friendly yet covetous of its tremendous wealth, Kuwait was always under the shadow of possible invasion from either of its neighbors. Moreover, the rapid growth of the population, because of the oil boom, brought in a huge variegated element of foreigners: Arabs, Asians, and Westerners, who potentially threatened Kuwait's stability.[50]

Kuwait's security was guaranteed by its status as a British Protectorate. However, the magnitude of its wealth, the state of the political climate in the Arab world and the general world international political situation made both Kuwait's and Britain's positions untenable. On June 19, 1961, Kuwait and Great Britain signed a new treaty which terminated the protectorate status and granted Kuwait full independence, including external affairs. To make sure, however, that Kuwait was not to be completely unprotected, a provision was inserted in the agreement that Great Britain should respond militarily whenever the Shaikh felt threatened and asked for help.

The fact that Kuwait was no longer protected, and the provision for a possible return of British military forces, prompted Abdel Karim Qasim of Iraq to claim, on June 25, 1961, Kuwait as an old integral part of the Iraqi Basrah province, and he threatened to march on Kuwait. On the following day Shaikh Abdullah as-Salim al-Sabah proclaimed officially the independence of Kuwait, and he turned to his fellow Arab rulers — with mixed feelings, for both Saudi Arabia and the UAR had ambitions of their own for the control of the shaikhdom — and hoped that they would restrain Iraq from proceeding with her threat. He failed with Arab leadership, and he called on Great Britain, on July 1, for military aid. The British responded and the case came before the United Nations, which soon became embroiled by international politics and inter-Arab rivalries and was incapable of any action. The case went to the Arab League, where some sort of collective military assistance was provided for Kuwait against Iraq.[51]

This episode cost Kuwait immediate membership in the United Nations. It applied for membership on June 30, 1961, and no action was taken on it because of the crisis; it was not until November 30, and at the special request of the United Arab Republic, that the Security Council took it up. Iraq objected, and the Soviet

Union supported Iraq in objecting even to placing the application on the agenda. In this, which was a procedural matter, the Soviet Union was defeated. But on the substance of the application itself the Soviet Union used its veto. [52]

In May 1963, the Security Council took up the application again; this time it was unanimously approved. The decision then went to the General Assembly, which elected Kuwait as the 111th member of the world organization. [53]

Meanwhile Shaikh Abdullah announced in 1961 the formation of a provisional government, and established a constituent assembly to draft a constitution and act as interim legislature. On November 11, 1962, the Shaikh approved the draft constitution, and elections for the first 50-seat assembly were held on January 23, 1963. Only Kuwait citizens could vote. The Cabinet was headed by a Prime Minister who was the heir apparent to the Shaikh's throne; and the Shaikh appointed the ministers. [54]

In 1965 Shaikh Abdullah died and was succeeded by his brother Sabah.

Kuwait's wealth might have been its weakness but it also served as the source of its strength. After Qasim was assassinated early in 1963, the new Iraq regime recognized the full sovereignty and independence of Kuwait; for which it was handsomely rewarded with an $84 million interest-free loan. Similarly, Egypt's efforts on behalf of Kuwait at the United Nations were rewarded with a $70 million grant. [55]

Kuwait demonstrated its devotion to the cause of Arab nationalism by first stopping oil production after the June 1967 Arab-Israel war and later prohibiting shipments to the United States and Great Britain, the homes of the parent companies of KOC. Subsequently at the Khartoum conference, though the ban of oil shipments to the Western countries was lifted, Kuwait, together with Saudi Arabia and Libya, underwrote an annual multimillion dollar contribution to the two defeated nations: Egypt and Jordan.

Kuwait, however, it would seem, was determined to keep the flow of oil out of political objectives. The Shaikhdom was economically dominated by the rich families who were conservative in their outlook, and the Shaikhdom achieved a very high rate of foreign investments. Nationalization of the oil industry was not part of Kuwaiti leadership thinking, nor were the Kuwaitis ready to permit the Arab advocates of nationalization to use oil as a weapon against the United States. Speaking at the Seventh Arab Petroleum Congress held in Kuwait in March 1969, Oil Minister Abdel Rahman al-Aliki warned the other delegates that the stoppage of the flow of oil would harm Arab interests more than American interests. [56]

It was perhaps the same motivation that prompted Kuwait in joining Saudi Arabia and Libya in January 1968 to set up OAPEC (Organization of Arab Petroleum Exporting Countries), membership qualification for which was that oil exports be the major income of the Government. This would have protected the basic economic interests of the oil producers and would have prevented the other Arab countries, which had no vital interest in the continued flow of oil to world markets, from interfering in the oil affairs of the region. [57]

One of the most frequently mentioned measures of the producing countries against the concessionaire companies had been curtailed production. In 1971, deputies in the Kuwait National Assembly proposed to limit crude oil production to 3 million barrels daily, and subsequently the Government imposed a limit increase of 1% over that of the previous year. [58]

Although it was stated above that Kuwait opposed nationalization, it must be kept in mind that Kuwait was a founding member of OPEC and worked steadily for better terms from the oil companies. Kuwait was part of the OPEC participation program from 25% to 51%; and, perhaps more than any other OPEC member, would be in a position to finance future development from its own resources.

NEUTRAL ZONE

The boundaries between the various shaikhdoms and the newly emergent states after World War I, practically nondefined, caused serious clashes between rival tribes claiming pasture rights between Iraq and Najd, and between Najd and Kuwait. To bring about some *modus vivendi,* a conference was held in 1922 at Uqair. [59] Sir Percy Cox, representing Iraq and Kuwait, met with Ibn Saud, representing Najd, and worked out a series of agreements which created two neutral zones, one between Iraq and Najd, and one between Kuwait and Najd along the coast south of Kuwait, covering an area of about 250 square miles. Because of the abundance of oil both in Kuwait and on the eastern coast of Saudi Arabia, the neutral zone between the two gained in importance.

The supplemental agreement between Aramco and Ibn Saud, signed in 1939, granted the Company the area covered by Saudi Arabia's half interest in the neutral zone. [60] In 1948 Aramco surrendered its interest, apparently in return for the extension by Ibn Saud of the Company's rights into Saudi Arabia's offshore areas in the Persian Gulf. [61] On February 20, 1949 King Ibn Saud signed a concession with the American Pacific Western Oil Corporation for the exploitation of oil in the neutral zone. The reported terms were: Western Oil to pay Ibn Saud at the time of the signing of the concession a lump sum of $9,500,000 plus the first year's minimum royalty of $1 million; the Company guaranteed to pay the minimum royalty for three years even if it cancelled the agreement within two years; the Company was to pay a royalty of 55 cents per barrel of oil produced, and to grant the Government a 25 per cent interest in the Company. [62]

In the meantime, on June 6, 1948 the American Independent Oil Company (Aminoil), registered in Delaware in August 1947 with an authorized capitalization of $100 million, and composed of eight American oil companies and Ralph K. Davies, former vice-president of Standard of California and Deputy Petroleum Administrator under Secretary Ickes, who headed the company, [63] obtained from the Shaikh of Kuwait a concession to exploit his share of the neutral zone. The terms provided for a down payment of $7.5 million in cash; a minimum annual royalty of $625,000 whether oil was discovered or not; a royalty of 34 cents a barrel on all oil produced, plus a 15 per cent interest in the Company and a gift of a $1 million yacht. [64] The two American companies worked out an agreement between themselves and late in 1949 began prospecting in the neutral zone. In March 1953 Aminoil struck oil; by 1954 the oil was being conveyed by pipeline to a loading port in Kuwait territory. The Company employed 270 people. [65]

On December 8, 1957 the Japanese Petroleum Company was granted a concession to exploit oil in Saudi Arabia's undivided share of the offshore area of the Neutral Zone, between Saudi Arabia and Kuwait. The Government was to receive 56 per cent of the profits; it was also entitled to acquire, at a minimal price, 10 to 20 per cent of the Company's shares. A basic departure from all previous concessions was the provision that the Government was to share with the producing company in the profits made outside as well as inside Saudi Arabia. The Japanese concessionaire was to set up a company to carry out all production, refining, transportation and marketing operations. It was to deliver to the Government free, in kind or in cash, a royalty of 20 per cent of its oil production, and in addition was to make available 20 per cent of its production at a price five per cent lower than to other customers. The Government was to name one-third of the members of the Board of Directors, and the Company renounced its "right of diplomatic recourse." The Company's management and books were to be in Saudi Arabia, and 70 per cent of the total labor force were to be Saudis. On the discovery of oil in commercial quantities, a forty-year exploration lease would be granted. The Company was to pay in advance an annual rental fee of $1,500,000, and on discovery an additional $1,000,000 a year from the effective date of the agreement until the discovery. The Company was to erect a refinery in Saudi Arabia. [66]

In February 1958 the Japanese Arabian Oil Company was formed in Tokyo. Its immediate step was to obtain the other half of the concession from Kuwait.

On May 11 of the same year the Shaikh of Kuwait granted the Company the concession for his half of the Neutral Zone offshore area. The terms in general were the same as those of the Saudi Arabian agreement, except that Kuwait was to receive 57% of the profits and a $5 million bonus when production reached 50,000 barrels daily. The Company was to establish and maintain a research institute for the study of Kuwait's natural resources. [67] The basis of calculating the profits was to have been on the operations of an integrated company functioning as producer, transporter, refiner and marketer.

The existing companies became alarmed on both scores: the percentage increase in the profit-sharing which they feared was a breach in the established 50/50% formula and the integrated company basis of calculating profits. To be sure, some analysts pointed out that the real profits were derived from the production of crude and not from the other operations; in fact, in some cases, they maintained, these operations incurred losses which the companies covered from their share of production. [68] At the time, it was implied, and not very subtly, that if integrated companies replaced the existing producing companies and the calculations of profits were based on all operations, the Government's share would be less. As for the increased percentage of the Government's share in the profits, the established companies believed that the newcomers would not be able to exploit their concessions to a point where they would challenge the existing pattern, since offshore exploration was very costly and very difficult, and, moreover, the new company did not have the necessary financial means for a successful operation.[69]

On the last score the established companies miscalculated. The Japanese Government made available to AOC $42 million to develop its concession. The Company was successful in discovering oil, and it began production in April 1961. It built a 200,000-barrel daily capacity terminal at Ras al-Khafji to handle crude, and laid submarine pipelines to transport the oil from the wells to the terminal. In 1962 production reached 21,900,000 barrels.[70]

To alleviate resentment in Saudi Arabia over the terms with Kuwait, the Arabian Oil Company signed an agreement on September 12, 1963, which raised Saudi Arabia's share in the profits to 57%; it was to expense royalties for tax calculation purposes; granted Saudi Arabia a $5 million bonus when daily pro-

Table **XXXVI**. *Neutral Zone Production, 1955–1971*

Year	Barrels
1955	8,605,155
1956	11,400,000
1957	23,259,000
1958	29,310,000
1959	41,544,000
1960	48,678,000
1961	70,520,000
1962	82,000,000
1963	116,000,000
1964	130,000,000
1965	132,240,056
1966	154,096,000
1967	151,458,771
1968	156,600,000
1969	163,000,000
1970	182,800,000
1971	198,400,000[a]

a. *World Petroleum Report,* 111, 209, January 15, 1957; –
XII, 80, March 15, 1966; *Petroleum Press Service,* XXXIII,
230, June, 1966; CFP, *Annual Report 1964,* 6; Saudi
Arabian Monetary Agency, *Annual Report 1964,* 6; Saudi
Arabian Monetary Agency, *Annual Report 1389–90 A.H.
(1970),* 87; *World Oil,* CXLVIII, 108, February 15, 1959;
– CL, 95, February 1960; – CLII, 111, February 15, 1961;
– CLIV, 120, February 15, 1962; CLVI, 125, February 15,
1963; – CLXIII, 177, August 15, 1966; – CLXIV, 144,
February 15, 1967; – CLXVII, August 15, 1968; –
CLXVIII, 114, February 15, 1969; – CLXXIV, 63, Feb-
ruary 15, 1972.

duction reached 50,000 barrels; was to erect a refinery at Ras al-Khafji, and was to contribute annually $200,000 to the faculty of the Petroleum and Minerals Institute of Saudi Arabia. [71]

Production in the Neutral Zone grew steadily as is indicated in Table XXXVI. But while that of the Arabian Oil Company grew from 7,400,000 barrels in 1961 to 131,000,000 barrels in 1971, that of Getty-Aminoil increased in the same period from 57,400,000 barrels to 67,400,000 barrels. [72]

Each of the three companies, Getty, Aminoil onshore and American Oil Company offshore, erected refineries of the topping variety. The Aminoil refinery was in Mina Abdullah in Kuwait, and in 1965 had a daily capacity of 110,000 barrels; Getty had its refinery in Mina Saud in the Neutral Zone with a daily capacity of 50,000 barrels; AOC built its refinery at Ras al-Khafji in the Neutral Zone, and in 1971 it had a daily capacity of 30,000 barrels. [73]

The oil reserves of the Neutral Zone were constantly revised upwards but at a much lower rate than either of its two owners, and so were the natural gas reserves.

In March 1964, Saudi Arabia and Kuwait decided to divide the zone between them, but the existing arrangements for the equal sharing of the natural resources of the zone were to remain the same. In 1970 the Zone was officially partitioned for administrative purposes and the official name became the Partition Zone. [74] In the same year the conessionaire companies in the zone agreed to raise the profit percentage of the Governments by 5 and increase posted prices in line with all the Persian Gulf companies. [75]

NOTES

1. C. U. Aitchison (ed.), *A Collection of Treaties, Engagements and Sanads Relating to India and Neighboring Countries*, XI, 264–265 (Delhi, 1933).
2. See above, Chapter IX.
3. For a short outline of Kuwait's vacillations between the Allied and Central Powers during World War I, see Kuwait Oil Company, Ltd., *Kuwait Past and Present*, 8–10.
4. US Senate, *International Petroleum Cartel*, 130.
5. Somehow both DeGolyer and Sanger make the point that only after Gulf retired from the Near East Development Corp. and thus escaped the restrictions of the Red Line Agreement could it pursue its concession in Kuwait. However, Para. 10, which was the basis for the Red Line Agreement, specifically excluded Kuwait. Gulf therefore would not have been debarred, under the agreement, from developing its concession in Kuwait. DeGolyer, *loc. cit.*, 79; Sanger, *The Arabian Peninsula*, 164. See also "Future of Middle East Development Clarified by Red Line Settlement," *World Petroleum*, XIX, 46–48, Dec., 1948.
6. *Foreign Relations, 1932*, II, 6–8.
7. *Ibid.*, 20.
8. *Ibid.*, 1–29.
9. US Senate, *International Petroleum Cartel*, 130–134; Anglo-Iranian Oil Com-

pany, *Proceedings of Annual Meeting June 27, 1935*, 4; US Senate. *American Petroleum Interests in Foreign Countries*, 24. As in the case of Bahrain, the State Dept. claimed credit for this arrangement. "The continued representation of our Government had secured equal American participation in this important field, which otherwise might be wholly British." *American Petroleum Interests in Foreign Countries*, 318–319.

10. US Senate, *American Petroleum Interests*, 379–380; Anglo-Iranian Oil Company, *Report and Balance Sheet at 31st December 1942* (henceforth cited AIOC, *Report* and year); Paul Edward Case, "Boom Time in Kuwait," *The National Geographic Magazine*, CII, 783, Dec., 1952.

11. AIOC, *Annual Report 1949*, 16; Kuwait Oil Company Ltd., *Oil and Kuwait*, 18–19; UN *Review 1949–50*, 60; – *Review 1951–52*, 56; – *Summary 1952–53*, 49.

12. US Senate, *Investigation of the National Defense Program. Additional Report of the Committee Investigating the National Defense Program. Report of the Subcommittee Concerning Investigations Overseas. Section 1 – Petroleum Matters* (Washington, 1948), 8.

13. DeGolyer, *loc cit.*, 83.

14. US Senate, *International Petroleum Cartel*, 6.

15. UN, *Review 1951–52*, 53.

16. *World Oil*, CXL, 205, Feb. 15, 1955; Case, *loc. cit.*, 784. Following is the upward revision of the oil reserve estimates of Kuwait in tons: 1949, 1,487,000,000; 1951, 2,172,700,000; 1953, 2,444,300,000, UN, *Review 1949–50*, 43, 60; – *1951–52*, 56; – *Summary 1952–53*, 49.

17. AIOC, *Report, 1951*, 22.

18. E. Lawson Lomax, "Kuwait Completes Model Oil Port," *World Petroleum*, XXI, 36–39, Oct., 1950.

19. Monroe, *loc. cit.*, 282. At the end of 1953 the total number of Company employees was 8,200; of these 3,000 were Pakistanis and Indians, 2,400 Kuwaitis, 1,800 Arabs, and 900 Americans and Britons. UN, *Summary 1952- 53*, 62–63.

20. See E. Lawson Lomax, "Kuwait Oil Company Provides Social Services and Training for Its Workers," *World Petroleum*, XXIII, 36–39, June, 1952. According to Lomax in 1952 the total medical staff of the Company was 288, of whom 16 were doctors and 73 male and female nurses.

21. John E. Cunningham, "Kuwait Boosts the Standards of Education in Record Time," *Middle East Report*, VI, May 3, 1954.

22. *Foreign Commerce Weekly*, May 11, 1953; Monroe, *loc. cit.*, 276–278. She reported that the Shaikh accepted one piece of advice (from whom she did not say) and decided to award the main jobs on a cost-plus basis to five first-class British contractors, each of whom took a Kuwaiti partner. The charge was cost plus 15 per cent, of which one-half went to the contractor and one-half to the Kuwaiti partner.

23. "Water for Kuwait," *The Economist*, CLXVII, 28–29, Apr. 4, 1953; Monroe, *loc. cit.*; Case, *loc. cit.*, 800–802.

24. "From him and the groups he represents, Shaikh Abdullah Salim gains valuable counsel in handling Kuwait's income and in administering the public-works program," Case, *loc. cit.*, 800. The decisive role of the British Government in

the destiny of Kuwait is illuminated by the following statement with which Elizabeth Monroe concludes her article: "Politically speaking, the unanswerable question of the future is whether so covetable a place can retain the independence that has in this century been preserved chiefly thanks to the unobstrusive but steady British protection." Monroe, *loc. cit.*, 284.

25. *Le Commerce du Levant,* Sept. 15, 1954.
26. KOC, *Oil in Kuwait,* 9; – *Annual Review 1959,* 7; *Petroleum Press Service,* XXIV, 225, Sept., 1956; *World Trade Information Service,* Pt. 1, 57–32, Mar., 1957; *World Petroleum Report,* XVII, 72–73, 1971.
27. KOC, *1959,* 13; – *Fact Sheet 1970; World Petroleum Report,* VIII, 44, Feb. 15, 1962.
28. BP, *Annual 1967,* 27; – *1968,* 30; KOC, *1968,* 19; *World Petroleum,* XXXIX, 26–27, July, 1968.
29. KOC, *1961,* 13, 14; – *1971,* 7, 9.
30. KOC, *1971,* 7, 18.
31. KOC, *Desert Epic,* 30.
32. BP, *Annual 1965,* 23; KOC, *1965,* 20, 54.
33. BP, *Annual 1966,* 25; KOC, *1966,* 12.
34. KOC, *1967,* 17, 18; – *1968,* 14; – *1969,* 11.
35. *World Petroleum Report,* XVII, 72–73, 1971; – XVIII, 59, 1972.
36. UN, *Economic Development 1961–1963,* 64; *World Petroleum Report,* XVIII, 59, 1972; *Petroleum Press Service,* XXXIX, 310, Aug., 1972.
37. UN, *Economic Development 1961–1963,* 64; *World Petroleum,* XXXIII, 58, May, 1963; *World Oil,* CLVII, 177–178, Aug. 15, 1963.
38. KOC, *Desert Epic,* 45; – *1961,* 24; – *1967,* 13; – *1971,* 7; *World Petroleum Report,* VIII, 44, Feb. 15, 1962; *Petroleum Press Service,* XXXV, 204, June, 1968; – XXXVIII, 194, May, 1971; – XXXIX, 132, Apr., 1972; *World Oil,* CLXXV, 126, Aug. 15, 1972.
39. KOC, *1967,* 13; *World Petroleum,* XXXIX, 29, July, 1968; *World Oil,* CLXV, 152, Aug. 15, 1967. The formula was the same as that of Iran outlined above.
40. KOC, *1967,* 11; – *1968,* 30; *Petroleum Press Service,* XXXVI, 131–133, Apr., 1969; – XXXIX, 390, Oct., 1972; SAMA, *Statistical Summary,* 16, Dec., 1968–Jan., 1969.
41. *World Oil,* CLXIII, 174, Aug. 15, 1966; – CLXV, 152, Aug. 15, 1968; *World Petroleum Report,* XVIII, 59, 1972.
42. *World Petroleum Report,* XV, 87, 1969; – XVIII, 59, 1972; SAMA, *Statistical Summary,* 15, Sept.–Oct., 1969.
43. *World Petroleum Report,* VII, 227, Feb. 15, 1961; – XVIII, 59, 1972; *World Oil,* CLI, 204, Aug. 15, 1960; *Petroleum Press Service,* XXVIII, 22, Jan., 1961.
44. *World Petroleum Report,* XI, 79, Mar. 15, 1965.
45. KOC, *1965,* 54; *World Petroleum Report,* XVII, 73, 1971; SAMA, *Statistical Summary,* 20, Sept., 1971.
46. William R. Polk, *The United States and the Arab World* (Cambridge, 1969), 239.
47. *Ibid.*
48. It can readily be appreciated how politically motivated withdrawals from London banks could jeopardize the entire financial structure of the British economy.

49. *World Petroleum Report,* VIII, 44, Feb. 15, 1962; *Petroleum Press Service,* XXIX, 109–110, Mar., 1963; – XXXVI, 131–133, Apr., 1969. The mission did not visit Iraq, for Iraq at that time was claiming Kuwait; and the UAR was not granted credits, for the Cairo authorities refused to submit the proposed projects for examination.

50. Even though censuses were held in Kuwait from time to time, the population figures given by different sources do not agree. The following picture, however, emerges. In 1940, it was given as under 100,000; in 1957 it jumped to 206,000; in 1965 it was given as 468,000 and in 1972 as 700,000. SAMA, *Statistical Summary,* 14, Nov., 1966; *New York Times,* Oct. 27, 1972. In 1970, 55% of the population were listed as foreigners. Of the 700 medical doctors in the shaikhdom in 1969, only 33 were Kuwaitis. Robert G. Landen, "State of Kuwait," in Abid Al-Marayati, *The Middle East: Its Governments and Politics* (Belmont, California, 1972), 282.

51. For a detailed analysis of events and political moves and countermoves, especially in inter-Arab politics, and how the crisis was ultimately overcome see: "The Kuwait Incident," *Middle Eastern Affairs,* XIII, 2–13, Jan. 43–53, Feb., 1962.

52. UN Security Council, *Official Records Sixteenth Year, 984th and 985th Meetings,* November 30, 1961.

53. UN, General Assembly, *Resolutions Adopted During Fourth Special Session 14 May–27 June, 1963,* Supplement No. 1, 7.

54. Landen, *loc. cit.,* 284–287. He noted that in 1970 only 40,000 out of a possible 700,000 inhabitants qualified as voters.

55. *Ibid.*

56. *World Oil,* CLXXI, 169, Aug. 15, 1970.

57. *Petroleum Press Service,* XXXV, 47, Feb., 1968.

58. *World Petroleum Report,* XVIII, 59, 1972; *World Oil,* CLXXV, 126, Aug. 15, 1972. In Jan., 1972, the daily production average was 3,486,609 barrels.

59. See above, Chapter XV.

60. US Senate, *Hearings Saudi Arabia,* 24959.

61. Roy Lebkicher, *Aramco and World Oil* (New York, 1952), 30.

62. *The New York Times,* Apr. 19, 1953; *The New York Herald Tribune,* Apr. 2, 1949; United States Department of Commerce, *Business Information Service. World Trade Series,* 9, Dec., 1952; UN, *Summary 1950–51,* 32. It subsequently became the Getty Oil Company.

63. 'Middle East Offers No Threat to International Oil Trade," *World Petroleum,* XVIII, 64–65, Sept., 1947.

64. Sanger, *The Arabian Peninsula,* 165; *The New York Times,* July 7, 1948; – Apr. 19, 1953. Early in Aug. 1947 it was reported that King Ibn Saud and the Shaikh of Kuwait had agreed to share the income from any oil discovered in the neutral zone. *The New York Times,* Aug. 6, 1947.

65. Monroe, *loc. cit.,* 275.

66. *Petroleum Press Service,* XXV, 46–48, Feb., 1958; *Le Commerce du Levant,* Dec. 14, 1957; "New Oil Agreements in the Middle East," *The World Today,* XIV, 141–142, Apr., 1958.

67. *Foreign Commerce Weekly,* June 12, 1958; *Petroleum Press Service,* XXV, 208, June, 1958; *International Oilman,* XII, 327, Nov., 1958.

68. "From the standpoint of investors, the most important development is the overall financial loss sustained in 1958 by the seven companies on their Eastern Hemisphere non-producing activities." William S. Evans, *Petroleum in the Eastern Hemisphere* (New York, 1959), 11.

The Arab specialists at the Arab Petroleum Congress presented figures showing that for the last ten years total company profits had exceeded the receipts of Middle East governments by $3,330,500,000; the 50/50 formula basis was therefore unrealistic. These experts claimed that in addition to the 80¢ profit per barrel from production, the companies had made 20¢ on transportation, $1 on refining, and $2 on marketing. Farouk El-Husseiny, "The Economics of the Petroleum in the Middle East," *L'Egypte Industrielle*, 6–23, Apr., 1959; "The First Arab Petroleum Congress," *World Today*, XV, 251, June, 1959.

69. For the reaction of the companies to the increased percentage share for the government in the Iranian case see Chapter VII, footnote 17. The change in the percentage basis came about in 1970, and not because of the innovation in the Neutral Zone.

70. UN, *Economic Development 1961–1963*, 45; *World Oil*, – CLVII, 178–180, Aug. 15, 1963.

71. *World Petroleum Report*, X, 76, Mar. 15, 1964.

72. See Table XXVI above.

73. *World Petroleum Report*, XII, 80, Mar. 15, 1966; – XVIII, 62, 1972.

74. UN, *Economic Developments 1961–1963*, 65; *World Oil*, CLXI, 180, Aug. 15, 1965; *World Petroleum Report*, XVII, 74, 1971; *Petroleum Press Service*, XXXVIII, 273, July, 351, Sept., 1971.

75. *Petroleum Press Service*, XXXVIII, 273, July, 1971.

Chapter XXII
OTHER PERSIAN GULF SHAIKHDOMS:
QATAR, OMAN, ABU DHABI, DUBAI

QATAR

Qatar is a peninsula situated south of the Bahrain islands, covering an area of about 4,100 square miles, and ruled by the Thani family in protectorate relations with Great Britain. Its capital is Doha and it had a population estimated in 1954 at 17,000. When it became known that the Standard Oil Company of California had been granted the concession in the Bahrain islands, Anglo-Persian obtained an exclusive license, in September 1932, for a two-year geological examination of Qatar. Although this was a violation of the Red Line Agreement, the Iraq Petroleum Company did not penalize Anglo-Persian, but on the contrary agreed that the Qatar license should remain in the name of Anglo-Persian as the nominee of IPC.[1]

After the geological examinations indicated favorable prospects, IPC authorized Anglo-Persian to negotiate a concession with the Shaikh of Qatar. On May 17, 1935 the Company was granted a 75-year concession covering all of the territory. About two years later an IPC subsidiary, Petroleum Development (Qatar) Ltd., took over the concession from the Anglo-Iranian Oil Company. The concession provided for a per ton royalty of 2 rupees.[2] Drilling operations began late in 1938, and the following year oil in commercial quantities was discovered in the Dukhan field. By 1940 about 4,000 barrels per day were being produced; however, operations were interrupted because of war conditions. They were resumed in 1947, and a 51-mile pipeline connecting the oil field with the eastern coast of the peninsula was completed. By 1950 Qatar was producing an average of 33,800 barrels per day; export of crude began December 31, 1949.

Table XXXVII. *Qatar Oil Production, 1950–1954*

Year	Tons
1950	1,636,000
1951	2,370,000
1952	3,296,000
1953	4,003,000
1954	4,704,423[a]

a. UN, *Summary 1952–53*, 47; AIOC, *Report, 1950*, 24; – *Report, 1951*, 23; – *Report, 1952*, 20; – *Report, 1953*, 19; *Le Commerce du Levant-Beyrouth Express*, Mar. 16, 1955.

431

Qatar's reserves were greater than those of Bahrain, although far from the fabulous potentialities of Saudi Arabia and Kuwait. In 1944 DeGolyer estimated the proven reserves at one-half billion barrels, while unofficial estimates were one billion barrels;[3] at the end of 1954 they were estimated at 1.5 billion. IPC's estimates were three billion barrels.[4] Estimates in tons were given in January 1949 as 67,600,000; in 1951 as 135,000,000; and in 1953 as 163,200,000.[5]

Falling in line with the rest of the area, on September 1, 1952 the Company signed an agreement establishing a 50/50 profit-sharing arrangement from the earnings of its operations within the peninsula. The Shaikh's income in 1950 was about $1 million; in 1952 in increased to $9 million, in 1953 it jumped to $16 million and in 1954 to $23,300,000. In addition to the income from Petroleum Development, Qatar also received income from its offshore areas. In June 1952 Royal Dutch-Shell was granted a concession for the offshore waters beyond the three-mile limit around the peninsula covering an area of about 10,000 square miles; that year the Shaikh received from Shell a bonus equivalent to about $730,000.[6] Ninety-five per cent of Qatar's budget was derived from oil revenue; only 25 per cent of the receipts were reserved for the Shaikh, and the rest of the royalties were spent on public utilities and new buildings and for the maintenance of welfare services.[7]

Developments 1955–1971

As will be seen from Table XXXVIII, production of crude oil was increasing, but at a rather slow rate until 1965, and from then to 1971 practically doubled because of the production of the Shell Company's Qatar offshore concession. Production from all areas quadrupled between 1955 and 1971. Similarly, the direct payments for oil produced gradually climbed until 1966 when they jumped by almost one-third over the previous year. In 1971 the direct payments were almost six times that of 1955.

The impact of the oil industry was far-reaching. Doha, the capital, which was a small sleepy village, became a bustling town of 30,000 inhabitants. streets were paved, hotels were erected and new attractive residential areas were developed. A large airport to accommodate jet airplanes was built, and drinking water was supplied by two distillation plants. Health services and education were free; a large hospital with the latest modern equipment was opened at Doha, and a women's hospital was added. All this was in addition to the health and medical facilities which the Company provided for its employees. In order to increase school attendance the Government paid $15 a month to the father of each boy who went to school.

The Shaikh, following the practice of the Shaikh of Bahrain, and no doubt with the guidance of the British advisor, established a reserve fund of about 40% of the oil revenue.[8]

Table XXXVIII. *Qatar Oil Production, 1955–1971*

Year	Tons
1955	5,362,000
1956	5,784,000
1957	6,505,000
1958	8,657,000
1959	8,150,000
1960	8,212,000
1961	8,382,000
1962	8,808,000
1963	9,095,000
1964	10,125,000
1965	10,961,000
1966	13,845,000
1967	15,479,000
1968	16,363,000
1969	17,000,000
1970	17,700,000
1971	20,500,000[a]

a. BP, *Annual Report 1956*, 27; – 1957, 30; – *Statistical Review 1971*, 18; *World Oil*, CXLVIII, 108, February 15, 1959; *Petroleum Press Service*, XXVII, 5, January, 1960; – XXXVI, 120, March, 1969.

Table XXXIX. *Qatar. Direct Payments, 1955–1971*

Year	Millions
1955	$ 35
1956	36
1857	45
1958	61
1959	53
1960	54
1961	54
1962	56
1963	60
1964	66
1965	69
1966	92
1967	102
1968	110
1969	115
1970	122
1971	198[a]

a. UN, *Economic Development 1958–1959*, 77; – *Economic Developments 1961–1963*, 83; *Petroleum Press Service*, XXXVIII, 326, September, 1971; – XXXIX, 322, September, 1972.

Qatar Petroleum Company Operations

The pattern of relations between the Government and the Qatar Petroleum Company was not unsimilar to those of the other Persian Gulf oil-producing countries. The company expanded production facilities; it relinquished a goodly portion of its concession area; it handed over the local distribution facilities to the Government, at first supplied the Government with the products of the topping plant and later, when the Government took over the plant, supplied crude oil. The Company increased the utilization of natural gas for its own needs, gave the Government substantial quantities of natural gas for electric power generation and for the water-distillation plants, and subsequently for other industrial uses.

The Company adopted the royalty expensing formula, increased profit-sharing to 55%, agreed to the OPEC requests for posted prices increases and the adjustments necessitated by inflationary pressures and the devaluations of the British pound and the American dollar. Finally, the Company agreed to the 25%–51% government participation schedule.

In 1965, the Company completed terminal expansion at Umm Said; and in 1969 it further extended the south berth at the terminal to accommodate the new giant-size ocean-going tankers. In the same year the company contributed £300,000 towards the cost of a new state road from Doha to Umm Said.[9]

Almost simultaneously with the unilateral Iraq Government action on the IPC concession area at the end of 1961, Qatar Petroleum Company relinquished 1,737 square miles of its concession; two years later it gave up an additional 1,237 square miles, and in August 1965 the Company returned another 1,104 square miles; in fact it relinquished to the Government the entire eastern zone and limited itself to the western part of the peninsula, concentrating on the Dukhan oilfield and retaining only ¼ of the original area.[10]

The Company built a topping plant, near the Umm Said terminal, to meet the local needs for non-octane oil products. In 1966 the topping plant supplied 2 million gallons of motor spirit, 1¼ million gallons of kerosene and 2 million gallons of gas oil. However, in 1968, the Government took over the topping plant, and the refining and distribution facilities were administered by the National Oil Distribution Company. In 1969 the Company supplied 13,824 tons of crude oil. To meet the refined-products needs the Government awarded in 1972 a contract for the construction of a 6,000-barrel daily capacity refinery.[11]

As mentioned above, the Company supplied the Government, through a gas pipeline from the Dukhan field to Doha, natural gas to fuel the electric-power station and the water-distillation plants. In 1966, it supplied the Government for these purposes 3.2 billion cubic feet of natural gas, while it utilized for its own needs 19 billion cubic feet. Government consumption went up in 1967 to 3.6 billion cubic feet and reached 5.24 billion cubic feet in 1971. In 1967, the Company experimented with injecting 505 million cubic feet of gas in the wells to keep up the pressure, but subsequently discontinued it as water injection was found to have been more effective. Beginning with 1969 the Company supplied annually some 700 million cubic feet of natural gas to the Qatar National Cement Company at Umm Bab.[12]

Early in 1971 the Company informed the Government that it would undertake to erect an installation to produce natural gas liquids for export with an annual capacity of 800,000 tons. The facilities, including a pipeline, jetty and special loading equipment, was to cost about £25 million and was to be completed in 1974.[13]

Other Operations

As mentioned above Shell Petroleum Qatar held an offshore concession. In 1969, two fields were producing: Idd al-Shargi with a 35,000-barrel daily capacity and Maydan Muhzan with a 100,000-barrel daily output. A terminal for Shell crude was built at Halul Island, 11 miles north of the first field and 12 miles northeast of the second. Production rose to 203,700 barrels daily in 1972. In the same year Shell discovered a third offshore field, Bul Hanine, with an estimated daily production rate of 30,000 barrels and an ultimate 100,000-barrel daily rate output. The terminal at Halul had a 2.5 million-barrel storage capacity.[14]

Meanwhile the Government granted an exploration permit covering some 3,000 square miles in onshore and offshore areas, including sections relinquished by QPC, to the American Continental Oil Company. In 1969, a Japanese group of companies and the Southeast Asia Oil Company of Houston, Texas, signed agreements with the Qatar Government. The American Company's area covered two offshore sections of some 900 square kilometers northwest of Qatar. After three years the kilometrage was to be reduced by ¼, after six years by ½, and after eight years to 30% of the original grant. Bonuses amounting to $13.5 million were to be paid by the Company on signature, on commercial discovery and on increased rate of production. The Company was to spend $24 million in the first 8 of the 30-year agreement, and it was to pay fully expensed 12½% royalty of posted prices, which was to go up to 15% with higher rate of production. Tax rate was to be 50% of profits. The Government was granted the right to acquire up to 50% participation on commercial discovery; and the Company was to invest 10% of its annual profits in feasible economic projects, at the discretion of the Government, in Qatar or abroad.[15]

Two years later, on August 12, 1971, the Qatar Government signed an agreement with the Belgian Oil Corporation for the exploration for oil and natural gas in a 12,000-square-kilometer area on- and offshore. The terms were practically, with minor variations, the same as those of the American Company.[16]

However, after the Qatar Government organized in April 1972 the Qatar National Petroleum Company — which was to undertake all kinds of oil operations in Qatar and overseas, excluding local refining and distribution— it cancelled the Belgian and American Companies' concessions.[17]

Abu Dhabi had granted an offshore concession to the Al-Bunduq Company owned by BP, CFP and Petroleum Development Company (Japan), each one 1/3.

However, Qatar claimed that Al-Bunduq was in its area. In 1969, Qatar and Abu Dhabi reached agreement on the demarcation line between them. The offshore demarcation median cut across Al-Bunduq field, and both countries agreed to share in the profits of the Al-Bunduq Company.[18]

After concluding a treaty with Great Britain terminating the protectorate, Qatar proclaimed its independence on August 14, 1971, and less than a month later applied for membership in the United Nations. On September 15, the Security Council approved the application, and on September 21, 1971, the General Assembly elected Qatar to membership in the world organization.[19] Qatar was a member of OPEC since 1961, and followed the policies of that organization. After independence it joined the Arab League.

Qatar has acquired some valuable experience and knowledge in the various branches of the oil industry, but its dependence on the revenue from the Qatar Petroleum Company was still overwhelming. The Company agreed, in line with the other companies in the region, to grant the Qatar Government 25 to 51% participation. Not being financially strong like Kuwait, participation would put a great strain on the Government and create serious problems.

OMAN

One of the late comers into the oil-producing Middle East area of the Persian Gulf was the Sultanate of Muscat and Oman. It's political history was turbulent and unstable. The Sultanate was made up of the long, almost 1000-mile coastal strip and the province of Dhofar in the interior. The formal relations between the Sultanate and Great Britain were not, like the other Shaikhdoms in the area, based on overall written agreements. However, small specific agreements and practical experience cast Britain in the role of protector, which she fulfilled loyally.

From 1920 to 1954, the interior section of the Sultanate was administered by the Imam of Oman in a semi-independent relationship with the Sultan. However, the new Imam, Ghalib bin Ali, attempted to enter into direct relations with Saudi Arabia and rebelled against the Sultan. The latter called upon the British for help, which came forth with land and air forces. On August 13, 1957, 11 Arab countries requested the United Nations Security Council to deal with British "armed aggression" in Oman. The Sultan sent a telegram to the Security Council protesting against the Arabs' letter and declaring that the issue between himself and the Imam of Oman was purely internal and was of no concern to the United Nations. On August 20, the Security Council took up the issue and rejected the Arab request that the United Nations deal with it.[20]

In 1958, Britain undertook to strengthen the Sultan's military forces, establish an air force and help with civilian administration, for which she was granted continued privileges of civil aviation, use of the airfields by the RAF and the naval facilities by the British Navy.[21]

With the decision of the British to withdraw from the Persian Gulf, and the new wealth that descended upon Oman through oil revenue, Qabus, son of the

Sultan, deposed his conservative old-fashioned father and seized power himself in July 1970. He declared that he would modernize his backward country and bring it in line with the other advanced newly independent shaikhdoms of the area. But the old resistance in Dhofar reemerged in the form of the Popular Front of Liberation of the Occupied Arab Gulf, and caused Qabus to mobilize his armed forces led by British officers and wage a prolonged battle. [22]

In spite of internal difficulties, Oman applied for membership in the United Nations. On September 16, 1971, the Security Council decided to refer the application to the Committee on Admissions, and some two weeks later it approved the application. On October 7, the General Assembly elected Oman to membership. [23]

Oil Developments

Among the specific treaties between the Sultan and Great Britain was one signed in 1923 which provided that the Sultan was not to allow the exploitation of his oil resources without permission of the British. In 1937, Petroleum Development (Oman), an affiliate of the IPC group, obtained a 75-year concession covering the Sultanate of Muscat and Oman with the exception of the district of Dhofar on terms similar to the standard IPC arrangements. In 1953, a petroleum concession was granted to two American companies: Cities Service and Richfield in the Dhofar area; it was to be for a period of 25 years, after commercial discovery, with an option for renewal for an additional 25 years. A large airport for the landing of supplies had been built near Salala, and exploration was to have begun within half a year of signature.

However, it would seem that oil was slow in coming to Oman. After four unsuccessful well drillings all the members of the IPC group dropped out; only Shell with 85% and Partex with 15% remained. When oil was discovered in 1966, CFP bought back a 10% interest from Partex.

On March 7, 1967, the company signed a revised agreement with the Sultan granting him a 50% share of the profits based on posted prices less allowances; 12½% royalty was to be expensed. [24]

Two fields, Fahud and Natih, far in the interior began producing in August 1967, for which a 172-mile 30/36/32-inch pipeline was built to a new terminal at Mina al-Fahal near Muscat on the Gulf of Oman. A 900,000-barrel storage tank was erected at the terminal and a large underwater pipe was installed for loading large tankers. Later two additional fields, Yibal and Al-Huwaisah, were discovered; by 1972 both were producing. Early in 1970, Petroleum Development (Oman) relinquished 1/3 of its concession, and agreed to two additional relinquishments in 1980 and 1990, which would reduce its holdings to about 2/9 of the original area. In 1968 the group acquired exploration rights in the Dhofar province covering some 40,000 square miles. [25]

Production grew steadily from 1967 to 1970, as is indicated in Table XL; however, it dropped sharply in 1971. Payments began in earnest in 1968 and rose steeply to 1970; these included not only royalty and share of profits but also the rentals for the exploration rights in the Dhofar district. Although Oman was not a member of OPEC the Company agreed to increase payments when the British pound was devalued, increase the percentage profit and raise the posted prices agreed upon in the Teheran agreement, and make dollar devaluation adjustment agreed to in Geneva. [26]

Table XL. *Oman. Production and Direct Payments, 1967–1971*

Year	Tons	Barrels	£
1967	2,900,000	23,029,000	1,366,625
1968	12,000,000	98,813,000	25,461,784
1969	16,400,000	117,200,000	38,518,356
1970	16,600,000	121,856,000	44,391,937[c]
1971	14,400,000[a]	111,318,000[b]	

a. BP, *Statistical Review of the World Oil Industry 1971*, 18.
b. *World Oil*, CLXVII, 216, August 1968; – CLXVIII, 114, February 15, 1969; – CLXX, 96, February 15, 1970; – CLXXII, 106, February 15, 1971; – CLXXIV, 63, February, 1972.
c. *Petroleum Press Service*, XXXVIII, 215–217, June, 1971.

Like all the other Persian Gulf countries Oman tried to exploit its offshore areas. On February 18, 1966, it granted a concession extending 300 miles along the northern coastline to Wintershall of West Germany, which was to be the operator; ownership was 59% Wintershall, 24% Shell, 7% Partex and 10% another German company. In 1967 a concession was granted in the territorial waters relinquished by Petroleum Development (Oman). In 1971, Oman granted Wendell Phillips Oil Company a concession extending 450 miles along the coast of southern Arabia. [27]

Oil had a great impact on the least developed and most turbulent of the Persian Gulf shaikhdoms. But Oman with an oil revenue of £44 million in the peak year, and a relentless war against a persistent rebel movement, covering a population of some 700,000 spread over vast territories, in spite of the great efforts of Sultan Qabus, could not achieve the stage of economic development of the other Gulf shaikhdoms, not did Oman advance in political development.

ABU DHABI

A backward shaikhdom of the Trucial Coast with a sparse population in ill-defined boundaries under a close relationship with Great Britain, Abu Dhabi emerged — after the four giants: Saudi Arabia, Kuwait, Iran and Iraq — as one of the great oil producers of the Middle East, with prospects for even greater output. Statistically, Abu Dhabi is supposed to have the highest per capita income in the world.

An exclusive agreement signed by the Shaikh of Abu Dhabi with Great Britain in 1892, and subsequent agreements, have established and steadily increased the role of Britain as the protector and controller of Abu Dhabi's government and affairs. The estimated 16,000 population roaming over an area of some 26,000 square miles of desert made a very small stir or none at all in the affairs of the Persian Gulf. In 1908 the total revenue of the Shaikhdom amounted to $750,000.[28]

When the scramble for oil resources in the Persian Gulf, right after World War I, was on, the British signed an agreement with the Shaikh on May 3, 1922, in which he obligated himself to grant oil concessions only to a "person appointed by the British Government."[29]

The ill-defined boundaries between the shaikhdom and its neighbor Saudi Arabia mattered very little until the oil potentialities of the area came to the fore. As a consequence the relations between Abu Dhabi–Great Britain and Saudi Arabia were bedeviled by the conflicting claims both onshore and offshore. The most celebrated controversy was that over the Buraimi Oasis.[30]

In line with the 1922 agreement, Abu Dhabi granted in 1935 an oil exploration option to the Anglo-Iranian Oil Company for two years. The option was not picked up, and in January 1939 Abu Dhabi granted Oil Development (Trucial Coast), an IPC affiliate, a 75-year concession. Shaikh Shakhbut strove to grant an offshore concession independently to IMOC. However, the Company relinquished it in 1952. In the following year he granted a new offshore concession to Abu Dhabi Marine Areas, which was owned 2/3 by BP and 1/3 by CFP. It was for a period of 65 years, and drilling was to commence within five years.

Petroleum Development Trucial Coast relinquished, in the late 'fifties, all its concessions except that of Abu Dhabi. It began drilling in 1952; in 1960 it discovered oil in commercial quantities in the Murban field. Two years later it changed its name to Abu Dhabi Petroleum Company.

Meanwhile, in September 1958, ADMA made its first discovery in the Umm Shaif field. Exports began in 1962 when a terminal was built at Das Island and a 22-mile, 18-inch pipeline laid to connect the wells with the terminal. The second field, Zakum, was discovered in 1963. After a 56-mile, 26-inch submarine pipe to Das Island was laid, production began.[31] In December exports commenced from the onshore Murban-Bu Hasa field, from which, although it is only 12 miles inland, crude had to be taken 70 miles westward by a 24-inch pipeline to Jebel Dhanna where there is sheltered deep water. A tanker terminal with two loading berths was built three miles out at sea. ADPC expanded facilities at the field to make possible the exporting of six million tons a year.[32]

Payments and Relinquishments

Both the ADMA and ADPC concessions provided for the payment of 20% of the profits to the Government. In 1965 ADPC signed a revised agreement with the Shaikh which raised the profit share to 50%; royalty was to be expensed in accordance with the OPEC formula, and the Company was to relinquish unexploited areas. A similar agreement was signed with ADMA on November 10, 1966. [33]

Meanwhile, in August 1965, Shaikh Shakhbut ibn Sultan an-Nahaiyan was deposed by his brother Shaikh Zaid. [34]

As the established companies relinquished sections of their concession areas the Shaikh granted concessions to new companies onshore and offshore.

Onshore concessions were granted to a group of Phillips (operator), AGIP and Aminoil, covering an area of some 13,000 square kilometers. It was to spend a minimum of $12 million on exploration and pay a cash bonus of $3 million on signature, annual rental until commercial discovery, and bonuses based on rate of production. Another onshore concession was granted in September 1969, to the Middle East Oil Company (Japanese). Terms were along OPEC lines plus bonuses. On commercial discovery the Shaikh had the option of 50% participation by paying half of the accumulated costs. When daily production reached 200,000 barrels, the Company was to build, within 3 years, a 30,000-barrel daily capacity refinery; and when daily production reached 300,000 barrels, the company was to plan for chemical and petrochemical projects to be jointly undertaken, or pay the Shaikh $2 million.

Offshore concessions were granted in 1968 to Abu Dhabi Oil Company (Japanese group). It was to pay a $1 million bonus on signature, and was to spend $13 million on exploration. In June 1970, the Shaikh granted a concession to a group consisting of Pan Ocean Corporation, Syracuse Oils and Wington Industries, covering an area 3,150 square kilometers in three offshore blocks. The group was to pay a $2.5 million bonus on signature, and other bonuses on commercial discovery and on rise in rate of production. It was to spend $19 million over an 8-year period on exploration and development. Royalties were to be 12½ to 16% of posted prices, depending on the rate of production. The Shaikh was to have the option of up to 50% participation on commercial discovery, paying half the costs, over a 10-year period. [35]

When oil was discovered in ADMA's field of El Bunduq a new separate company was organized in which BP sold ½ of its holding to a group of Japanese companies. These companies were to market BP's 1/3 of the oil in Japan. After an agreement was reached with Qatar, as mentioned above, the Company continued to operate the field. [36]

The relations between the Abu Dhabi Government and the producing companies were proper. In 1967, Abu Dhabi joined OPEC, and one year later the Government signed new agreements with both companies, eliminating OPEC allowances off posted prices. Three years later ADPC began supplying, from the Murban field, natural gas to the Government Abu Dhabi town power generation plant. [37]

When the departure of the British from the Persian Gulf became a certainty, Abu Dhabi sought union with the other Trucial shaikhdoms rather than independence. The Union of Arab Emirates emerged in December 1971, comprising seven shaikhdoms which assumed the name The United Arab Emirates. The President of the Union was Shaikh Zaid of Abu Dhabi, the Vice President Shaikh Rashid of Dubai. There could be no doubt that oil production and its level in the two shaikhdoms determined their respective positions in the Union.

In spite of the Union, Abu Dhabi apparently did not feel secure — especially with the ever increasing revenue from oil — in the face of the rising power of the Arab socialist countries. It therefore followed in the footsteps of Kuwait, and set up the Abu Dhabi Fund for Arab Economic Development. The capital of the Fund was about $105 million and was to be provided by the Government in stages. [38]

In line with the practice of the other Persian Gulf oil-producing countries, Abu Dhabi established in 1971 the Abu Dhabi National Petroleum Company, with a capital of BD 20 million contributed entirely by the Government. Its functions were to be: local refining and distribution, and taking up the Government's option of participation in the oil companies. [39]

Oil and Development

Production of oil after 1962 advanced very rapidly from both offshore (ADMA) and onshore (ADPC), as is evidenced from Table XLI. Even disregarding 1962, when production began late in the year, and taking 1963 as the first year of production, the increase from that year to 1971 — nine years — was almost tenfold. The increase in payments from 1963 to 1971 was indeed staggering, more than 71-fold. [40] The increases in reserves were substantial, from 6.5 billion barrels in 1963 to almost 19 billion barrels in 1971. But the share in the total world or even total Middle East reserves must be considered very modest. [41]

Table XLI. *Abu Dhabi. Production and Direct Payments, 1962–1971*

Year	Long Tons	Barrels	Dollars
1962	800,000	5,997,000	3,000,000
1963	2,600,000	17,800,000	6,000,000
1964	9,000,000	68,456,000	12,000,000
1965	13,500,000	103,113,000	33,000,000
1966	17,300,000	131,196,000	100,000,000
1967	18,300,000	138,912,000	105,000,000
1968	24,000,000	189,149,000	153,000,000
1969	28,900,000	219,657,000	191,000,000
1970	33,400,000	253,233,000	231,000,000
1971	44,900,000[a]	327,287,000[b]	431,000,000[c]

a. BP, *Statistical Review of the World Oil Industry 1971,* 18; *New York Times,* January 14, 1973.

b. *World Oil,* CLVIII, 137, February 15, 1964; – CLXI, 187, August 15, 1965; – CLXII, 120, February 15, 1966; – CLXIV, 144, February 15, 1967; – CLXVI, 123, February 15, 1968; – CLXVIII, 114, February 15, 1969; – CLXX, 96, February 15, 1970; – CLXXIV, 63, February 15, 1972; *World Petroleum Report,* 64, 1971.

c. *Petroleum Press Service,* XXXVIII, 326, September, 1971; – XXXIX, 322, September, 1972.

With the growth in production Abu Dhabi undertook, in the spirit of the times, its first Five-Year Plan for social services: education, health and social welfare, and development of towns, harbors and an airport. With the growth of the population these services would of necessity increase. Before oil developed the population was estimated at less than 16,000; in 1971 it was estimated to have jumped to 70,000.

Even though Abu Dhabi did not itself produce crude oil, the Government apparently expected to obtain crude for export, and it was reported — late in 1972 — that the Government had signed an agreement with Rumania for the supply of crude oil for Rumanian equipment and services.[42]

Although Abu Dhabi was a relatively great producer of oil in the Middle East, it was neither a leader in political-military power nor in Middle East oil production. It had a number of foreign companies operating on- and offshore, and it followed the lead of the great producers by agreeing to the 25–51% participation schedule. From its high revenue it could, to some extent, finance part of its participation. But Abu Dhabi, regardless of all the institutional efforts at establishing a National Oil Company, would depend, to a very large extent, on the revenue-producing operations and the marketing facilities of the foreign companies.

DUBAI

The latest newcomer to the Persian Gulf oil-producing family was Dubai. This shaikhdom on the Trucial Coast has an area of some 1,500 square miles with a population estimated in the late 'sixties as 80,000, of which 15,000 were Iranians and mostly concentrated in the town of Dubai. Geographic determinants made Dubai a commercial center, and it was designated as the future capital of the United Arab Emirates.

The development of oil in Dubai was late in coming. The first concession was granted to Dubai Marine Areas in 1952. It was an offshore undertaking owned 2/3 by BP and 1/3 by CFP. Some eleven years later Dubai Petroleum Company, a wholly owned subsidiary of the American Continental Oil Company, acquired a concession covering the mainland and territorial waters. In August 1963, Continental bought a 50% interest in the Dubai Marine Areas, and was to be the operator of the concession.[43]

BP withdrew from DUMA, and CFP increased its share to ½, and entered into an agreement with the Spanish state-owned Instituto Nacional de Industria for a half share in its holding. The operator of the concession, DPC, discovered the Fateh field located 55 miles off the coast and began production in September 1969. By that time Continental sold 20% of its holdings: 5% to the German Wintershall Company, 10% Deutsche Erdöl (Texaco) and 5% Sun Oil Company.

To expedite exports DPC installed a $7 million underwater oil storage facility of a unique design in the Persian Gulf. It was a 500,000-barrel tank 205 feet high, 270 feet in diameter and had no bottom; the sea floor served as the base for oil storage.[44]

Production proceeded regularly: half a million tons in 1969, 4.3 million tons in

1970, and 6.5 million tons in 1971. Reserve estimates rose from one billion barrels in 1969 to 1.5 billion barrels in 1971.[45]

Other recorded concessions in the territories of the United Arab Emirates were: John W. Macom and Pure Oil Company acquired a 9,380 square mile oil concession – onshore and offshore – in the Shaikhdom of Sharjah. Shell obtained in January 1967 a 2,199 square kilometer concession, and relinquished it at the end of 1971. In January 1970, Buttes/Clayco were awarded a 40-year 562 square mile concession in the offshore and territorial waters of the Shaikhdom.

On March 20, 1964, Union Oil of California (80%) and Southern National Gas Company (20%) obatained a concession of 1,800 square miles, onshore and offshore, from the Shaikh of Ras al-Khaimah. In the middle of 1972, Union Oil reported its first offshore discovery. On March 15, 1969, Shell Hydrocarbons was granted a 40-year onshore concession. The Company was to begin exploration within 18 months, and drilling within 3 years, and was to spend $3.5 million on exploration and development within 8 years. $9.5 million in bonuses were to be paid on signature, on commercial discovery and on achieving specific production rates. In 1970, Shell relinquished its concession.

In November 1969, the Umm al-Qaiwain Government signed an agreement with the American Occidental Petroleum Corporation for a 40-year 755 square mile offshore oil concession. The Company was to pay a total of $10.5 million in bonuses from signature to daily production of 200,000 barrels; it was to spend a minimum of $5 million on exploration in the first four years, and pay the Government 12½% royalties and 50% of profits. In 1969, Shell obtained a concession, and in 1971 it relinquished it.

The Shaikh of Ajman awarded a 600 square mile concession on- and offshore to Occidental Oil Company in January 1970. Payments of $2.4 million were to be made over the ensuing four years.[46]

NOTES

1. US Senate, *International Petroleum Cartel*, 86. The obtaining of the license by Anglo-Persian was interpreted as a "preclusive measure in order to keep the area out of the hands of Standard Oil Company of California." This was also given as the reason for the benevolent attitude of IPC to Anglo-Persian's violation of the Red Line Agreement. DeGolyer declared: "The India Office refused to permit Americans to negotiate for concessions in Qatar, Trucial Oman coast, and these shaikhdoms fell like ripe fruit into the lap of Petroleum Concessions Ltd." DeGolyer, *loc cit.*, 79.
2. AIOC, *Report, 1950*, 24; US Senate, *International Petroleum Cartel*, 86; James Terry Duce, *Middle East Oil Developments*, 10; UN, *Review 1951–52*, 61; Sanger, *The Arabian Peninsula*, 124.
3. *World Oil*, CXL, 205, Feb. 15, 1955.
4. US Senate, *International Petroleum Cartel*, 87.
5. UN, *Review 1949–50*, 60; – *Review 1951–52, 53;* – Summary 1952–53, *40.*
6. AIOC, *Report, 1952*, 20; UN, *Review 1951–52*, 61; – *Summary 1952–53*, 42.

In connection with the offshore concession, Sanger reported the following. In July 1949 Shaikh Abdullah Ibn Qasim al-Thani received a down payment for the concession rights to the submerged areas of Qatar. He refused to share this new income with his relatives, and a split developed in the family which started with demonstrations and ended in a shooting war. The British Political Agent intervened, and on August 20 Shaikh Abdullah abdicated in favor of his son Ali, "thus bringing to an end the first revolution in the Arabian Peninsula to be caused by oil." Sanger, *The Arabian Peninsula,* 124.

7. UN, *Summary 1952–53,* 63.
8. Aramco, *Middle East Oil Development* (1956), 16; *World Oil,* CXLVIII, 108, Feb. 15, 1959; *Aramco World,* XIV, 16–17, Oct., 1963.
9. Qatar Petroleum Company,*Review for 1969;* henceforth cited QPC, and year.
10. UN, *Economic Developments 1961–1963,* 64–67; CFP, *Exercice 1961,* 13; – *Exercice 1963,* 9; – *Annual Report 1965,* 5; *Petroleum Press Service,* XXXII, 13–14, Jan., 394, Oct., 1965; BP, *Annual Report 1965,* 23.
11. QPC, *1966; – 1967; – 1968; – 1969; Petroleum Press Service,* XXXIX, 331, Sept., 1972.
12. QPC, *1966; – 1967; – 1968; – 1969; – 1971.*
13. QPC, 1971; *Petroleum Press Service,* XXXIX, 330, Sept., 1972.
14. *World Petroleum Report,* XII, 80, Mar. 15, 1966; *World Petroleum,* XXXVI, 22–27, July, 1965; *Petroleum Press Service,* XXXIX, 320, Sept., 1972. ENI bought in 20% in the Shell Qatar Company.
15. *World Petroleum Report,* X, 76, Mar. 15, 1964; *Petroleum Press Service,* XXXVII, 173, May, 1970.
16. SAMA, *Statistical Summary,* 17–18, Sept., 1971; *Petroleum Press Service,* XXXVIII, 394, Oct., 1971.
17. *World Oil,* CLXXIV, 21, May, 1972; *Petroleum Press Service,* XXXIX, 330–331, Sept., 1972.
18. *Petroleum Press Service,* XXXV, 304, Aug., 1968; – XXXVI, 153, Apr., 1969; – XXXIX, 330, Sept., 1972.
19. UN, Security Council, *Official Records Twenty-Sixth Year. Resolutions and Decisions 1971,* 12; – General Assembly, *Official Records Twenty-Sixth Session,* Supplement No. 29, 2.
20. UN, Security Council, *Official Records Twelfth Year,* S/PV. 783–784, Aug. 20, 1957. It should be noted that the United States abstained in the vote. See also the debate on the subject in the House of Commons, *PDC,* 574, cols. 870–877, July 29, 1957.
21. Alexander Melamid, "Economic changes in Yemen, Aden and Dhofar," *Middle Eastern Affairs,* V, 88–91, Mar., 1954; Rupert Hay, "Great Britain's Relations with Yemen and Oman," *Middle Eastern Affairs,* XI, 142–146, May, 1960. Behind the purely local differences between the Sultan and the Imam was the Buraimi oasis which reportedly was rich in oil reserves, and claimed by Saudi Arabia, which would have meant additional reserves for Aramco, and Oman, which would have meant additional resources for the IPC group. Caught between were Standard of New Jersey and Mobil Oil Corporation, which had shares in both groups.
22. *World Petroleum,* XLI, 14, Oct., 1970. The writer declared that Qabus' move

in deposing his father had the tacit approval of the oil company. *World Petroleum Report*, XVII, 73, 1971; *New York Times*, Feb. 14, Nov. 14, 1971.

23. UN, Security Council, *Official Records. Twenty-Sixth Year. Resolutions and Decisions 1971*, 12–13; — General Assembly *Official Records. Twenty-Sixth Session*, Supplement No. 29, 2.

24. *World Petroleum*, XXXVIII, 22, July, 1967; — XLI, 14, Oct., 1970; *Petroleum Press Service*, XXXVII, 331–332, Sept., 1970; — XXXVIII, 215–217, June 1971, CFP, *Annual Report, 1967*, 7.

25. *Petroleum Press Service*, XXXV, 407–409, Nov., 1968; – XXXVII, 331–332, Sept., 1970; – XXXVIII, 215–217, June, 1971.

26. *Petroleum Press Service*, XXXVIII, 215–217, June, 1971; – XXXIX, 109, Mar., 1972.

27. *World Oil*, CLXXII, 10, June, 1971; *Petroleum Press Service*, XXXV, 407–409, Nov., 1968.

28. J.G. Lorimer, *Geographical and Historical Gazetteer of the Persian Gulf, Oman and Central Arabia* (Calcutta, 1908), I, 409 Quoted in Clarence Mann, *Abu Dhabi. Birth of an Oil Sheikhdom* (Beirut, 1964), 112.

29. Mann, *op. cit.*, 79.

30. For background of the Buraimi issue see: Alexander Melamid, "The Buraimi Oasis Dispute," *Middle Eastern Affairs*, VII, 56–63, Feb., 1956; Mann, *op. cit.*, 91–100; J.B. Kelly, "Buraimi Oasis Dispute," *International Affairs*, XXXII, 318–26, July, 1956; — "Sovereignty and Jurisdiction in Eastern Arabia," *International Affairs*, XXXIV, 16–24, January, 1958.

31. CFP, *Exercice 1962*, 10; – 1963, 10; *World Petroleum Report*, X, 69, Mar. 15, 1964; *World Petroleum*, XLI, 63–65, Apr., 1970.

32. *Petroleum Press Service*, XXXI, 28, Jan., 1965; *World Petroleum Report*, XI, 68, Mar. 15, 1965.

33. BP, *Annual Report 1965*, 23; – 1966, 25; CFP, *Annual Report 1965*, 5; *World Petroleum*, XXXVIII, 44, July, 1967; *World Petroleum Report*, XIII, 66, 1967.

34. *World Petroleum*, XXXVII, 162, Oct., 1966.

35. *World Petroleum Report*, XIII, 66, 1967; *World Petroleum*, XXXVIII, 44, July, 1967; *Petroleum Press Service*, XXXV, 24, Jan., 386, Oct., 1968; – XXXVIII, 245–246, July, 1970, SAMA, *Statistical Summary*, 16, Sept., 1970.

36. *Petroleum Press Service*, XXXVII, 382, Oct., 1970.

37. *Petroleum Press Service*, XXXV, 304, Aug., 1968; *World Oil*, CLXVII, 214, Aug. 15, 1968; *World Petroleum Report*, XVII, 64, 1971.

38. *Petroleum Press Service*, XXXVIII, 313, Aug., 1971. The motivation for establishing the Fund was frankly stated in a United Arab Emirates advertisement in *The New York Times* of Jan. 14, 1973. "The UAE has been quick to make contacts with the republican, socialist Arab world which in turn has looked to it for the financial handout of the sort which helped to establish the security of the traditional regime in Kuwait."

39. *World Petroleum Report*, XVIII, 52–53, 1972.

40. This high ratio of increase in payment was due to the high cost of production at the early stages of development and the low basis of payments. In 1963, the average income of all Middle Eastern producers was 77.7 cents per barrel, that of Abu Dhabi was only 36.4 cents; in the following year it was even lower,

18.2 cents, compared to the Middle East average of 78.3 cents. It was not until 1968 that Abu Dhabi caught up with the rest of the Middle East. With the new increases in percentage profits and in posted prices in 1971, it surpassed the Middle East average. *Petroleum Press Service*, XXXIX, 322, Sept., 1972.

41. *World Oil*, CLIX, 94, Aug. 15, 1964; Aramco, *Facts and Figures 1971*.
42. *Petroleum Press Service*, XXXIX, 390, Oct., 1972.
43. *World Oil*, CLIX, 177, Aug. 15, 1964; CFP, *Exercice 1966*, 38.
44. CFP, *Annual Report 1969*, 9–10; *Petroleum Press Service*, XXXVIII, 109–110, Mar., 1971; *World Petroleum*, XL, 26–27, July, 1969; *World Petroleum Report*, XVII, 76, 1971.
45. *Petroleum Press Service*, XXXVII, 6, Jan., 1970; *Petroleum Times*, LXXIV, 27, July 3, 1970; – LXXV, 22, July 30, 1971; *World Oil*, CLXXI, 71, Aug. 15, 1970; – CLXXIV, 63, Feb. 15, 1972, *New York Times*, Jan. 14, 1973. The last named source (paid advertisement) gave an estimate for 1972 the same as the production figure for 1971.
46. *Petroleum Press Service*, XXXVI, 146–147, Apr., 1969; SAMA, *Statistical Summary*, 13, Dec., 1969–Jan., 1970; *World Petroleum Report*, X, 76, Mar. 15, 1964; – XVIII, 64, 1972: *World Oil*, CLXXV, 128, Aug. 15, 1972.

Section Six

MEDITERRANEAN

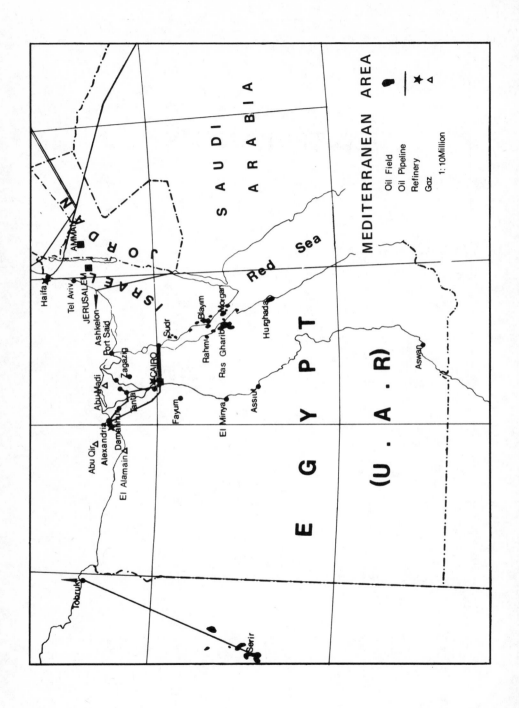

MEDITERRANEAN AREA

Oil Field
Oil Pipeline
Refinery
Gaz

1:10Million

SAUDI ARABIA

JORDAN

ISRAEL

Red Sea

Haifa
Tel Aviv
JERUSALEM
AMMAN
Ashkelon
Port Said
Sudr
Blaym
Morgan
Rahmi
Ras Gharib
Hurghada
Zagazig
CAIRO
Abu Madi
Tanta
Fayum
El Minya
Assiut
Aswan

EGYPT

(U . A . R)

Abu Qir
Alexandria
Damain
El Alamain

Tobruk
Serir

Chapter XXIII
ISRAEL, JORDAN, LEBANON, SYRIA

The impact of the oil industry on the Mediterranean countries — Turkey, Syria, Lebanon, Israel, Jordan and Egypt — had been pronounced but of less importance than on the Persian Gulf area countries. All except Jordan and Lebanon have been successful in discovering oil (Egypt first, followed by Turkey, Israel and Syria) in various quantities — in some insufficient even to meet their own internal needs — in their territories. Oil has affected them not only in terms of consumption but also in terms of pipelines, terminals and refineries, and those, in turn, have had repercussions in all the countries that are part of the overall pattern of the Middle East.

ISRAEL

The background of the oil story in Israel goes back to pre-World War I days when the territory was part of the Ottoman Empire. On February 3, 1914 three Turkish subjects, Ismail Hakki Bey al-Husseini, Sulaiman Nassif Bey and Charles Ayoub Bey, obtained from the Turkish Government seven licenses for mineral prospecting in Palestine; these were confirmed on March 26 by the Imperial Department of Commerce and Agriculture. The three transferred their licenses to W. E. Bemis and Oscar Gunkel, trustees of the Standard Oil Company of New York, before the competent Ottoman authorities in Jerusalem and Constantinople, in May 1914. That same year Standard also obtained eleven licenses directly from the Ottoman Government, for prospecting in the vicinity of Beersheba. The activities of the Company were suspended during the war, but when the British forces occupied Palestine, representatives of Standard Oil requested permission to proceed with the development of their licenses. The British refused, and on September 3, 1918 the military authorities in Jerusalem ordered the Standard Oil Company's representative to allow Lieutenant Goodrick to inspect maps and other documents of its concessions. Goodrick examined them, borrowed them, and later returned them.

Standard appealed to the American Government against the action of the British authorities in Jerusalem. After discussing the matter informally with the American Chargé d'Affaires in London, the British Foreign Secretary, Lord Curzon, declared that until the mandate over Palestine was established and war conditions no longer existed, all claims must await clarification. The State Department was not willing to accept this decision, especially in view of the activities of British concessionaires in Mesopotamia, also an Ottoman territory that was to be constituted as a mandate. This was the beginning of a controversy between the United States and Great Britain over the oil resources of the Middle East which did not end until the American companies had been granted a share in the Turkish Petroleum Company.

Both sides took steps to strengthen their positions. On August 9, 1919 Lord

Curzon asked Colonel French in Cairo to send information to the British Peace Delegation about concessions in either Syria or Palestine in which British interests participated, as well as about concessions of any other kind. On November 1, Colonel Meinertzhagen replied from Cairo that he was unable to trace any concessions in Syria or Palestine in which British interests had a share. "As far as can be ascertained there are no pre-war concessions held by British subjects in Palestine." He supplied the Foreign Secretary with a list of concessions, among which was an oil concession in Palestine to Standard Oil Company.[1]

As to the actual claims of Standard, Colonel French reported to Lord Curzon on August 13, 1919 that the Company had satisfied the British administration that there was no doubt as to the rights which the Company had obtained from the late Ottoman Government. The claims were divided into three categories: 1) operations approved by the Government of Turkey and on which work had actually been begun, but no drilling;[2] 2) operations approved by the Council of State of the Turkish Government, but on which no work had been undertaken; 3) operations approved by the departments of the Government of Turkey but which had not been submitted to the Council of State. The Company asked for permission to continue work on the first category, for which it had full documentary evidence of its rights, which evidence it was willing to submit to the British administration in Jerusalem. Colonel French asked the Foreign Office for a decision. At the end of the month Lord Curzon replied: "Permission cannot be granted until the question of the Mandate has been settled."[3]

The Americans on their part also tried to establish the facts. On July 7, 1919 Acting Secretary of State Breckinridge Long instructed the American Ambassador in London to find out whether there were other claims, and whether the military authorities in Palestine had permitted holders of concessions granted by the Ottoman Government to operate there since the occupation of the territory by British forces. About two months later he expressed to the American Commission to negotiate peace his grave doubts about British intentions. The Secretary was afraid that the British would squeeze Standard out, and a report prepared by Captain William Yale, who was with the King-Crane Commission, indicated that Standard's apprehensions were justified.[4] Long informed the Commission that the State Department intended to take "as strong a position as possible in regard to rights of our nationals to concessions obtained or in process of consummation before or during the war" and would request the British to facilitate the American concession in Palestine.[5]

The American Consul in Baghdad informed the State Department on February 4, 1920 that the Civil Commissioner had refused a license to investigate the land in Mesopotamia to the representative of the Standard Oil Company, E. S. Sheffield, but had granted a license to the representative of Shell. The Consul expressed the opinion that the Shell Oil Company would be able to obtain the desired information regarding oil lands which had been denied to the Americans. The State Department thereupon instructed the American Ambassador in London to protest against the discrimination in Mesopotamia between Standard and Shell. The Department was fully aware that it had no legal ground for representation to the British authorities to obtain the right to explore the territory during military occupation, but the American Chargé in London was told: "It is believed important, however, to maintain pressure so Foreign Office will realize vital concern of this Government and understand that during negotiations Turkish Treaty the United States will look to the British Government to see that throughout such negotiations the vested rights of its nationals under the Turkish law are protected and guaranteed."[6]

The strongest weapon of the United States was its refusal to recognize the legality of the Turkish Petroleum Company, under which the British claimed their oil concession in Mesopotamia. The suspicion that the British were attempting to gain control of the oil resources of the area and prevent Americans from getting a foothold there was strengthened by the British refusal to permit the Standard Oil Company of New York to proceed with its concession in Palestine. On June 18, 1920 the American Ambassador told Lord Curzon that the British were discriminating, as regards oil exploitation, between British and Americans in their administration of Palestine and Mesopotamia.[7]

The dispute between the two countries over the oil in Mesopotamia increased in intensity, especially after the publication of the San Remo Agreement, and agitation against the British in the United States mounted. The British were ready for some compromise. On August 12, 1921 L. I. Thomas, vice-president of the Standard Oil Company, asked the Secretary of State to obtain from the British permission for his Company to make a geological survey of its petroleum concessions in Palestine; the Company would guarantee that the activities of the geologists would not in themselves form a basis for claim; that the Company be permitted to find out whether these concessions should be developed. Thomas left for London, where he arrived on September 2, and with the help of the American Ambassador he pressed his demands on the Foreign Office. He also conferred with Sir John Cadman, Anglo-Persian's chief technical adviser. The upshot was that the Foreign Office asked for a formal request. On September 15 Ambassador Harvey made the formal request to the Foreign Office; on October 20 the American Consul in Jerusalem reported to Secretary of State Charles Evans Hughes that Standard's request for permission to prospect had been referred to Jerusalem for recommendation and Jerusalem had reported favorably. On October 26 Lord Curzon informed the American Ambassador that informal permission would be accorded to Standard's representative by the Palestine Government to conduct researches within the limit of the areas over which the Company was known to claim concession rights.[8]

In the meantime Sir John Cadman, British oil troubleshooter, had visited the United States and worked out an understanding between the American and British companies on the development of Middle East oil resources; in this arrangement the Palestine issue completely disappeared. Standard gave up its claims, no doubt as the price for becoming one of the seven companies which comprised the American group negotiating with the Turkish Petroleum Company for a share in the Iraqi concession.[9]

Exploration

Until the final settlement in 1929 between the American group and the Turkish Petroleum Company, no attempt was made either by American or British companies to explore the oil possibilities of Palestine; after the signing of the Red Line Agreement it would have been up to the Iraq Petroleum Company to seek concession rights in that area. In the meantime, the Palestine Mining Syndicate Ltd. had been organized and registered in Palestine in 1924 and had applied for exploration licenses, especially in the vicinity of the Dead Sea. In 1933 permits were issued to the IPC and to the Palestine Mining Syndicate to search for oil in parts of Palestine and Transjordan.[10] IPC, however, was not satisfied; it wanted exclusive concessions. In 1938 a new mining law was enacted by the Government of Palestine.[11] The following year the Palestine Mining Syndicate was granted two oil prospecting licenses covering about 1,000 square kilometers on the southwestern shore of the Dead Sea. These were subsequently transferred to the Jordan Exploration Company Ltd., registered in Palestine. After World War II a number of exploration drillings were undertaken, but with the war of the Arabs against Israel the Company's activities came to an end.[12]

At the end of February 1939 the Palestine Government granted IPC eleven prospecting licenses, each for about 500 kilometers, covering the total shore area from the Lebanese to the Egyptian frontiers; these were transferred in the middle of March 1940 to the IPC subsidiary, Petroleum Development (Palestine) Ltd. In July 1939 eighteen more prospecting licenses were granted to the subsidiary, involving a total of 13,645 kilometers. No drilling was undertaken during World War II; in 1946 IPC began drilling near the village of Huliqat, about twenty kilometers north of Gaza. On February 7, 1948, after reaching a depth of 3,464 feet, operations were suspended, according to the Company because of the politically tense situation. With the passage of the new Israeli Mining Law, all the old concessions were cancelled.[13]

On August 31, 1952 the Knesset passed the Petroleum Law, based on the recommendations of two American oil consultants, Max and Douglas Ball. It provided for three stages: preliminary permits, exploration licenses, and leases. It limited the leases to thirty years, and the area to not more than 250,000 dunams. The country was divided into petroleum districts and the licenses and leases in each district as well as in the entire country were limited. Royalty was established at 12 1/2 per cent, to be paid to the state, which was to obtain the surface rights from the owners of the land that might be involved in the companies' operations; profits were to be subject to income tax. The following February the Government published the Petroleum Regulations, providing in detail for all the activities of companies engaged in exploring for or developing the oil resources of the country. A number of American, Swiss, Canadian and other companies, as well as Israeli companies, either singly or in combination, applied for and obtained licenses and began drilling for oil in various districts.[14]

Pipeline and Terminus

On January 5, 1931 J. Skliros, managing director of the Iraq Petroleum Company, obtained from the High Commissioner for Palestine an agreement for constructing a pipeline from Iraq across Palestine, with a terminus at Haifa.[15] The best description of the character of this agreement and similar agreements signed with Transjordan, Syria and Lebanon, is given by one who represented IPC on many occasions and participated in many of the negotiations in the area. "While favorable to the Company, the terms which they embodied were in no way unfavorable to the States. The agreements were for 70 years. They gave almost complete customs and taxation exemption and permitted the Company to use, on payment, all local services and resources, to construct all necessary works, and to use all public communications. The Government was to assist the Company in obtaining the necessary lands and wayleaves. No charge was made for the right of transit as such."[16] Longrigg listed the conditions favorable to the Company, but, except for the general benefit, perhaps, of some employment in the construction of the line and its maintenance, he failed to list any returns for the privileges and exemptions which the Company would enjoy.

Construction of the line began in the spring of 1932 and it was completed by the end of 1934.[17] Haifa was to be the terminal. A farm of ten 93,000-barrel tanks was built on the bay of Acre. A submarine line was laid on the sea bed to a loading berth a mile from shore. In 1934 the Company signed an agreement with the Palestine Government covering loading regulations and other services, for which the Company paid 2 pennies per ton. Five more tanks were later added, and in 1935 an oil dock, especially built for the Company, was completed; the charge for loading at the dock was 5 pennies per ton.[18]

Refinery

On October 18, 1933 the Anglo-Iranian Oil Company obtained transit rights across Transjordan and Palestine.[19] Five years later it reached an agreement with the Iraq Petroleum Company and the High Commissioner for Palestine to handle and load petroleum products in Palestine. This made it possible for AIOC to proceed with the building of a refinery, for Palestine was, for refining purposes, within the area allotted by the various companies operating in the region, by the Consolidated Refineries Ltd., a company owned half by AIOC and half by Shell Oil. The Haifa refinery, completed at the end of 1939, had an annual capacity of one million tons, two years later another unit was added which doubled the capacity; in 1944, because of war needs, the refinery was further extended to bring it to a four million ton capacity. Since the supply of crude was limited by the pipeline capacity, oil from the second terminal — Tripoli in Lebanon — was brought to Haifa by tanker.[20]

In May 1948 operations at the refinery came to a standstill as Iraq, warring with

Israel, stopped the flow of oil through the southern branch of the pipeline. By special arrangement with Consolidated Refineries Ltd., the Israel Government re-activated the refinery in 1950 with oil imported from overseas, mostly from the Western Hemisphere, since the Arab countries of the Middle East banned the supply of crude oil to Israel. In 1952, 843,000 tons were processed in the refinery, which employed a staff of 700 persons, about one-third the total required for full-scale operations. [21]

Because Palestine was a British mandate, it benefited very little from the privileges granted IPC for constructing the pipeline, terminal and refinery. The Company was exempted from all taxation and levies, and for landing services it paid only 2 pennies or 5 pennies per ton. On the other hand, because of the high level of the country's development, the internal consumption of petroleum and petroleum products steadily increased. As long as oil flowed through the IPC pipeline, that is, up to 1948, the country's needs were met from Iraq oil, the refinery worked to full capacity, and more than three-fourths of its output was exported. When the Iraq Government stopped the flow of oil at the southern branch of the line, the refinery had to be completely shut down. This created difficulties for IPC and for the entire European Recovery Program, as well as financial problems for the new State, since the demand for petroleum was increasing very rapidly. [22]

1955–1971

On September 22, 1955 oil was struck at Heletz (Huliqat) at a depth of 4,906 feet. The quality was reported to be a 30.5 API density (American Petroleum Institute index rating). This is approximately the same quality as Iraq and Kuwait oil. The strike was made at the very site where in 1947 an IPC subsidiary had drilled to a depth of 3,464 feet. About three weeks later a second strike was made at Heletz: a new oil layer was discovered 45 feet deeper in the same hole. Reserves at the Heletz field were estimated as between 50- and 60,000,000 barrels. At the end of 1956 production was 700 barrels a day from five wells. In June 1957 oil was struck at Bror Hail, near Heletz, and by the end of 1958 the Heletz-Bror fields were producing 2,700 barrels a day.

In order to encourage exploration, the Petroleum Law was amended in August 1956. Oil operators were freed from import, purchase and excise taxes on machinery, equipment, installations, and other necessities. Depletion allowances, equal to 27 per cent of the gross income, were granted. [23]

Because of the boycott of the Arab countries, Israel signed a contract with the Soviet Union for the purchase of 350,000 to 400,000 tons of crude oil annually; Russia was to buy bananas, oranges and lemons. Israel's rate of consumption went steadily upward — from some 420,000 tons in 1948, to 1,150,000 tons in 1954, and to 1,327,000 in 1956. [24] The Suez crisis and the subsequent unilateral cancellation of the contract by the Soviet Union raised the discussion of a pipeline from Elath to Haifa into a practical realizable issue. Some Western governments viewed such a pipeline as a possible potential alternative to the Suez Canal. With the opening of the Gulf of Aqaba to Israel shipping, the Israel Government began laying the pipeline, and on April 15 oil began to flow from Elath to Beersheba, at the rate or 70 tons an hour; by August 3, 1958 it was flowing directly into the Haifa refinery. From Elath to Beersheba the pipeline was eight inches in diameter; from Beersheba to Haifa, sixteen inches. The oil from Heletz-Bror was carried in the sixteen-inch pipeline.[25] In 1960 the entire line was expanded to sixteen-inch and the flow from Elath to the Haifa refinery increased to 1,700,000 tons in addition to 120,000 tons from Heletz.

Late in 1958 the Pazit Company of Israel bought Consolidated Refineries Ltd. from the British Petroleum and Shell Companies.

During the early 'sixties it looked as if Israel might produce a substantial part of its own needs. However, by the late 'sixties no important oil reserves were found.

Table XLII. *Israel Production, 1959–1971*

Year	Barrels	Tons
1959	927,000	130,000
1960	936,000	132,000
1961	978,000	140,000
1962	967,000	138,000
1963	1,082,000	149,530
1964	1,439,425	200,000
1965	1,469,079	204,500
1966	1,352,954	187,380
1967	974,000	155,000
1968	823,191	114,305
1969	725,000	101,381
1970	555,000	76,131
1971	540,000[a]	72,000[b]

a. *World Oil*, CL, 95, February 15, 1960; – CLII, 111, February 15, 1961; – CLIV, 120, February 15, 1962; – CLVIII, 137, February 15, 1964; – CLXI, 162, August 15, 1965; – CLXII, 120, February 15, 1966; – CLXV, 134, August 15, 1967; *Hadashoth Haneft*, 1, February 3, 1969; – 6, March 15, 1970; – 10, January 25, 1971.

b. *Petroleum Press Service*, XXVII, 5, January, 1960; – XXIX, 4, January, 1962; – XXXV, 114, March, 1968; – XXIX, 10, January, 1972; UN, *Economic Development 1961–1963*, 44–46; *Hadashoth Haneft*, 1, February 3, 1969; – 6, March 15, 1970; – 10, January 25, 1971. Production to the end of 1971 was given as 13,942,000 barrels, or 1,954,000 tons; and natural gas total production to the end of 1971 was the caloric equivalent of 779,460 tons of liquid gas. *Hadashoth Haneft*, 1, March 12, 1972.

Both the Government and private investors made strenuous efforts and invested substantial sums of money to discover additional reserves of oil and natural gas onshore and offshore. But so far their efforts have not been crowned with success. Table XLII shows that 1965 was the peak year of oil production, and even in that year the country produced about 7% of local needs. [26] From that year on production dropped rapidly, and in 1971 local production supplied less than 2% of home consumption. In 1962 the estimated reserves were 15 million barrels, two years later they went up to 27 million barrels, but they soon began to drop; in 1967 they went down to 18 million barrels, and in 1970 they came down to about 14 million barrels, less than half of that year's consumption. [27]

Table XLIII. *Israel Local Consumption, 1957–1971*

Year	Tons
1957	1,392,000
1958	1,524,000
1959	1,683,000
1960	1,761,000
1961	1,978,000
1962	2,165,000
1963	2,400,000
1964	2,578,000
1965	2,700,000
1966	3,000,000
1967	3,251,700
1968	3,975,300
1969	4,310,000
1970	5,061,900
1971	5,540,000[a]

a. *World Petroleum Report*, X, 70, March 15, 1964; *Petroleum Press Service*, XXXII, 110, March 1965; – XXXVI, 86–87, March, 1969; *Hadashot Haneft*, 10, March 19, 1969; – 7, February 1, 1970; – 8, March 12, 1972. The later totals include the occupied areas and natural gas.

However, because of geography, international political circumstances, advanced technological knowledge, availability of highly skilled and trained human resources, and that intangible quality, determination, Israel became an important factor in some aspects of the oil complex of the Middle East.

Refining

As was mentioned above, the BP-Shell Haifa Refinery was taken over in 1958 by Pazit. Later the Haifa Refinery Company became a government concern. After the 1956 Suez crisis, Elath on the Gulf of Aqaba became a crude-oil unloading center. With the steady growth of local consumption and the increase of supplies of crude, the Haifa refinery was constantly expanded, modified, improved and extended. The refinery output in 1963 was 2,360,000 tons, in the following year it went up to 3,000,00 tons, of which 1/3 was exported. By 1968 the refinery processed 5 million tons of which 2,300,000 were exported. Two years later the refinery's capacity went up to six million tons. The throughput that year was 5,880,000 tons; since local consumption of refined products was 4.5 million tons, very little was left for exports. [28]

The Haifa refinery could not be expanded any more, and it was decided to build a second refinery with more up-to-date facilities at Ashdod with an initial output of 2.7 million tons of oil products. It was to be completed in 1973, at a cost of $37.5 million. Even before completion it was expanded to 3.5 million-ton capacity. To partially finance the construction of the new refinery, the Government sold 26% of its holdings in the Haifa Refinery Company to the Israel Corporation for IL 42.1 million. In 1971 the Ashdod refinery throughput was extended to 4.25 million tons. [29]

Pipeline and Development

The consequence of the 1956 Suez crisis was the opening of Elath, which became the gateway to Israel's oil development; the consequences of the 1967 Arab-Israel war were the occupation of the Sinai Peninsula and the building of the Elath-Ashkelon pipeline.

Belayim field in Sinai, which began producing in 1955, was considered by the Egyptians as one of their greatest oil fields. In 1958, they announced that they would be able to export 500,000 tons annually from the Sinai fields. Since then an Italo-Egyptian Company made new discoveries in the peninsula. [30] The occupation of Sinai gave Israel an additional temporary source of oil. In 1967, two Egyptian fields — Ras Sudr operated by El Nasr Oilfields and Mobil Oil and Belayim operated by Egyptian Government Petroleum Corporation and ENI — produced daily 50,000 barrels. The Israel Government Company Netivei Neft took over the fields and operated, expanded and developed them; by 1971, 17 wells offshore and 100 wells onshore produced a total of 6 million tons. This was a substantial contribution to Israel's oil resources; in fact it was equivalent to that year's total home consumption. [31]

Pipelines

But the more important consequence of the war which resulted from the closure of the Suez Canal was the laying of a 162-mile, 42-inch pipeline from Elath on the Gulf of Aqaba to Ashkelon on the Mediterranean. The grand design of the pipeline was not only as a temporary alternative to the long haul around the Cape of Good Hope as long as the Suez Canal was closed, but as a permanent solution to the problem of the modern mammoth tankers lifting the oil in the Persian Gulf. The new tankers could not pass through the Canal even when it would be opened; nor were most of the European consumer countries' ports capable of receiving the giant tankers. The pipeline was offered as a solution. Elath port was to be expanded to be able to receive the giant tankers for rapid unloading; and from Ashkelon smaller tankers would ship the oil to the consuming countries. The pipeline company was also to function as a service company and would export oil products.

On February 15, 1970, Israel announced that the pipeline had been completed at a cost of $136 million. Although the line was originally built by a private company, the Israel Government bought out the shares, and the pipeline became a Government concern. The initial annual capacity of the line was to have been 12 million tons, with an ultimate throughput of 60 million tons, through the construction of additional pumping stations. In July 1970, it was reported that the initial capacity had been doubled by completing an additional pumping station in the Negev, and that work had begun on the second pumping station to bring the pipeline to its ultimate capacity.

From Ashkelon a 10-mile pipeline from the terminal to the Ashdod refinery was to be completed in 1973. It was to have an annual capacity of 4.2 million tons, which could be raised to 7 million tons by pumping stations. Elath port was expanded at a cost of $12 million to be able to pump through 60 million tons and to accommodate the unloading of giant tankers of up to 300,000-ton capacity; with some modification it could unload even 500,000-ton tankers in 24 hours. The tank-farm at Elath consisted of three 100,000-ton storage units, and the Ashkelon terminal tank-farm consisted of seven 100,000-ton units. Ashkelon terminal was provided with single-berth moored piers that could accommodate the loading of 150,000-ton tankers in 24 hours and double-berth piers that could accommodate two 80,000-ton tankers.[32]

The major source of the crude oil arriving at Elath was, according to the *Petroleum Press Service,* Iran. In 1969, the Economic Council of the Arab League decided to boycott all the international oil companies which intended to use the Israeli pipeline. It also called on the Arab Governments to plead with Iran and with other countries not to make use of the line. But the efforts of the Arab League were obviously unsuccessful. For the throughput of the line in 1970 was 11 million tons; in 1971 it rose to 19 million tons; of these 13 million were shipped onward from Ashkelon and the balance went to the Haifa refinery. In 1972 the oil liftings from Ashkelon terminal during the first six months of the year rose by some 80% over 1971. The 19 million tons of crude in 1971 were unloaded by 302 tankers in

Elath, and 389 tankers took on some 13 million tons at Ashkelon. In March 1972, it was reported that the pipeline was operating at an annual capacity of 30 million tons, and that the year would end up with a total of 26 million tons. This would surpass Tapline, which had a 25 million ton capacity.[33]

Tankers

The building of the Elath-Ashkelon pipeline, the terminal at Ashkelon and the refinery at Ashdod presupposed the availability of tankers which would make the disposal of the oil possible. Israel, therefore, steadily built up its tanker fleet. At the end of 1966, the entire tanker fleet consisted of 11 vessels making a total of 330,000 tons capacity; at the end of 1970 it went up to 25 vessels with 1,750,000 tons capacity; and only one year later it rose to 28 vessels with 1,900,000 tons capacity. Plans for 1974 called for a fleet capable of handling 20 million tons of oil coming through the pipeline.[34]

To make available products and crude to the home consuming areas pipelines were built from Haifa to a distribution center near Tel-Aviv; from the new refinery at Ashdod a pipeline was built to Jerusalem, and other branches from the main pipelines were built to industrial centers that use crude oil. Similarly, pipelines from the natural gas fields were laid to the potash-works at Sodom on the Dead Sea and to other plants.

The availability of oil by-products at the refinery, as well as various chemicals from the Dead Sea, the advance technical-engineering knowledge, and highly skilled manpower as well as administrative and managerial resources made possible the development of a wide range of chemical, petrochemical, electrochemical and fertilizer industries, concentrated in the Haifa bay area and in Arad, which over the years supplied local needs and exported products. Together with oil they had become an important source for the country's foreign exchange supply.[35]

Although Israel itself is not a substantial producer even to meet a small percentage of its own needs, it established itself as a factor in the Middle East oil industry in spite of Arab opposition, nay hostility. Obviously strong economic forces make the main pipeline useful and important to the total picture of Middle East oil development. If production should increase in accordance with anticipated future demands, as predicted by both producing and consuming countries, existing pipelines, even with the opening of the Suez Canal, could not meet the demand for the oil, and more pipelines would have to be built, and indeed were being planned. The Elath-Ashkelon pipeline must be seen within that framework, assisted by the growing Israel tanker fleet.

TRANSJORDAN

Transjordan was part of the Palestine mandate and in 1931 the government of Emir Abdullah signed the same contracts with IPC as the Palestine Government. The pipeline ran for 205 miles in Transjordan territory. The first arrangement with

Tapline was discussed above.[36] In 1952 the agreement was amended, granting Jordan $600,000 annually in fees and the crude needed for internal consumption, about 100,000 tons, at the lowest of the competing crude oil prices posted in eastern Mediterranean ports.

IPC organized the Petroleum Development (Transjordan) Ltd. in 1938 for operations in Transjordan. In May 1947 the Government granted the Company a 75-year exclusive right to explore and develop oil and gas resources in the entire territory. The agreement provided for an initial payment of £50,000 gold; an annual payment beginning with £15,000 (gold) and increasing until after twelve years it reached £80,000; royalty was to be 4 shillings (gold) per ton. The concession could not be cancelled for ten years unless the Company was released by the Transjordan Government.[37] In April 1948, just prior to the expiration of the Palestine Mandate, the High Commissioner for Palestine granted the Transjordan Petroleum Company free passage through Palestine to the Mediterranean for any oil discovered in Transjordan.[38] In the middle of 1954 the Company decided to give up its concession and it began to negotiate with the Jordan Government for release; in November the Government signed an agreement with the Company to terminate the concession. The agreement provided for a final payment of £400,000, bringing the total amount paid by the Company since it was granted its concession in 1947 to £717,000.[39]

JORDAN 1955–1971

Concessions, transit fees and refinery were the three issues in Jordan during the period. Some success was achieved in the last two, and very little success was registered for the first. In November 1955 the Government granted an exploration concession covering the entire country to Edwin A. Pauley, an independent American petroleum operator. After eight months of exploration, one-third of the territory was to be selected for drilling. Pauley was to pay an annual fee of $50,000 for the first three years, $100,000 for the next three years, and $200,000 for each year thereafter. When oil was discovered, the Government was to have an equal share in the profits.[40]

Two years later Georges Izmiri of Guatemala was granted an exploration concession that covered about one-third of the country. The following May a concession for the remaining third was granted to the Arab Trading and Development Company. In October 1958, with the consent of the Jordan Government, Pauley transferred his concession to the Phillips Company.[41]

A Jordanian refining company, with a capital of JD 4,000,000, was organized in November 1957. It obtained a 50-year concession to operate a refinery with a 330,000-ton annual capacity. The Government was to supply it with the crude it received from Tapline as transit dues.

Early in 1959 the Government cancelled the concession of Georges Izmiri because he had failed to proceed with exploration. At the end of the year Izmiri's acreage was granted to Phillips Petroleum Company at an annual rental of $60,000.[42] After spending over $5 million in drilling six dry holes, Phillips surrendered its acreage which it obtained from the Pauley concession. In 1962, Phillips relinquished its own concession area. Two years later the 31,333 square kilometers were granted to John M. Mecom in association with a German company for a 40-year period.[43]

Iraq Petroleum Company closed its operations in Jordan in 1961, and under the agreement reached with the Government all the Company's facilities passed to the Jordanian authorities. Two years later IPC cancelled its pipeline concession which was signed with Abdullah on January 3, 1931, as mentioned above.[44]

This left Tapline the only oil transit concessionaire in Jordan. Transit fees for the oil flowing through the pipelines had been a constant issue between the oil companies and the territories through which the pipelines ran: Jordan, Syria and Lebanon.

On March 13, 1962, Jordan signed an agreement with Tapline raising the payment for the oil transited through the 110-mile stretch of pipeline in Jordan. Annual revenues, hitherto $850,000, were to increase, effective as of November 1961, to $4 million on a daily throughput of 350,000 barrels. Jordan was also to receive a lump sum of $10 million in settlement of all past claims.[45] Four years later, as a result of an adjustment in the boundary between Saudi Arabia and Jordan, Tapline added $125,000 to the annual payment for the new mileage in Jordanian territory.

Because of the interruption in the flow of oil in 1970 and 1971 – about which later – Jordan lost about $1 million each year in transit revenue.

Finally in March 1971, Tapline agreed to increase Jordan's annual transit revenue from about $4.5 million to about $7.25 million. The company also agreed to pay, in installments, a sum of $9 million for all past claims.[46]

The Jordan Refinery Company plant was built in 1961 at Zarqa with an annual capacity of 325,000 barrels at a cost of $9.5 million. Part of the cost was contributed by the Jordan Government, the rest was raised by selling shares throughout the Arab world. Crude oil was supplied by Tapline through a 43-kilometer, 8-inch pipeline. In 1970, the daily throughput was increased from 7,000 to 15,000 barrels. In the same year Saudi Arabia offered to remit half of its annual subsidy – granted since after the Arab-Israel 1967 war – in crude oil.[47]

LEBANON

Lebanon dealt, over the years, with three oil issues: search for oil reserves, pipelines and loading fees and refineries. The efforts in the first area have, so far, not yielded any returns, and Lebanon, like Jordan, is still a non-producer. It reaped a constantly growing revenue from the two pipelines, IPC and Tapline, crossing her territory, and she had two expanding local refineries, and was about to gain a third one.

Exploration

In March 1938 IPC was granted a license covering the area of Jebel Terbol; the outbreak of the war prevented any operations. In May 1947 the Lebanon Petroleum Company (a subsidiary of Petroleum Concessions Ltd.) began drilling operations, but it later abandoned its license. In January 1954 it was reported from Beirut that oil experts who had been prospecting for oil in Lebanon for a number of years had come to the conclusion that there was no possibility of finding rich oil deposits in the country.[48]

An exploration and producing concession was granted in July 1955 to Compagnie Libanaise des Pétroles. It was for a period of 75 years and covered about one-third of the country. The basis of government profit-sharing was on a sliding scale: from 15 per cent on annual production of less than 1,000,000 tons of crude oil, to 50 per cent on production of 8,000,000 tons and over.[49]

Pipelines

Lebanon signed an agreement with IPC in 1931 under the same conditions as the other countries. The first pipeline ran through her territory for only seventeen miles. The northern branch of the line terminated at Tripoli where a farm of fifteen 93,000-barrel tanks was built. The loading fee at Tripoli was 2 pennies per ton.

In 1947 Lebanon signed the agreement with Tapline mentioned above, and that same year IPC agreed to pay Lebanon transit fees amounting to £45,000 annually. A new agreement with IPC was signed in 1952, raising the transit fees to about £450,000 annually. A new agreement was also signed with Tapline in 1952; it provided for a basic increase in transit fees, for $3 per ton up to 200,000 tons for petroleum products consumed in the country; an increase in the loading fee at Sidon; and annual contributions of $21,000 for the maintenance of the public highways used by the Comapny's vehicles and $17,000 to the municipalities of Beirut and Sidon in consideration of services.[50]

The Chamber of Deputies refused to ratify the agreement and unanimously adopted a resolution demanding that the Government undertake immediate negotiations with all oil companies for the revision of oil agreements.[51] The companies refused to respond to the demands of the Government. At the end of December 1953 the Justice Ministry informed Premier Abdallah al-Yafi that it was impossible to take any legal action against Tapline and IPC to oblige them to discuss the question of an increase in royalties.

Attempts were made to present a united front of Syria, Lebanon and Jordan in their demands for higher royalties, but apparently with little success. At the end of May 1954 it was reported from Beirut that the Syrian and Lebanese Governments had agreed to send a joint note to the companies asking for an increase in the royalties.[52]

Lebanon's main preoccupation, as far as oil was concerned, was the question of royalties. Both Tapline and IPC have pipelines and terminals in Lebanon. Lebanon's share in a profit-sharing arrangement became an issue not only between the Government and the companies, but also between Lebanon and her neighbors. For lengthwise, Lebanon's share of the pipeline is the smallest, and Beirut insisted that the apportionment be on a country basis, which would entitle Lebanon to one-fourth of the 50 per cent profits. The others involved insisted on a mileage basis.

In July 1955 Lebanon reached an agreement with IPC which more or less followed the unratified 1952 agreement, except that the basis of payment was increased from LL 1,000,000 to about LL 4,000,000, and a lump sum of LL 6,500,000 was granted for back payments from May 15, 1952 to January 31, 1955. The Company was also to increase its refining facilities by 25 per cent.

Earlier that year, in April, IPC notified Lebanon that it intended to build a new 24-inch pipeline, with an annual capacity of 9,000,000 tons, to terminate in Tripoli. At the end of the year, IPC and Syria agreed to a new system of royalty payments, and the agreement with Lebanon, in accordance with its provisions, was automatically reopened. The Company offered Lebanon the same terms as Syria. Lebanon refused. On June 29, 1956, Parliament enacted a law imposing an income tax, retroactive to January 1, 1952, on the oil pipeline companies, which had previously been exempted from all taxation by special agreement. IPC challenged the legality of the act and the Lebanese Government announced that it was accepting an offer made by the United States to mediate the dispute. The mediation failing, the Government proclaimed that the new law had gone into effect. IPC called for arbitration; Lebanon refused. The Government demanded taxes and fines amounting to $7,500,000 and threatened to take over the company's properties. The Company stood firm.

Under Tapline's profit-sharing proposal, Saudi Arabia was to share in the arrangements, as the line passes also through her territory. Since the Lebanese law affected Tapline as well as IPC, the former protested that the law imposing taxation was a violation of the convention that Lebanon had signed with it. Late in August it was reported from Damascus that Syria, Lebanon, Jordan and Saudi Arabia had accepted in principle the Tapline offer to share its profits on an equal basis. [53]

Refineries

In 1940 the French High Commissioner for Syria and Lebanon built an emergency refinery at Tripoli under an agreement with the Iraq Petroleum Company that he was to own and operate during the war emergency but that was to be turned over to IPC at the end of hostilities for the crude oil which the Company was to supply. The transfer was made on January 1, 1946. Three years later the refinery's quarter of a million barrel capacity was doubled. [54]

A second refinery in Lebanon, the Mediterranean Refinery, was inaugurated in February 1955. It is located at Sidon, built at a cost of about $20 million, and has a capacity of about 6,000 barrels daily. It is owned by the Mediterranean Refining

Company, which is in turn owned 50% by Caltex and 50% by Mobil Oil.[55] This refinery was to supply Lebanon with the major part of its benzine and kerosene requirements.

The throughput of the two refineries at Tripoli and Sidon went up steadily: 1955 — 893,945 tons; 1956 — 1,007,757; 1957 — 1,051,535.[56]

1959–1971

The differences between Lebanon and IPC over transit and loading fees were finally settled in the middle of 1959. The Company agreed to raise the annual payments from about $1 million to about $3,458,000. In addition the company was to pay a lump sum of $16.5 million in settlement of past claims. The Company also agreed to raise the output of the Tripoli refinery by 25%, and that the 20% of the refined products reserved for local consumption were to be assigned to local distributors nominated by the Government. In addition, the Company undertook to finance the establishment of a technical trade school in Tripoli, and to provide LL (Lebanese pounds) 35,000 annually to be distributed among the municipalities in whose territories the Company operated.[57]

Three years later the Government signed a new agreement with Tapline which increased the annual payments from $1.2 million to $5.3 million, and the Company agreed to pay a lump sum of $12.5 million in settlement of past claims.[58]

In 1967, following an agreement with Syria, IPC agreed to increase the annual transit rates and loading fees, retroactive to January 1, 1966, from about $3 million to about $6 million annually.[59]

The next increment came in the agreement with Tapline of April 1971, which provided for an annual payment of $8.6 million instead of the previous $5.1 million, and a supplementary payment of $900,000 when the pipeline reached maximum throughput. Tapline also undertook to pay $9 million in five installments, over a two-year period, in settlement of past claims.[60]

In the negotiations with IPC, Lebanon asked for a 50% increase in transit rates and loading fees. It also asked for a change of method in calculating the price of crude entering the Tripoli refinery. The outcome of these negotiations for Lebanon depended on the outcome of negotiations with Syria.[61]

When Iraq nationalized IPC in June 1972, and Syria followed through by nationalizing the pipeline running through its territory, Lebanon was not willing to follow suit, for it was both against her economic ideology, and politically mischievous, to tamper with a Western company. Iraq did not permit the flow of oil through the Lebanese branch of the pipeline. It was reported that the Arab Fund for Economic and Social Development had offered Lebanon a long-term loan to purchase or lease IPC properties. Meanwhile, oil for the Tripoli refinery came from Abu Dhabi. In December 1972, Iraq reported that an Iraqi delegation was conducting talks with Lebanese officials on the question of transit fees. After long negotiations between Iraqi and Lebanese officials, Lebanon decided on March 5, 1973, to take over the pipeline section, the loading facilities and the refinery of the

Iraq Petroleum Company, for which Iraq permitted 40% of its oil to flow to the Tripoli terminal. The transit and loading fees from the Syrian-Lebanese border to Tripoli were set at 11 cents per barrel. Iraq was to supply to Lebanon annually up to 1.5 million tons of crude oil, to be refined for local consumption, at a price equal to that charged the Homs refinery.

The agreement was for a period of 15 years, but the transit fees and the crude oil price were to be renegotiated at the end of 1975. [62]

Refineries

The two established refineries were constantly increasing their output; and in 1970 the daily throughput of the Tripoli refinery reached 24,000 barrels and that of Medreco 16,000 barrels. At the end of 1971 Lebanon reached agreement with IPC on new methods of calculating the price of crude entering the refinery. In settlement of past refinery claims, the Company undertook to pay, in installments, the sum of £3.8 million. [63]

Apparently looking for an outlet for its crude oil as well as investment possibilities, Saudi Arabia began in the middle of 1969 to press Lebanon for building a third refinery. At first Lebanon tried to resist; however, after Saudi Arabia intimated that she might not renew her commercial treaty, Lebanon relented. In May 1971, Lebanon signed a preliminary agreement with Petromin for the construction of a jointly owned refinery which would process Saudi Arabian crude. The Saudi Arabian Government would put up the capital, oil and equipment, which it was to get from Rumania under a barter agreement. Lebanon's share of the capital was to be paid out of her part in future profits. [64]

SYRIA

Syria, like Jordan and Lebanon, though she made efforts was until the middle 'sixties a non-producer, and her oil income was primarily derived from pipeline transit rates and loading fees and from refining.

Exploration and Exploitation

Syria having been included in the Red Line Agreement, IPC obtained, early in 1934, a four-year exploration permit in the Shamiyah desert; this was later extended to other areas. In 1936 the subsidiary Petroleum Concessions (Syria and Lebanon) Ltd. was organized. Two years later the Syrian Government signed an agreement with this Company granting it a 75-year concession covering all the Syrian territory (except the Alexandretta district) north of Damascus. It provided for an immediate payment of £50,000, a rental rising from £15,000 to £80,000 in ten years, and a lump sum of £100,000 when oil began to be exported. Royalty was to be 4 shillings per ton. The Company was to build a refinery to supply Syria's internal petroleum needs. If oil was not discovered within ten years, or no commercial export was under way within fifteen years, the concession was to be cancelled. The agreement was not ratified until 1940, but in the meantime the Company carried on exploration and drilling. The outbreak of World War II interfered with operations, which were resumed in 1947. The Company gave up its concession in August, 1951.[65]

In 1954 it was reported in Damascus that the annual internal consumption of petroleum was about 350,000 tons and that the Government was negotiating with Tapline and IPC to build two refineries near Damascus, the cost of which was to be deducted from the Government's revenue from the two companies.

The Société des Pétroles Concordia, a Syrian company affiliated with the Deutsche Erdöl of Hamburg, obtained a concession in October 1956 which covered 5,800 square miles, and the Company agreed to invest $5,000,000 in exploration during the ensuing three years.[66]

An exploration permit covering some 5,625 square miles in the southern part of the country was granted on May 16, 1955 to the James W. Menhall Drilling Company; and in October 1956, it was reported that oil had been discovered at Karatchak, close to the Iraqi border.

In February 1958, Syria was merged with Egypt as the United Arab Republic; this combined with the general turbulence of Syrian politics gave oil development and policies a new direction. In October the Government cancelled the Menhall concession and took it over. A year later, the Minister of Industry for Syria announced in Damascus that an agreement had been reached with the Soviet Union to conduct oil exploration in Syria. On August 12, 1959 it was officially announced that a new oilfield had been discovered at Suwaidiyah, some 12 miles from Karatchak.[67]

In 1960, the Karatchak concession was taken over by the Syrian branch of the General Petroleum Authority; while the Concordia Company reported the discovery of a new well at Suwaidiyah,. However, at the end of 1963, oil reserves were estimated to be only 300 million barrels.

In December 1964, the Syrian Government decreed that only the Government agency, the General Petroleum Authority, was to be permitted to produce oil in Syria. Four months later nine domestic and foreign marketing firms operating in the country were nationalized.[68] Two years later it was reported that the Soviet Union had agreed to drill 20 field wells in Syria to an average depth of 6,500 feet.

However, the Karatchak and Suwaidiyah fields could not produce until pipelines were built to the refinery for home consumption and to an outlet on the Mediterranean for export. The Government, therefore, awarded a contract to the Italian ENI subsidiary SNAM to build the pipeline. In 1968 it was officially announced that oil was flowing through a 560-kilometer, 20/22-inch pipeline from the fields to the Homs refinery, and through a 90-kilometer 18-inch section from Homs to the terminal at Tartus. The building of the line and the terminal was supervised by Soviet advisors.[69]

Since that year production of crude grew and the estimates of reserves were marked upwards. In 1968, Syria produced 1,000,000 tons; in 1969, it went up to

3,500,000 tons; in 1970, it rose to 4,500,000 and in 1971, it reached the 5,500,000-ton mark. The Minister of Oil, Ahmad Yusuf Hassan, predicted late in October 1969 that within 4 or 5 years Syria would be producing at the rate of 15 million tons a year. He claimed that his country's oil income had reached $14.4 million, exclusive of the revenue from IPC and Tapline. The Minister of Economics and Commerce declared early in the year that all the oil expected to be produced in 1969 had already been sold to foreign purchasers. Under a general export-import agreement with the USSR, Syria was to export five million tons of crude to that country over the period 1970–1975 on a rising scale beginning with 500,000 tons in 1971. It was also reported that Syria had signed contracts with West European countries to export 3 million tons of crude. [70]

Estimated reserves by the end of 1967 were given as 1.050 billion barrels; by the end of 1969 they rose to 1.320 billion barrels; and two years later jumped to 7.3 billion barrels. [71]

It would seem that, although by comparison with Middle East producers, Syria's output was of small consequence, she had succeeded, to be sure with Soviet assistance and guidance, in running her oil industry on a nationalized basis; and the Government boasted of this achievement publicly and officially and indeed tauntingly to the other Arab producing countries. However, no information was provided as to the real extent of the export, to whom it was sold and under what terms.

Pipelines

The major battles of the Syrian Government were waged, and waged frequently and strenuously, against the owners of the two pipelines running through its territory.

The IPC pipeline from Iraq to the Mediterranean, built in the early 'thirties, ran about 267 miles in Syria, and as in the case of Palestine and Transjordan, the agreement signed in 1931 between the Government of Syria and the IPC provided for no payment for transit rights. After the war, when Syria emerged as an independent country, her Premier, Jamil Mardam, announced in Damascus in the middle of May 1947 that the IPC had agreed to pay the Syrian Government LS 750,000 annually (about $336,000) for the right of passage and protection of the pipeline. [72] This was increased in 1950 to £108,000. [73] The 1949 agreement between Syria and Tapline is outlined above. [74]

Premier Fawzi Selo stated at the end of 1952 that the annual royalties from Tapline totaled about $490,000 and that the Syrian Government found that sum inadequate as compared with the advantages the Company enjoyed in Syria. The Government had therefore asked the Company for an increase, and after negotiations a new agreement was signed in May 1952 which raised the royalty to about $1 million, and which provided that the parent companies of Tapline were to supply Syria with increased amounts of crude oil to meet part of her internal need, at the most favorable prices prevailing in the eastern Mediterranean area. [75]

Soon the Syrian Government felt that even the increased royalties and crude supplies were not sufficient and it demanded from IPC and Tapline one-half of all savings from the transit of oil through Syria, based on what the cost would be if the oil had to be moved by tanker. It also demanded that the crude oil be supplied free to the Government. [76]

At the end of 1955 the Syrian Parliament ratified the agreement with IPC of November 29, under which Syria was to receive $30 million annually for the oil passage and loading fees. In settlement of past claims the Company agreed to pay $17,000,000. In 1956, after the Suez Canal crisis, Syria sabotaged the pipeline, and for a considerable time the oil was not flowing. Indeed the oil revenue for 1957 was cut in half.

With Tapline, the situation was more difficult. The Company paid a flat annual rate of $1,250,000, and Syria demanded higher royalties. Early in 1956 Tapline recognized Syria's demands in principle. In May it agreed to become a commercial profit-making company and to share the profits on an equal basis with all the governments through whose territories its pipeline ran. However, it left it up to the governments concerned to agree on the share of each in the 50 per cent profit. [77]

After six years of negotiations Tapline signed a new agreement with Syria on February 25, 1962. It raised the annual transit fees from $1,250,000 to $3,000,000. In addition Tapline was to pay a lump sum of $10 million for settlement of all past claims. [78]

As was mentioned above, in the section on Iraq, the negotiations between Syria and IPC about transit rates and loading fees began in the later part of 1966. Although the total payments in 1965 amounted to £8.6 million, the Syrian Government insisted that the basis of payment be a share of the profits, as the value of the oil increased after it passed the pipeline in the Syrian section. [79] On August 28, 1966, Premier Zuayn announced that the Government asked for immediate negotiations with IPC. On September 10, negotiations began; however, only ten days later amid a violent Government and press campaign against the Company, the negotiators returned to London to fetch some documents and for consultations. On November 8, negotiations resumed and ended in a stalemate. Iraq became disturbed about the possibilities of an oil stoppage, and on November 22, an Iraqi delegation paid a one-day visit to Damascus in an effort to settle the dispute. It obviously failed, for three days later the Syrian Government threatened IPC that it would take "measures already decided upon" to safeguard its rights. On December 8, Syria impounded all the moveable and immoveable assets of IPC in the country until the Government's claim to back royalties and higher transit rates had been paid. On December 13, the flow of oil to both Banias and Tripoli was stopped. The Government rejected a Company request for arbitration. Early in January 1967, IPC offered to pay the Government an advance of $10.4 million against any arbitral or negotiated settlement in return for the opening of the pipelines.

Meanwhile Iraq received no revenue and it was pressing the Company to settle with Syria or it would have to pay the revenue anyway. In the middle of February,

Premier Najib Talib of Iraq arrived in Damascus to help solve the conflict.

On March 2, 1967, the Government and IPC reached agreement: Syria was to return the Company property impounded and open the pipelines; the Company was to increase transit fees by 50%. The new rates were to be applicable as of January 1, 1966, which would grant the Government an additional £4.9 million, and the new income was estimated at about £15 million annually. [80]

In July 1971, after a threat by Syria to shut down the pipelines, IPC raised its payments to Syria from 14.96 to 22.5 cents per barrel for transit, and an extra 0.5 cents a barrel until the Suez Canal was reopened; and the loading fees at Banias terminal were raised from 3.7 cents to 4 cents per barrel. IPC also undertook to pay $33.6 million, in installments over a two-year period, in settlement of past claims. It was calculated that with a maximum throughput of the pipelines Syria was to receive close to $80 million from IPC. [81]

On May 3, 1970, Tapline was sabotaged in the Golan Heights and the Syrian Government did not permit the line to be repaired. The pipeline was out of commission until January 29, 1971, when the Syrian Government permitted the Company to make the necessary repairs, [82] for which Syria received an annual increase in transit payments from $4.5 million to $8.5 million, and $9 million as a settlement bonus. [83]

It would seem, however, that the question of pipeline rates and loading fees were not to be solved even after nationalization. As mentioned above, Iraq nationalized IPC on June 1, 1972 and Syria immediately nationalized the pipeline in her territory; difficulties were anticipated between the two Arab countries. For as final plans were worked out for the Nationalization Act the Iraqi Minister of Oil, Sadoun Hamadi, paid a visit to Damascus on May 24, 1972, and met with President Hafez al-Asad to discuss Syria's transit share in the new setup. At first, according to Iraqi sources, President al-Asad expressed the hope that Syria would obtain not less than she received under IPC. But later the demands rose. On July 12, an Iraqi delegation headed by the Foreign Minister visited Damascus again, and one of the issues under discussion was Syria's share from the pipeline and Banias loading terminal. Syria demanded double what she was getting under IPC. All the arguments of the Iraqis in the name of Arab solidarity, in the name of the cause of nationalization, and their financial difficulties as a result of nationalization did not help. The Iraqis expressed willingness to accept any formula or procedure in force between any Arab or foreign states to serve as a basis for action to be taken. But Syria refused. Resort was made to two Arab oil specialists: Abdullah Tariqi, former Saudi Arabian Minister of Oil, and Nikolas Sarkis were asked to study the case and submit recommendations. Moreover, Iraq was willing to pay Syria an advance on the transit share until the issue was resolved. In February 1973, it was reported that a compromise was worked out under which Syria was to receive 50% more than what she obtained from IPC for a period of fifteen years, with a possible revision of the level of dues at the end of 1975. [84]

Refining

In 1956, the Syrian Government asked for tenders for the construction and equipment of an oil refinery at Homs. In March 1957, it awarded the contract to the Czechoslovak Technoexport Company. In June 1959, the 10,000,000-ton annual capacity refinery was completed, and Syria produced oil products from crude supplied by IPC and Tapline.[85]

Ten years later, for the first time, Syria began to process in her Homs refinery locally produced oil.[86]

CONCLUSION

All the four Eastern Mediterranean countries — Israel, Jordan, Lebanon and Syria — have become determining factors in the operation in the Middle East oil industry primarily as transporters, although Israel to some extent and Syria to a much larger extent were also producers, and both, although from different sources, were also exporters. Not only had the revenue or the earnings of the pipelines in each become an important source of revenue and foreign exchange, but they collectively and individually demonstrated their vital role in making the oil available to the consuming world public. Syria in particular had proven many times her ability to seriously interfere with the flow of oil to world markets; and each time she came out the winner.

These transporting countries do not have the great crude resources of the Middle East producers but, since the oil must reach the Mediterranean on its way to the consuming countries, they had become an integral part of the Middle East oil pattern complex. And there is no alternative. Even should the Suez Canal be opened, the Persian Gulf could not possibly handle all the exports from the area. Nor would the proposed new pipelines via Turkey diminish the role of the four

Table XLIV. *Transit Revenue, 1955–1971 (Selected Years)*
Million Dollars

Year	Jordan	Lebanon	Syria	Total
1955	1.0	1.4	2.9	5.3
1957	1.0	1.4	9.1	11.5
1959	1.0	13.5	23.8	37.3
1963	4.2	8.5	24.9	37.6
1968	5.1	14.5	57.6	77.2
1970	1.7	11.4	53.9	67.0
1971	11.9	16.8	75.6	104.3[a]

a. UN, *Economic Developments 1961–1963,* 116; *Petroleum Press Service,* XXXIX, 322, September, 1972.

countries under discussion. For should the new proposed pipelines be built, they would simply serve to handle the additional production anticipated at the time they would be completed. And in the final analysis, Turkey would become the new member of this oil transportation group and would play the same role as these four have played.

All four were fully equipped with their own oil refining facilities not only to meet home demands but also to export. But all in turn, except to a limited extent Syria, depended on the producing countries for the crude oil for their refineries, and this formed part of the mutual dependence of the producers and the transporters; and it would make no difference if the producers were companies or countries.

NOTES

1. Great Britain, Foreign Office, *Documents on British Foreign Policy, 1919–1939,* First Series, IV (London, 1952), E. L. Woodward and Rohan Butler (eds.), 337, 504 (henceforth cited *Documents*).
2. Standard had planned to drill in the vicinity of Beersheba and a road had been built and drilling material and equipment imported from the United States, but with the outbreak of the war the material had been unloaded at Alexandria.
3. *Documents,* First Series, IV, 352, 366. On Sept. 12 Col. Meinertzhagen asked Lord Curzon: "Is geographical survey included in refusal resume operation?" To which Curzon replied in the affirmative. *Ibid.,* 382–406. It should be noted that on Sept. 8, 1919 the French chargé d'affaires in London informed Lord Curzon that a group headed in England by Lord Inchcape and in France by the Banque Demarchy had been formed for the purpose of searching for and obtaining oil concessions in Palestine and in the region of Aqaba. While the French Government fully understood that until the status of the territory was established no concessions could be granted, it nevertheless was favorably disposed to permit this newly formed group to proceed with preliminary studies and ultimately to grant concessions at the appropriate time, and would be happy to receive the British reaction. About two weeks later Lord Curzon told the French Ambassador that the British Government had consistently refused to permit any geological or other survey in the territories of Syria and Palestine, and still adhered to the decisions without exception. *Ibid.,* 372–391.
4. On June 17, 1919 Gen Gilbert Clayton of the Arab Bureau in Cairo reported to Lord Hardinge: "Captain Yale is now working with the Commission [King-Crane]. His connection with Standard Oil Company is known to us and will not be lost sight of." *Ibid.,* 278.
5. *Foreign Relations 1919,* II (Washington, 1934), 250–254. The American Consul in Jerusalem reported on Sept. 30, 1919 that the Standard Oil Co. had spent $89,018 in securing and working the concessions and had about $125,218 worth of materials stored in Alexandria.
6. *Foreign Relations 1920,* II (Washington, 1936), 650–651.
7. *Ibid.,* 652.
8. *Foreign Relations 1921,* II, 94–106; *PDC,* 150, col. 1012, Feb. 15, 1922.

9. See above, Chapters II, X and XI.

10. *PDC,* 299, col. 1000, Mar. 19, 1935.

11. "Oil Mining Ordinance," *Official Gazette of the Government of Palestine,* Supplement No. 1 to No. 793 of July 7, 1938, 49—80.

12. Curt Nawratzki, "Oil Prospecting in the Middle East and in Israel," *Economic News* (Tel Aviv), V, 40—41, 1953.

13. *Ibid.,* 42; Stephen Hemsley Longrigg, *Oil in the Middle East* (London, 1954), 90, 247—248 *et passim.* The Petroleum Commissioner of the Israeli Ministry of Development stated in 1953 that in the course of the overall negotiations on the various IPC rights in Israel, the Company had finally surrendered the exploration permits it had obtained from the previous regime and freed the Israel Government from all moral commitments. I. S. Kosloff, "Oil Development in Israel," *Economic News,* V, 3, 1953.

14. Both the Petroleum Law of 1952 and the Petroleum Regulations of 1953 are reproduced in English in *Economic News,* V, 75—108, 1953. An analysis of both and their interpretation are given by M. D. Schlosberg and L. Kuenstler, *ibid.,* 49—52. The full debate in the Knesset on the Petroleum question will be found in *Divrei Haknesset,* XII, 1st Session, No. 43, 3121—3130, Aug. 25, 1952; see also "Israel Looks to Oil Possibilities," *World Petroleum,* XXIII, 97, 112, Sept., 1952; "Israel's New Petroleum Law," *Petroleum Press Service,* XIX 319—320, Sept., 1952. An encouraging report on the possibilities is presented in "Israel's Oil Prospects Look Better Than Ever," *The Oil Forum,* VIII, 92—94, Mar., 1955.

15. "Convention Regulating the Transit of Mineral Oils of Iraq Petroleum Company Ltd. through the Territory of Palestine," *Official Gazette of the Government of Palestine,* No. 276, 75—85, Feb. 1, 1931.

16. Longrigg, *Oil in the Middle East,* 87.

17. For the various aspects of the construction of the pipeline, see *World Petroleum,* VI, 55—94, Feb., 1935; Iraq Petroleum Company, *An Account of the Construction in the Years 1932 to 1934 of the Pipe-line of the Iraq Petroleum Company Limited from its Oilfield in the vicinity of Kirkuk, Iraq to the Mediterranean Ports of Haifa (Palestine) and Tripoli (Lebanon)* (London, 1934).

18. For a description of the pipeline, see above, Chapter XI.

19. See the provisions of the Long-Berenger draft agreement discussed in Chapter IX above. "Convention Regulating the Conveyance of Mineral Oils by the Anglo-Persian Oil Company Ltd. through the Territory of Palestine," in *Report by His Majesty's Government in the United Kingdom of Great Britain and Northern Ireland to the Council of the League of Nations on the Administration of Palestine and Transjordan for the Year 1933,* Colonial No. 94 (London, 1934), 301—311.

20. For the role of the Haifa refinery during World War II, see Ernst Aschner, "Haifa Refinery Played Important Part in Supporting Allied Operations in Mediterranean Theatre," *World Petroleum,* XVI, 44—46, December, 1945. "The Haifa refinery took its place by 1944 as the largest single industrial asset in Palestine." Longrigg, *Oil in the Middle East,* 141.

21. "The Haifa Refinery," *Economic News,* V, 68, 1953; "The Haifa Oil Refineries," *Palestine Affairs,* III, 93—94, Aug., 1948; Longrigg, *Oil in the Middle*

East, 89. For the financial difficulties created in Israel by the need for importing oil from overseas, see "Israel's Struggle," *Petroleum Press Service,* XXI, 41–45, Feb., 1954.

22. In 1948 Israel consumed 420,000 tons of petroleum, in 1953, 980,000 tons, and in 1954, 1,150,000 tons. *Petroleum Press Service,* XXI, 43, Feb., 1954.

23. UN, Dept. of Economics, *Economic Development of Under-Developed Countries. The International Flow of Private Capital 1957* (UN (mimeographed) 1958), 76; *World Oil,* CXLIV, 166–169, Feb. 1, 1957; "Israel," *World Petroleum Report,* III, 124–125, Jan. 15, 1957; *Davar,* Mar. 13, 1957.

24. *Davar,* Mar. 20, 1957.

25. *The New York Times,* Aug. 4, 1958.

26. *World Oil,* CLXIII, 166, Aug. 15, 1966.

27. *Davar,* Oct. 31, 1962; *World Oil,* CLXVII, 199, Aug. 15, 1968; – CLXXIII, 116, Aug. 15, 1971; *World Petroleum Report,* X, 70, Mar. 15, 1964; writing in *Beolam Hadelek,* 31, July, 1969, on the occasion of twenty years of search for oil in Israel, I. R. Kosloff concluded that from a strictly business point of view Israel's search for oil and gas had been a failure. For the growth of home consumption see Table XLIII.

28. *Petroleum Press Service,* XXXI, 430, Nov., 1964; – XXXVI, 85, Mar., 1969; – XXXVII, 126–127, Apr., 1970; *World Petroleum,* XXXV, 45, July, 1964; *World Petroleum Report,* X, 70, Mar. 15, 1964; *Hadashoth Haneft,* 5, June 5, 1971; – 4, Feb. 8, 1972.

29. *Petroleum Press Service,* XXXVII, 127, Apr., 1970; – XXXVIII, 176, May, 1971; – XXXIX, 297, Aug., 1972; *Hadashoth Haneft,* 5, Feb. 1, 1970; – 5, June 8, 1971; – 6, Aug. 27, 1971.

30. See below, Chapter XXIV.

31. *Petroleum Press Service,* XXXVIII, 394, Oct., 1971; – XXXIX, 298, Aug., 1972; *World Oil,* CLXVII, 199, Aug. 15, 1968, *World Petroleum Report,* XVII, 72, 1971.

32. *The New York Times,* Jan. 6, Feb. 16, 1970; *Oil and Gas Journal,* 68, Aug. 14, 1972; *Petroleum Press Service,* 127, Apr., 269, July, 1970; – XXXIX, 298, Aug., 1972; *World Petroleum Report,* XVI, 77, 1970; *Beolam Hadelek,* 32, July, 1969; *Hadashoth Haneft,* 9, Mar. 15, 1970; – 6, Aug. 10, 1970; – 10, Sept. 27, 1970.

33. *Oil and Gas Journal,* 68, Aug. 14, 1972; *Petroleum Press Service,* XXXIX, 298, Aug., 1972; *Hadashoth Haneft,* 6, Mar. 15, 1971; – 7, June 6, 1971; – 7, Feb. 8, 1972; – 5, Mar. 10, 1972.

34. *Petroleum Press Service,* XXXIX, 297, Aug., 1972; *Hadashoth Haneft,* 10, Mar. 15, 1971; – 13, Aug. 23, 1971; – 8, Oct. 5, 1971.

35. *Petroleum Press Service,* XXXVIII, 176, May, 1971; – XXXIX, 298, Aug., 1972; – *Hadashoth Haneft,* 6, Feb. 1, 1970; – 2, Mar. 15, 1970.

36. See above, Chapter XVII.

37. Longrigg, *Oil in the Middle East,* 250; *The New York Times,* May 13, 1947.

38. This was later repudiated by the Israel Government. For an official British summary of the various concessions granted by the British Administration in Palestine and Transjordan, see Palestine, *A Survey of Palestine Prepared in December 1945 and January 1946 for the Information of the Anglo-American Committee of Inquiry,* II, 969–972.

39. *Middle Eastern Affairs*, V, 408, Dec., 1954. Some American experts did not share the pessimistic view of the IPC about oil possibilities in Jordan. See, for instance, L. S. Gardner, "Oil Possibilities in The Hashemite Kingdom of Jordan," *The Oil Forum*, VIII, 448–450, 454, November, 1954.
40. UN, *Economic Developments 1955–1956*, 45; *The New York Times*, Nov. 23, 1955.
41. *International Oilman*, XII, 36, Feb., 1958, 132, May, 1958; *Le Commerce du Levant*, Nov. 4, 1957, Oct. 13, 1958.
42. *World Petroleum Report*, VI, 82, Feb. 15, 1960; *International Oilman*, XIII, 390, Dec., 1959.
43. *Petroleum Press Service*, XXVIII, 74, Feb., 1961; *World Petroleum Report*, XII, 77, Mar. 15, 1966.
44. *World Petroleum Report*, X, 70, Mar. 15, 1964; *The New York Times*, July 10, 1961.
45. *World Petroleum Report*, X, 70, Mar. 15, 1964.
46. *World Petroleum Report*, XVII, 71, 1971; – XVIII, 58–59, 1972; *Petroleum Press Service*, XXXVIII, 153, Apr., 1971; *The New York Times*, Mar. 2, 1971.
47. *Petroleum Press Service*, XXXVIII, 273, July, 1971; *World Petroleum Report*, VIII, 43, Feb. 15, 1962; – XVII, 71, 1971.
48. *MENA* (Cairo), 20, Jan. 30, 1954.
49. "Lebanon," *World Petroleum Report*, III, 138, Jan. 15, 1957; *World Trade Information Service*, Pt. 1, No. 56–60, June, 1956.
50. UN, *Review of Economic Conditions in the Middle East 1951–52*, 59; *Le Commerce du Levant*, May 21, 1952; *The New York Times*, May 16, 1952; Longrigg, *Oil in the Middle East*, 243; *The Christian Science Monitor*, Dec. 19, 1952. Direct payments by IPC and Tapline in 1952 amounted to £15,500,000; local expenditure by Tapline was £17,600,000 and by IPC about £10,000,000. UN, *Economic Developments, 1945 to 1954*, 168.
51. The magnitude of the flow of oil through the terminals of IPC and Tapline in Lebanon is illustrated by the 1953–54 figures. The total for both terminals in 1953 was 22,519,831 tons; in 1954, 22,905,708 tons. In the latter year Saudi Arabian oil amounted to 15,694,761 tons and Iraqi oil to 7,210,947 tons. *Le Commerce du Levant-Beyrouth Express*, Mar. 14, 1955.
52. *Mideast Mirror* (Cairo), 19, May 22, 1954.
53. Charles J. V. Murphy, "Oil East of Suez," *Fortune*, 258–260, Oct. 1956; *Foreign Commerce Weekly*, Aug. 8, 1955; *World Trade Information Service*, Pt. 1, No. 56–66, June, 1956; "Lebanon," *World Petroleum Report*, III, 35, Jan. 15, 1957; *The New York Times*, Feb. 10, 1957.
54. US Senate, *International Petroleum Cartel*, 97.
55. *World Petroleum Report*, XVII, 73, 1971.
56. Republique Libanaise, Ministere de l'Economie Nationale, Service de Statistique Generale, *Bulletin Statistique Trimestriel*, IX, 49, 1st Trimestre, 1958; *World Petroleum*, XXVII, 83, 148, 164, July 15, 1956.
57. *World Petroleum Report*, VI, 84, Feb. 15, 1960; *The Times* (London), June 4, 1959; *Iraq Petroleum*, VIII, 37, July-Aug., 1959.
58. *New York Herald Tribune*, Aug. 8, 1962; *The Financial Times*, Aug. 9, 1962.
59. *Petroleum Press Service*, XXXIV, 184, May, 1967.

60. *Petroleum Press Service,* XXXVIII, 193—4, May, 1971; *World Petroleum Report,* XVIII, 60, 1972.
61. *World Oil,* CLXXIII, 7, Aug. 15, 1971.
62. INOC, *Weekly Bulletin,* 6, Dec. 16, 1972; — 1—3, Feb. 24, 1973; — 14—16, Mar. 10, 1973; *The New York Times,* Mar. 6, 1973.
63. *World Petroleum Report,* XVIII, 60, 1972; IPC, *Review 1971.* For the rate of increase of throughput in the Tripoli refinery between the years 1962—1971 see Table XIX above.
64. *Petroleum Press Service,* XXXVIII, 193, May, 1971; *Hadashoth Haneft,* 9, Nov. 17, 1969.
65. A. H. Hourani, *Syria and Lebanon* (London, 1946), 155; Longrigg, *Oil in the Middle East,* 91—93.
66. UN, *Economic Developments 1955—1956,* 45; "Syria," *World Petroleum Report,* III, 246, Jan. 15, 1957; *The Middle East Economist and Cotton Recorder,* 32, Nov., 1958; *Le Commerce du Levant,* Oct. 10, 1956.
67. *Petroleum Press Service,* XXVI, 362, Sept., 460—461, Dec., 1959; *World Petroleum Report,* VI, 88, Feb. 15, 1960.
68. UN, *Economic Development 1961—1963,* 103; *World Oil,* CLXI, 162, Aug. 15, 1965.
69. *World Oil,* CLXVII, 200, Aug. 15, 1968; *Petroleum Press Service,* XXXV, 52—53, Feb., 1968.
70. *Petroleum Press Service,* XXXVII, 6, Jan., 230, June, 1970; — XXXIX, 10, January, 1972; *Hadashoth Haneft,* 9, Oct. 15, 1969; — 15, Feb. 3, 1969.
71. *Petroleum Press Service,* XXXV, 52—53, Feb., 1968; *World Oil,* CLXXI, 71, Aug. 15, 142, Oct., 1970; Aramco, *Facts and Figures 1971.*
72. *The New York Times,* May 13, 1947.
73. Longrigg, *Oil in the Middle East,* 243. Actually it was no increase, only an adjustment for the devaluation of the £.
74. See above, Chapter XVII.
75. *Middle Eastern Affairs,* IV, 20, Jan., 1953.
76. Longrigg, *Oil in the Middle East,* 244; *Le Commerce du Levant-Beyrouth Express,* Mar. 5, 1955. It was calculated by the Syrian press that the oil companies saved through the pipelines over $220 million, while the payments to Syria were no more than LS 6 million.
77. See Lebanon, above. UN, *Economic Developments 1954—1955,* 67; *Christian Science Monitor,* Feb. 6, 1957; *The New York Times,* Nov. 28, 1955; *New York Herald Tribune,* May 8, 1956.
78. *Petroleum Press Service,* XXIX, 142, Apr., 1967; *The New York Times,* Mar. 10, 1962.
79. *Petroleum Press Service,* XXXIII, 444—445, Dec., 1966.
80. *Petroleum Press Service,* XXXIV, 126, Apr., 1967; *World Oil,* CLXVII, 202, Aug. 15, 1968; IPC, *Review for 1967.* In April the Syrian Government invited Tapline to discuss increases in transit fees.
81. IPC, *Review 1971; Petroleum Press Service,* XXXVIII, 307—8, Aug., 1971; *World Oil,* CLXXII, 7, Mar., 1971; *World Petroleum Report,* XVIII, 64, 1972.
82. *Aramco 1970,* 4; — *1971,* 3; *World Petroleum Report,* XVI, 84, 1970; — XVII, 75, 1971.

83. *World Petroleum Report,* XVIII, 64, 1972.
84. INOC, *Weekly Bulletin,* 1–7, Oct. 21, 1972; – 6, Jan., 1973; *Petroleum Press Service,* XXXIX, 463, Dec., 1972; – XL, 65, Feb., 1973.
85. *Petroleum Press Service,* XXVI, 362, Sept., 1959; *World Petroleum Report,* VI, 88, Feb. 15, 1960.
86. *Hadashoth Haneft,* 9, Oct. 15, 1969. See the drop off in IPC's dispatch of crude to Homs after 1969 in Table XIX above.

Chapter XXIV
EGYPT AND TURKEY

EGYPT 1911–1954

In terms of oil developments, Egypt presented a different picture from the other countries of the Middle East. Although she actually has had as long a production period as Iran, she did not achieve expanded operations until the late 'sixties. After a short time most of her oilfields were exhausted and production either stopped completely or diminished to an insignificant trickle. Over the years production, with a few ups and downs, has progressed, and until the middle of the last decade did not reach a level which could meet the expanding home demand, let alone exporting on a large or even small scale. Also in contrast to the other countries, Egypt had never granted one oil company exclusive rights. Perhaps, in a reverse order from Syria, Egypt had always been primarily a transporter of oil rather than a producer.

The presence of oil was known in Egypt as early as 1869, in the very field — Gemsa — where the first commercial oil well was subsequently discovered, but early searches revealed no commercial quantities of oil. Three years after the establishment of the British regime in 1882, the Minister of Works invited foreign scientists to investigate the area, but the results were not too encouraging.[1] Between 1904 and 1911 different companies obtained exploration licenses in different parts of the country; in the latter year the Anglo-Egyptian Oilfields Ltd., a Shell subsidiary, was registered and acquired the field which in 1909 had showed sufficient quantities of oil to be exploited commercially. In 1912 the Company's capital consisted of £1 million; "A" shares were held by the Anglo-Saxon Petroleum Company (the same that had a 25 per cent share in the Turkish Petroleum Company); "B" shares were given to the companies that sold the concessions. A year later 100,000 "C" shares of £1 each were turned over to the Egyptian Government, thus giving it a 10 per cent limited equity in the Company. The management was in the hands of the Anglo-Saxon Company.

In 1912 Anglo-Egyptian Oilfields obtained a lease from the Egyptian Government which provided for a royalty of 6 pennies per hundred gallons of crude produced. The first to produce was the Gemsa field, located on the African side of the Red Sea at the mouth of the Gulf of Suez. In 1911 it produced about 2,793 tons. Hurghada, about thirty miles to the south of Gemsa, the second field, was discovered in 1913, and for about twenty-five years was the major producing field in Egypt. The leases for this field, granted to Anglo-Egyptian Oilfields in 1914 and 1923, carried the same terms as those for Gemsa.[2]

The third major field, Ras Gharib, about 150 miles southeast of Suez,[3] discovered in 1938, became the most prolific in the country and accounted for about half the oil production. The lease for this field, also granted to Anglo-Egyptian

Oilfields, was for thirty years, with an extension provision for another fifteen years; royalty was 14 per cent of production, either in kind or in cash, at the discretion of the Government; the latter also had the right to buy additional amounts of crude, up to about 20 per cent, at favorable prices. Ninety per cent of the labor force and 50 per cent of the staff were to be Egyptians.

The discovery of oil on the Asiatic side of the Sinai Peninsula did not take place until 1946 when the Sudr field, thirty miles south of Suez, was discovered. Production started in 1947 with 118,306 barrels per year, and rose to some 5 million barrels in 1952, and this from about eight wells. Ten miles from Sudr the second field in the Peninsula, Asl, was discovered, in 1948, and a year later commercial production commenced.[4] In 1951 seven wells were yielding an annual output of 4,249,298 barrels.[5]

The leases for the Sudr and Asl fields, jointly owned by Anglo-Egyptian Oilfields and Socony-Vacuum, provided the same terms as those for Ras Gharib, except that 14 per cent royalty applied to the half belonging to Anglo-Egyptian and 15 per cent to the half belonging to Socony-Vacuum.[6]

In the meantime, the Egyptian Government took steps which seriously restricted the freedom of the foreign companies. On January 20, 1947 a companies law was enacted which required that 40 per cent of the boards of directors must be Egyptian nationals; that at least 75 per cent of the employees must be Egyptian and that their salaries must be at least 65 per cent of total salaries paid; that the number of laborers must be not less than 90 per cent of the total engaged and their wages not less than 80 per cent of total wages paid; and that 29 per cent of the shares of a company must be held by Egyptians. One year later the law was amended; it stipulated that 51 per cent of the shares be held by Egyptians. On August 12, 1948 the Mines and Quarries Law was enacted; it provided that only Egyptian companies could obtain oil mining licenses; royalty payments to the Government were to be 15 per cent for one-half, and 25 per cent for the other half, of an area under exploration license qualifying for a lease; royalty for the renewal period of a lease was to be 25 per cent for the entire area.[7] The foreign companies operating in Egypt refused to accept the new conditions; Standard of New Jersey abandoned its license and withdrew from Egypt in 1949; Socony-Vacuum and Anglo-Egyptian suspended all exploration operations.[8]

Refineries

In 1913, soon after the Gemsa field began producing crude in commercial quantities, the Anglo-Egyptian Oilfields erected a refinery at Suez and the output was taken over by the Asiatic Petroleum Company, a Shell subsidiary. In 1919, in anticipation of production from the Hurghada field, the refinery was enlarged. Three years later the Egyptian Government erected its own hundred-ton daily capcity refinery at Suez to refine the crude which it received as royalty, and in 1938 the capacity of the Company refinery was doubled. At the beginning the Company refinery operated only part time on native crude and the rest of the time on imported crude; after 1938 the increase in local production restricted its operations to Egyptain crude. In 1950 the throughput was 14,511,000 barrels.[9] The refineries of both the Anglo-Egyptian Oilfields and of the Government were further expanded in 1954 to a combined annual capacity of 3,300,000 tons.[10]

Production

The growth of Egyptian oil is illustrated by Table XLV.

The reason for the irregularities in the curve of production was that the fields became exhausted after a number of years and were then abandoned. Gemsa was abandoned in 1927; the total production during its active life was 193,000 tons. Hurghada reached the end of its life in 1954, with total production about five million tons. Ras Gharib was still producing regularly at the beginning of 1954, up to which time it had a total output of 17 million tons. Sudr had passed its peak in 1954, having produced a total of a little over three million tons, and Asl, which began to produce only in 1948, reached its peak at the end of 1953, with a total production until then of only about three million tons.[11]

Reserves

Estimates of Egyptian reserves were raised from time to time but they never reached the figures that the other producing areas in the Middle East attained. In 1949 the reserves were estimated as 17 million tons; at the end of 1951 the figures were up to 23,400,000 tons, and in 1953 they were given as 28,600,000 tons.[12]

Exploration

Internal petroleum consumption since World War II has been much higher than the crude produced. An official Egyptian Government estimate for the annual average for the period 1930–1939 gave home consumption as about 100,000 tons; for 1953 it jumped to 3,200,000 tons. This has been primarily due to the rapid industrial expansion in Egypt, as well as to the increased use of petroluem products for home cooking and lighting. According to Egyptian statistics, the 9.5 gallon per capita consumption in Egypt was the highest in the world.[13] During the period 1949–1953, in spite of the increased production from the Sudr and Asl fields, about one million tons of petroleum products had to be imported annually to meet the home demand, and even in the peak year of 1952, home production supplied only 68 per cent of the demand.

The Egyptian Government realized that if the country possessed oil, it would have to be discovered and produced by foreign companies. This meant modification both of the Companies Law and the Mines and Quarries Law. One of the first acts of General Naguib's regime after the ousting of King Farouk was to amend the Companies Law to permit foreigners to hold up to 51 per cent of the shares in Egyptian companies (July 20, 1950).[14] At the end of 1953 a new Mines and

Table XLV. *Egyptian Oil Production, 1911–1954*

Year	Barrels	Tons
1911	21,000	2,793
1912	200,000	27,960
1913	93,000	12,620
1914	715,000	103,000
1915	210,000	34,000
1916	415,000	55,000
1917	980,000	135,000
1918	2,016,000	272,000
1919	1,658,000	232,000
1920	1,058,000	156,000
1921	1,308,000	181,000
1922	1,237,000	169,000
1923	1,093,000	150,000
1924	1,165,000	160,000
1925	1,287,000	175,000
1926	1,237,000	169,000
1927	1,315,000	180,000
1928	1,923,000	264,000
1929	1,944,000	267,000
1930	1,996,000	277,000
1931	2,038,000	280,000
1932	1,895,000	260,000
1933	1,663,000	228,000
1934	1,546,000	211,000
1935	1,301,000	173,000
1936	1,278,000	174,000
1937	1,196,000	168,000
1938	1,581,000	225,000
1939	4,666,000	660,000
1940	6,053,000	926,000
1941	7,659,000	1,221,000
1942	8,190,000	1,182,000
1943	8,994,000	1,285,000
1944	9,125,000	1,053,000
1945	9,406,000	1,340,000
1946	9,070,000	1,250,000
1947	8,627,000	1,320,000
1948	13,172,648	1,900,000
1949	15,802,221	2,270,000
1950	16,111,135	2,365,000
1951	16,299,191	2,368,000
1952	16,373,493	2,377,000
1953	16,501,000	2,351,000
1954	13,806,000[a]	1,758,000[b]

a. US Senate, *Special Committee Investigating Petroleum Resources. Petroleum Requirements – Postwar. Hearings* (Washington, 1946), 40; *Economic News* (Tel Aviv), V, 47, 1953; Valentine R. Garfias, *Petroleum Resources of the World* (New York, 1923), 111–112; "Egypt Offers Encouragement to Bolster Declining Production," *World Petroleum*, IX, 96, July, 1938; De-Golyer and MacNaughton, *Twentieth Century 1954*, 9; *World Oil*, CXL, 198, Feb. 15, 1955.

b. UN, *Review 1951–52*, 55; *– Summary 1952–53*, 47; Longrigg, *Oil in the Middle East*, 23, 40, 95, 258; *Annuaire de la Fédération Egyptienne de l'Industrie*, 1952–1953, 81–82; *L'Egypte Industrielle*, 42, June, 1953; – 38–39, June, 1954.

Quarries Law was promulgated, permitting foreign companies to explore for oil and obtain exploitation leases. The companies which had operated in Egypt were not satisfied with these modifications and negotiated with the Government for terms which would enable them to resume the operations of their old licenses as well as obtain new ones. New foreign companies, however, were willing to undertake exploration in Egypt under the new laws.

On February 3, 1954, the American Conorada Company (a combine of the Ohio Company, the Continental Company, and the Petroleum Corporation of America),[15] signed an agreement with the Government for exploration rights in the western desert covering an area of 185,000 square kilometers, divided into 456 blocks, for a period of thirty years, and renewable for another thirty years. The Company was not required to pay any rental on exploration blocks for the first twelve years, but was required to spend on exploration $1 million a year during the first three years, $1.5 million during the fourth and fifth years, $2 million in the sixth year, and between $50,000 and $60,000 for each block retained during the subsequent six years. Rental after the twelfth year was to be $25,000 for each exploration block retained.

Royalty was to be 15 per cent of the total quantity of oil produced for a maximum of ten years' production, or up to the expiration of the thirty years' period of the agreement, after which time and during the renewal period it was to be 25 per cent. The Government had the choice of taking the royalty either in kind or in cash; it also had the right to purchase up to 20 per cent of the crude produced by the Company in any calendar year at a price 10 per cent less than the price provided for settlement of royalty when paid in cash. The Company was to use Egyptian refineries up to their capacity.[16]

At the end of May the Government signed an agreement with the Egyptian Oil Exploration Company (owned by the same four companies as the Egyptian-American Oil Company) for exploration and exploitation in the eastern part of the western desert. The agreement was for thirty years and was renewable for another thirty years. The royalty terms were the same as those in the Conorada agreement. The Company was to spend about $1 million in prospecting operations during the first three years.[17]

At about the same time the International Egyptian Oil Company, owned by American and Swiss interests and registered in Panama, was reported to have obtained an exploration concession covering an area of about three million acres on both sides of the Gulf of Suez and in part of the western desert. Neither the terms nor the period of concession were made public. The National Petroleum Company of Egypt, an operating body of the International Egyptian Oil Company, began drilling operations in September 1953 under leases held by the Cooperative Centrale des Pétroles. National brought in a well in the Wadi Firan area in the Sinai Peninsula late in May 1954, with a 2,500-barrel daily capacity.[18]

The old foreign companies were meanwhile negotiating with the Government regarding exploration and exploitation terms. On June 26, 1954 an agreement was signed between the Egyptian Government and Anglo-Egyptian Oilfields, Shell Oil

Company of Egypt, Socony-Vacuum Company of Egypt, Esso-Standard, and the Société Egyptienne des Pétroles, amending the 1913 lease as regards royalty payments, and settling all differences on exploration licenses and exploitation leases.

Anglo-Egyptian Oilfields was permitted to double its capital to LE 4 million; Hassan Marei, Minister of Commerce and Industry, announced on June 29, 1954 that the Government owned 360,000 shares or about 10 per cent of the Company's capital as a result of an agreement signed earlier in the year. In September 1954 Anglo-Egyptian Oilfields and Socony-Vacuum signed a thirty-year agreement with the Government for the exploration of the Ras Matarma area near the Sudr oil fields.

About a month later the Minister of Commerce and Industry signed an oil prospecting contract with the Vereinigte Gewerkschaften Borgholzhausen of Hanover, covering an area of 30,000 square kilometers in the desert to the northeast of Sinai, for a period of thirty years and renewable for another fifteen years. Royalty was to be 15 per cent during the first half of the agreement, 25 per cent during the second half and the renewal period. The annual rent was fixed at LE 10 for each unit of land [19] for the first year, LE 100 for the second year, and LE 2,500 for each succeeding year. At the same time, the Government signed an exploration agreement with the National Petroleum Company covering areas in the eastern desert of Sinai and the territorial waters in the Red Sea. [20]

Through these various agreements the new Egyptian regime succeeded in dotting the eastern and western deserts and the coasts along the Gulf of Suez with many oil exploration units. Through the carefully drawn contracts, with their steeply rising rental fees and obligations on the part of some of the companies to spend specific sums on exploration, the Government guaranteed a dynamic search for oil in Egypt. [21]

EGYPT 1955–1971

Exploration and Exploitation

Because of the nature of most of the Egyptain oil wells, the peak year of 1953 was not surpassed until 1958. Sudr field began to decline seriously in 1956, as did Asl. On the other hand, Belayim, which had begun to produce early in 1955, increased sharply, and in 1958 was considered the most important field in Egypt. Its estimated reserves of 100,000,000 tons far surpassed the estimates of Ras Gharib, which were given as 14,000,000 tons. In June 1957 a new field was discovered at Abu Rudais; that year it produced some 60,000 tons. Consumption during 1955 and 1956, however, ran far ahead of production (it reached 3,650,000 tons in 1955 and 3,835,000 tons in 1956), yet it was reported that the General Petroleum Authority, a government agency set up in April 1956 as an advisory body on petroleum matters to the Ministry of Industry, had announced on April 5, 1958 that in May it would have available for export — for the first time in the country's history — 500,000 tons of Egyptian crude from the Sinai fields. [22]

Meanwhile three major developments took place which affected Egypt and the oil question in her territory, with repercussions of many ramifications. Because of financial difficulties in building the Aswan High Dam, and the mishandling of the issue by the Western diplomats, President Nasser nationalized the Suez Canal on July 26, 1966. This act brought on the war between Egypt and England, France and Israel. One of the major consequences of the war was the prolonged closure of the Suez Canal, which created an oil supply crisis in Western Europe; it also caused economic difficulties for Egypt through the loss of the revenue of the Canal, a major source of her income.

The second event was the merging of Syria and Egypt in February 1958 into the United Arab Republic as the prototype of the modern national Arab renaissance and transformation under the leadership of Egypt, an attempt which was quickly dissolved by Syria's breakaway late in 1961.

The third, whose effects were felt long afterwards, was the Arab-Israel war of June 1967. It was hoped in Egypt that it would be a triumphant achievement for Arab nationalism, of which Egypt was the dynamic progressive leader if not protector. The outcome, however, was a military defeat, and the Suez Canal was closed and remains closed until this very day.

These three events affected drastically not only the internal political and economic patterns of the country but also the relations between the Egyptian Government and the foreign companies which were developing the oil resources of Egypt, relations with the great oil-consuming countries, and relations with other Arab countries.

At the end of January 1957, President Nasser decreed the Egyptianization of all foreign-owned banks and commercial and export companies. This brought about the sequestration of some of the foreign oil companies, especially Anglo-Egyptian Oilfields. In May, negotiations began between the British and Egyptian governments, and in October between the French and Egyptain governments, and the outlook for the return of the oil companies to their original owners seemed hopeful. Finally, in February 1959, Abdul Muneim al-Kaissouni, Minister of Economy, announced that an agreement settling outstanding issues between the Egyptian Government and Anglo-Egyptian Oilfields Ltd. had been reached, and a new ten-year exploration concession was granted the Company.

In December 1957 the Petroleum Authority began exploration operations in the desert region of Sinai and the Red Sea; in January 1959 the Italo-Egyptian Petroleum Company was granted a thirteen-year exploration concession in the Sinai desert and offshore areas in the Red Sea.[23] A new refinery at Alexandria was completed in July 1956 by the Société Egyptienne pour le Raffinage et le Commerce de Pétrole, and in March of the following year it began to operate. The annual throughput was 200,000 tons, which could be expanded to 500,000 tons.

On July 21, 1961, as Egypt was moving further toward socialism, the Government issued a decree nationalizing all companies; this affected Anglo-Egyptian Oilfields, which yielded in 1960 about 1.2 million tons of oil; and the Egyptian Independent Oil Company (renamed Nasr Oil Company). A new government agency, the Egyptian General Petroleum Corporation (EGPC), was established and it took over direction of the seven government-owned oil companies. It became the policymaker and controller of the oil industry in Egypt.

The major operating company at the time was the Compagnie Orientale des Pétroles d'Egypt (COPE),[24] jointly owned by the Italian ENI and the Egyptian EGPC. COPE was granted two new concessions covering an area of 26,300 square kilometers. ENI gave the company $20 million to be spent on exploring and developing the two new areas within 12 years. All costs were to be borne by ENI should no oil be discovered; however, if oil were discovered in commercial quantities, the Government was to receive 75% of the net profits; 50% as royalty and half of the remainder based on its half interest in COPE. The Government's half of the costs was to be deducted from its share of the profits from the crude oil produced.

At the end of 1961, COPE discovered a well in the Gulf of Suez, 9 kilometers offshore from its field in Sinai. In that year Belayim produced 65% of Egypt's 3.7 million tons total output.[25]

It would seem that beginning with 1963 Egypt's oil development policy assumed a new direction in dealing with foreign companies. Although she nationalized the marketing firm Shell Oil Company of Egypt, which was jointly owned by Shell, BP and Nasr Oilfields, she signed an agreement on September 25 with Phillips Oil Company for prospecting for oil in an area of 37,000 square miles between the Rosetta branch of the Nile and the Libyan border, including coastal and territorial waters. The terms were practically the same as those of the last grant to ENI.

In the same year Pan American UAR Oil Company, a subsidiary of Standard Oil Company of Indiana, obtained exploration and development rights covering an area of 28,125 square miles south of Cairo and west of the Nile. It was granted for a period of 30 years and renewable for an additional 15 years.

COPE was granted additional exploration rights in the Nile Delta and on the African coast of the Gulf of Suez. It discovered the Gharah offshore oilfield in the Gulf of Suez 50 miles south of Belayim field. Early in 1964 Pan-American was awarded a license to explore offshore areas from the town of Suez southward to the Red Sea, excluding the sections previously granted to COPE. The terms were the same as those for the western desert concession, except that the Company was to spend a larger minimum amount on exploration.[26]

In March 1965, Pan-American made its first discovery – the El Morgan field in the Gulf of Suez, 150 miles southeast of Suez. The Government and Pan-American thereupon organized the joint-company Gulf of Suez Petroleum Company (GUPCO). It began producing in April 1967, and the major part of the crude was to have been exported to world markets.

The successful production in El Morgan field and the impending promising developments in the other concession areas prompted President Nasser to declare that he expected production from established oilfields to rise to 30 million tons by 1970; this would provide Egypt with an annual oil revenue of $280 million. Taher

Hadidi, general manager of the General Petroleum Corporation, went out on a limb by stating that his country's oil revenues might rival those of Saudi Arabia in three years.[27]

In the same year Phillips and General Petroleum completed a well in the El Alamain field in the western desert. A new joint-company, Western Desert Operating Petroleum Company (WEPCO), was formed. Both parent companies were to provide equally for development costs. The initial daily output was to be at the rate of 40,000 barrels, and the crude was to have been refined in Alexandria.[28]

Production from El Morgan soon reached proportions of a major Middle East field, and the Minister of Industry, Petroleum and Mining, Aziz Sidky, reported that of the 20 million tons which Egypt was to produce in 1969 (which she did not achieve) El Morgan was to contribute 15 million tons. It was reported early in the year that the field produced at the rate of 10.2 million tons, and that the oil output of the first six months was worth $130 million. The Company built new platforms and submarine pipelines to increase output.

So successful was the GUPCO undertaking that on September 30, 1969, the Government signed a new agreement with Pan-American UAR for an 8-year exploration, 30-year development concession covering an area of 28,000 square kilometers in the western desert and the southern part of the Nile valley. Pan-American was to spend between $15.5 and 20 million on exploration. Terms were the same as of the previous concession: income tax 50% and expensed royalty 15%. Both royalty and profits were to be based on realized sales prices to non-affiliated companies. Each of the two partners was to be responsible for the marketing of its own share of production.[29]

Because some of the oil produced in Egypt could not be refined for home needs, the country had always exported as well as imported oil, and some imports were refined and exported as products. The Government General Petroleum Corporation dealt with both. In 1969, the Corporation signed six contracts for exporting crude and products amounting to $10 million. 2.9 million tons of crude from the El Morgan field were contracted by Spain, 370,000 tons by East Germany, 50,000 tons by the Soviet Union, 730,000 tons by Hungary, 100,000 tons by China and 140,000 tons by Sudan.[30]

Although the huge production figures predicted by President Nasser and Ministers were far from being realized, production in 1970 continued to expand and reached a daily average of 430,000 barrels; El Morgan (GUPCO) between 350,000 and 400,000 barrels, with storage facilities at the Ras Shukeir terminal; El Alamain (WEPCO) 27,000 barrels, with the oil piped to the tanker terminal Ras el-Shegig on the Mediterranean. The rest came from the Egyptian General Petroleum Corporation fields on the western shore of the Gulf of Suez — the older fields Ras Gharib and Bakr produced at a daily average of 28,000 barrels, Karim contributed 2,500 barrels, and the newer fields Amin, Shukeir, Uyun and Umm el-Yusr averaged 5,000 barrels.[31]

Refining

By 1963, three refineries, all state owned, operated in Egypt, two in Suez and one in Alexandria with a total throughput of 6,560,000 tons. Egypt imported in that year 4,323,000 tons of crude and exported 3,419,000 tons. [32]

In the late 'sixties, during the war of attrition between Egypt and Israel across the Suez Canal, the Suez refineries were heavily if not totally damaged. They had to be rebuilt, and the Alexandria refinery was extensively enlarged. In 1972 it was reported that the Suez refineries were restored, enlarged and extended. [33]

Three natural gas fields were discovered in Egypt by the new concession com-

Table XLVI. *Egyptian Oil Production, 1955–1971*

Year	Barrels	Tons
1955	12,966,000	1,807,777
1956	12,174,000	1,729,393
1957	16,150,000	2,337,546
1958	21,718,000	2,792,000
1959	21,720,000	3,000,000
1960	24,528,000	3,600,000
1961	25,550,000	3,700,000
1962	32,774,000	4,300,000
1963	38,759,000	5,720,000
1964	47,040,000	6,720,000
1965	44,320,000	6,334,000
1966	43,298,000	6,279,000
1967	44,418,000	5,580,000
1968	52,338,000	8,995,000
1969	87,857,000	12,348,000
1970	119,229,000	16,404,000
1971	114,646,000[a]	15,500,000[b]

a. *World Oil,* CXLII, 186, Feb. 15, 1956; – CXLVI, 195, Feb. 15, 1958; – CXLVIII, Feb. 15, 1959; – CL, 95, Feb. 15, 1960; – CLII, 111, Feb. 15, 1961; – CLIV, 120, Feb. 15, 1962; – CLVI, 125, Feb. 15, 1963; – CLIX, 144, Aug. 15, 1964; – CLXII, 120, Feb. 15, 1966; – CLXIV, 144, Feb. 15, 1967; – CLXVI, 123, Feb. 15, 1968; – CLXVIII, 114, Feb. 15, 1969; – CLXX, 96, Feb. 15, 1970; – CLXXIV, 63, Feb. 15, 1972.

b. *L'Egypte Industrielle,* 37, April, 1957; – 23, Mar. 1958; *L'Egypte Contemporaine,* XLVII, 83, Oct., 1957; *Petroleum Press Service,* XXVII, 5, Jan., 1960; – XXIX, 4, Jan., 1962; – XXX, 4, Jan., 1963 – XXXVII, 6, Jan., 1970; – XXXIX, 10, Jan., 1972; *Petroleum Times,* LXXI, 264, Feb. 17, 1967; – LXXII, 991, July 5, 1968; *World Petroleum,* XXXIV, 519, June, 1963; *World Petroleum Report,* XI, 68–70, Mar. 15, 1965.

panies. Phillips found one in the Abu Qir area 9 miles offshore in the Mediterranean northeast of Alexandria in 1965; COPE discovered a field in 1967 in the Abu Madi area in the northern Delta, and Pan-American found one at Abu Gharadig in the western desert.[34]

The rate of production was steadily climbing from some 13 million barrels in 1955 to some 115 million in 1971, almost a ninefold increase, as is seen from Table XLVI. At the same time home consumption was continuously rising. While Egypt became ultimately a net exporter, it was far from becoming an exporter on the scale of the major producers in the region; the production curve dropped in 1971 by some 5 million barrels. The rate of production at El Morgan declined, and the Soviet Union drillings in the Siva oasis have yielded no results. Both the Egyptian Government and the operating companies no longer spoke in glowing terms about great promising future oil developments as they had done in the late 'sixties.[35]

Sumed Pipeline

The extended closure of the Suez Canal after June 1967 posed serious economic difficulties for Egypt. As the prospects for the opening of the Canal grew dimmer and the new mammoth tankers began operating from the Persian Gulf's newly expanded ports, it became obvious that Egypt might lose the oil transport even should the Canal be reopened, for it could not accommodate the new giant vessels. Moreover, Israel was planning to build a pipeline from Elath to Ashkelon which would be not only an alternative to the Canal but also a solution to the huge-sized tanker problem. To overcome these dangers a plan was worked out to build a pipeline parallel to the Canal, from a point in the Gulf of Suez to Alexandria. The projected line would accomplish all the objectives the Israelis sought to achieve with the Elath-Ashkelon line. Indeed, the Economic Council of the Arab League approved the Egyptian plan for the pipeline as a counter program to the Israel line.[36]

The building of the pipeline presented a complex of problems, the major one of which was financing. Egypt would have to be the owner and builder of the line, and award contracts to foreign companies. Not being able to finance it herself she would have to obtain credit from foreign bankers. In order to cover the risk of the loans the bankers would want to see legally binding commitments from international oil companies to ship their oil through the pipeline. The international oil companies in turn would want to know in advance of their commitments what the rates of transit would be, and whether they would be economical. Egypt on her part must charge rates that would amortize the investment, pay the interest and make a profit; these might be higher than the tanker freight around the Cape. This complexity, in fact, became the viscious circle in the negotiations between Egypt and a European consortium headed by French companies.

On July 13, 1969, a tentative contract to lay the pipeline at a cost of $175 million was signed. France obligated herself to grant Egypt a long-term loan amounting to 25% of the cost in hard currency. Italy was contemplating, reportedly, to grant a similar loan, should Italian companies participate in the project. Early in 1969 it was believed that it would take about 2½ years to construct the line, but after 19 months a goodly portion of the contemplated quantity of crude could be flowing through. The length of the pipeline was given as 330 kilometers and diameter as 42 inches. In order to ensure the safety of the line it was provided that at no place should it be nearer than 64 kilometers from the Canal. The port near Suez was to be able to receive tankers of up to 312,000 tons. The pier at Alexandria was to accommodate tankers between 50,000 and 250,000 tons.

Negotiations continued and a new modified agreement was signed on July 31, 1971 in Cairo for the construction of the Sumed (Suez–Mediterranean) crude oil pipeline, between the Egyptian Government and the international consortium. The line was to have consisted of two parallel 320-kilometer, 42-inch pipes from a receiving terminal at Ain Sukhna, 40 kilometers south of Suez, to a loading terminal at Alexandria. The initial annual capacity throughput with a pumping station at Suez was to be 80 million tons; with an additional pumping station near Cairo it could be boosted to 120 million tons. The cost of construction went up to $280 million plus $20 million for unforeseen expenses. Foreign credits from the participating countries in the consortium were to amount to $225 million, $44 million was to be obtained in loans from Middle Eastern countries, and the two prospective users of the line, Amoco (Pan-American) and Mobil, were to provide $15 million.

The Sumed Pipeline International Consortium (SPIC) was to be managed by a board of directors from seven nations representing the companies in the countries involved in the construction of the line.

Early in 1971 it was reported that Amoco and Mobil would undertake to put through annually 5 million tons of crude oil; that Shell, BP and CFP had tentatively committed themselves to put through annually 7½ million tons each for a period of 10 years with subsequent five-year extensions; and the French Elf-ERAP and German Gelsenberg would put through 3 million tons each, while Hispanoil would ship 6 million tons. Although the total of all these tentative commitments was far from the initial 80 million tons, it was a very substantial achievement to encourage the beginning of construction.[37]

Yet at the end of 1972, it was reported that while the project was not completely on its way to realization the Egyptians were making headway in their efforts. The major obstacle to the consummation of the project was the reluctance, for various reasons, of the international producing companies to commit themselves to put through their oil over prolonged periods, while these commitments had to serve as guarantees for the consortium. Early in 1973 it looked as if the pipeline would be built.[38]

Reflections

Egypt had functioned in the oil development of the region as producer but more as a transporter. Since the middle 'sixties it also had become an exporter as well as an importer. The closure of the Suez Canal since June 1967 eliminated the function of transporter; with the building of the Sumed line that function would be restored, at least partially. Should the Canal be reopened the function would be restored fully. The role of producer was greatly broadened but its extent would depend on the continuity of output of the existing fields as well as on discovery of new fields. Egypt had been and is an integral part, more perhaps as transporter but also as producer, of the Middle East oil industry.

Egypt apparently has found her relations with the foreign companies in the joint-ownership type of arrangements rather satisfactory. Whether the wave of nationalization through 51% participation would also affect Egypt only time will tell; but it would seem that Egypt would have little to gain from such a move.

One thing Egypt had definitely learned from the 1967 closure of the Canal was that the world petroleum industry and the consuming countries could and did get along without the Suez Canal. While the 1956–57 experience gave the Canal serious if not vital control over the supply to Western Europe, ten years later this importance had totally vanished. [39]

TURKEY

Turkey is perhaps one of the lesser oil-producing countries in the Middle East. Like Egypt, she has been exerting great efforts to discover oil to meet part if not all of internal demand for petroleum products. The rapid development of the country required huge amounts of petroleum and this has taxed Turkey's foreign exchange heavily. [40] At the beginning of 1955 there were a half dozen foreign companies, mostly American, searching for oil in various parts of the country under permits from the Turkish Government. From a governmental point of view this has been a complete reversal of economic policy.

The emergence of the new Turkey in the early 'twenties, with its emphasis on the State, brought with it general nationalization of all economic resources. In 1929 a special law covering exploration of petroleum was enacted. A search for oil was undertaken and in 1933 a law was enacted which created the Maden Tetkik ve Aramu Enstitusu (Mining Research Institute), popularly known in Turkey as MTA, as the agency for oil research, exploration and exploitation. It began drilling in 1934, concentrating in the Ramandag, some sixty miles from Diarbekr; the various attempts, however, brought no sizeable results and for a while the efforts were given up. [41] Another attempt in 1948 did bring results; six wells produced between 50 and 100 tons a day. A refinery at Batman, the railway station in Ramandag, with a 200-barrel per day capacity, was erected; in the following year it was expanded to 350 barrels per day capacity. [42] In 1951 a second field was discovered at Garzan in the same region; it produced about 500 barrels daily and the Government considered building a new refinery at Batman. Between 1941 and 1951 the Government spent some $10 million on petroleum research, drilling, refining and other facilities, but had very little to show for it.

With the victory of the Democrat Party in 1950, étatism, the cornerstone of the Ataturk regime, was weakened and the question of petroleum was seriously raised. On November 12, 1952 the Government announced that it had decided to denationalize the oil industry and would prepare a new petroleum law. Some eight international oil companies were permitted to undertake geological survey work in anticipation of the new law. The following May, Max W. Ball of Washington, D. C. was retained as consultant to prepare the draft of a new petroleum law to be presented to the Grand National Assembly. [43]

After considerable discussion and debate, during which the proposed law became a hot political issue, the National Assembly adopted, on March 7, 1954, the new law, which was based on the original proposals submitted by Ball. The country was divided into nine petroleum districts; no exploration area was to exceed 50,000 hectares, and no single person or concern was to be granted more than eight different exploration permits; the permit rights in any one particular area were to run for a maximum of six years. In order to guarantee the financial stability of the applicants and their earnest to explore, a high graduated rental per hectare was prescribed, but 80 per cent of that rental was deductible from future royalties.

Exploitation units were restricted to 25,000 hectares, and also limited to eight for any one person or concern. Concessions were limited to forty years, with a possible extension of another twenty years; royalty was to be 12½ per cent of production; companies were subject to corporation tax and special fees, and after the company recovered its initial investment, profits were to be subject to a special surtax which would give the State a total of about 50 per cent of the profits. In calculating profits the companies were to be granted all the allowances which the United States Government granted to oil producers in the United States for tax purposes. [44]

By May 1954 about seven companies, among them Socony-Vacuum, American Overseas Petroleum Company (an affiliate of Standard Oil of California and the Texas Company), Esso-Standard of Turkey (an affiliate of Standard Oil of New Jersey), and the Royal Dutch-Shell had been granted exploration permits under the provisions of the new law.

At the same time the Petroleum Law was enacted, the National Assembly voted to establish the Turkish National Petroleum Company in which the State was to own 51 per cent of the shares. It was to have exclusive rights over the existing oil wells and refineries, to control refining throughout the country, and it could also undertake exploration and exploitation. The Company was soon organized and undertook the expansion of the Batman refinery to a 330,000-ton annual capacity. The Turkish Government had thus combined State ownership and control of the existing oil wells and refineries with foreign private investment concentrated on exploration for new petroleum resources. Turkey succeeded in interesting the great international oil companies to search for oil.

The discrepancy between production and consumption had become alarming. The rapid increase in consumption and the figures for production are indicated by the following table:

Table XLVII. *Turkish Internal Consumption and Production, 1937–1954*

Year	Consumption (In Tons)	Production (In Tons)	Production (In Barrels)
1937	90,000		
1938	126,300		
1947	280,000		
1948		3,000	13,000
1949	432,000		95,000
1950	413,000	17,000	108,000
1951		19,000	133,000
1952		22,000	146,000
1953	963,000[a]	28,000	179,000
1954			309,000[b]

a. UN, *Summary 1950–51,* 27; *Foreign Commerce Weekly,* XXXV, 35, Apr. 25, 1949; *Mideast Mirror,* 19, May 15, 1954.
b. UN, *Summary 1952–53,* 47; *Le Commerce du Levant,* May 15, 1953; *World Oil,* CXL, 200, Feb. 15, 1955.

The outlook in Turkey was not too good. Nevertheless, being on the fringe of the rich Middle East oil fields, and still nursing a grudge against Great Britain and Iraq for depriving her of the Mosul oil fields, Turkey was preoccupied with the problem of oil, and oil had influenced her national thinking.

TURKEY 1955–1971

To make exploration more attractive to foreign companies, the Government amended the Petroleum Law in 1955, and a number of additional exploration licenses were granted. By the end of the year, twelve companies were carrying on exploration. The law was amended again in 1957, and in 1958 twenty-one companies were in the field.

Production

Production of crude was steadily but slowly increasing during the later 'fifties to the early 'sixties, and then suddenly began, relatively of course, to climb upward, reaching a peak in 1970, and then dropped by more than 1/3 in 1971, as shown in Table XLVII. Although the 1963 production increased over the previous years, three foreign companies, Deilmann, Pauley, and Tidewater, gave up their exploration licenses covering an area of 22,000 square kilometers.[45]

In the following year the country had four producers, TPAO (the national company), which produced approximately 68% of the total 6,990,000 barrels, and Mobil, Shell and Esso affiliates in Turkey which produced the rest. Exploration in Turkey was very expensive and unrewarding. In 1960, for instance, Mobil Explo-

Table XLVIII. *Turkish Production, 1955–1971*

Year	Thousands of Barrels
1955	1,276
1956	2,200
1957	2,135
1958	2,350
1959	2,617
1960	2,588
1961	3,083
1962	4,131
1963	4,725
1964	6,990
1965	9,770
1966	14,172
1967	19,516
1968	22,000
1969	24,911
1970	26,376
1971	17,918[a]

a. UN, *Economic Development 1954–1955*, 57; *World Petroleum Report*, III, 254, January 15, 1957, *New York Times*, January 13, 1959; *World Oil*, CL, 95, February 15, 1960; – CLII, 111, February 15, 1961; – CLVII, 172, August 15, 1963; – CLXII, 120, February 15, 1966; – CLXIV, 144, February 15, 1967; – CLXVII, 201, August 15, 1968; – CLXVIII, 114, February 15, 1969; – CLXX, 96, February 15, 1970; – CLXXII, 106, February 15, 1971; – CLXXIV, 63, February 15, 1972.

ration Mediterranean, affiliate of Mobil International, struck oil in the Bulgardag area, fifty miles from Mersin, but it cost $16 million. Esso Standard (Turkey) withdrew from the country after having spent $20 million on dry holes. Most of the fields were concentrated in Diyarbekr district in the Sirit province, and it had to be brought west by pipeline. In January 1967, a 310-mile, 18-inch pipeline with a daily capacity of 700,000 barrels was completed. It runs from the oilfields to Dörtyol on the Gulf of Iskanderun. The pipeline made possible the activation of the fields which previously could find no outlets. In that year daily production rose to 54,000 barrels, making a total of 19,516,000 barrels; this amounted to a 47% increase over the previous year.[46]

The oil reserves estimates were constantly increasing until 1969; while the official estimates at the end of 1963 were given as 2.4 billion barrels, actual recoverable reserves were given as between 300 and 400 million barrels. One year later they were given as 440 million barrels. In 1969 they dropped to 200 million, and in 1971 to 181 million barrels.[47]

Refining

Local consumption was constantly rising; and since home production could meet only part, and at times a very small part, of the local needs, the Government concentrated on building refining capacity, and imported crude.

In December 1955, the new refinery at Batman was completed; its daily capacity was 6,250 barrels. In January 1965, it began to produce commercial gasoline. Early in 1957 the Government signed an agreement with British Petroleum, Shell, Socony-Mobil and Caltex to build a refinery with a 3.2 million-ton throughput, to cost $48 million and to begin production in 1960. By 1967, three refineries were operating in Turkey: 1) Mersin, in the southern part of the country. It was owned by ATAS — 56% Mobil, 27% Shell and 17% BP — and its daily capacity was between 25,000 and 90,000 barrels. 2) Izmit, near Istanbul; owned by IPRAS — 51% TPCO (Government corporation) and 49% Caltex; the daily capacity was between 44,000 and 110,000 barrels. 3) Batman, expanded to a daily capacity of 13,500 barrels.

The crude for the Izmit refinery was totally imported, as was part of the crude for the Mersin refinery. Before the closing of the Suez Canal crude was imported from Qatar and from Libya. After the closing of the Canal when Mediterranean oil became very expensive, the crude was imported from northern Iraq. In 1970, Turkey imported 3.8 million tons, almost 1 million more than in the previous year.[48]

If the drop of production from 1970 to 1971 was part of a pattern, then Turkey would become a heavy importer of crude. Of all the countries in the region Turkey was the least involved in the oil developments of the area. There were, however, intimations that new pipelines from the producing countries might debauch on the Turkish coast; should these plans materialize Turkey might join the other transport countries discussed in the previous chapter.

NOTES

1. A helpful but not too optimistic report was presented by C. E. Stewart, *Report on the Petroleum Districts Situated on the Red Sea Coast* (Cairo, 1888).
2. For the Government oil policy, see Ministry of Finance, Egypt, *Note on the Programme and Policy of the Government with Regard to the Investigation and Development of the Petroleum Resources of Egypt,* by E. M. Dowson, Under-Secretary of State for Finance (Cairo, 1921).
3. "Ras Gharib Discovery Brings Rush of Prospecting Activity to Egypt," *World Petroleum*, X, 52—53, June, 1939.

4. For a detailed table of the annual production of each field to the end of 1953, see *The Egyptian Economic & Political Review*, I, 52, Mar., 1955 (henceforth cited *EE & PR*).

5. With the discovery of oil in Saudi Arabia, the international companies applied for prospecting licenses in Egypt. By the end of 1938, 846 licenses had been applied for, covering an area of 84,600 square kilometers in the western desert and the Sinai Peninsula. Among the companies were Shell, Caltex, Standard of New Jersey, and Socony-Vacuum.

6. Nawratzki, *loc. cit.*; Longrigg, *Oil in the Middle East*, 22–24, 94–97, 254–255; *EE & PR*, I, 29, Nov., 1954.

7. *EE & PR*, I, 26–27, Nov., 1954.

8. "Egyptian Ambitions," *Petroleum Press Service*, XVI, 261–263, Nov., 1949; Rodger L. Simons, "Egypt's Oil Industry Faces Uncertain Future," *World Petroleum*, XXI, 29, Aug., 1950.

9. UN, *Summary 1950–51*, 24.

10. *EE & PR*, I, 24, 33, Nov., 1954.

11. *EE & PR*, I, 34, Nov., 1954; Nawratzki, *loc. cit.*, Garfias, *Petroleum Resources of the World*, 111–112.

12. UN, *Review 1951–52*, 53; – *Summary 1952–53*, 40.

13. *EE & PR*, I, 125, Nov., 1954. This was contradicted by the figures compiled by the United Nations. In 1950 the per capita consumption for the world as a whole was 240 kilograms, while in the Middle East it was about 55 kilograms. UN, *Summary 1950–51*, 25–26.

14. *Middle Eastern Affairs*, III, 256, Aug.–Sept., 1952.

15. The Egyptian-American Oil Co. was later formed to take over the concession from the Conorada Co. The Petroleum Corp. of America withdrew and the Egyptian-American Oil Co. was owned by the Continental Oil Co. and the Ohio Co., 27½ per cent each, and Cities Services Co. and Richfield Oil Corp., 22½ per cent each. *EE & PR*, I, 29, Nov., 1954; *Newsweek*, XLIV, 49–50, Oct. 25, 1954.

16. *Middle Eastern Affairs*, V, 103, Mar., 1954; *EE & PR*, I, 41–44, Nov., 1954.

17. *EE & PR*, I, 29, Nov., 1954; *Mideast Mirror*, 15, May 29, 1954.

18. "Vigorous Oil Search in Sinai," *The Oil Forum*, IX, 48–49, 59, Feb., 1955.

19. A unit was not to be more than 1,000 square kilometers.

20. *EE & PR*, I, 29–30, Nov., 1954.

21. A very helpful map indicating the areas granted to the various companies for exploration is given in *EE & PR*, I, 28, Nov., 1954.

22. *The New York Times*, Apr. 6, 1958. In Mar., 1959 a new General Petroleum Authority was established; its function was defined as central planning for petroleum policy in the Egyptian and Syrian regions. Among its privileges were the right to import and export all crude oil and derivatives needed by the United Arab Republic, to fix selling prices of petroleum products, to undertake all prospecting, production, refining, purchase, sale, transport and distribution operations; purchase, merge with, and annex public and private organizations carrying out work similar to its own, and advise on the granting of oil prospecting and mining leases. A two-thirds vote of the members of the Authority could override the Minister of Industry, whom the Authority was to advise. *Foreign Commerce Weekly*, 12, Mar. 16, 1959; *L'Egypte Contemporaine*,

XLVIII, 77, Oct., 1957, *Le Commerce du Levant,* Sept. 9, 1958.

23. *International Oilman,* XIII, 4, Jan., 38, Feb., 1959; *Petroleum Press Service,* XXV, 12, Jan., 1958; *The New York Times,* Dec. 29, 1958.
24. In English sources it is referred to as the Oriental Petroleum Company and in Arabic sources as As-Sharqiyah Company.
25. *Petroleum Press Service,* XXIX, 11–14, Jan., 1962; *World Petroleum Report,* VIII, 228–229, Feb. 15, 1962; *World Oil,* CLVII, 168, Aug. 15, 1963; *The New York Times,* Nov. 26, 1961; Radio Cairo, Nov. 15, 1961.
26. *World Oil,* CLIX, 144–148, Aug. 15, 1964; *World Petroleum Report,* XI, 68–70, Mar. 15, 1965; *Petroleum Press Service,* XXXI, 463–465, Dec., 1964.
27. *World Petroleum,* XXXVIII, 36–37, July, 1967.
28. *World Oil,* CLXIV, 69–72, Mar., 1967; *Petroleum Press Service,* XXXV, 243–245, July, 1968.
29. SAMA, *Statistical Summary,* 15, Dec., 1969–Jan., 1970; *Wall Street Journal,* Oct. 22, 1969; *Hadashoth Haneft,* 10, July 31, 1969.
30. *Hadashoth Haneft,* 14, 21, Feb. 3, 1969.
31. *World Petroleum Report,* XVII, 54–55, 1971; *Petroleum Press Service,* XXXVIII, 95–96, Mar., 1971.
32. *Petroleum Press Service,* XXXI, 463–465, Dec., 1964.
33. *Petroleum Press Service,* XXXIX, 366, Oct., 1972.
34. *Petroleum Press Service,* XXXVIII, 468, Dec., 1971.
35. *Ibid.*
36. *Hadashoth Haneft,* 18, Feb. 3, 1969.
37. *Petroleum Press Service,* XXXVIII, 25, Jan., 84, Mar., 313, Aug., 1971; *World Petroleum,* XLII, 13–14, Nov., 1971.
38. *The New York Times,* Dec. 26, 1972.
39. For an analysis of the two closures from the point of view of those affected see *World Petroleum,* XLI, 94–106, June, 1970; – XXIX, 25, 30, July, 30, Aug., 1968.
40. For a description of the financial difficulties which Turkey has had to face as a result of her great demand for foreign oil, see *The New York Times,* Oct. 18, 1954. See also E. A. Bell, "Mineral Oil in Turkey," *World Petroleum,* XXIII, 94, 112, Sept., 1952.
41. William Gillman, "No Commercial Results from Million Dollar Campaign in Turkey," *World Petroleum,* IX, 82–83, July, 1938.
42. Kemal Apak, "Turkey's Growing Role in Mid-East," *The Oil Forum,* VI, 380, Nov., 1952; Longrigg, *Oil in the Middle East,* 238; *The Christian Science Monitor,* Mar. 14, 1951. The general reserves of Ramandag have been estimated as 80 million tons, *News From Turkey,* IV, 3, Mar. 27, 1950, and as between 50 and 100 million barrels, Apak, *loc. cit.;* at the end of 1954 it was estimated at only 25,000,000 barrels. *World Oil,* CXL, 205, Feb. 15, 1955.
43. UN, *Summary 1952–53,* 43; *Middle East Report,* VI, Aug. 23, 1954; *Mideast Mirror,* 17, May 30, 1953.
44. *The New York Times,* Mar. 8, 1954; B. Orchard Lisle, "Turkey Offers Self-Determination of Capital Writeoff," *The Oil Forum,* VIII, 163, 172, May, 1954; *Foreign Commerce Weekly,* LII, 19, Aug. 16, 1954.
45. *World Petroleum Report,* X, 78, Mar. 15, 1964.
46. *Petroleum Press Service,* XXXV, 377–379, Oct., 1968; *World Oil,* CLXVII,

201, Aug. 15, 1968. The pipeline belonged to the Government company.
47. *World Petroleum Report,* X, 78, Mar. 15, 1964; — XI, 86, Mar. 15, 1965; —
XII, 84, 1966; *World Oil,* CLXI, 166, Aug. 15, 1965; — CLXXV, 60, Aug. 15,
1972.
48. *Petroleum Press Service,* XXXV, 377—379, Oct., 1968; — XXXVIII, 194, May,
1971.

Section Seven

CONCLUSIONS

Chapter XXV

MIDDLE EAST COLLECTIVE INSTRUMENTALITIES

Given the generally underdeveloped nature of the area, the somewhat conservative and primitive nature of the society, the lack of political experience and administrative skills, and on the other hand a sudden influx of sophisticated technological innovations with a simultaneous outpouring of wealth in monetary units calculated in astronomical figures, when within one generation the region had practically emerged in a number of separate sovereignties, it was bound to happen, at some stage in the development, that attempts would be made to unite against the outsiders. No matter how much wealth and prosperity the outsiders might have brought with them, they were always under suspicion, for they must have amassed for themselves much more than they gave up and have cheated and wronged the natives.

The most plausible answer to this problem of being cheated would naturally be unity. If all the members of the area had united against the foreign companies they could have obtained their maximum rights and prevented the foreigners from exploiting, for their own ends, the God-given treasures. But unity presupposed a number of factors: that there was a sense of common objective, that there existed a willingness to make sacrifices for the sake of a common cause, that the members of the union possessed the ability for sustained planned effort to achieve the distant goal, and that there existed suitable, able and skillful leaders who were recognized by all members as the personification of the ideal of unity and were trusted by all. These prerequisites did not always exist in the different moves for unity.

Since 1959, three collective instrumentalities were set up in the Middle East for the over-all purpose of uniting against the oil companies: the Arab Petroleum Congress, the Organization of Oil Exporting Countries (OPEC) and the Organization of Arab Oil Exporting Countries (OAPEC). Each had different specific objectives and as a result employed different methods and structures in efforts to attain its goals.

ARAB PETROLEUM CONGRESS

First in chronological order was the Arab Petroleum Congress. Ever since Middle East oil became an important issue in world affairs, a decisive if not vital commodity for the Western Powers and hence of great international political value, the professional leadership of the Arab League had tried to bring Arab oil under the League's political wings. Had they succeeded, the League would have become a significant factor in world affairs. The countries that had the oil, however, stubbornly and consistently refused to allow it to become a political football of Arab politicians. Although the League had a committee on oil, it never functioned because the oil-producing centers — Saudi Arabia and Iraq at the time — jealously kept the Arab League leadership, Egypt, out of the oil business.

Beginning with the Iraq revolution in July 1958, a number of changes took place. Iraq felt, at first, that through united Arab efforts she might perhaps get better terms from the Iraq Petroleum Company. In Saudi Arabia, Amir Feisal became the dominant figure and a rapproachement between his country and the leader of the United Arab Republic took place. And, Abdullah at-Tariqi, Saudi Arabia's Director General of Petroleum and Mineral Affairs, began to agitate for greater Arab participation in the petroleum industry and to call for greater concessions from the foreign companies; finally there was the cut in oil prices, which amounted to a 10% decrease in revenue. These factors opened a limited area for Arab League activity in the oil arena.

As early as January 15, 1959, Iraq called for some united Arab action. It was pointed out that the various concessionary companies — although supposedly competing with one another — were actually pursuing a uniform policy towards the Arab governments, while the latter were following separate and different policies towards the companies, policies that worked against Arab interests. To remedy the situation, Iraq suggested that the Arab countries should, on the basis of broad-line agreement, arrive at a uniform policy for the exploitation of their resources. Said Radio Baghdad: "The Iraqi Government proposes the discussion of this subject in the Arab Economic Council at its current session. It wishes this subject to be included in the agenda with a view to studying the possibility of forming a committee of legal, technical and economic experts, who would draft the proposed agreement in light of the oil economy in the Arab world, and whatever new developments have taken place in the relations between oil companies and producing countries."

The Arab League Secretariat invited the member-states, and Kuwait, Bahrain and Qatar to name representatives to attend the sessions of the technical committee. On April 16, the chief of the Arab League's petroleum bureau, Muhammad Salman, opened the first Arab Petroleum Congress. Five hundred and thirty-five delegates attended. They represented the members of the League, Qatar, and Kuwait; Venezuela and Iran sent observers; thirty oil companies were represented. Jordan and Tunisia were absent, and Iraq paradoxically did not attend because of her tension with the UAR.

The Congress closed after seven days, with the release of an official communiqué summarizing the results of the meetings.[1] The experiences of both the Arab and oil company delegates were instructive and sobering; no major changes either in royalty arrangements or in the basis of calculation were recommended.

Interestingly enough, not one of the Arab delegates raised the threat of nationalization, nor were any threats made of unilateral modification of the concession agreements. The only exception was the American legal adviser to the Saudi Arabian Directorate of Petroleum and Mineral Affairs, Frank Hendryx, who read a paper entitled "A Sovereign Nation's Legal Ability to Make and Abide by a Petroleum Concession Contract," which sought to prove that the Arab governements could legally nullify unilaterally some of the provisions in the agreements. While this thesis greatly upset the companies, especially Aramco, for whose benefit the paper was obviously intended, it disturbed even more the Arab delegates, who were the severest critics of Hendryx, and they disavowed him.

The Congress' proposals were of two categories: those affecting relations between the governments and the companies and those affecting inter-government relations. The first category demanded greater representation for the governments in the actual management of the companies; more Arabs trained for administrative and executive positions; relinquishment of areas not exploited and for which no rent was paid; no drop in the price of oil without prior consultation with the governments concerned.

The second category recommended the formation of national oil companies which would embrace all the phases of the industry, "side by side with the private oil companies operating the states," and a unified policy on the exploitation and preservation of oil resources.[2]

Because of the very nature of the Congress the recommendations were just that — recommendations. At best the Congress could be described as a kind of discussion not binding on the participants.

For even after the Arab League Petroleum Experts Committee met in Jidda in October, and worked the recommendations of the Cairo Congress into recommendations of the League Economic Council, they were not binding on the League members, and Kuwait, a major oil producer, was not a member.

Second Congress

In the meantime the Organization of Oil Exporting Countries came into being. The Arab League, however, was not willing to give up its established base and mode of operation. At the League meetings all members participated as well as delegations from the companies and from foreign countries; the meetings were open and could serve as a great platform for the promotion of League objectives. The League, therefore, convened the second Arab Petroleum Congress in Beirut, and it met from October 17 to 22, 1960.[3] It was attended by some 600 delegates and observers the latter from the companies operating in the area, from some South American governments, from Eastern European and Eurpoean countries, and from Indonesia and Japan. Iran, because of the tension between her and the UAR, did not send observers.[4]

The major items of the agenda were about the same as at the first Congress. The burning issue was, of course, the second cut in prices, made by the companies without previous consultation with the governments concerned, in defiance of the resolution adopted by the first Congress.

Saudi Arabia, through Abdullah at-Tariqi and Frank Hendryx, lashed out against the companies, the former pressing for the integrated company. To back up at-Tariqi's demands, Hendryx repeated his doctrine of unilateral modification of concession terms by governments. Company representatives took serious objection to at-Tariqi's charges and the Hendryx doctrine. The Arab delegates were divided on the doctrine; some maintained that Islamic law forbade the breaking of contracts, others held that since the doctrine was based on considerations of public interest, it was within Islamic law.

The Congress produced two surprises. On October 20, E.P. Gurov, chairman of the Soviet Oil Export Organization and head of the Soviet delegation, replied to a paper, "Russia's Potential in Future Oil Markets," read by Theodore Shabad of Paris, which charged that western markets were being flooded with cheap Russian oil. Gurov stated that while it could not be said that Soviet oil was flooding Western European markets (during the period 1956–59 Russian had sold Western Europe only 24 million tons, which was about 4% of that region's total imports), the USSR was determined to recapture its pre-war place among the oil-exporting nations. He told the Congress that the Soviet Union would compete in the Middle East's traditional oil markets in Western Europe and that it would not help with the maintenance of artificially high prices.[5] Gurov's statement was as much a shock to the Arab oil-procuding countries as it was a threat to western companies. Price cuts were now inevitable and the future looked menacing indeed.

The second surprise was the nature of the four resolutions adopted. These were, of course, not binding on anyone, including the countries by whose delegates they were adopted; they were merely expressions of the chiefs of the Arab delegations. The first resolution declared that the Congress upheld the demands of the Arab countries in their efforts to improve the terms of the petroleum concessions. The second disavowed "the resort of the petroleum companies to a reduction of prices of crude oil and its products without the consent of the Arab producing countries." The third recommended that the Arab governments double their efforts for greater Arab participation in the activities of the Petroleum Congress. The fourth resolution proposed that the Congress be convened "every year on Monday of the third week of October," and that the third Congress be held in Cairo.[6]

Third Congress

In spite of the assertion by the semi-official Egyptian *Al-Gumhuriyah* on October 20, 1961, that "one of the main aims sought by the Arab peoples in which the Arab oil Congresses are expected to exert coordinated efforts to accomplish is that Arab oil be freed from the domination of monopoly, imperialism and reactionism" the third Arab Petroleum Congress held in Alexandria from October 16 to 21, 1961, was, comparatively speaking, the friendliest as regards the treatment of the westerm concessionaire companies, of any of the previous Congresses. The threat which the Soviet Union represented to Middle East oil hung heavily over the Congress and although no formal resolution or recommendation on it was adopted, it became the main issue of the proceedings; all other considerations were secondary.

Some six hundred delegates participated. The major topics were issues between the producing countries and the concessionaire companies, and technical methods of prospecting for and transporting and refining oil.

The general atmosphere of the Congress was one of cooperation rather than hostility. The western oil companies' delegates described the Congress as a "forum and not a battleground," and the Egyptian chairman of the Congress described it as a straight interchange of views between the countries and the oil companies. The Soviet Union was the major target of the Congress; it was attacked by a number of delegates, and especially by Lebanon's Emile Bustany, for dumping oil in the Western European market and thus depressing the prices of Arab oil. The Russians replied with a paper on the development of their oil industry and their attempts to regain their rightful position in the international oil market.

On the last day the Congress adopted ten resolutions, among them one backing the demand of the Arab countries for the restoration of the price cuts.

Fourth Congress

The fourth Congress was held on November 5–12, 1963 in Beirut. Company representatives announced at the very outset of the proceedings that they would not participate in open-panel discussions dealing with royalty payments and division of profits as these were, at the time, under negotiation between themselves and the governments. Among the five resolutions adopted at the conclusion of the session on November 12, one supported OPEC's demands on royalties and the restoration of crude oil prices to their pre-August 1960 level.

Two noteworthy points were made at the Congress. One was by the Venezuelan delegate, Francisco Para, who argued for administrative prices for crude oil to be fixed by the OPEC guidelines; he maintained that it was for the governments and not the commercial companies to operate a price cartel. He pleaded that prices should be free from violent fluctuations and should be reasonably predictable for a number of years. The other point was made by the companies' delegate B.A.C. Sweet-Scoot of British Petroleum, who called attention to the fact that until some 10 years previously the companies formed most of their investment capital from retained profits; by the end of the 'fifties there was a sharp drop in the earnings, and recourse had, therefore, to be made to the capital markets for development investment resources.[7]

Fifth Congress

The fifth Congress met in Cairo on March 16–23, 1965; 928 delegates and observers attended. Of the total number of participants 400 were Egyptians. The tone of the Congress was much more rhetorical and political than of the previous Congresses. It became clear that whatever tangible achievements could be attained would have to come from OPEC and that the Congress was to be utilized for propaganda purposes. Ahmad Kamel el Badri, chairman of the Congress, declared:

"Arab petroleum is today, as it has ever been, the axis and object of all the conspiracies plotted by the alliance of colonialism and Zionism . . . .The weapon that Arab petroleum represents can be redirected towards the heart of Zionism and colonialism, should they ever be tempted to commit any new acts of aggression."

The Congress resolutions supported the basic demands of OPEC: expensing royalties and restoration of the price cuts, and all the other regular demands formulated by the first Congress. It also recommended that "petroleum and other economic potentials should be utilized as an effective weapon in the struggle of the Arab nation to liberate Palestine, and that the relations of the Arab states with all other states be determined in the light of their attitude to the cause of Palestine." Other recommendations were the establishment of an Arab Petroleum Organization, an Arab Petroleum Research Institute, and an Arab Petroleum Company. Abdullah at-Tariqi. no longer Saudi Arabian Oil Minister, but a private oil consultant in Beirut, delivered a paper, "Nationalization of Arab Petroleum Industry is a National Necessity."[8]

Sixth Congress

The sixth Congress opened on March 6, 1967, in Baghdad and lasted 7 days. The tone of this delayed Congress was much more moderate than its predecessor; it was dominated by Arab professional oil men wrestling with real practical problems. Almost all the papers read were by Arabs, and they were the major participants in the discussions. Technical problems of the industry in addition to the economic and financial issues were integral parts of the Congress' operation. The resolutions passed were along the old lines, with stress on Arab participation in and control of the technical phases of the industry. Of course, OPEC demands were supported.[9]

Seventh Congress

Although originally scheduled to be held in the first quarter of 1969, the seventh Congress was convened in Kuwait only in March 1970, and it sat from the 16th to the 23rd.

By then some of the Middle Eastern oil-producing countries had made considerable progress in the areas of new concession agreements, relinquishment of unexploited areas, establishment of national oil companies, local refining and distribution, utilization of natural gas and the development of the petrochemical industry; in addition, the practice of expensing royalties was almost universally adopted. What was the Congress to demand? The recommendations adopted sought greater gains in all these areas for the producing countries. They were summarized in three categories: 1) support for Arab national oil companies; 2) progressive relinquishment in all concession agreements, and greater participation by Arabs in administrative, technical and professional positions; and 3) relating prices of petroleum and its products to the prices of imported manufactured goods.[10]

Eighth Congress

The eighth Congress was convened in Algiers from May 28 to June 3, 1972, and the major preoccupation of the delegates was, naturally, the nationalization of IPC by Iraq. While the ranking level of the delegates was much lower than that of the previous Congresses, the nationalization advocates had their day. Abdullah at-Tariqi made his old request, but this time with greater assurance, and, to wild applause from his audience, he proclaimed: "Foreigners have no rights on our territories or to our national resources which are the exclusive ownership of our peoples." It was a triumphant Congress for Arab achievements. Most of the recommendations adopted dealt with nationalization.[11]

This was a long and successful way from the first Congress when nationalization was not even mentioned as a threat. But were all or even part of the achievements of the Middle East producing countries, since the first Arab Petroleum Congress, the result of the united efforts of the Arab collective instrumentality? To be sure, the Congress was an excellent public relations platform from which the Arab radical articulate leadership could stage its propaganda slogans and watch the reactions of the producing companies and of the political world beyond. The concessionaire companies' mode of operation made the defense of the position of the producing countries at these Congresses imperative and that gave these public conferences a real battleground atmosphere. The real negotiations, however, were not between the companies and the Congresses but between the individual companies and the individual governments, and later through OPEC. The Congress had no executive body, no permanent continuity and no collective executive power. It was technically a body of the Arab League Economic Council, and could make recommendations. But the Arab oil-producing countries never submitted their oil interest to the jurisdiction of the Arab League.

The Congress began as and remained a public forum where oil issues could be raised and aired without any real responsibility for the nature of the developments of these issues. The Congress functioned within the inherent limitations, and as objective conditions changed in favor of the producing countries, the Congress functioned more demonstratively.

ORGANIZATION OF OIL EXPORTING COUNTRIES

The birth and successful growth of the Organization of Oil Exporting Countries, which had ultimately become a real force in the international oil industry, was brought about by purely economic considerations, and its life and strength were dominated by economic determinants.

On February 13, 1959, the British Petroleum Company announced a cut of 18 cents per barrel on its Middle East oil, the other companies operating in the region soon followed suit. This price reduction amounted to more than a 10% decrease in the total income from royalties, as compared with 1958.[12]

On August 8, 1960, Esso Export Corporation, an affiliate of Standard Oil Company (New Jersey), announced cuts in posted Middle East prices ranging from 4 to 14 cents a barrel, averaging about 10 cents. Three days later British Petroleum Trading, an affiliate of BP, announced a reduction in prices of refined products at the Abadan refinery; on the following day Royal Dutch-Shell reduced its posted prices for Middle East oil to bring them in line with the prices of the other two companies. By the end of the month most of the companies operating in the Middle East had cut their prices, on an average of 6%.

Iran, attempting to regain, at the time, her position of Middle East producer which she had held prior to nationalization, and experiencing financial budgetary difficulties, was greatly exercised over the price cuts. The Shah felt that although these might have been unavoidable, the companies should have consulted with the government before announcing them. Kuwait, the greatest Middle Eastern oil producer at the time, with the highest revenue from oil and the smallest population, was the harshest critic of the cuts. Her Oil Affairs Bureau declared on August 24: "The Government has great interest in this matter and is dissatisfied that such steps were taken by oil distributors without reference to producing countries. Companies selling oil should not look to their own interest alone but should adopt a co-operation policy with producing countries." [13]

The companies maintained that to sell in the international market the Middle East oil they produced, they must cut prices to meet competition; the exigencies of the market did not allow for previous consultation and negotiations. The major producing countries failed to see any justification for the price cuts and their reaction was soon forthcoming.

On August 25th the Iraqi Oil Ministry invited representatives of Saudi Arabia, Kuwait, Qatar, Iran and Venezuela to meet in Baghdad on September 10 to adopt a "unified attitude" against oil companies that reduced prices "unilaterally and without consultations." [14] All accepted the invitation; Qatar, however, came only as an observer. The meeting lasted from the 10th to the 14th and adopted the following resolutions, which were to become effective after they had been approved by the respective governments: cancel the price cuts; companies to consult with governments concerned before cutting prices, should this again become necessary; devise a method, with proper safeguards, for controlling production in order to maintain prices; members to stand united against any efforts of the companies to divide them; establish, with themselves as founding members, a permanent organization with a secretariat; unify oil policies of member-countries and work out the best means for safeguarding the specific interests of each member.

By the end of October all the member-governments had ratified the resolutions and the Organization of Petroleum Exporting Countries (OPEC) came into formal being. However, the inherent difficulties of converting the objectives of the resolutions adopted into an operative organization were not easily overcome. If prices were to be maintained at the pre-cut level, production would have to be curtailed; this curtailment and future production would have to be based on a pro-rating method. The question would then arise: what criterion should be employed? Was it to be petroleum resources? This would have given a high priority, at the time, to Kuwait and a very low priority to Venezuela. Was it to be population? This would have given priorities to Iran and Venezuela against Iraq and Kuwait. Was it to be the share of oil revenue in national budgets? This would have favored Kuwait and Saudi Arabia against Iraq and Iran. What means could the Organization command to enforce, should it adopt, pro-rating decisions?

If pro-rating of production should be adopted in order to maintain high prices for crude, exploration and extension of exploitation or reserves would practically come to a standstill in the pro-rated countries and, conversely, would expand in the non-pro-rated countries, even though the lands of OPEC members contained a very high percentage of the world's reserves.

Second Conference

The second Conference of OPEC met in Caracas on January 16, 1961. President Romulo Betancourt of Venezuela, the host, declared at the opening session that the Conference's principal aims were cancellation of the price cuts and devising of means to regulate production. These were, in fact, the same objectives that had occupied the members of the Organization at its first conference in Baghdad. However, while the Middle Eastern producing countries were unanimous on restoration of prices, the regulation of production was primarily, if not exclusively, a Venezuelan objective.

Before OPEC could tackle its objectives, or even fully agree on them, it had to set up organizational and administrative machinery. This was the main item on the agenda. After adopting a series of resolutions, the Conference closed on January 21. Some three weeks later, the resolutions were made public. [15]

It was, of course, inevitable that most of the resolutions should deal with the organizational structure. OPEC was to consist of a Conference which was to meet twice a year; a Board of Governors (each founding member was to have one Governor) to meet at least four times a year; and a Secretariat. The seat of the body was to be at Geneva (later moved to Vienna). The chairman of the Board of Governors was also to serve as Secretary General, and the Secretariat was to consist of departments of Technology, Marketing, Production and Finances, Administration, Public Relations and an enforcement section. The chief Iranian delegate, Fuad Rouhani, was elected first chairman. Qatar was voted in as a new member.

The substantive aspects of the Organization, since there was no unanimous agreement on them, were only touched upon. Resolution II–12 called on each member-country to compile within two months and send to the Board of Governors a statement, supported by proper documents, of "its position in the matter of determination of prices at which its petroleum is paid for by the exporting companies." The Board of Governors was to refer these statements to competent legal advisers for consideration and recommendations for possible steps to be taken, in each case, for restoration of prices.

Another resolution called for a study by the Board of Governors of the prevailing oil prices "in order to arrive at a just pricing formula supported by a study of international pro-rating, should that prove essential."

At the first meeting of the Board of Governors, which took place early in May 1961, at Geneva, the question of Iraq-IPC negotiations was raised. According to an Iraqi report, a spirit of understanding among the OPEC members prevailed and there was full agreement in viewpoints on all issues on the agenda. The Iraq representative submitted the question of the negotiations, and the Venezuelan and other delegates supported the Iraqi request to place the subject on the agenda of the forthcoming Conference which was scheduled to be held in Teheran on August 19, 1961. However, on August 10, the chairman of the Board, Fuad Rouhani, announced in Teheran that at Iraq's request it had been decided to postpone the Conference to October.[16]

Third Conference

OPEC's third Conference opened in Teheran on October 28 with Iraq absent. The meeting closed on November 1, and the resolutions were made public on November 15. An analysis of the resolutions would reveal that the member-countries came to the conclusion that they could not attain their objectives quickly, that they were woefully uninformed on the intricate factors involved in the determination of prices and production, and above all, on the behavior of the international oil market. The Organization, therefore, decided to settle down to the task of getting information and transmitting it to the members, after which consideration, based on intelligent understanding and solid data, could be given on what action should be taken.

Fourth Conference

The fourth Conference convened on April 5, 1962 in Geneva, with Iraq again absent. The basic issues of OPEC were the same as they were when the Organization first came into being except that the arguments advanced were supposed to have been based on better information and on practical studies made by experts of the Organization. The fourth session was primarily devoted to a discussion of the studies prepared by the consultants on oil prices and profits, as compared with the profits of other industries. On April 9, the Conference was adjourned to June 4, to allow the delegates to consult with their governments. It reconvened on June 4, and adjourned on the 11th after adopting seven resolutions, most of them based on the reports of the consultants. One requested the member-states to enter into negotiations with their oil companies for the restoration of payments based on prices prevailing prior to August 1960.

Another resolution called for talks between governments and companies to establish a uniform royalty payment which would not be deductible from income tax.[17] A third urged members not to contribute towards the companies' expenses on marketing operations.[18] The Conference accepted Libya and Indonesia as new members.[19]

Fifth Conference

Because of unsuccessful negotiations between the producing countries and the companies about expensing royalties, the fifth Conference did not convene in Riyadh until November 22, 1962, and after two days adjourned to March 30, 1963. However, the Conference did not open on the set day for the same reason, and the opening was postponed to November 1963. The talks between Fuad Rouhani, Chairman of OPEC and Chairman of NIOC, and the Consortium and Aramco made no progress. A three-member committee was named, and the meeting was again postponed to December. In spite of the strong resistance of the companies it was decided to continue negotiations rather than take unilateral action. The companies estimated that if they complied with the request of the governments to expense royalties it would cost them about $300 million annually, "a sum equivalent, perhaps to one quarter of the international oil companies' total profits from all the integrated operations in the eastern hemisphere."[20]

The negotiations were shifted from the three-man committee to individual country-company units. The issue was solved by a compromise worked out by the companies and governments. All, except Iraq, agreed to the compromise, and the seventh Conference meeting at Djakarta in November–December 1964 dropped the issue from the agenda to permit members to conclude agreements with their respective companies. An extraordinary Conference was held at the end of the year in Geneva; on December 28, OPEC announced that five Middle Eastern countries, except Iraq, had accepted the latest OPEC royalty proposals.[21]

Ninth Conference

The major purpose of OPEC — restoration of the price cuts — was still to be achieved. The ninth Conference meeting in Tripoli from July 7 to 13, 1965, attempted to resort to controlled production. OPEC did not actually propose any curtailment of production; it merely formulated limited percentage increases for 1966 on an experimental basis: Venezuela 4%, Indonesia 4%, Kuwait 5%, Qatar 6%, Saudi Arabia 9%, Iraq 9%, Iran 16% and Libya 33%.[22]

Needless to say, the countries concerned did not abide by this modest proposal nor could OPEC enforce its terms. In fact, on February 14, 1966, Ahmad Zaki Yamani, the Saudi Arabian Oil Minister, declared that OPEC's crude oil production controls were unacceptable to his country. The actual percentage increases for 1966 were: Iran 11%, Iraq 6%, Kuwait 5%, Qatar 25% and Saudi Arabia 18%. The total percentage increase for the Middle East over 1965 was 12, while the five-year 1961/1966 average was 10.5.[23] By June 1967, OPEC gave up the attempt of production regulation.

After the major Middle Eastern oil-producing countries had established national oil companies, OPEC leaders realized the potential danger which these companies represented to the international oil-price structure. As the national companies became producers searching for new markets they were bound to enter into competition with one another and even with the established foreign companies through price cuts. An effort was, therefore, made to get together all the national companies and prevent them from resorting to unbridled competition.

In September 1966, delegates from the national oil companies of Iran, Iraq, Saudi Arabia, Kuwait, Indonesia and Venezuela met in Caracas. They all agreed to coordinate the marketing efforts of their companies, to exchange non-confidential information, and to meet once a year and consult together before taking any decisions which might be prejudicial to the interests of other national companies.[24]

After the 1967 Arab-Israel war and the subsequent closure of the Suez Canal, and the general increase in world demand for oil, the basic relationship pattern between the governments and the companies changed, and as a result the role of OPEC changed. At the thirteenth Conference, which met in Rome from September 15 to 17, 1967, the major issue was Iraq and Libya's demand for higher posted prices for their Mediterranean oil; Persian Gulf oil was costlier because of increase in tanker rates. OPEC pledged support to both. Abu Dhabi and the Sultanate of Muscat and Oman joined the Organization.[25]

The issue of royalty expensing was only partially resolved. While the companies conceded the principle of expensing they exacted for it a sliding discount allowance.[26] The 1966 allowance was 6½%; and no provision was made for the subsequent years. An extraordinary Conference was, therefore, convened in Beirut on January 8- 9, 1968, to deal with the deadlocked negotiations. Two days before the Conference opened, the companies submitted a new offer, and the Conference decided that member-countries should accept the new proposal.[27]

Sixteenth Conference

The sixteenth Conference met in Vienna in June 1968. The major item under consideration was the Governments' profit share. OPEC issued a Declaratory Statement of Petroleum Policy which recommended the principle of renegotiating agreements in cases of excessively high net earnings by the companies. The other issue was participation of producing countries in the ownership of concessions. The most ardent advocate of participation was Saudi Arabia's Ahmad Zaki Yamani.[28]

The third meeting of the national oil companies was held in Teheran in October 1968. The objective was avoidance of price disruption by some form of coordination and liaison among themselves.[29]

In 1969, Algeria joined OPEC, which increased the oil output of all OPEC members to 48% of world production and 59% of total free-world production.

The twentieth Conference was held in Algiers from June 24 to 26, 1970. The three outstanding items on the agenda were: participation, production program and the integration of the oil industry in the national economies of the producing states.

OPEC's report declared that 1970 marked a turning point in the history of the oil industry. It witnessed major shifts in the price structure and in bargaining positions between governments and companies. According to OPEC, economic and political forces had changed the international buyers' market to a sellers' market.

As regards production, OPEC saw two immediate factors which affected the situation: Libya's curtailment of production and the closure of Tapline; these two events reduced seriously the supply of oil to Europe from the Mediterranean, the consequence of which was an increase in posted prices for all Middle Eastern producing countries; and the government's share of profits was increased by 5%. The basic aims of OPEC were being realized. [30]

The twenty-first Conference was held in Caracas on December 9–12, 1970. It adopted resolution XXI–120, which declared 55% as the governments' share of profits; provided for general uniform increase in posted prices, the elimination of allowances granted to oil companies from posted price and prescribed methods of achieving these ends. The fruits of this resolution were the Teheran agreement of February 1971 and the subsequent Tripoli accord. [31]

After the Teheran agreement for the Persian Gulf area and the Tripoli agreement for the Mediterranean outlets established the higher posted rates, and the governments' higher profit share, OPEC launched its next campaign for government participation in company ownership. In an interview with the editor of *Petroleum Times*, before the scheduled Conference, Nadhim Pachachi, the Secretary-General of OPEC, stated, in reply to a question as to what the major immediate objective of OPEC was: "We, the producing states, believe that, at some future date, we should become more involved in the downstream operations of the oil industry. The achievement of this objective is currently the major part of the work of OPEC. So far we have made no progress in achieving this objective but we have been studying the subject fairly strenuously and there is a ministerial committee composed of the Ministers of Iran, Iraq, Kuwait, Libya and Saudi Arabia, who will meet before the next annual conference which is to be held in Vienna on July 12, to decide on what form our participation in downstream operation might take. This objective is the most important aim of OPEC, now that the posted prices issue has been settled." [32]

In July 1971, the twenty-third and twenty-fourth Conferences were held in Vienna. In addition to the issue of participation the Conference took up the question of planned production, the details of which were not made public. It reportedly provided for compensation to producers which would reduce output for the main marketing regions in times of price weaknesses.

Nigeria was admitted to membership at the twenty-third Conference. Saudi Arabian researchers estimated that at the end of 1970 the reserves of OPEC countries, including Nigeria, amounted to 70% of total world reserves; production was 49.1 of total world output.

Resolution XXIV–135 requested the implementation of the participation principle, and appointed a ministerial committee consisting of Iran, Iraq, Kuwait, Libya and Saudi Arabia to draw up a practical plan for implementation to be submitted to the extraordinary Conference to be held on September 22, 1971. [33]

Meanwhile the devaluation of the dollar became an acute issue and a subject of long negotiations between governments and companies. An extraordinary twenty-fifth Conference was convened in Beirut on September 22, 1971, but it lasted only two hours, and it adopted two resolutions, one on government participation and the other on dollar parity in oil revenues.

Some of the leaders of OPEC argued that the principle of changed circum-stances *(res sic stantibus)* justified nationalization, and that should be the im-mediate objective. Three distinct positions were taken on this issue. Abu Dhabi, Iran, Iraq, Qatar and Saudi Arabia favored a 20% shareholding in operating com-panies; Libya and Algeria called for a minimum of 51% or full nationalization; Indonesia, Nigeria and Venezuela did not support either position. Hence the un-usual brevity of the specially convened Conference.

Resolution XXV–139 called on member-countries to institute negotiations with their respective companies individually or in groups to attain participation.

Resolution XXV–140 urged all members to negotiate individually or in groups with the oil companies to achieve adjustments in payments in view of the devalu-ation of the dollar. [34]

Hectic negotiations began between the governments and the companies. First the Iranian Finance Minister met a delegation of the companies in Teheran on October 30, and they continued discussions on November 8 and 9, without reaching a conclusion. The negotiations then moved to Vienna on November 22. The companies were beset by a sense of insecurity, for only a number of months previously they concluded the Teheran and Tripoli agreement which they thought guaranteed them supplies and stability for at least five years. [35]

The twenty-sixth Conference opened in Abu Dhabi on December 7, 1971, with the two issues of the twenty-fifth Conference unresolved. It therefore resolved to meet at Geneva on January 12, 1972, with representatives of the companies to resolve the two issues. The position of the governments stiffened and the Secretary-General of OPEC, Pachachi, declared: "Our recent demand for government participation in existing oil concesssions is based on our conviction that such direct participation would reinforce and render more effective our right of permanent sovereignty over our resources." [36] There must have been some sort of understanding between the OPEC leadership of the Gulf states and the companies prior to the Geneva meeting. For on January 20, 1972, it was announced in Geneva that the two sides had arrived at a solution. [37]

The question of participation was not so easily resolved. The Secretary-General of OPEC intimated a possible stoppage of the flow of oil should the companies persist in their unyielding position. After the Geneva accord was reached he de-clared: "Prompt implementation of our right to take active participation in the existing concessions would be not only in the interest of our people but also in the interest of the companies concerned, as well as the countries whose economies depend on uninterrupted flow of OPEC oil." [38]

The pressures began to bear fruit, and the OPEC goal of participation was on its way to realization. For on March 12, 1972, while the extraordinary Conference of OPEC was being convened in Beirut to deal with the participation question, Saudi Arabian Oil Minister, Ahmed Zaki Yamani, announced that Aramco had accepted the principle of a minumum of 20% participation by the Saudi Government. The Conference subsequently urged all oil companies to follow the lead of Aramco. Later it was reported that the Iraq and the Kuwait companies agreed to government participation. Meanwhile OPEC was called upon to act on Iraq's nationalization of IPC. An extraordinary Conference convened in Beirut in June 1972 and adopted a unanimous resolution supporting Iraq's move. [39]

The principle of participation was accepted by the companies but implementation became the subject of the twenty-ninth Conference, which met in Vienna in the latter part of June 1972. A number of aspects of participation had to be clarified: the percentage which the governments would obtain, the basis of payment for their shares, and, of no less importance, the methods of disposing of the governments' share of crude after participation. The Conference rejected the companies' proposals, and it declared that it was "determined to achieve participation, and in case of failure of negotiations definite concerted action" would be taken by member-countries. [40]

Ahmad Zaki Yamani, who negotiated participation for the Persian Gulf countries, warned the companies that the alternative to participation was nationalization, for which the Arab public was ready. [41]

OPEC was from the very beginning, in its methods of operation and in its institutional structure, neither political nor a propaganda instrumentality. Its primary aim was economic, but the targets changed with changing conditions. These changing conditions made its achievements possible. Its meetings were closed and its resolutions were not made public until they were approved by member-states. Its research facilities were adequate, and its negotiations were persistent. For a long time it could make no progress in its primary aim because objective conditions made the companies strong vis-à-vis the governments. But as soon as the buyers' market changed to a sellers' market OPEC began to advance in royalty expensing, in increases in posted prices, in increases in profit sharing and in participation.

ORGANIZATION OF ARAB PETROLEUM EXPORTING COUNTRIES

The Organization of Arab Petroleum Exporting Countries (OAPEC) is the youngest of the united-effort instrumentalities, and its aim is not easily defined. It was organized in January 1968 in the face of the consequences for Middle East oil of the Arab-Israel war in 1967. The conveners of the first meeting in Beirut, Saudi Arabia and Libya, were the greatest victims of the stoppage of oil production during the war and apparently attempted to remove the oil from political interference in any future events. The requirements for membership ruled out Egypt and Algeria, and eliminated all the Arab countries of the Arab League which were either non-producers or minor exporters. Iraq did qualify but was not interested in joining, which because of her radical orientation would not have been welcome anyway.

The headquarters of OAPEC was to be Kuwait. OPEC, reacting to charges that the new organization might negate the old one, described it as a significant and potentially far-reaching development of 1968, and declared that "the establishment of OAPEC does not in any way affect the status of its members in OPEC." [42]

The first meeting was held in Kuwait on September 9–11, 1968. Admission of new members had to be approved unanimously by the founder-members. It declared that it was to be non-political, and its aim was to strengthen cooperation between and protection of member-states. It hoped to unify efforts to insure the flow of oil to its consumption markets at fair and reasonable terms, and thus create a favorable climate for the capital and enterprise investment in the oil industry of member-states. The new organization was empowered to conclude agreements for commercial transactions and for the establishment of joint-venture undertakings. It had all the earmarks of an effort on the one hand to protect its members from non-producing Arab countries, and on the other hand guarantee the western companies not only the uninterrupted flow of the oil but also the protection of the necessary investments for the development of the industry.

The structure of OAPEC was to have been a council composed of the oil ministers of the member-states, an executive bureau, a secretariat and a legal committee or court to deal with disputes. [43]

However, with the political instability of the Arab Middle East, pressure against the new organization began to build up. The qualifications of membership were eased and in May 1970 Algeria, Dubai, Bahrain and Abu Dhabi joined. Membership was expanded to eight; Iraq still refused to join. [44]

In June 1971 the council of oil ministers met in Kuwait, and a serious dispute arose over the membership application of Iraq, which threatened to break up the Organization. Since OAPEC had been organized a basic change had taken place in Libya, from a conservative monarchy to a radical republic. Libya and Algeria sought to include Iraq in the Organization, which would enable the three radical states to dominate the policies of OAPEC, while Saudi Arabia sought to keep Iraq out; the small Persian Gulf shaikhdoms were standing on the sidelines. Saudi Arabia, at least at the time, was adamant. Ahmad Zaki Yamani expressed the hope that the Organization would "continue to fulfill the goals for which it had been created."

From Beirut it was reported that the real difference between the two sides was using Arab oil as a means of pressure to deter the West from supporting Israel advocated by Libya and Algeria and vigorously opposed by Saudi Arabia, which maintained that the basic purpose of the Organization was to keep the oil question out of politics. [45]

However, Saudi Arabia could not hold out indefinitely and at the end of 1971 relented. The projected joint-undertakings were an Arab tanker fleet, a dry dock in the Gulf and a general petroleum service company for ancillary services for oil-operating states. Later an agreement was concluded to establish a joint-tanker fleet which would have the nationality of all members. [46]

On June 11, 1972, an emergency meeting of the Organization took place in Beirut and adopted a resolution supporting Iraq. The Finance Ministers met in Baghdad, and recommended that the Arab countries extend financial aid to Iraq and Syria in their difficulties with the nationalization of the IPC and the pipelines. Iraq was to obtain £53,900,000 for a period of three months after nationalization, and Syria £6,800,000. However, since OAPEC had no means of its own it could only have recommended that individual members extend loans to these two Arab states. Accordingly Kuwait granted Iraq a KD 11.6 million loan and Syria a KD 1.5 million loan. [47]

While the objective of OAPEC was clear when it was established, the changed circumstances have, in a sense, invalidated its aims, and, at best, it could only compete with OPEC. As to its Arab collective functioning, the determination to work together, under pressure to find markets and outlets for ever-increasing supplies for each member, would indeed be very challenging.

NOTES

1. The full text is reproduced in *Middle Eastern Affairs*, X, 238, June–July, 1959.
2. *Ibid.*
3. Against the belief expressed by many observers in the Middle East that OPEC would weaken the Arab League's oil efforts, Muhammad Salman declared on Oct. 7 that OPEC would work "side by side" with the League and would "in fact strengthen the League's oil role." *New York Times*, Oct. 8, 1960.
4. *Arab News and Views* of Nov. 1, 1960, reported that Iran participated in the second Congress as an observer. However, on July 28, Radio Cairo had broadcast the following: "In an official note sent to the governments of its member states on July 27, the Arab League has accused Iran of being the main source of oil supplies for Israel for many years. The note made it clear that Iran would not be accepted, even as an observer, at the Arab Oil Congress which will be held in Beirut on October 12. In this note the League stated that Iran had asked to take part in this Congress, but due to its recent recognition of Israel it had become impossible for Iran to attend the Congress."
5. The similarity of assessment of prices by Gurov and the editorial "Oil Sense," in the *Egyptian Political & Economic Review* of November–December, 1960, is more than striking and would indicate a direct connection between the two. At that time Egypt was a heavy importer of oil. See footnote 13 below.
6. *Arab News and Views*, IV, Nov. 1, 1960; *Petroleum Press Service*, XXVII, 399–406, Nov., 1960. Muhammad Salman tried to give the impression, early in November, that the oil companies had seen the error of their ways and had promptly complied with Arab demands. He was reported to have said at a press conference in Beirut on Nov. 2 that the companies had agreed to seek approval

of producing countries before announcing future reductions in the price of
crude oil, and that they had promised to try to protect the royalties of pet-
roleum-producing countries from loss incurred in any future cuts. *New York
Times*, Nov. 3, 1960.

7. UN *Economic Development 1961–1963*, 63; "Issues at the Arab Congress,"
   Petroleum Press Service, XXX, 443–446, Dec., 1963; *World Petroleum*,
   XXXIV, 44–45, Dec., 1963.

8. *Petroleum Press Service*, XXXII, 166–170, May, 1965; Ruth Sheldon Knowles,
   "Arab Countries Continue to Drive for Bigger Role in Oil," *World Petroleum*,
   XXXVII, 52–54, 56, 94, June, 1965; Arab League Secretariat, *Resolutions Fifth
   Arab Petroleum Congress*. The recommendation of the establishment of an
   Arab Petroleum Company was apparently an attempt to bring Arab oil resources under one centralized control which would logically be the Arab
   League. Saudi Arabia could not countenance such a possibility . On May 28,
   1965, the Saudi Arabian Governor of OPEC, Fahd Khayyal, stated that his
   country would not participate in the proposed Arab Petroleum Company, and
   it even had reservations about the financing of the Arab petroleum research
   institute.

9. *Petroleum Press Service*, XXXIII, 274, July, 1966; – XXXIV, 124–126, Apr.,
   1967; Arab League Secretariat, *Recommendations of the Sixth Arab Petroleum
   Congress*.

10. *Petroleum Press Service*, XXXVII, 153, 177–179, Apr., 1970; *Arab Oil Re-
    view*, 20, Jan.–Mar., 1970; Arab League Secretariat, *Recommendations of the
    Seventh Arab Petroleum Congress*.

11. *Petroleum Press Service*, XXXIX, 239–240, July, 1972; *New York Times*, June
    11, 1972, INOC, *Weekly News Bulletin*, 18, June 8, 1972; Arab League Secre-
    tariat, *Recommendations of the Eighth Arab Petroleum Congress*.

12. It had been estimated that the cut in price reduced royalties in Iraq by $24
    million; Iran by $27 million; Kuwait $46 million; and Saudi Arabia $35 mil-
    lion. *World Petroleum Report*, IV, 71, Feb. 15, 1960.

13. In this connection it might be of interest to note an editorial "Oil Sense,"
    mentioned above, which appeared in the November–December, 1960 issue of
    the *Egyptian Economic & Political Review*, a publication close to both the
    UAR government and the Arab League secretariat. The editorial saw price cuts
    of petroleum, because of increased supplies from North Africa and the Soviet
    Union, as inevitable and resulting in reduced revenue to the Persian Gulf
    governments. However, it viewed the latter as a welcome means for reforms
    long needed in the area. "This [reduced revenue] should mean fewer Cadillacs
    in Arabia, and notwithstanding the effort of the Baghdad-sponsored organiz-
    ation of exporting countries to keep up the present price level, the trend seems
    likely to go the other way. An unkind critic might suggest that for all the good
    that the high prices may have brought with them, they might have been con-
    siderably lower and had a no less useful effect on the exporters' economies.
    The irrationl spending which is a feature of the economic policies of oil
    countries disposes of sums well above those that might have been lost through
    price fluctuations. More progressive governments would, therefore, have little
    difficulty through the observance of mild forms of restraint to use substantially

smaller revenues to considerably better effect than has so far been the case. One can feel little sympathy for the dilemma of some Gulf governments, and in the same order of things one can welcome a fear of the future which must sooner or later inject a great sobriety in the character and degree of spending."

14. *Middle Eastern Affairs*, XI, 222, Aug.–Sept., 1960.

15. Reproduced in *Middle Eastern Affairs*, XII, 179–84, June–July, 1961.

16. It may be recalled that Iraq was embroiled at the time in her claim to Kuwait.

17. Details are give in Chapter VII above.

18. *Ibid.*

19. UN, *Economic Development 1961–1963*, 61–63; Geoffrey Drayton, "The Travails of OPEC," *World Today*, XIX, 485–491, Nov., 1963.

20. *Petroleum Press Service*, XXXI, 44, Feb., 322–324, Sept., 1964.

21. *Petroleum Press Service*, XXXI, 470, Dec., 1964; – XXXII, 2–4, Jan., 1965.

22. "OPEC's Recipe," *Petroleum Press Service* XXXII, 328–331, Sept., 1965. These percentage increases, except for Iraq and Iran, were lower than the average percentage increases for the period 1961/66.

23. BP, *Statistical Review 1966*, 7; OPEC, *1967 Review and Record*, 4; *Petroleum Press Service*, XXXIII, 42–44, Feb., 1966, Schurr, *Middle East Oil*, 124.

24. *Petroleum Press Service*, XXXIII, 403, Mar., 1966.

25. *Petroleum Press Service*, XXXIV, 362–363, Oct., 1967.

26. Aramco and Saudi Arabia agreed to eliminate the 6½% discount off posted prices on oil shipped via pipeline to the Mediterranean in view of the differential advantage of the price of the oil. *Petroleum Press Service*, XXXIV, 363, Oct., 1967.

27. OPEC, *1968 Review and Record*, 4–5; *Petroleum Press Service*, XXXV, 44, Feb., 1968.

28. OPEC, *1968 Review and Record*, 4–5; *Petroleum Press Service*, XXXV, 363–364, Oct., 1968.

29. *Petroleum Press Service*, XXXV, 470, Dec., 1968. In 1966, at the first meeting of the national companies, the then Secretary General of OPEC and delegate of Kuwait, Ashraf T. Lutfi, proposed the establishment of a joint marketing company of the seven national oil companies, but the proposal fell on deaf ears. *Petroleum Press Service*, XXXIII, 402–3, Nov., 1966. For an insider's assessment of the achievements of OPEC during the first seven years of its existence see Ashraf T. Lutfi, *OPEC Oil* (Beirut, 1968).

30. OPEC, *1970 Review and Record*, 2.

31. *Ibid.*, 6, *World Oil*, CLXXII, 7, Feb. 1, 1971.

32. *Petroleum Times*, LXXV, 10, May 7, 1971.

33. *Petroleum Press Service*, XXXVIII, 305, Aug., 322, Sept. 1971; SAMA, *Statistical Summary*, 17, Sept., 1971.

34. *Petroleum Press Service*, XXXVIII, 403, Nov., 1971; *Petroleum Times*, LXXV, 34, Oct. 22, 1971; *World Oil*, CLXXIII, 7, Dec., 1971; *World Petroleum*, XLII, 30–31, 54, Dec., 1971; *New York Times*, Sept. 22, 1971.

35. *Petroleum Press Service*, XXXVIII, 463, Dec., 1971.

36. *Petroleum Press Service*, XXXIX, 23, Jan. 1972; *Petroleum Times*, LXXV, 9, Dec. 17, 1971. It should be noted that in September at the twenty-fifth Conference he was reported to have said that the primary objective of OPEC in

seeking ownership in the existing concessions had been to acquire technical and economic know-how. *Ibid.*, 34, Oct. 22, 1971.

37. *Petroleum Press Service*, XXXIX, 64, Feb., 1972.

38. *World Oil*, CLXXIV, 10, Mar., 1972.

39. *Petroleum Times* LXXVI, 6, Mar. 24, 1972; INOC, *Weekly News Bulletin*, 3–5, June, 1972. A provision in the resolution requested member-countries not to permit oil companies in their territories to increase output to make up the loss of supply from IPC. A similar resolution was adopted at the eighth Arab Petroleum Congress; see above.

40. INOC, *Weekly News Bulletin* 14, July 1, 1972. See Chapter XXVIII, below.

41. *Petroleum Press Service*, XXXIX, 272, July, 382, Oct., 1972; INOC, *Weekly News Bulletin*, 2, 14, July 1, 1972.

42. OPEC, *1968 Review and Record*, 3.

43. *Petroleum Press Service*, XXXVI, 46–47, Feb., 393, Oct., 1968.

44. SAMA, *Statistical Summary*, 9, Dec., 1968–Jan., 1969; – 15, Dec., 1970; *Petroleum Press Service*, XXXVII, 301, Aug., 1970; *World Petroleum*, XLI, 13, June, 1970.

45. *Petroleum Press Service*, XXXVIII, 268, July, 1971; *New York Times*, June 13, 1971.

46. *Petroleum Press Service*, XXXIX, 29, Jan., 229, June, 1972.

47. INOC, *Weekly News Bulletin,* 10, June 17, 1972; – 6, June 24, 1972; *Petroleum Press Service*, XXXIX, 390, Oct., 1972.

Chapter XXVI
THE MIDDLE EAST AND OIL

1920–1958

The Middle East as an oil-producing region as well as an area of great reserves has become increasingly significant. In 1920, the potential reserves were estimated at 5,820,000,000 barrels;[1] in 1945, De Golyer estimated them at about 18 billion barrels; three years later he scaled up the estimate to 32 billion. In 1951, the estimates were 48 billion barrels; in 1953, 61 billion;[2] and at the end of 1954, 83 billion.[3] The American Institute of Mining, Metallurgical and Petroleum Engineering estimated the reserves in 1956 as 117 billion barrels.

These figures were not only of absolute importance for the region as such, but they also indicated the steady rise of the region's share in total world reserves. By 1949, when the reserve figures had advanced staggeringly from those of 1920, the Middle East was considered to contain 41% of the total world reserves, while North America's at that time were considered to be about 38%; five years later the estimates for the Middle East went up to 56%. According to an American Institute of Mining, Metallurgical and Petroleum Engineering report early in 1956, they were 68%.[4]

Not only in estimated reserves, however, but also in actual production has the Middle East emerged as an important factor in the world oil markets, as indicated by Table XLIX.

Since 1960 the Middle East has become the major single supplier of oil in the international market. The movement of oil between the Eastern and Western Hemispheres changed completely. While in 1938 the Western Hemisphere exported 34 million tons to the Eastern Hemisphere, by 1950 this was reduced to 15 million tons, and the United States, which had been an exporter of oil, had become an importer.[5] The two major exporting regions in 1950 were the Caribbean and the Middle East, the former exporting to the Americas, West Africa, and to a limited extent Western Europe, and the latter supplying Europe, most of Asia, East Africa, and to some, though a limited, extent the Western Hemishpere. Europe had become more and more dependent on the Middle East as the source of its petroleum supply, and in 1955, 85% of its petroleum imports came from that region.[6]

By 1956, the Middle East formed a closely connected oil region. Iran, Iraq, Saudi Arabia, Kuwait, Bahrain, and Qatar were the producers; Syria, Lebanon and Jordan (Israel was temporarily out of the picture) were the transporters through which the oil was shipped by pipeline to the Mediterranean, along with Egypt, through whose Suez Canal the oil moved by tanker.

Although the refining capacity had expanded with new refineries in Kuwait, Saudi Arabia, Lebanon and Iraq, and the expansion of existing refineries in Egypt, all were for home consumption; and the Bahrain refinery, Ras Tanura in Saudi

Table XLIX. *World and Middle East Oil Production*
Selected Years

Year	World Production (in tons)	Middle East Production (in tons)	Per Cent of Column 3 to Column 2
1930	195,000,000	6,340,000	3.3
1935	225,000,000	11,490,000	5.0
1940	289,000,000	13,900,000	4.8
1945	356,000,000	26,550,000	7.5
1948	465,000,000	56,910,000	12.2
1950	518,000,000	86,600,000	16.8
1951	582,000,000	96,190,000	16.5
1952	605,000,000	104,440,000	19.0
1953	654,000,000	121,620,000	20.1
1954	681,000,000	136,000,000	20.0
1955	768,790,000	162,468,000	21.26
1956	841,650,000	172,549,000	20.90
1957	882,200,000	177,900,000	20.20
1958	905,200,000	214,300,000	23.66[a]

a. Longrigg, *Oil in the Middle East*, 276–277; *Le Commerce du Levant*, June 3, 1954; Mar. 9, 1955; – UN. *Review 1949–50*, 26; – *Summary 1950–51*, 17; – *Review 1951–52*, 53, 55; – *Summary 1952–53*, 39–40; – *Economic Developments 1955–1956*, 47; *World Oil*, CXL, 198, Feb. 15, 1955; *Petroleum Press Service*, XXIV, 44, Feb., 1957; *World Oil*, CXLVI, 196, Feb. 15, 1958; – CXLVIII, 108, Feb. 15, 1959.

Arabia, and British Petroleum's in Aden were for export. The general trend had been reversed. While before the war Europe imported mainly refined petroleum products, the major portion of its petroleum imports – and a great part even of Asia's imports – had been crude.[7]

The development of the oil industry, both in terms of production and in terms of transportation, introduced an economic element in the Middle East which had far-reaching consequences. The industry brought with it modern technology and gave employment to a force, at the end of 1955, of some 150,000 natives and foreigners.[8] It trained natives for positions which, in accordance with the policies of the companies, could be filled by them. The absolute number of such employees in the total population of the area might have been very small, but the impact of their technological training combined with all the different operations of the various companies, had had a decisive effect on the existing social patterns and economic structures. The companies' investments in exploration, exploitation, refining and transportation, and the other indirect activities ran into hundreds of millions of dollars and affected the different states and territories in different ways.[9]

The greatest and most obvious impact had come from the direct payments to the governments for the oil produced. In the period 1952–1958 the size of the payments had been staggering indeed. It had been estimated that during the three-year period 1950–1952 the area, excluding Iran, had received over $809 million in direct payments, and that in 1954 alone it received about £200 million ($560 million). During the following four years, in spite of the Suez Canal crisis, the total direct payments to the governments rose rapidly: in 1955 – $880 million, 1956 – $940, 1957 – $1,016, and 1958 – $1,200 million. [10]

The figures were impressive in themselves, but their importance was heightened if one kept in mind the population in each of the respective countries, and the ratio of oil royalties to other income. In Iran, for instance, 40% of the Government's 1955 budget was covered from oil revenue, and in Iraq 54%, while in Saudi Arabia it covered 71%, and in Kuwait 97%. [11] Furthermore, the oil-producing countries had benefited indirectly from the oil industry – in terms of investment expenditure on local products, employment, wages, and economic and other projects undertaken by the oil companies, which cumulatively had run into billions of dollars. The companies had also supplied these countries with most of their foreign exchange needs.

Moreover, the non-producing countries of the area also had derived incomes from the oil industry – Syria, Lebanon and Jordan from the pipelines running through their territories and Egypt from the Suez Canal tolls paid by oil tankers. The growing importance of petroleum cargo going through the Suez Canal was illustrated by the following statistic. In the years 1933–1937 the annual average tonnage, other than petroleum, in both directions was 23 million, and petroleum accounted for an additional 5.5 million tons; in 1955, general cargo was 39 million tons and petroleum 69 million tons. The Middle East had thus formed a more or less close petroleum unit between the producers and transporters.

What had been the result of this great outpouring of capital?

By all standards, the Middle East was backward and the different populations had a very low, though varying from one country to another, standard of living. With the exception of Israel, the region suffered from illiteracy, poverty, very bad health and sanitary conditions, poor housing, and a general mere subsistence level. On the other hand, while it was not too rich in natural resources other than oil, it had huge supplies of land, water and potential energy from natural gas. By utilizing the direct income from its oil for the development of water resources and cultivation of the enormous land areas, the Middle East could have become one of the promising areas of the world.

Yet, in spite of some fifty years, at the time, of activity of the oil industry, not a single Middle Eastern country was itself in a position, either technically, financially or administratively, to operate the oil industry. Even Iran, the oldest of the oil-producing states, could not activate the industry after she nationalized it, and had to resort to the International Consortium.

Nor had the Arab countries developed a sense of regional cooperation and responsibility; the rulers and the peoples had not reached, then, the stage where they would be willing to pool their resources to develop the region for the good and welfare of all the Arabs. Indeed, even the willingness to develop their countries for the good and welfare of their own population had not been very successful during the period under review as was evidenced in the discussions of the various economic plans detailed in the respective chapters above.

Sensible arrangements for utilizing at least part of the oil revenues for the general welfare were to be found in the then British protectorates of Bahrain, Kuwait and Qatar. There, although the ruling Shaikhs had appropriated fantastic amounts of the income for their own needs (between 25 and 30%), the balance at least had been employed, under the "guidance" of British advisers and political agents, for the general welfare of the population, for education, health, housing, and other improvements. But even in these cases the natives had not been made to realize the need, nor given the means of producing by their own efforts, new sources of wealth, which would enable them to maintain the higher standard of living and the social and governmental services they were enjoying and which had been made possible only because of the income from the oil.

Looking at the region as a whole, it would seem that the yearly oil revenues had been spent, during the period under review, at least in great measure, on non-productive undertakings, and thus the major natural resource of the area had been uselessly squandered. But as a result of the development of the oil industry, the region was stirring. The activities of the companies, the technological training of the employees, the standard of living introduced both for the native employees and the foreign staff, the extravagancies of the ruling families, and the economic projects undertaken by the public authorities had not only brought about a desire for the better things in life, but a restlessness and discontent that had potential social and political explosiveness.

Many social and economic patterns were breaking down, and the old fatalism, resignation and obedience to authority were being seriously questioned. While the social and cultural stirrings were not the result of the oil industry alone, the industry had hastened the process. Unfortunately for the region and for its peoples, the rulers and governments had not foreseen the dangers in the abandonment of old loyalties and traditions and had not prepared, sufficiently and constructively, for the emergence and growth of a new society.

Great as had been the amounts the rulers and governments had received from the oil resources of their countries, the profits of the companies had been many times greater. The West, through the eight major international petroleum companies – Anglo-Iranian Oil Company (later British Petroleum Company), Royal Dutch-Shell Co., Compagnie Française des Pétroles, Gulf Oil Corp., Socony Mobil Oil Co., Standard Oil Co. (New Jersey), Standard Oil Company of California and Texas Oil Co. – had been exploiting the oil resources of the area in an ever-increasing measure. The ease of production, the quality of the oil, the plenitude of the wells, the cheap labor, and the comparatively low royalties, at the time, had brought the companies enormous profits. In a report to the shareholders, the President of AIOC stressed the high quality of the Iranian oil and the ease of its production as the reason for the huge profits the company was making.

The Chairman of the Board of Directors of the Bahrain Oil Company testified that the cost of producing a barrel of oil in Bahrain had been 25 cents, including the royalty to the Shaikh. The extent of the accessibility of Middle East oil was demonstrated by the fact that the entire 1949 production of over 70 million tons, about 12½% of total world production, came from only 287 wells. [12] In a statement on January 19, 1948, to a special Committee of the House of Representatives, the late Secretary of Defense James V. Forrestal declared that the 1947 Middle East crude oil production, 10 per cent of total world production, came from only 199 wells. By way of contrast, the position of the United States in world crude production was achieved as the result of the output of some 440,000 wells. [13]

Furthermore, the companies were also exempted from all taxation, imposts and levies. Compared with profits, royalties were very low, and it was only after long and bitter struggles that the companies increased royalties and finally arrived at a general practice of sharing production profits on an equal basis with the governments.

For obvious reasons no attempt could be made to calculate the extent of the profits of the companies operating in the Middle East, but one small fact might reveal the possible size of their revenues. The agreement between the Arabian American Oil Company and Saudi Arabia called for a 50/50 profit-sharing arrangement on the production of crude. King Saud's royalties in 1955 were $338.2 million. Aramco's profit from production was at least the same amount. In addition there were the profits on transportation, refining and marketing. A rough estimate of the gross profits from Middle Eastern operations of the eight major companies for the year 1955 had been given as $1,599,826,000.

The companies, motivated exclusively by profit considerations, had directed all their activities and based their attitudes and positions on that objective. Even their undertakings in the fields of education, health and economic projects had been as much for their advantage as for the improvement of the life of the natives, and the cost of those undertakings had been insignificant in comparison with the profits reaped.

The United Nations' *Review of Economic Conditions in the Middle East 1949–1950* succinctly summarized the situation: "The terms of their concessions give the foreign companies a freedom of action which substantially insulates them from the economy of the Middle East countries. Output is determined by consideration of world rather than local conditions. Moreover, it is the companies which provide and own the means of transport, whether pipelines or tankers, to carry Middle Eastern oil to its markets, and it is they who secure these markets, both in Western Europe and in other parts of the world. The foreign exchange derived from sales of oil accrues to the petroleum companies and is in large measure retained by them." [14]

1959–1971

Right after the 1967 Arab-Israel war, with the changed international-political situation in the inter-bloc struggle, and with the later increase in world demand for oil, as well as the general political, technological, ideological and revolutionary developments in the area, the patterns of relations in the oil industry began to change, and the changes affected the region and the world outside the region.

The early years of the period under review operated in favor of the companies because of two causes: 1) The drop in world demand reduced the profits of the companies, and as much as the producing countries protested they had no alternative to the companies. 2) The Soviet Union was increasing its exports to Western Europe and undercutting the price of Middle East shipments. This brought the producers closer to the companies against the Soviet efforts.

However, in the later 'sixties the Soviet Union was making great inroads both politically and economically in the Middle East, and the threat of her exports fizzled out. The developments outlined in the chapters above revamped the entire pattern of oil affairs of the region. Since 1962 most of the governments have organized national oil companies which developed the local oil industries — from production to marketing — and the producing countries have gained full control of local refining and distribution. They contracted new concessions which made them the masters of the oil potential of their countries. But even the old concessions were unilaterally or bilaterally modified by relinquishments of unexploited areas; by royalty expensing, first after discount allowances and later with no allowances; by a 5% increase in their share of profits; by substantial increases in posted prices; by obtaining a price premium for the oil coming through the Mediterranean as long as the Suez Canal was closed, and by acquiring participation in the ownership of the companies, 25% in 1973 and 51% in nine years.

During the period under review Israel became a very small and declining producer, and entered the ranks of the major transporters. A number of the Trucial Coast Shaikhdoms became producers, and some of considerable magnitude. Persian Gulf offshore developments added great quantities of oil to the total production and reserves of the region; and loading facilities, storage capacity and pipelines were constantly added. The region became a great natural gas producer, and a center of the petrochemical industry. Egypt became a substantial producer, and although the Suez Canal was closed since 1967, it was reentering the transporter's group through the planned Sumed pipeline. Syria became a producer, and its role as transporter was enhanced by its nationalization of the IPC pipeline running through its territory. Iraq had nationalized the IPC and MPC and was struggling to operate the enterprise and find markets for the crude produced.

The region, more than ever, had become inextricably enmeshed in the oil industry with all its local, regional and international ramifications. The consequences of these developments, economically and politically — internally and externally — were subject to future action and policies by the different forces involved.

Production and Reserves

During the period under review production was steadily increasing. As shown in Table L it almost quadrupled between 1959 and 1971 and the percentage of the region's share in world production went up from 23.6% to 32.4%. While in 1963 the Middle East share was still below that of the United States, in 1971 the United States share was only 19.1%. [15]

But even more impressive was the dazzling rise in the estimated reserves presented in Table LI, from 181.365 billion barrels in 1959, which was impressive enough, to 366.8 billion barrels in 1971, double what it was at the end of 1959 in spite of the constant heavy accelerated production since 1959. Curiously enough the share of the Middle East in total reserves did not increase since 1963; in fact it might have dropped somewhat. It should be noted that the United States share of world reserves dropped from 9.1% in 1963 to 6.8% in 1971, and of the whole Western Hemisphere from 20.4% to 13.4%. The slight drop in the Middle East share was primarily the result of the increment of reserves of the Communist countries and of Africa. Africa's share accounted for 4.7% in 1963 and that of the Com-

Table L. *Middle East Production, 1959–1971*

Year	Barrels (Thousands)	Tons (Thousands)	% of World Production
1959	1,644,294	227,900	23.6
1960	1,955,475	264,092	25.1
1961	2,061,062	279,000	25.2
1962	2,281,000	311,000	30.4
1963	2,483,000	338,300	25.0
1964	2,768,000	380,500	25.8
1965	3,074,450	416,100	26.4
1966	3,450,431	462,600	27.2
1967	3,712,263	495,700	27.3
1968	4,176,929	559,100	28.1
1969	4,598,073	616,200	28.6
1970	5,158,748	687,600	29.3
1971	5,992,710[a]	803,700[b]	32.4[c]

a. *World Oil.* CL, 95, Feb. 15, 1960; – CLII, 111, Feb. 15, 1961; – CLIV, 120, Feb. 15, 1962; – CLVI, 125, Feb. 15, 1963; – CLXII, 120, Feb. 15, 1966; – CLXIV, 144, Feb. 15, 1967; – CLXVI, 123, Feb. 15, 1968; – CLXVIII, 114. Feb. 15, 1969; – CLXX, 96, Feb. 15, 1970; – CLXXII, 106, Feb. 15, 1971; – CLXXIV, 63, Feb. 15, 1972; CFP, *Annual Report 1964.* 6.

b. CFP, *Exercice 1960,* 11; – 1962, 2; BP, *Statistical Review 1971,* 18.

c. BP, *Statistical Review 1963,* 7; – *1964,* 7; – *1965,* 7; – *1966,* 7; – *1967,* 7; – *1968,* 7; – *1969,* 6; – *1970,* 6; – *1971,* 6; UN, *Economic Development 1959–1961,* 49; *World Petroleum,* XXXIV, 30, July, 1963.

munist countries was 8.8%; in 1971 Africa's share increased to 8.9% and that of the Communist countries to 15.4%; Western Europe's share of reserves rose from .8% in 1963 to 2.3% in 1971. [16]

Regardless of the drop of a few percentage points in the last two years, the outstanding fact remained that the Middle East had been for a long time the single huge oil reservoir of the world, and it was constantly expanding in absolute numbers of barrels as well as percentages of the other world reserve centers. It long surpassed the United States in reserves, it passed it also as a producer and it was destined to become the greatest producer in the world.

Table LI. *Middle East Reserves, 1959–1971*

Year	Barrels (Million)	% of World	Tons (Million)	% of World
1959	181,365		24,345	
1960	183,085		24,475	
1961	188,130	60.9	25,270	
1962	193,875		26,045	
1963	207,020	61.7	27,825	61.9
1964	211,500	60.9	28,400	61.1
1965	214,900	60.1	28,900	60.3
1966	234,600	59.9	32,100	60.2
1967	248,500	59.5	33,900	59.8
1968	270,100	58.1	36,800	58.7
1969	332,800	61.6	45,400	62.0
1970	343,900	55.4	47,000	55.9
1971	366,800	57.6	50,100	57.6[a]

a. BP, *Statistical Review 1963*, 5, 19; – *1964*, 5; – *1965*, 5; – *1966*, 5; – *1967*, 5; – *1968*, 5; – *1969*, 4; – *1970*, 4; – *1971*, 4; UN, *Economic Development 1959–1961*, 130; – *Economic Development 1961–1963*, 103; *World Oil*, CLIX, 94, Aug. 15, 1964; – CLXXI, 71, Aug. 15, 1970.

The Middle East had been obtaining steadily and increasingly enormous amounts of revenue from its oil. Table LII shows the unbelievable climb in revenue from $1.274 billion in 1959 to $7.088 billion in 1971, a more than fivefold jump. Great portions of this enormous revenue were invested in economic projects which have intensified the economic development of the region. An interesting indicator of this development growth was the substantial increase of consumption of oil and oil products, as indicated in Table LIII, from some 14 million tons in 1959 to 54 million tons in 1971, a fourfold jump in 12 years. It should, however, be noted that total consumption never surpassed 2% of total world consumption.

The Middle East had seen some expansion of refining capacity during the period under review, as shown in Table LIV. Most of the refineries were producing for local consumption, some refineries for export of products. But the total never went beyond 5.6% of world refining capacity, and in 1971 went down to 4.3%.

While the total amount of revenue, if it were equally distributed throughout the region, would be inadequate for the needs of economic development, the fact was that some countries in the area remained distressingly underdeveloped and were starved for investment funds, others had reached and some were about to reach the saturation point in the absorption of their oil revenue, and were to become the potential world money-lenders or investors.

Table LII. *Middle East Revenues, 1955–1971*
$ Million

1955	898
1956	968
1957	1,023
1958	1,238
1959	1,274
1960	1,364
1961	1,498
1962	1,649
1963	1,861
1964	2,131
1965	2,342
1966	2,682
1967	2,898
1968	3,370
1969	3,666
1970	4,214
1971	7,088[a]

a. UN, *Economic Development 1958–1959*, 77; – 1959–1961, 56; *Petroleum Press Service*, XXXVIII, 326, Sept., 1971; – XXXIX, 322, Sept. 1972. The average Government receipts per barrel of oil throughout the region for the period 1957–1971 were: 1957 – 85.7 cents; 1958 – 84.8; 1959 – 79.8; 1960 – 77.7; 1961 – 75.8 cents; 1962 – 75.7 cents; 1963 – 77.7 cents; 1964 – 78.3 cents; 1965 – 79.5 cents; 1966 – 81.5 cents; 1967 – 82.2 cents; 1968 – 85 cents; 1969 – 84.3 cents; 1970 – 86.3 cents; 1971 – 124.5 cents. Schurr, *Middle East Oil*, 103, 122; SAMA, *Statistical Summary*, 22, Sept.–Oct., 1969; *Petroleum Press Service*, XXXIX, 322, Sept., 1972.

Table LIII. *Middle East Domestic Consumption, 1959–1971*
Selected Years

Year	Tons	% of Total World Consumption
1959	14,270,000	
1962	29,000,000	2
1963	31,000,000	2
1964	32,000,000	2
1965	33,000,000	2
1966	36,000,000	2
1967	42,000,000	2
1968	41,000,000	2
1969	45,000,000	2
1970	51,000,000	2
1971	54,000,000	2[a]

a. UN, *Economic Development 1958–1959*, 79; BP, *Statistical Review 1963*, 9;
– *1964*, 9; – *1965*, 9; – *1966*, 9; – *1967*, 9; – *1968*, 9; – *1969*, 8; – *1970*, 8;
– *1971*, 8.

Table LIV. *Middle East Refining, 1963–1971*

Year	Tons (Million)	% of World Refining Capacity
1963	81	5.6
1964	81	5.2
1965	83	5.1
1966	91	5.1
1967	96	4.7
1968	104	4.8
1969	110	4.7
1970	114	4.6
1971	118	4.3[a]

a. BP, *Statistical Review 1963*, 11, 13; – *1964*, 11, 13; –
1965, 11, 13; *1966*, 11, 13; – *1967*, 11, 13; – *1968*,
11, 13; – *1969*, 10, 12; – *1970*, 10, 12; – *1971*, 10,
12. Other sources give different percentages because
their basic numbers are differently compiled.

Nevertheless, the region as a whole, some countries more while others less, has been deeply and irrevocably involved if not committed, economically, psychologically, technologically, financially, culturally, socially and politically, to the oil industry of the area.

NOTES

1. David White, "Our Future Oil Supply," *Engineering and Mining Journal*, 951–955, June 4, 1921.
2. James Terry Duce declared in a speech at the University of Michigan on July 29, 1952, that the probable ultimate production of the area lies in the vastly spectacular realm of over one hundred billion barrels. He also stated that an estimate by Standard of New Jersey gave the proved reserves of the Near and Middle East as 71,632,000,000 barrels. *Op. cit.*, 4.
3. *World Oil*, CXL, 205, Feb. 15, 1955.
4. US Senate, *International Petroleum Cartel*, 6; UN, *Review 1949–50*, 60; – *Summary 1952–53*, 39–40; Lees, "The Middle East," in *World Geography of Petroleum*, 161; De Golyer and MacNaughton, *Twentieth Century Petroleum Statistics 1954*, 2. At the end of 1954 UN experts estimated the share of the Middle East in world reserves as 60%. UN, *Economic Developments 1945 to 1954*, 2. Another estimate gave the percentage as 64, that of the Western Hemisphere as only 27.6. "Oil and Social Change in the Middle East," *The Economist*, CLXXVI, 8, July 2, 1955.
5. In 1948, about 23 million barrels of crude moved from the Middle East to the US and Canada. US Senate, *International Petroleum Cartel*, 16.
6. UN, *Review 1951–52*, 57–58; John M. Cassels, "Meeting the Challenge of World Demands," *Oil Lifestream of Progress*, III, 14, Oct., 1953.
7. In 1950, even before Abadan was shut down, the Middle East refined only 48% of its crude production. UN, *Summary 1950–51*, 20. In the following three years the percentage went down further.
8. UN, *Economic Developments 1954–1955*, 5.
9. Duce estimated that the gross investments in property, plants and equipment before depreciation was, by the end of 1951, $1.9 billion. *Middle East Oil Development*, 19. In 1954, it was estimated that the cumulative gross interest in properties, plant and equipment, before depreciation, was approximately $2.2 billion. UN, *Economic Developments 1954–1955*, 2. The cumulative investment capital, including the refineries and pipelines, was estimated in 1955 and $2.6 billion. UN, *Economic Developments 1954–1955*, 5.
10. S.J. Langley, "Oil Royalties and Economic Development in the Middle East," *Middle East Economic Papers 1954* (Beirut, 1954), 92. The following table gives a vivid picture of the rapid increase in direct payments (in US dollars) to the region as a whole:

1940	–	$ 26,600,000
1946	–	55,000,000
1948	–	97,900,000
1949	–	142,300,000
1950	–	193,300,000
1951	–	260,200,000
1952	–	440,100,000

UN, *Review 1951–52*, 59. UN experts estimated in 1954 that direct revenues of the oil-producing countries amounted to about $2,500 million in the period between 1946 and 1954; of this, approximately $55 million accrued in 1946, the amount increasing to $700 million in 1954, with Iraq, Kuwait and Saudi Arabia receiving over 93 per cent of the latter amount. *Economic Developments 1945–1954*, 6; – *1955–1956*, 58; *The New York Times*, Dec. 7, 1958.

11. *The New York Times*, Nov. 18, 1956; UN, *Economic Developments 1954–1955*, 5.

12. US Senate, *International Cartel* 9. The average per well output per day was 5,143 barrels. This compared with a 200-barrel per day average in Venezuela and 11 in the United States. It had been estimated that each foot of exploratory drilling located more than 13,000 barrels of oil in the Middle East, about 700 in Venezuela, and only about 30 in the United States. *Petroleum Press Service*, XXIII, 239, July, 1956. In 1957 the daily production per well in Kuwait was over 24,000 barrels. See above, Chapter VII, footnote 5.

13. US Senate, *Hearings Saudi Arabia*, 25450.

14. UN, *Review 1949–50* 25. After calculating the direct oil income of the oil-producing countries during the period 1946–1954 as $2,500 million, the UN experts declared that only "a limited part of this income has been utilized for economic development." *Economic Developments 1945 to 1954*, 6.

15. BP, *Statistical Review 1963*, 7; – *1971*, 6.

16. BP, *Statistical Review 1963*, 5; – *1971*, 4.

Chapter XXVII
GREAT POWER POLICIES

In dealing with the complexity of relations between the producing countries and the operating companies, between the producing countries and the home governments of the companies and between the Great Powers among themselves we should identify, although they are interwoven and affect one another or both work as one, two different sets of factors: one is an aggregate of economic and financial elements and the other is a network of political-international aspects of regional and global strategy. The former would primarily affect the producing countries and the concessionaire companies, while the latter would entail the relations among the different individual Powers involved, relations between the global groupings and relations between the individual Powers in their groupings as they bear on the region as a whole.

Policy, by its very definition, must be flexible and adjustable to changing circumstances. In analyzing developments in regard to Middle East oil we should expect modifications and changes of policy, attitudes and approaches determined by the evolvement of events. Nevertheless, we may discern two major periods of general continued trends.

1927–1965

The first period ended in the middle 'sixties. During this era the companies, to a very large extent, were in a strong position to determine the terms of the concessions, the rates of royalties and taxes and the levels of production. For they controlled not only the investment resources necessary to develop and operate the oil industry, but also possessed the technological as well as practical means of production, refining, transporting and marketing of both crude and products. Moreover, the companies were assumed, in different ways and in various degrees, to have, and they did, the unconditional backing of their respective governments, against which the producing countries were, in a sense, helpless and powerless. To be sure, from time to time difficulties arose, as in the 'thirties and again in the 'fifties with Iran, but they were subsequently smoothed out. The relations of the companies and producing countries, which almost always revolved on monetary considerations, did not determine the Great Powers' position towards the oil question of the region. The outstanding aspect were the policies and objectives of the Great Powers in relations among themselves.

The Western Great Powers' oil policies had not been as consistent as they were usually presented, nor had the different companies worked as harmoniously as some anti-cartel writers would have wanted us to believe. The British recognized from the very beginning the national imperial significance of oil resources and supplies from territories other than their own. The British government became a

531

partner in the Anglo-Persian Oil Company as early as 1914, and it had always sought to protect and preserve for the British as much of the world's oil resources as possible. The British were greatly disturbed by the fact that while the United States possessed huge oil resources in its own territory, they had very little anywhere in the empire. The Middle East became their preserve and they resented penetration into the area by anyone else, especially the Americans. Until the end of World War I they succeeded in dominating the region.

The Americans on their part showed no great desire to search for oil far away in the Middle East until the end of the First World War, when they realized the importance of oil for the military machine and saw how much American oil had been consumed during the conflict. From then until World War II and afterwards, a continuous struggle had been waged between the United States and Great Britain for the exploitation of the oil resources of the Middle East. Although arguing in the name of abstract general principles the Americans were determined not to let the British outsmart them. The victories of the United States — some due to stubborn persistence and some to the oversight or ineptness of the British — have been impressive indeed. American companies have obtained a quarter share in the Iraq Petroleum Company, the entire concessions of Saudi Arabia and Bahrain, one half of the Kuwait concession, and a 40 per cent share in Iranian oil. They have also built many refineries and laid the longest pipeline in the area — from Saudi Arabia to the Mediterranean.[1]

Into this Anglo-American struggle for petroleum resources entered naturally the question of strategy and the importance of the Middle East as an oil center in time of war. The strategic value of the area was demonstrated in the two World Wars. After the second, the control of Middle East oil assumed a new importance, for oil had become a decisive factor in the recovery of Western Europe and it was the backbone of the NATO structure. Moreover, it was of prime importance that the oil resources should not fall into enemy hands, and the enemy must not be able to cut off supplies to the West; these objectives had become a cornerstone of Western policy.

The dependence of the Western nations on Middle East oil, both in terms of home consumption and in terms of revenue, differed from Power to Power. Great Britain depended very heavily on the oil for home consumption and equally as heavily on the revenue derived from the sale of Middle East oil in world markets. Middle East oil accounted for 43% of French oil imports for home consumption. The United States, on the other hand, although its companies owned the largest share of Middle East oil concessions, did not depend on the oil of the area for home consumption until the early 'seventies. Nevertheless, the American oil companies, because of their enormous profits, have had a far-reaching influence on American Middle East policy.[2]

Two major issues have bedeviled the oil question in particular and the Middle East problem in general, during the period under review. One was the question of oil supply. The American public and the world at large had been told over and over that the great American oil reserves had been dwindling and that the world would become more and more dependent on Middle East oil. To substantiate this assertion, it was pointed out that the United States was importing Middle East oil, and that Western Europe had been importing about 80% of its petroleum needs from the Middle East. In order to perserve the dwindling resources of the United States and guarantee the supply of oil in time of peace as well as in time of war, the point was constantly stressed that the availability of Middle East oil must become the cardinal aspect of American oil policy and American foreign policy. This implied that not only must concessions in the Middle East be obtained to guarantee supplies for Western use, but that American companies must obtain these concessions exclusively, even to the extent of limiting British interests.

The cry of America's dwindling supplies were raised right after World War I and during and after World War II. But the nearer the United States seemed to be approaching the end of its petroleum resources the greater had the American reserves become. This of course was no paradox and no mystery. The new discoveries and the better and more scientific methods employed to determine reserves and the exploitation of these reserves had resulted in a constant up-marking of reserves.

A primary issue was the extent of American reserves. The American Petroleum Institute, the most conservative of estimators, gave the figure of proved reserves on December 31, 1954, as 29.6 billion barrels. In 1947, however, L.G. Weeks estimated that, exclusive of the continental shelves, reserves in the United States would total 110 billion barrels. In 1952, P.R. Schultz estimated that the ultimate oil reserves of the continental shelves would add another 30 billion barrels. Since with the then prevailing methods of production only between 30 and 40 per cent of the oil in the ground was recovered, with newer and more scientific methods which were constantly being perfected greater recoveries would be made possible, and estimates of proven reserves would have to become larger. In fact, in 1955 Paul D. Torrey declared that the method of obtaining oil through fluid injection into the wells increased recoverable reserves from existing United States fields by 11 billion barrels. Torrey stated that significant improvements in the recoverability process would place the ultimate reserves of the United States at 300 billion barrels.

These estimates did not include the vast petroleum resources which could be derived from shale. According to one authority, the United States has shale resources which could produce 1,000 billion barrels of petroleum: 125 billion barrels could be economically recoverable with improved techniques and experience. Another authority declared that 45 billion barrels were definitely recoverable with existing techniques.[3]

It then appeared that the cry that the United States was about to run out of oil, and that Middle East oil must be made availabe to guarantee adequate supplies, was far from justified; and the motivation for the campaign was the huge profits made by the American companies from their concessions in the Middle East.

For the British and the French the situation had been quite different. They did not have any alternative resources to the Middle East; their economies could not permit them to depend on American dollar oil; neither would they have wanted to be exclusively dependent on the United States for their oil, nor would they have been willing to lose the revenue which they had derived from the Middle East oil. Hence the divergent policies of the three great Western Powers.

The 1956 Suez Crisis

Different approaches operated in the 1956 Suez crisis. The British saw in the rise of the ruler of Egypt, Gamal Abd an-Nasser, the greatest danger to their very existence. They felt that they had to guarantee the flow of oil from the Middle East; that they could not allow the situation to deteriorate to such a state that they would be placed at the mercy of the whim of an Arab leader. They felt that they were first losing their position in the Middle East under the relentless pressure of the Arabs, who were in the some measure supported by the Americans, and that with the rise of Nasser, under the guidance of the Kremlin, they would lose their oil. They therefore embarked on a policy which had two objectives: to weaken Nasser and thus reduce the Soviet influence and hold on the Middle East; and on the other hand, to build Iraq — one of the major British oil sources in the area — as the leader in the Middle East against Egypt. Though the Baghdad Pact was supposed to have been the brainchild of United States Secretary of State John Foster Dulles and was meant to protect the area against Russian encroachment, it had been adopted by the British as the best means of achieving their two objectives. In the final analysis, the British had believed that the immediate major danger to their Middle East oil supplies had been the ruler of Egypt.

The French, on their part, while vitally interested in the oil supplies from the area, and in reducing the ruler of Egypt whom they accused of supporting the rebels in their North African possessions, had no political interests and objectives. They did not want to see the Baghdad Pact emerge as a new Arab front which would only serve British interests, and which would complicate their own difficulties in North Africa, and the international situation by directly challenging the Soviet Union.

The Americans, primarily concerned with profits and with keeping the Soviet Union from entering the area, had still another approach. Their feeling was that the danger in the Middle East did not stem from Russian penetration through Arab susceptibility to Communist wooing, but from Western policies and measures that antagonized the Arab leaders and were driving them into the Soviet orbit. They believed that the Arab countries had been basically well disposed to the West and that every measure adopted had to be directed toward conciliating them so that they would stay with the West. They were, therefore, determined not to join officially the Baghdad Pact, which in the eyes of Egypt, Syria and Saudi Arabia had not been a means of protection against the Soviet Union, but an instrument for promoting and protecting Iraq as the leader of the Arab world against King Saud, the dynastic enemy of the ruling house of Hashim, and against the rising power of the ruler of Egypt, as well as for facilitating new Western entrenchment in the area. In the opinion of the Americans the only way to guarantee the flow of Middle East oil would have to be by granting political concessions to the Arabs and avoiding any measures which aroused suspicion and antagonism.

In the nationalization of the Suez Canal by Egypt, the British and French had seen a clear determination on the part of Egypt's ruler to make himself master of the area. They were convinced that if President Nasser had been susscessful in nationalizing the Canal, and if he had become the sole arbiter of commercial movements through it, he would have started to press, not without Soviet guidance and encouragement, for the nationalization of the Middle East oil resources. The United States thought otherwise. It had seen in Egypt's act an effort to rid herself of Western influence and Western "colonialism"; here had been an issue which could and should have been solved peacefully, by negotiation, although Dulles' efforts in the Suez Canal nationalization crisis were inconsistent and inexplicable. Because of the basic American system of foreign economic exploitation through non-political control, the American oil companies and policymakers believed that they could retain their concessions and make their profits even if Egypt nationalized the Canal.

Soviet Union Oil Exports

A great deal of speculation has been advanced about Russia's Middle East oil objective. A study of the USSR's Middle East oil policies, during the period under review, would have indicated that the Soviet aim had been a negative one: to deny the oil to the West both in time of peace and in time of war rather than to have acquired the oil for its own purposes. Russia had successfully prevented Western penetration into northern Iran, but did not attempt to exploit it herself.

The physical difficulties of transporting oil from the Middle East to the Soviet Union, the question of world markets, and the cash requirements for royalty payments and development must have ruled out a direct Soviet Union takeover of the Middle Eastern oil fields to operate them commercially.

The Soviet Union's policy seemed to have been to arouse the Middle Eastern countries and create such conditions in the area as to make it impossible for the Western Powers to exploit the oil, thereby causing difficulties in Western Europe as well as chaotic conditions in the Middle East which could only redound to Russia's advantage.

Beginning with the middle 'fifties and continuing to the middle 'sixties the Soviet Union exported oil to Western countries in ever-increasing quantities. At the time, there was considerable speculation as to the motives for the steady and rapid growth of the exports. There were some analysts who maintained that the Soviet Union had a surplus of oil and was in desperate need of foreign exchange for the machinery and equipment for its industrial development, and oil was the best and surest means for obtaining hard currency. The Western oil companies, which were in control of the international oil price structure, saw in the Russian move a danger to their profits, and argued that Soviet exports at lower prices had been primarily political rather than economic in purpose. They justified their argument by the fact that the Soviet Union charged its own Communist satellites higher prices than it charged non-Communist purchasers. In 1961, for instance, Italy obtained a barrel of Soviet oil at $1.68, while Iraq oil cost $2.68 a barrel. Poland was charged by the Soviet Union in 1960 60% higher than the price of Iraqi oil. In 1962, the price of Soviet oil to Eastern Europe was 19.5 and 27.5 rubles per ton respectively for crude and products compared to 8.6 and 13.0 rubles to Western Europe.[4] The phenomenal achievement in the Soviet oil exports was attributed to "price cutting and other aggressive tactics." A French petroleum expert declared in 1962 that more than ever Russian oil was becoming the direct competitor of Arab oil.[5]

The Chairman of the Board of Directors of the American Petroleum Institute and at the time President of Standard Oil Company (New Jersey), M.J. Rathbone, declared on March 14, 1960, that Russia was using oil to upset the economy of the Western world. Although the absolute amount of Russian exports was not a very significant percentage of the total world production, it was enough to affect dangerously an already glutted international market.[6]

The Middle Eastern producing countries which were struggling, at the time, with the operating companies against cuts in the posted prices, were not at first disturbed by the increase in Soviet exports. However, later on when Soviet exports rose sharply they began to express resentment. The Soviet Union risked the possible alienation of the Arabs by defending its position, and publicly and almost defiantly informed the Arab countries at their own oil Congress, that it would increase exports to attain its position of exporter which it held prior to the outbreak of World War II.

At the time (early 'sixties) Soviet oil entering non-Communist nations had three aspects: 1) impact on the European Common Market countries in general and on the relations between members of the market; 2) impact on the Middle East producing countries; and 3) impact on the Western oil companies and the latter's efforts to eliminate it as non-economic competitive oil.

The Common Market in 1962 presented a most muddled picture. Before oil was discovered in Algeria, practically all market-members were resentful at the Anglo-American monopoly over oil resources; they would have been eager to find an alternative to that source. However, with the Algerian oil coming to market, France wanted the Common Market as one of the outlets for the new oil. Germany and Belgium, great coal producers, were somewhat hesitant to become dependent on Soviet oil, lest the Soviets convert from coal to petroleum and endanger the coal markets. Only Italy, which has neither oil nor coal, was eager for Soviet oil, which would relieve her of the Anglo-Americn oil monopoly, and she would obtain the oil at much cheaper prices. However, the overall fear of the Common Market was that it might become dependent on the Russian source, for the Soviet Government, for its own reasons, might, in moments of crisis, cut off the supply. The Parliamentary Assembly of the Market adopted a resolution in March 1962 urging member-states to adopt a policy that would restrict import of fuels from states that were not in a position to guarantee "long-term stability of supply."[7]

The crucial question, at the time, was the amount of oil the Soviet Union had available for export. Was it of such a magnitude as to be able to replace the other sources on which Western Europe had relied or was the available supply limited only to affect the oil price structure for the Common Market?

To give a clearcut answer was not too easy. A study prepared by the Library of Congress for the subcommittee of the Senate Committee on the Judiciary investigating the Administration of the Internal Security Act came to the conclusion that it appeared likely "that there will be continuing surpluses of oil in the USSR through 1970. We have seen all kinds of guesses of future Soviet petroleum demand, but they are no more than just that – guesses." Nevertheless, the study group, after making its own calculations and doing its own guessing, stated: "Our calculations indicate that the USSR will then [1965] have an exportable surplus of 1,000,000 barrels daily . . . . USSR–satellite agreements point to an export level of 400,000 barrels daily to the satellite countries, leaving 600,000 barrels daily available for free world markets. There is nothing sacred about this splitup of the extra 1,000,000 barrels daily. As a matter of fact, it could very well be that economic growth of the satellites will require larger drafts on Russian supply."[8]

The impact of the Soviet exports on Middle East producing countries was decisive. Prices dropped, and the revenue of the governments was reduced. Moreover, with lower prices and no increase in demand the rate of increased production also dropped. For this, however, the Middle Eastern countries blamed the Western companies rather than the Soviet Union, and they pressed, though unsuccessfully until 1971, the companies for the restoration of the price cuts.

The impact on the Western oil companies involved more than price structure: the security of the Western world, according to them, was at stake. They maintained that the Soviet Union had been using oil as one of its weapons in the world struggle. The Senate Committee on the Judiciary investigated Soviet oil as a factor in the cold war and held public hearings.[9]

The witness Samuel Nakasian stated that the great sedimentary basins of the Soviet landmass capable of producing oil were three times greater in size than the sedimentary basins of the United States, and he declared: "On the basis of present trends and opportunities available to the Soviet Union, it is possible to estimate that Soviet oil exports will rise from 222 million in 1961 to 365 million barrels in 1965 and 730 million barrels in 1972. Practically all such increase will be exported to Europe and Japan."[10]

The American companies' alarm impressed the United States Government, and on November 28, 1961, the Secretary of the Interior asked the National Petroleum Council (the industry's advisory group to the Secretary) to make a study of the impact of increased Soviet oil exports on the American petroleum situation. The Council named a 23-member committee. In 1962 the committee presented a two-volume report.[11]

Among its major conclusions were: 1) "Soviet-bloc petroleum exports to the free world have grown and will continue to grow unless action is taken by the West"; 2) From 1955 to 1961 oil exports to the free world increased from 116,300 to 610,000 barrels per day; 3) "Soviet oil exports have reduced the revenue of the

free world's oil-producing countries and oil industry." An estimate of the cumulative loss of direct income by the Venezuelan and Middle Eastern governments
over the period 1956–61 was shown to have reached $490 million. The estimate
was given "without considering the depressing effect on prices of the dumping of
Soviet oil in the free world"; 4) "Though the degree of responsibility is not demon-
strable there can be no doubt that cut-price Soviet-bloc exports contributed to the
reduction in the Middle East posted prices which took place in February 1959 and
August 1960." [12]

All the alarms and prognostications slowly but surely fizzled out, for the
exports began to decline in 1965 and dwindled down to a point where the Soviet
Union and its satellites became importers of Middle East oil.

Arab Denial of Oil to West

The second issue which affected the oil policies of the Western Powers was the
constant fear that the Middle Eastern states, if Western policy became offensive to
them, would deny their oil to the West, and transfer the oil to the Soviet Union as
the West's rival.

This fear, it would seem, was based on many errors and misleading assumptions. To begin with, it must be remembered that Iranian oil could not have been
included; for the Iranians have never been a part of the Arab League and never
disputed the West's general Middle East policy. Should ever a permanent break
between the Arabs and the West take place, Iran would, as she has in past temporary crises, continue to supply oil, and no doubt in even greater quantities, to the
West.

Furthermore, when the statement was made that the Arabs might deny the oil
to the West, it was never specified which Arabs would be involved, as if they were
all unified and of one mind to achieve political objectives through the utilization of
the entire region's oil. It must be ovbious to all students of the Middle East that oil
has been a devisive element in the Arab camp.

Oil Crises

Both major issues – the question of oil supply to the West by the Middle East
producing countries and the threat to that supply from the Arab transporting
countries – have been tested in three previous crises in the region: Suez Canal
closure in 1956; Iraq revolution in 1958, and the Arab-Israel 1967 war and subsequent closure of the Canal. On October 31, 1956, following the Anglo-French
attack on the Suez Canal, Egypt closed and later blocked the Canal, which stopped
all Suez-bound oil tankers from the Middle East. Three days later the Syrian authorities sabotaged the Iraq Petroleum Company's pipeline which passes through their
country from northern Iraq to the Mediterranean. Both these actions combined to
cause a critical oil shortage in Western Europe. Although it was primarily a transportation problem – the oil-producing countries were not cooperating with Egypt

in her clash with the Western Powers — it nevertheless created a very dangerous situation in Europe and posed a real challenge to the West.

Prior to October 31, the Middle East produced on the average some 3,700,000 barrels of oil daily. Of the total exported, 43% moved by way of the Suez Canal and 23% by international pipelines to the Mediterranean. The total daily export in 1955 was 2,840,000 barrels, of which 27,000 went to Canada, 272,000 to the United States, 70,000 to Latin America, 75,000 to Africa, 500,000 to the Far East, and 1,880,000 to Europe. The closing of the Canal and the cutting of the pipeline affected the movement of only 200,000 barrels daily, for oil was still flowing through Tapline and some tankers moved by way of the Cape of Good Hope.

The test was at hand. Could the West meet its petroleum needs from other than Middle Eastern sources, and could it overcome the transportation difficulty to undo the effect of the stoppage? It was clear even before the Canal was closed that it would be up the United States to accept such a challenge. The answer would lie in increased production in the Western Hemisphere to replace the oil it imported from the Middle East, in additional production to supply Europe's needs, and in diverting to Europe the tankers en route to America from the Middle East.

On August 14, 1956, the office of Defense Mobilization in Washington set up a Committee consisting of representatives of the thirteen United States oil companies engaged in foreign oil operations, to be known as the Middle East Emergency Committee. Its functions, in case of need, were: 1) to arrange for the most efficient use, without regard to ownership, of terminal, storage and transportation facilities of the various participating companies; 2) to arrange for exchange between companies of crude oil and petroleum products by sale, loan, or otherwise for distribution to meet domestic requirements and requirements in foreign countries or areas affected by the emergency; 3) to alter the rate of production of crude oil or the manufacture of refined products in the foreign areas to reduce the transportation difficulties.

The anticipated emergency called for an additional production of 800—900,000 barrels daily to meet British and European needs, if the Suez Canal were closed, and 1,100,000 barrels daily if the pipelines were also cut.[13] The overall result of the crisis was that the American Middle East Emergency Committee, with the close cooperation of the Organization of European Economic Cooperation, met the Middle East oil challenge successfully.[14]

Many important consequences followed the Suez crisis. The Western Powers learned that they could, at least in an emergency of up to six months' duration, find a substitute for Middle East oil, and therefore the political price they had to pay for Middle East oil would be determined by themselves and not by the Arab producers. They also fully realized that the weak link in the Middle East supply was transportation. The oil-producing countries of the region, which had been seriously affected by the crisis, arrived at two conclusions: 1) Egypt and Syria were grave dangers to their prosperity, if not to their very existence; 2) the West could do without their oil; they must, therefore, not allow the oil to become an issue in their relations with the West.

In July 1958 came the second crisis – Iraq. The President of Egypt was on the march: his aim, control of the oil-producing states of the region; the means to that end, Lebanon and Jordan. He challenged the West in the two latter countries; and when the coup d'état took place in Iraq, the West accepted the challenge in Lebanon and Jordan and was on its way to stop Nasser in Iraq. But the Iraqi revolutionaries, though they supported Nasser's quest for Arab nationalism, had ambitions of their own. Why should they play second fiddle to Egypt? They were the real Arabs; they had a rich country; they had oil and water; their country could and should fulfill its own independent role in the Arab world and not be subservient to Egypt, newly converted to Arab nationalism. General Abdul Karim al-Qasim and his colleagues knew what Nasser was after, and they knew what the West was after. By announcing to the West and to the world that they would continue to honor the Western oil concession, they would achieve a double purpose: they would stop the West from marching on Iraq and would give unequivocal notice to Egypt that they were masters in their own land. The objective was achieved.

The third crisis followed the Arab-Israel war of early June, 1967. The subsequent closure of the Suez Canal created a transportation question, but more serious was the shutting off of all production by Arab producing countries and the subsequent ban on shipments to the United States, Great Britain and West Germany. Iran, however, not only continued her shipment of oil but the increase of production in 1967 over 1966 was double that of 1966 over 1965 and less than half of 1968 over 1967. Soon all the Arab producing countries removed some or all the shipping restrictions. The Arab Summit Conference in August 1967 in Khartoum permitted each oil-producing country to develop and apply its own oil-sales policy, which meant a return to normalcy.

The closure of the Suez Canal did not create a transportation crisis as it did in 1956, and the European countries resisted the efforts of the United States to reactivate the Middle East Emergency Committee. The storm was easily weathered by the Europeans themselves.[15]

1967–1971

It would seem, and very likely it was only coincidental, that following the Arab-Israel 1967 war the Middle East picture and the Great Powers policies have changed; but these changes were the direct results of events and developments in the world's regional and international scene.

USSR

The Soviet Union's objective has not changed but its methods have; and the methods seem, at least on the surface, to have been successful. The policy of exporting oil to non-Communist countries which resulted in antagonizing Middle Eastern producing countries was changed to one of extending assistance – through agreements and other means — to the producing countries in developing their own oil resources. Through barter arrangements and credit grants, the Soviet Union and its satellites contracted for Middle East crude oil and natural gas in exchange for equipment, machinery and services. The Communist-bloc countries have heavily penetrated practically all the oil-producing countries of the area through advisors, technicians, contractors and experts. This encouraged and made possible not only development of as yet untapped oil and gas resources but also the development of North Rumaila field and the nationalization of IPC and MPC in Iraq. Similar developments took place in Syria and Iran; and even conservative monarchical Saudi Arabia entered into barter agreements with Communist-bloc members. Whether this change of policy was the result of the reduced production and reserve ratios of the Soviet Union, or merely a stratagem for penetrating the area and alienating the producing countries from the West or even changing their dependence on the West to ultimate dependence on the Soviet Union, is subject to different speculations and conclusions by various students of the subject.

France

France's policy did not change radically from what it had been ever since the end of World War I. She lost position after position in the area, and yet hoped that she could regain influence by adopting a more friendly attitude towards the Arabs at everybody else's expense. More than ever France was determined to increase her Middle East oil resources. In 1967 she attempted to obtain the North Rumaila field which was expropriated from IPC. After she lost the field to the Russians and Iraq nationalized the IPC fields, she contracted to receive her quarter crude on the same terms as before nationalization. France also entered, through her various state-owned oil firms, many joint-venture and especially contract-service arrangements with a number of Middle Eastern oil-producing states as detailed in chapters above.

Great Britain

While Great Britain was still concerned with the safety and supply of Middle East oil for her domestic needs and for the revenue which oil companies derived from the oil[16] and the strategic defense of the area, she was just concerned. Great Britain was anxious, if not determined, to bring the Middle East Arab-Israel crisis to an end on the basis of mutual compromise, and to reopen the Suez Canal for her oil and for her general trade.[17] England was first determined to and then did withdraw from the Persian Gulf, and with that came to a close a long and colorful chapter in modern Middle Eastern history. England was no longer the determining factor in Middle East developments. She suffered a number of blows and was impotent to do anything about it. IPC companies were expropriated in 1961 of over 99% of their concession areas, and in 1972 IPC was nationalized. The Iranian Government exacted concession after concession from the Consortium. But England saw in Iran the only possible strong power in the region and supported Iran's effort to secure the safety of the oil passage in the Gulf by occupying three disputed little islands. The occupation was bitterly resented by some Arab countries and greatly intensified the conflict between Iran and Iraq. It also cost England a British Petroleum concession in Libya.

United States

The United States companies, though late-comers to the area, had the greatest share of Middle East oil. Both the major international companies — Standard of New Jersey (now Exxon), Mobil Oil, Standard of California, Texaco — and the comparatively smaller ones like Standard of Indiana (Amoco), Phillips, Continental, Getty and many others, have been producing Middle East oil in prodigious quantities. The United States was and is the most concerned about the strategic importance of the region as such and from a global point of view. This was magnified and intensified by the British total withdrawal.

In the last half a dozen years two phenomena have greatly disturbed the United States, and the resultant state of mind approached panic. Oil production and estimated reserves in the United States have been steadily and rapidly declining; at the same time both were constantly climbing upwards in the Middle East. The various elements dealing with the question of fuel energy supply seemed to have agreed that within 10 or 15 years, if not sooner, the United States would have to draw heavily on the Middle East for its expected supplies.

Paradoxically, the closure of the Suez Canal by Egypt, which the United States was so energetically striving to reopen, had postponed, temporarily, the problem of the defense of the region. For it practically prevented the Soviet Union from rushing in to fill the vacuum created by the British withdrawal. It enabled the United States, in the meanwhile, to help Iran assume the role of protector by supplying her, through grants and purchases, with some $2 billion worth of ultramodern sophisticated weapons, admittedly not necessary for internal security.[18]

But whether with all these weapons Iran would be able to resist a Soviet move, should this ever take place, even as a holding action, remains doubtful. But perhaps the military build-up of Iran was meant as a guarantee of intra-regional security.

NOTES

1. While the struggle had been going on, the major British and American companies, together with Royal Dutch-Shell and Compagnie Française des Pétroles, organized world markets and controlled production for their own advantage. For a detailed treatment of the issue, see US Senate, *International Petroleum Cartel*.

2. Herbert Feis, who worked closely in the State Department with representatives of the American oil companies, wrote: "The course of our past and present diplomacy indicates that either the oil of the Middle East or the oil companies of the Middle East had had much, and perhaps excessive, influence on American policy in the region. It would be healthy not only for themselves but also for

the Arab states if the Western governments took adequate measures to bring home to one and all that they could, if need be, get along without most, if not all, of the Middle East oil. Such a course might well improve rather than impair future prospects of the desert enterprise." *New York Herald Tribune*, Oct. 28, 1958.

3. US Congress, Joint Committee on Atomic Energy, *Peaceful Uses of Atomic Energy Background Material* (Washington, 1956), 81–106.

4. *Middle Eastern Affairs*, XII, 227–228, Oct., 1962; *ACEN NEWS* 41, May–July, 1963; Robert W. Campbell, *The Economics of Soviet Oil and Gas* (Baltimore, 1968), 230. The Soviet exports of crude oil to the non-Communist world had risen from 1.3 million tons in 1956 to 13.4 million in 1961, a tenfold increase in five years.

5. *Petroleum Press Service*, XXX, 123–126, Apr., 1963; Georges Spillmann, "Le Pétrole en 1962", *L'Afrique et L'Asie*, No. 3, 27–32, 1963.

6. *The New York Times*, Nov. 15, 1960.

7. *The New York Times* Mar. 4, 1962. Italy continued to import great quantities of Soviet crude.

8. US Congress, Senate Committee on the Judiciary, *Soviet Oil in the Cold War. A Study prepared by the Library of Congress at the Request of the Subcommittee to Investigate the Administration of the Internal Security Act and other Internal Security Laws* (Washington, 1961), 18–19.

9. US Congress, Senate Committee on the Judiciary, *Problems Raised by the Soviet Oil Offensive, Study Prepared for the Subcommittee to Investigate the Administration of the Internal Security Act and Other Internal Security Laws* (Washington, 1962); – *Soviet Oil in East-West Trade, Hearings before the Subcommittee to Investigate the Administration of the Internal Security Act and other Internal Security Laws. Testimony of Samuel Nakasian*, July 3, 1962 (Washington, 1962); – *Export of Strategic Materials to the USSR and other Soviet Bloc Countries Hearings . . . Part 3, Problem Raised by Soviet Oil Development. Testimony by George T. Piercy*, Oct. 26, 1962 (Washington, 1962). Note the study mentioned in footnote 8 above.

10. See above footnote on Nakasian's testimony, page 16.

11. National Petroleum Council, *Impact of Oil Exports from the Soviet Bloc*, 2 V. (Washington, 1962). George T. Piercy, a leading member of the committee, testified before the Senate Committee on Oct. 26, 1962, and stated: "I would like to show you the petroleum balance for the bloc. The entire bloc production will go from 3,762,000 barrels to 5,911,000 barrels per day in 1965. Consumption is not keeping with the growth in production, so the net exports the bloc will have to ship to the free world will go from 596,000 in 1961 to 1,020,000 barrels per day in 1965. This is what will be available for export. Whether this much is exported to the free world, of course, will depend upon what action is taken by the free world to control it." See footnote 9 above on Piercy's testimony, 377.

12. National Petroleum Council, *Impact of Oil*, I, 24, 28, 29.

13. *The New York Times*, Aug. 15, 1956.

14. UN, *Economic Developments 1955–1956*, 43–44, 50–51, *et passim*; Chase Manhatten Bank, *A Monthly Review of the Petroleum Situation*, July, 1956; US Congress, Senate Committee on the Judiciary, *Petroleum, The Anti-Trust*

Law and Government Policies, Report of the Subcommittee on Anti-Trust and Monopoly (Washington, 1957); Organization for European Economic Co-operation, *Europe's Need for Oil. Implications and Lessons of the Suez Canal* (Paris, 1958); *The New York Times*, Mar. 22, 1957.

15. For explanations of this development see: *World Petroleum*, XLI. 94—106, June, 1970; BP, *Annual Report 1967*, 5—6. *A World Petroleum* editorial on July 1968 commented: "Quite possibly the most significant thing to come out of the entire affair was the demonstration by the world petroleum industry that it could indeed handle, and handle very well, the disruptions. Furthermore, it is continuing to demonstrate that it can handle, very likely for an indefinite time, the blockage of the Suez Canal." XXXIX, 25, July, 1968.

16. Aside from taxation BP reported in 1971 that it contributed £170 million to the United Kingdon's balance of payments. *Annual Report 1971*, 10.

17. As early as Sept., 1967, the British National Institute of Economic and Social Research (NIESR) estimated the net cost to the United Kingdom balance of payments from the Arab-Israel war at £130 million. *Petroleum Times*, LXXI, 1269, Sept. 15, 1967. In reply to a question, early in 1968, in the House of Commons, a Government representative stated: "The loss resulting from the Middle East situation as a whole was £10 million a month in July—September, 1967; and it may have been running at some £20 million a month in October—December, 1967. Most of this loss probably arose from the cost of transporting oil round the Cape, but this cannot be separated from other increases in oil cost arising from the crisis." *PDC*, 759, col. 291, Feb. 27, 1968.

18. *The New York Times*, Feb. 22, 25, 1973.

Chapter XXVIII
OUTLOOK

The prolonged and almost never-ending struggle between the producing countries and the concessionaire companies, which primarily, if not exclusively, revolved around economic considerations, began to bear fruit in 1970 in favor of the producing nations. Because of the increase in world demand for oil it progressed triumphantly to the end of 1972 when four out of the five Persian Gulf countries reached agreement with the companies for an initial 25% government participation in company ownership; a percentage which was to increase annually by 5%, reaching 51% on January 1, 1982. This would give the producing countries absolute control of the oil industry in their territories. Iraq was the only Arab country holding out with Basrah Petroleum Company while invloved in the full nationalization of IPC and MPC.[1]

Iran, which nationalized her oil industry in 1952, with the oil produced, refined and marketed by the international Iranian Consortium, was not, as mentioned above, included in the participation negotiations. In June 1972, when the Shah was in London, it was reported that the two sides arrived at an amicable accord in which the Consortium acceeded to Iran's demands on increased production, on great extension of facilities, on turning over the Abadan refinery to NIOC and on building a new refinery, probably on Kharg island. There was apparently a tacit understanding for extending the arrangements after the expiration date in 1979, for the three additional periods of five years each as provided in the 1954 agreement.

Meanwhile the other Persian Gulf producers were making headway in their negotiations with the companies on participation under very favorable terms. Iran, therefore, began to renegotiate with the Consortium companies. On January 23, 1973, Iran gave the Consortium an ultimatum: either surrender the producing and manufacturing assets to the government and sign a new agreement which would provide for the purchase of crude oil and products on favorable terms over a period of 20 to 25 years, or insist on compliance with the terms of the existing agreement until the end in 1979, after which the agreement would not be renewed and no supplies would be available. The demands were not subject to negotiations.[2]

On March 16, 1973, on the occasion of the dedication in Isfahan of a new steel mill built with Soviet assistance and in the presence of Soviet Premier Alexei Kosygin, the Shah announced that the Consortium companies had "surrendered totally" and handed over "full control" of oil operation to Iran with the ownership of all installations, for which detailed arrangements were being drafted in Teheran.[3]

These rapid and astonishing developments are rooted in a number of facts which should clearly be kept in mind in assessing the present situation and the outlook for the future.

In addition to all the factors operating in the region which were discussed in various chapters in the book, the following are the immediate components of the complex. The Middle East is, and more than likely will remain for some time to come, the greatest oil producer, and contains the greatest ordinary oil reserves in the world. The improved payment rates, the contemplated production increases and the inevitable higher posted prices and greater government share in the profits, would make the producing countries, at least the major ones, the financial bankers of modern investment needs. This would seriously and menacingly affect the international monetary structure and system.

Indeed the international monetary crisis of the first three months of 1973, culminating in the closing of the stock exchanges in Western Europe and in Japan in the middle of March, was greatly affected by the huge dollar surpluses of the Middle Eastern major oil-producing countries. A senior vice president of the New York Chemical Bank estimated that by the end of the decade Middle East governments would have $175 billion in their treasuries, representing the largest hoard of foreign-held money ever accumulated. He also predicted that the balance-of-payment drain resulting from importation of crude oil for the United States by the end of the decade would be about $21 billion annually; the European drain would amount to $25 billion, and that of Japan to $15 billion. He estimated the annual revenue that would accrue to the Middle East producing countries would amount to $30–40 billion, with Saudi Arabia receiving half.

The Organization for Economic Cooperation and Development in a working paper to a group of monetary officials meeting in Paris in February 1973, warned that the new Middle East oil wealth would keep the financial markets in constant turmoil.[4]

Most of the conventional major oil reserves are reportedly declining (the USSR reserves are still an unknown quantity), absolutely and relatively, most pronouncedly in the United States. The entire Western world, especially the United States, is in a state of panic — open and unconcealed — over future fuel energy supplies.[5]

It was originally understood by some members of OPEC, especially Venezuela, that the only effective way to raise the price of oil was by controlling and restricting production. OPEC encompassed the greater part of world producers, and therefore could limit production. Although all past experiences of OPEC with output regulation ended in dismal failure, under the new conditions of participation there might develop possibilities of collective curtailment of production. Ahmad Zaki Yamani, the main OPEC negotiator for the Persian Gulf producers, declared that participation would give OPEC great political weight in the balance of world powers. He added tactfully that because the member-states appreciated the power of their weapon they would use it "to build and not destroy, for peace and not for war, for cooperation not confrontation."

THIRD PARTY

In all the discussions and considerations between the producing countries and the concessionaire companies, and even in the Great Powers' relations and objectives, the third most important party, namely the oil consumers, were completely overlooked if not ignored.

It must have been obvious that the price structure, whether entirely determined by the companies or by the companies and the producing countries, was always at the expense of the consumers. The high profits of the companies and the huge revenues of the producing countries were exacted from the consumers by the cartel, direct or indirect, arrangements of the marketing agencies.

Oil supplies were never subjected to a free competitive market which could determine real prices. Of course, the companies, interested primarily in profits, squeezed out as much as they could from the oil. The producing countries, with all their attacks on the Western monopolists, would not be eager to kill the goose that laid the golden eggs. What they were struggling for, and in a very great measure successfully accomplished, was to obtain a much greater share of the golden eggs, and ultimately, if possible, all of them. This situation leaves the companies and their home governments with two options.

TWO OPTIONS

One option would be to go along with the producing countries and maintain the cartel's artificial price structure and squeeze the consumer even more, for he would have no alternative. But it would guarantee, at least it would be hoped, steady supplies in the next ten year. The order would, of course, be reversed: the lion's share of the profits would go to the producers and a smaller share to the companies. At the end, however, the companies themselves would be entirely squeezed out. It must be clear to them as it must be clear to the producing countries and the consuming nations that with participation the process of raising posted prices and higher government profits share would continue, and perhaps even at a more accelerated rate. But in spite of the blinding effect of the profits, the companies must realize that the producing countries have worked out the nine-year schedule of progressive participation for two objectives: technological takeover of the industry and the acquirement of the international marketing facilities with their cartel price structure. Once they have achieved these two aims the companies would be completely discarded.

Operating on the assumption that the Middle East was the only major source of oil for the consuming countries, the West would be subject to the political and financial whims of the producing countries.

There is, however, another option. It may be drastic, it may create temporary severe difficulties, dislocations and shortages, and to a very large extent may demand self-sacrifice, but in the long run may be a much better and safer solution for all concerned with the question of energy supply.

Let all the international companies invite all the producing countries to totally nationalize the industry with reasonable compensation while the companies are still in control of the dowstream operations and marketing outlets. Let the companies join the consumers as buyers of Middle East oil in a free, realistically competitive market. This would, to some extent, reduce company profits, but it would free the oil from the artificial price structure, and, above all, would produce some sane balance in the world's future energy, financial and investment patterns.

This option inevitably raises the threat of controlled production. Realizing, as they clearly must and do, that they possess the major source of world oil supply, the Middle Eastern countries might unite either under OPEC or OAPEC or both and ration production, which could raise prices to levels never experienced before. Such a possibility must be anticipated, even though the objective observation of past experiences may cast a heavy shadow of doubt about its realization. To meet such a possibility all consumer countries would have to mobilize all their energies to bring into operation alternatives to Middle East oil. To judge from past experience the success of such an undertaking could be more than a certainty.

Middle East oil could be replaced by a number of other sources of energy. To mention but a few: the enormous shale deposits in the United States, the tar sands of Canada, the petroleum reserves in Alaska, atomic energy and others could be developed. The only reason these alternatives were not hitherto worked in proper proportions was that the production of Middle East oil was economically much cheaper for the companies. In view of the new circumstances it would be advisable, indeed imperative, to turn to these resources and activate them both as alternatives to, and co-suppliers with, the Middle East oil. Not only would they bring down Middle East oil to natural competitive prices, but would free the West from absolute dependence on the Middle East.

In the discussion about the startling developments — hardly expected by the producers — in the Middle East and the transfer of cartel practices from the companies to the producing countries, the issue of ideal interdependence was injected. It was sweetly argued that since the West needs the oil, and the Middle East needs the markets which the West could supply, the arrangements which the new proposals would bring about would guarantee peaceful mutually beneficial coexistence.

Under normal circumstances where a well-balanced equilibrium on both sides existed interdependence would have been a highly desirable and profitable condition. However, closer scrutiny would reveal that the mutuality of interdependence was far from balanced. For the Western highly industrialized societies Middle East oil would be a basic necessity for their economic existence and the guarantee of their national security. The price they would pay for that oil would take out a very big chunk of their national earnings without getting any reciprocal benefits for their economies. For most of the Middle Eastern great producers, the revenue from the oil would simply be additional accumulation of surpluses which would give them political and financial manipulative power. The mutuality of interest would be heavily weighted against the Western countries.

Moreover, and it should not easily and lightly be dismissed, the internal political nature of the individual Middle Eastern states and of the region as a whole would not increase the viability even of this imbalanced interdependence. Nor for that matter would the external security of the area enhance the reliability of this interdependence.

Finally, in the second option, the role of the Soviet Union — whatever the real facts of its reserves of both oil and natural gas and its production capacity — would be greatly diminished, if not eliminated, in the complex Middle East problem — which would be the greatest boon to peace.

NOTES

1. SAMA, *Statistical Summary*, 15, Dec., 1972, *Petroleum Press Service*, XL, 44–46, Feb., 1973. Very elaborate details were worked out on prices to be charged and percentages of the oil to be disposed to ease the process of transfer from company to government and the continuity of supply to world markets. Some aspects of the industry were not included.
2. *Ibid.*, 47–48.
3. *The New York Times*, Mar. 17, 1973.
4. *The New York Times*, Mar. 16, 1973.
5. An indicator of this panic was the announcement by United States Secretary of Commerce Peter Peterson that the United States was seriously considering a plan which would make the Soviet Union a major source for Unites States natural gas supplies, which would surpass the Middle East and Algeria. The imports of Soviet gas would reduce American dependence on OPEC and would affect favorably the United States balance of payment, for the credit arrangements would provide for Russian imports of American products. To which *The New York Times*, Feb. 26, 1973, reacted editorially: "National security, basic foreign policy and elementary economics combine into a formidable argument against the huge Soviet-American co-venture to exploit the natural gas reserves of Siberia, first disclosed last November." It was a $3.7 billion project.

BIBLIOGRAPHY

The bibliographical listing is comprehensive but by no means exhaustive. The division into Official Documents and Books and Articles is more descriptive than qualitative.

OFFICIAL DOCUMENTS

Anglo-Iranian Oil Company Limited (until 1934 Anglo-Persian Oil Company), *AIOC, Operations in Iran,* London, 1951.
— *The Anglo-Iranian Oil Company and Iran,* [London, 1951].
— *A Brief Survey of a World-Wide Organization,* London, 1952.
— *Education and the Training for the Oil Industry in Abadan,* London, 1950.
— *50 Years of Oil. A Survey of the World-Wide Activities of the Anglo-Iranian Oil Company,* London, [1951].
— *Medical and Health Services of the Anglo-Iranian Oil Company Ltd. in Iran,* London, 1951.
— *Persia Past and Present,* London, 1950.
— *The Persian Oil Industry: An Account of its Origin and Development,* London, 1927.
— *[Proceedings] Ordinary General Meeting of the Anglo-Persian Oil Company, Limited,* 15th meeting — November 25, 1924.
— *Ordinary* . . . 18th meeting — November 2, 1927.
— *Ordinary* . . . 21st meeting — June 17, 1930.
— *Ordinary* . . . 22nd meeting — June 16, 1931.
— *Ordinary* . . . 24th meeting — July 11, 1933.
— *Ordinary* . . . 25th meeting — June 12, 1934.
— *Ordinary* . . . 26th meeting — June 27, 1935.
— *Report of the Directors and Balance Sheet to 31st March, 1910,* London, 1910.
— *Report . . . to 31st March, 1911,* London, 1911.
— *Report . . . to 31st March, 1912,* London, 1912.
— *Report . . . to 31st March, 1913,* London, 1913.
— *Report . . . to 31st March, 1914,* London, 1914.
— *Report . . . to 31st March, 1915,* London, 1915.
— *Report . . . to 31st March, 1916,* London, 1916.
— *Report . . . to 31st March, 1917,* London, 1917.
— *Report . . . to 31st March, 1918,* London, 1918.
— *Report . . . to 31st March, 1919,* London, 1919.
— *Report . . . to 31st March, 1920,* London, 1920.
— *Report . . . to 31st March, 1921,* London, 1921.

—— *Report . . . to 31st March, 1922,* London, 1922.
—— *Report . . . to 31st March, 1923,* London, 1923.
—— *Report . . . to 31st March, 1924,* London, 1924.
—— *Report . . . to 31st March, 1925,* London, 1925.
—— *Report . . . to 31st March, 1926,* London, 1926.
—— *Report . . . to 31st March, 1927,* London, 1927.
—— *Report . . . to 31st March, 1928,* London, 1928.
—— *Report . . . to 31st December, 1928,* London, 1929.
—— *Report . . . to 31st December, 1929,* London, 1930.
—— *Report . . . to 31st December, 1930,* London, 1931.
—— *Report . . . to 31st December, 1931,* London, 1932.
—— *Report . . . to 31st December, 1932,* London, 1933.
—— *Report . . . to 31st December, 1933,* London, 1934.
—— *Report . . . to 31st December, 1934,* London, 1935.
—— *Report . . . at 31st December, 1935,* London, 1936.
—— *Report . . . at 31st December, 1936,* London, 1937.
—— *Report . . . at 31st December, 1937,* London, 1938.
—— *Report . . . at 31st December, 1938,* London, 1939.
—— *Report . . . at 31st December, 1939,* London, 1940.
- — *Report . . . at 31st December, 1940,* London, 1941.
—— *Report . . . at 31st December, 1941,* London, 1942.
—— *Report and Balance Sheet at 31st December, 1942,* London, 1943.
—— *Report . . . at 31st December, 1943,* London, 1944.
—— *Report . . . at 31st December, 1944,* London, 1945.
—— *Report . . . at 31st December, 1945,* London, 1946.
—— *Report . . . at 31st December, 1946,* London, 1947.
—— *Report . . . at 31st December, 1947,* London, 1948.
—— *Report and Accounts as at 31st December, 1948,* London, 1949.
—— *Annual Report and Accounts as at 31st December, 1949,* London, 1950.
—— *Annual Report . . . as at 31st December, 1950,* London, 1951.
—— *Annual Report . . . as at 31st December, 1951,* London, 1952.
—— *Annual Report and Accounts for the Year ended 31st December, 1952,* London, 1953.
—— *Annual Report and Accounts for the Year ended 31st December, 1953,* London, 1954.
—— *A Short History of the Anglo-Iranian Oil Company,* [London, 1948].
—— *Social Services,* London, 1950.
Arabian American Oil Company, *Arabian Oil and its Relation to World Shortages,* n.p., [1947].
—— *Arabian Oil and World Oil Needs,* [New York, 1948].
—— *15 Years. A History of Achievement 1933–1948,* n.p., [1948].
—— *Handbook. Oil and the Middle East,* Dhahran, Saudi Arabia, 1968.
—— *Middle East Oil Developments* (1956).

—— *Report of Operations to the Saudi Arab Government for the Year 1948*, n.p., n.d.
—— *Report of Operations to the Saudi Arab Government 1949*, n.p., n.d.
—— *Report of Operations to the Saudi Arab Government 1950*, n.p., n.d.
—— *Report of Operations to the Saudi Arab Government 1951*, n.p., n.d.
—— *Report of Operations to the Saudi Arab Government 1952*, n.p., n.d.
– *Report of Operations to the Saudi Arab Government 1953*, n.p., n.d.
—— *Report of Operations 1954 to the Saudi Arab Government*, n.p. 1955.
—— *Report of Operations to the Saudi Arab Government 1955*, n.p., n.d.
– – *Report of Operations to the Saudi Arab Government 1956*, n.p., n.d.
- – *Report of Operations to the Saudi Arab Government 1957*, n.p., n.d.
—— *1958 Report of Operations to the Saudi Arabian Government*, Dhahran, 1959.
—— *1959 Report of Operations to the Saudi Arabian Government*, Dhahran, 1960.
—— *1960 Report of Operations to the Saudi Arab Government.*
—— *Aramco 1961: A Review of Operations.*
—— *Aramco 1962: A Review of Operations.*
—— *Aramco 1963: A Review of Operations.*
—— *Aramco 1964: A Review of Operations.*
—— *A Review of Operations 1965.*
—— *Aramco 1966: A Review of Operations.*
—— *Aramco 1967: A Review of Operations.*
—— *Aramco 1968: A Review of Operations.*
—— *Aramco 1969: A Review of Operations.*
—— *Aramco 1970: A Review of Operations.*
—— *Aramco 1971: A Review of Operations.*
—— *Middle East Oil Development, 1958.* Fourth edition.
—— *National Gas For Sale in Saudi Arabia,* 1962.
—— *Summary of Middle East Oil Developments,* n.p. 1948.
Bahrain, *Census 1950* (mimeographed), Bahrain, 1950.
Bahrain Petroleum Company, *Annual Report to the Ruler of Bahrain, 1953,* 1954.
—— *Annual Report to the Ruler of Bahrain, 1954,* 1955.
—— *1959 Annual Report to the Ruler of Bahrain and its Dependencies.*
—— *1960 Annual Report,* Bahrain.
—— *1961 Annual Report,* Bahrain.
—— *Annual Report 1962,* Manama, Bahrain.
—— *1963 Annual Report,* Bahrain.
—— *1964 Annual Report,* Bahrain.
—— *1965 Annual Report,* Bahrain.
—— *1966 Annual Report,* Bahrain.
—— *1967 Annual Report,* Bahrain.
—— *1968 Annual Report,* Bahrain.
—— *1969 Annual Report,* Bahrain.
—— *1970 Annual Report,* Bahrain.

—— *1971 Annual Report,* Bahrain.

—— *Bahrain,* 1952.

British Petroleum Company Limited, *Annual Report of Accounts for the Year Ended 31st December, 1955,* London, 1956.

—— *Annual Report of Accounts for the Year Ended 31st December, 1956,* London, 1957.

—— *Annual Report of Accounts for the Year Ended 31st December, 1957,* London, 1958.

—— *Annual Report and Accounts for the Year Ended 31st December, 1959,* London, 1960.

—— *Annual Report and Accounts for the Year Ended 31st December, 1964,* London.

—— *Annual Report and Accounts for the Year Ended 31st December, 1965,* London.

—— *Annual Report and Accounts for the Year Ended 31st December, 1966,* London.

—— *Annual Report and Accounts for the Year Ended 31st December, 1967,* London.

—— *Annual Report and Accounts for the Year Ended 31st December, 1968,* London.

—— *Annual Report and Accounts for 1969.*

—— *Annual Report and Accounts for 1970.*

—— *Annual Report and Accounts for 1971.*

British Petroleum, *Our Industry,* London, 1970.

—— *Statistical Review of the World Oil Industry 1956,* London, 1957.

—— *Statistical Review of the World Oil Industry 1957,* London, 1958.

—— *Statistical Review of the World Oil Industry 1963,* London.

—— *Statistical Review of the World Oil Industry 1964,* London.

—— *Statistical Review of the World Oil Industry 1965,* London.

—— *Statistical Review of the World Oil Industry 1966,* London.

—— *Statistical Review of the World Oil Industry 1967,* London.

—— *Statistical Review of the World Oil Industry 1968,* London.

—— *Statistical Review of the World Oil Industry 1969,* London.

—— *Statistical Review of the World Oil Industry 1970,* London.

—— *Statistical Review of the World Oil Industry 1971,* London.

Compagnie Française des Pétroles, *Exercice 1959.*

—— *Exercice 1960.*

—— *Exercice 1961.*

—— *Exercice 1962.*

—— *Exercice 1963.*

—— *Annual Report 1964.*

—— *Annual Report 1965.*

—— *Exercice 1966.*

—— *Annual Report 1967.*

—— *Annual Report 1968.*
—— *Annual Report 1969.*
—— *Annual Report 1970.* Paris, 1971.
—— *Annual Report 1971,* Paris, 1972.
Ente Natzionale Idrocarburi, *Annual Report and Statement of Accounts,* Rome, 1960.
—— *Annual Report and Statement of Accounts,* Rome, 1961.
—— *Energy and Hydrocarbons in 1959,* Rome, 1960.
—— *Energy and Hydrocarbons in 1960,* Rome, 1961.
France, Ministere des Affairs Etrangeres, *Actes Signes à Lausanne le 30 Janvier et le 24 Juillet 1923. Lettres et Accords en date du 24 Juillet 1923 relatifs à diverses Causes de ces Actes Convention et Protocole en date du 23 Novembre 1923 relatifs aux Dommages subis en Turquie, Paris, 1923.*
—— *Documents Diplomatiques Conference de Lausanne,* Paris, 1923.
France and Petroleum, Ambassade de France, New York, 1961.
Germany, Auswartigen Amt, *Die Grosse Politik der Europäischen Kabinette 1871–1914* (Vol. 37, I), Berlin, 1926.
Great Britain, Colonial Office, *Report by His Majesty's Government in the United Kingdom of Great Britain and Northern Ireland to the Council of the League of Nations on the Administration of Palestine and Transjordan for the Year 1933,* London, 1934. Colonial No. 94.
—— *Report of His Britannic Majesty's Government on the Administration of Iraq for the period April 1923–December 1924,* London, 1925. Colonial No. 13.
—— *Special Report by His Majesty's Government in the United Kingdom of Great Britain and Northern Ireland on the Progress of Iraq during the period 1920–31,* London, 1931. Colonial No. 58.
Great Britain, Department of Overseas Trade, *Iraq. Review Commercial Conditions,* London, 1945.
—— *Persia. Review of Commercial Conditions,* London, 1945.
Great Britain, Foreign Office, *Agreement between His Britannic Majesty's Government and the Persian Government, Signed at Tehran August 9, 1919,* London, 1919. Cmd. 300, 1919. Persia No. 1 (1919).
—— *Agreement on Petroleum between the Government of the United Kingdom and Northern Ireland and the Government of the United States of America,* London, 1945. Cmd. 6683, 1945. United States No. 3 (1945).
—— *Agreement on Petroleum between the Government of the United States of America and the Government of the United Kingdom of Great Britain and Northern Ireland,* London, 1944. Cmd. 6555, 1944. United States No. 1 (1944).
—— *Correspondence between Her Majesty's Government in the United Kingdom and the Iranian Government, and Related Documents, Concerning the Joint Anglo-American Proposals for the Settlement of the Oil Dispute August 1952 to October 1952, London, 1952.* Cmd. 8677, 1952. Iran No. 1 (1952).
—— *Correspondence between His Majesty's Government and the French Govern-*

ment Respecting the Angora Agreement of October 20, 1921, London, 1922. Cmd. 1570, 1922. Turkey No. 1 (1922).

—— *Correspondence between His Majesty's Government and the United States Ambassador Respecting Economic Rights in Mandated Territories*, London, 1921. Cmd. 1226, 1921. Miscellaneous No. 10 (1921).

—— *Correspondence between His Majesty's Government in the United Kingdom and the Persian Government, and related Documents concerning the Oil Industry in Persia, February 1951 to September 1951*, London, 1951. Cmd. 8425, 1951.

—— *Correspondence Respecting the Affairs of Persia December 1906 to December 1908*, London, 1909. Cmd. 4581, 1909.

—— *Despatch from His Majesty's Ambassador at Paris Enclosing the Franco-Turkish Agreement signed at Angora on October 20, 1921*, London, 1921. Cmd. 1556, 1921. Turkey No. 2 (1921).

—— *Despatch to His Majesty's Ambassador at Washington, a Memorandum on the Petroleum Situation*, London, 1921. Cmd. 1351 (Miscellaneous No. 17), 1921.

—— *Documents on British Foreign Policy 1919–1939*, First Series, Volume IV, edited by E. L. Woodward and Rohan Butler, London, 1952.

—— *Iraq. Treaty between the United Kingdom and Iraq signed at London, December 14, 1927*, London, 1927. Cmd. 2998, 1937.

—— *Iraq. Treaty with King Feisal*, London, 1922. Cmd. 1757, 1922.

—— *Lausanne Conference on Near Eastern Affairs 1922–1923*, London, 1923. Cmd. 1814, 1923.

—— *Memorandum of Agreement between M. Philippe Berthelot, Directeur des Affaires Politiques et Commerciales au Ministère des Affaires Etrangères, and Professor Sir John Cadman, Director in charge of His Majesty's Petroleum Department*, London, 1920. Cmd. 675, 1920.

—— *Protocol of the Proceedings of the Berlin Conference, Berlin, 2nd August, 1945*, London, 1947. Cmd. 7087, 1947.

—— *Protocol of 30 April 1923, and Agreements Subsidiary to the Treaty with King Feisal*, London, 1924. Cmd. 2120, 1924.

—— *Report by His Majesty's Government in the United Kingdom of Great Britain and Northern Ireland to the Council of the League of Nations on the Administration of Iraq for the period January to October 1932*, London, 1933.

—— *Treaty between Great Britain and Iraq, October 10, 1922*, London, 1922. Cmd. 2662, 1922.

—— *Treaty between Great Britain and Iraq and Turkey Regarding Settlement of Frontier between Turkey and Iraq, June 5, 1926*, London, 1926. Cmd. 2679, 1926.

—— *Treaty between His Majesty and the King of Hejaz and of Nejd and Its Dependencies, May 20, 1927*, London, 1927. Cmd. 2951, 1927. (Treaty Series, 1927 No. 25).

—— *Treaty between the United Kingdom and Iraq, Signed at London 14 December 1927*, London, 1927. Cmd. 2587, 1927.

—— *Treaty of Alliance between His Majesty in Respect of the United Kingdom and His Highness the Amir of Trans-Jordan,* London, 1946. Cmd. 6779, 1946.

—— *Treaty of Alliance between the United Kingdom and the Soviet Union and Iran, Tehran, January 29, 1942,* London, 1942. Cmd. 6335, 1942.

—— *Treaty of Peace with Turkey and other Instruments Signed at Lausanne on July 24, 1923,* London, 1923. Cmd. 1929, 1923. Treaty Series No. 16 (1923).

—— *Treaty with King Feisal signed at Baghdad, 13 January 1926, with Explanatory Note,* London, 1926. Cmd. 2587, 1926.

Great Britain, Navy, *Agreement with the Anglo-Persian Oil Company, Limited,* London, 1914. Cmd. 7419, 1914.

Great Britain, Overseas Economic Survey, *Iran Economic and Commercial Conditions,* London, 1948.

—— *Iraq,* London, 1949.

Great Britain, Parliament, *Parliamentary Debates House of Commons,* Fifth Series, Volumes: 2, 4, 5, 7, 8, 10, 18, 22, 23, 30, 32, 35, 41, 45, 49, 50, 63, 64, 65, 70, 71, 73, 76, 122, 123, 125, 126, 127, 128, 129, 130, 131, 132, 133, 134, 135, 136, 138, 139, 140, 142, 143, 144, 145, 147, 150, 151, 152, 153, 154, 155, 156, 157, 160, 161, 162, 163, 164, 167, 168, 169, 170, 171, 172, 176, 177, 180, 183, 184, 188, 189, 191, 193, 198, 199, 204, 208, 209, 210, 211, 213, 214, 215, 216, 217, 218, 219, 220, 222, 223, 224, 225, 226, 227, 230, 232, 235, 236, 237, 238, 239, 244, 247, 248, 249, 250, 251, 252, 253, 261, 263, 267, 268, 270, 272, 273, 274, 275, 276, 278, 280, 287, 288, 289, 290, 292, 293, 297, 298, 299, 300, 301, 302, 307, 310, 311, 319, 321, 324, 326, 330, 332, 338, 342, 343, 344, 355, 360, 364, 370, 371, 373, 374, 376, 377, 378, 380, 387, 396, 397, 399, 402, 404, 407, 408, 409, 413, 414, 418, 419, 420, 421, 422, 423, 424, 425, 426, 427, 430, 431, 432, 433, 434, 435, 436, 437, 441, 443, 444, 445, 446, 448, 450, 452, 453, 454, 457, 460, 463, 467, 469, 470, 472, 473, 475, 476, 477, 478, 484, 485, 486, 487, 488, 489, 490, 491, 493, 494, 495, 497, 498, 499, 501, 502, 504, 505, 507, 508, 509, 510, 511, 512, 513, 518, 521, 522, 523, 524, 526, 527, 528, 529, 530, 531.

—— *Parliamentary Debates House of Lords,* Fifth Series, Volumes: 42, 46, 61, 71, 81, 138, 139, 154, 170, 171, 172, 173, 176, 177, 178, 179, 180, 184, 185, 186.

India, *A Collection of Treaties, Engagements and Sanads Relating to India and Neighbouring Countries,* XI, edited by C. U. Aitcheson, Delhi, 1933.

International Bank for Reconstruction and Development, *The Economic Development of Iraq. Report of a Mission Organized by the International Bank for Reconstruction and Development at the Request of the Government of Iraq,* Baltimore, 1952.

—— *The Economic Development of Kuwait. Report of Missions Organized at the Request of the Government of Kuwait,* Baltimore, 1965.

International Court of Justice, *Pleadings, Oral Arguments, Documents, Anglo-Iranian Oil Co. Case (United Kingdom v. Iran) Judgment of July 22nd, 1952,* n.p., n.d.

International Court of Justice, Reports and Judgments, Advisory Opinions and Orders, *Anglo-Iranian Co. Case Request for the Indication of Interim Measures of Protection (United Kingdom v. Iran). Order of July 5th, 1951,* Leyden, 1951.

—— *Anglo-Iranian Oil Co. Case (United Kingdom v. Iran) Preliminary Objection Judgment of July 22nd, 1952,* Leyden, 1952.

International Labour Office, *Labor Conditions in the Oil Industry in Iran,* Geneva, 1950.

International Labor Organization, *Manpower Problems. Vocational Training and Employment Service,* Geneva, 1951.

Iran, Iranian Embassy, Washington, *The Nationalization of the Oil Industry in Iran,* [Washington, 1951]

—— *Some Documents on the Nationalization of the Oil Industry in Iran,* Washington.

Iran Today, Iranian Information Center, New York.

Iranian Oil Operating Companies, *Review of 1956.*

—— *Review of 1957.*

—— *Review of 1958,* London, 1959.

—— *1959 Review,* Teheran, 1960.

—— *1960 Annual Review.*

—— *1961 Annual Review.*

—— *1962 Annual Review.*

—— *1963 Annual Review.*

—— *Annual Review 1964.*

—— *Annual Review 1965.*

—— *Annual Review 1966.*

—— *Annual Review 1967.*

—— *Annual Review 1968.*

—— *Annual Review 1969,* Teheran.

—— *Annual Review 1970,* Teheran.

—— *Annual Review 1971,* Teheran.

Iranian Oil Services, Limited, *The Iranian Oil Operating Companies: The First Year of Operations,* London, 1955.

Iraq, *An Inquiry into Land Tenure and Related Questions. Proposals for the Initiation of Reform,* by Sir Ernest Dowson, Letchworth, England [1937].

—— "B.O.D. Company Limited. Convention made with the Government of Iraq on the 20th April, 1932," *Iraq Government Gazette,* Annexure 1 to No. 27 of 2/7/32, July 3, 1932.

—— "Basrah Petroleum Company Limited. Convention Made with the Government of Iraq on the 29th July, 1938," *Iraq Government Gazette,* No. 49, 686–706, December 4, 1938.

Iraq, Iraq Petroleum Company Limited, Mosul Petroleum Company Limited, Basrah Petroleum Company Limited, *Agreement with the Government of Iraq made on 3rd day of February, 1952 together with Law No. 4 of 1952 Ratifying Agreement* [n.p., 1952].

Iraq, "Law ratifying Two Agreements Amending the Turkish Petroleum Company's Concession," *Iraq Government Gazette,* No. 21a, 380–390, May 21, 1931.

Iraq, *Oil & Minerals in Iraq,* 1970, (First Annual Book).

Iraq, Development Board, *Annual Report for the Financial Year 1953–1954,* Baghdad, 1954.

Iraq, Development Board of the Ministry of Development, *Annual Report for the Financial Year 1954–1955,* 1957.

Iraq, Ministry of Development, Board of Development, *Development of Iraq Second Development Week March 1957.*

Iraq, Ministry of Economics, Principal Bureau of Statistics, *Statistical Abstract, 1939; 1940; 1941; 1942; 1943; 1944–1945; 1946; 1947; 1948.*

Iraq, Ministry of Oil and Minerals, *The Discriminatory Policies of Foreign Oil Companies Operating in Iraq and the Stand of the Iraqi Government [Baghdad].* 1972 Report Prepared for the 8th Arab Petroleum Congress, Algiers, May 28–June 3, 1972.

Iraq, Ministry of Oil, *Law Defining the Exploitation Areas for Oil Companies,* Baghdad, 1961.

Iraq, National Bank, Directorate of Statistics and Research, *The Annual Report of the National Bank of Iraq,* Baghdad, 1951.

Iraq National Oil Company, *Direct & National Exploitation of Iraqi Crude Oil,* Baghdad, 1972.

Iraq Petroleum Company, Limited, *An Account of the Construction in the Years 1932 to 1934 of the Pipe-line of the Iraq Petroleum Company Limited from its Oilfield in the vicinity of Kirkuk, Iraq to the Mediterranean Ports of Haifa (Palestine) and Tripoli (Lebanon),* London, 1934.

—— *Iraq Oil in 1951,* London, 1952.

—— *Iraq Oil in 1952,* London, 1953.

—— *Iraq Oil in 1953,* London, 1954.

—— *Iraq Oil in 1954,* London, 1955.

—— *Iraq Oil in 1957,* London, 1958.

—— *Iraq Oil in 1960.*

—— *Iraq Oil in 1961.*

—— *Iraq Oil in 1962.*

—— *Iraq Oil in 1963.*

—— *Review for 1964.*

—— *Review for 1965.*

—— *Review for 1966.*

—— *Review for 1967.*

—— *Review for 1968.*

—— *Review for 1969.*

—— *Review for 1970.*

—— *1971.*

The Kuwait Oil Company, *Annual Review of the Operations 1959.*

—— *Annual Review 1961.*

—— *Annual Review 1965.*

—— *Annual Review 1966.*

—— *Annual Review 1967.*

—— *Annual Review 1968.*

—— *Annual Review 1969.*

—— *1970 Review of Operations.*

—— *1971 Review of Operations.*

—— *Desert Epic. Commemorating the Silver Jubilee of Oil Exports. 1946–1971.*

—— *Inauguration of Mina al Ahmadi Refinery,* 1958.

—— *Kuwait Past and Present.*

—— *Oil in Kuwait. A Short Account of the Operations of Kuwait Oil Co.,* London.

—— *The Story of Kuwait,* London, 1955.

—— *The Story of Kuwait,* 1957.

—— *The Story of Kuwait,* London, 1959.

League of Nations, *Official Journal,* 5th Year, 1924; 6th Year, 1925; 7th Year, 1926; 9th Year, 1928; 10th Year, 1929; 11th Year, 1930; 14th Year, 1933, 15th Year, 1934.

League of Nations, *Question of the Frontier Between Turkey and Iraq. Report Submitted to the Council by the Commission Instituted by the Council Resolution of September 30th, 1924,* Geneva, 1925. c. 400. M. 147. 1925. VII.

National Iranian Oil Company, Teheran, 1962.

—— *Annual Report 1968,* Teheran.

—— *A Chronology of the Iranian Oil Industry,* [1970].

—— *NIOC in 1969,* Teheran.

—— *NIOC in 1970,* Teheran.

—— *1971,* Teheran.

—— *Oil and Economic Development of Iran,* [1970].

—— Affiliated Companies Affairs & International Relations Group, *The Economic Impact of Petroleum Industries on Iran,* 1972.

—— Affiliated Companies Affairs & International Relations Group, *Iranian Oil. A Decade of Success,* 1972.

Organisation for European Economic Co-Operation, *Europe's Need for Oil. Implications and Lessons of the Suez Crisis,* Paris, 1958.

Organization of the Petroleum Exporting Countries, *Annual Review and Record 1967,* Vienna.

—— *Annual Review and Record 1968,* Vienna.

—— *Annual Review and Record 1969,* Vienna.

—— *Annual Review and Record 1970,* Vienna.

—— *Background Information,* Geneva, 1964.

—— *Explanatory Memoranda for the OPEC Resolutions,* Geneva, 1962.

—— *Offshore Oil Concession Agreements in OPEC Member Countries,* 1965.

—— *OPEC and the Oil Industry in the Middle East,* 1962.

—— *Resolutions Adopted at the Conferences of the Organization of the Petroleum Exporting Countries.*

—— *Statute of the Economic Commission.*

—— *The Statute of the Organization of the Petroleum Exporting Countries.*

Palestine, "Convention Regulating the Transit of Mineral Oils of Iraq Petroleum Company Ltd. through the Territory of Palestine," *Official Gazette of the Government of Palestine,* No. 276, 75–85, February 1, 1931.

—— "Oil Mining Ordinance," *Official Gazette of the Government of Palestine,* Supplement No. 1 to No. 793 of July 7, 1938 49–80.

Permanent Court of International Justice, *Collection of Advisory Opinions No. 12 Article 3 Paragraph 2, of the Treaty of Lausanne,* Leyden, 1925, Series B. No. 12, November 21st, 1925.

—— *Treaty of Lausanne, Article 3, Paragraph 2,* Leyden, 1926, Series C. No. 10. Acts and Documents Relating to Judgments and Advisory Opinions given by the Court. Documents Relating to Advisory Opinion No. 12.

Qatar Petroleum Company, *Review for 1966.*

—— *Review for 1967.*

—— *Review for 1968.*

—— *Review for 1969.*

—— *Review for 1970.*

—— *1971.*

Saudi Arabia, General Petroleum and Mineral Organization (Petromin), *Annual Report 1969.*

Saudi Arabian Monetary Agency, Research Department, *Annual Report 1380 A.H. (1961).*

—— *Annual Report 1381–82 A.H. (1962).*

—— *Annual Report 1382–83 A.H. (1963).*

—— *Annual Report 1383–84 A.H. (1964).*

—— *Annual Report 1384–85 A.H. (1965).*

—— *Annual Report 1385–86 A.H. (1966).*

—— *Annual Report 1386–87 A.H. (1967).*

—— *Annual Report 1387–88 A.H. (1968).*

—— *Annual Report 1388–89 A.H. (1969).*

—— *Annual Report 1389–90 A.H. (1970).*

—— *Statistical Summary,* December, 1963.

—— *Statistical Summary,* March, 1964.

—— *Statistical Summary,* November, 1965.

—— *Statistical Summary,* November, 1966.

—— *Statistical Summary,* October, 1967.

—— *Statistical Summary,* December, 1968–January, 1969.

—— *Statistical Summary,* September–October, 1969.

—— *Statistical Summary,* December, 1969–January, 1970.

—— *Statistical Summary,* September, 1970.

—— *Statistical Summary*, December, 1970.

—— *Statistical Summary*, September, 1971.

—— *Statistical Summary*, December, 1971.

—— *Statistical Summary*, September, 1972.

—— *Statistical Summary*, December, 1972.

Standard Oil Company (New Jersey), *Libya Today*.

—— *Standard Oil Company (N.J.) and Middle Eastern Oil Production*, New York, 1947.

Trans-Arabian Pipeline Company, *Kilometer 2113: Tapline Today*, n.p., 1959.

Turkey, *La Question du Mossoul de la Signature du Traitè d'Armistice de Moudros (30 Octobre 1918) au 1er Mars 1925*, Constantinople, 1925.

United Nations, Conciliation Commission for Palestine. *Final Report of the United Nations Economic Survey Mission for the Middle East*, 2 Vols., Lake Success, 1949. AAC. 25/6.

—— Department of Economic Affairs, *Public Finance Information Papers. Iran*, New York, 1951, ST/ECA/SER. A/4.

—— *Public Finance Information Papers, Iraq*, New York, 1951, ST/ECA/SER. A/5.

—— *Review of Economic Conditions in the Middle East 1949–50*, New York, 1951. Supplement to World Economic Report, 1949–50. E/1910/Add. 2/Rev. 1, ST/ECA/9/Add.2.

—— *Review of Economic Conditions in the Middle East 1951–52*, New York, 1953. Supplement to World Economic Report. E/2353/Add.1, ST/ECA/ 19/Add.1.

—— *Summary of Recent Economic Developments in the Middle East 1950–1951*, New York, 1950. Supplement to World Economic Report 1950–1951. E/2193/Add.3 (ST/ECA/14/Add.3).

—— *Summary of Recent Economic Developments in the Middle East 1952–53*, New York, 1954. Supplement to World Economic Report. E/2581, ST/ECA/ 25.

—— Department of Economic and Social Affairs, *Economic Developments in the Middle East, 1945 to 1954*, New York, 1955. Supplement to World Economic Report, 1953–54. E/2740 (ST/ECA/32).

—— *Economic Developments in the Middle East 1954–1955*, 151, New York, 1956. E/2880 ST/ECA/39.

—— *Economic Developments in the Middle East 1955–1956*, 135, New York, 1957. E/2983 ST/ECA/45.

—— *Economic Developments in the Middle East 1956–1957*, 163, New York, 1958. E/3116 ST/ECA/55.

—— *Economic Developments in the Middle East 1957–1958*, United Nations, 1959 (Supplement to World Economic Survey, 1958), E/3256 ST/ECA/61.

—— *Economic Developments in the Middle East 1958–1959*, New York, 1960. Supplement to World Economic Survey, 1959. E/3384 ST/ECA/64.

—— *Economic Developments in the Middle East 1959–1961*, New York, 1962. Supplement to World Economic Survey, 1961. E/3635 ST/ECA/69.

—— *Economic Developments in the Middle East 1961–1963,* Supplement to World Economic Survey, 1963, New York, United Nations, 1964. E/3910 ST/ECA/85.

—— Economic and Social Office in Beirut, *Studies on Selected Development Problems in Various Countries in the Middle East,* 1967, New York, United Nations, [1967]. E/4361.

—— *Studies on Selected Development Problems in Various Countries in the Middle East, 1968,* New York, United Nations, 1968. E/4511.

—— *Studies on Selected Development Problems in Various Countries in the Middle East, 1969,* New York, United Nations, 1969. E/4638.

—— *Studies on Selected Development Problems in Various Countries in the Middle East, 1970,* New York, United Nations, 1970. ST/UNESOB/7.

—— *Studies on Selected Development Problems in Various Countries in the Middle East,* New York, United Nations, 1971. ST/UNESOB/8.

—— Secretariat Economic Commission for Europe, *The Price of Oil in Western Europe,* Geneva, 1955. E/ECE/205.

—— Security Council, *Official Records,* First Year: First Series, No. 1, London.

—— *Official Records,* First Year: First Series, No. 2, New York.

—— *Official Records,* First Year: First Series, Supplement No. 1, London.

—— *Official Records,* First Year: First Series, Supplement No. 2, New York.

—— *Official Records,* Sixth Year, 559th Meeting, October 11, 1951, S/PV.559; 560th Meeting, October 15, 1951, S/PV.560; 561st Meeting, October 16, 1951, S/PV.561; 562nd Meeting, October 17, 1951, S/PV.562; 563rd Meeting, October 17, 1951, S/PV.563; 565th Meeting, October 19, 1951, S/PV.565.

—— *Official Records,* Sixth Year, Supplement for October, November and December 1951, New York, 1952.

United States, Economic Cooperation Administration, Special Mission to the United Kingdom, *The Sterling Area: An American Analysis,* London, 1951.

—— *Report to the Secretary of the Interior, for the Director of the Voluntary Agreement Relating to Foreign Petroleum Supply, as Amended May 8, 1956 concerning the Activities of the Foreign Petroleum Supply Committee under the Voluntary Agreement and the Activities of the Middle East Emergency Committee and Its Subcommittees under the Plan of Action for the Period April 1, 1956 Through June 30, 1957.*

—— Congress, *Congressional Record. Proceedings and Debates,* the following Vols.: 58, part 4; 59, parts 6, 7; 60, part 1; 89, parts 5, 6, 8; 90, parts 1, 2, 3, 6, 8, 11; 91, parts 1, 4, 8; 92, parts 6, 10, 11; 93, parts 2, 3, 4, 6, 7, 9, 10, 11; 94, parts 1, 2, 3, 4, 6, 9, 11; 95, parts 1, 9, 10, 12, 13; 96, part 15; 97, parts 13, 14; 98, parts 4, 8, 11; 99, parts 1, 10.

—— Congress, House of Representatives, *United States Aid Operations in Iran. First Report of the Committee on Government Operations,* Washington, 1957. 85th Congress, 2nd Session, House Report No. 10, January 28, 1957.

—— Congress, House of Representatives, Foreign Affairs Committee, *Economic and Military Cooperation with Nations in General Area of Middle East. Hearings,*

Washington. 85th Congress, H. Joint Res. 117, June 7–22, 1957.

— Congress, Joint Committee on Atomic Energy, *Peaceful Uses of Atomic Energy Background Material for the Report of the Panel on the Impact of the Peaceful Uses of Atomic Energy,* Washington, 1956.

— Congress, Senate, *Diplomatic Protection of American Petroleum Interests in Mesopotamia, Netherlands East Indies and Mexico,* Washington, 1945. 79th Congress, 1st Session, Senate Document No. 43.

— *Oil Concessions in Foreign Countries,* Washington, 1924. 68th Congress, 1st Session, Senate Document No. 97.

— *Restrictions on American Petroleum Prospectors in Certain Foreign Countries,* Washington, 1920. 66th Congress, 2nd Session, Senate Document No. 272.

— Congress, Senate, Committee on the Judiciary, *Petroleum, the Antitrust Laws and Government Policies, Report of the Subcommittee on Antitrust and Monopoly. Pursuant to Senate Resolution 57,* 85th Congress, Washington, 1957.

— Congress, Senate, Committee on the Judiciary, *Hearings before the Subcommittee to Investigate the Administration of the Internal Security Act and other Internal Security Laws. Problems Raised by Soviet Oil Development,* Eighty-seventh Congress, Second Session, 1963.

— Congress, Senate, Committee on the Judiciary, *Hearings before the Subcommittee to Investigate the Administration of the Internal Security Act and other Internal Security Laws. Soviet Oil in East-West Trade,* 87th Congress, Washington, 1962.

— Senate, Committee on the Judiciary, Eighty-seventh Congress, First Session, *Soviet Oil in the Cold War: A Study Prepared by the Library of Congress,* Washington, U.S. Government Printing Office, 1961.

— Congress, Senate, Foreign Relations Committee, *Anglo-American Oil Agreement. Report to Accompany Executive H,* Washington, 1947. 79th Congress, 1st Session, July 1, 1947.

— *Petroleum Agreement with Great Britain and Northern Ireland. Hearings,* Washington, 1947. 80th Congress, 1st Session, June 2–25, 1947, Executive H.

— Congress, Senate, Special Committee Investigating the National Defense Program, *Additional Report of the Subcommittee Concerning Investigations Overseas. Section 1 – Petroleum Matters,* Washington, 1944. 78th Congress, 2nd Session, Report No. 10, Part 15.

— *Hearings. Part 41 Petroleum Arrangements with Saudi Arabia,* Washington, 1948. 80th Congress, 1st Session, March 28, 29; May 8; October 29, 30, 31; November 1, 3, 4, 1947 and January 24, 29, 30, 1948.

— Congress, Senate, Special Committee Investigating Petroleum Resources, *American Petroleum Interest in Foreign Countries. Hearings, Petroleum Resources United States,* Washington, 1946. 79th Congress, 1st Session, June 27 and 28, 1945.

— *Investigation of Petroleum Resources, Hearings,* Washington, 1945, 79th Congress, 1st Session, June 19–25, 1945.

—— *Petroleum Requirements — Postwar. Hearings,* Washington, 1946. 79th Congress, 1st Session, October 3 and 4, 1945.

—— Congress, Senate, Special Committee to Study Problems of American Small Business, *Problems of American Small Business, Hearings.* Parts 21–24, 25–28, 33–34, 36, 38, and 40–44, October 15, 1947–January 10, 1948.

· — Congress, Senate Subcommittee of the Committee on the Judiciary and Committee on Interior and Insular Affairs, *Joint Hearings on the Emergency Oil Lift Program and Related Oil Problems,* Washington, 1957.

—— Department of Commerce, "Economic Development in Iraq 1955," *World Trade Information Service,* Part 1, No. 56–27, March, 1956.

—— "Economic Development in Kuwait 1954," *World Trade Information Service,* Part 1, No. 55–26, March, 1955.

—— "Economic Development in Kuwait 1955," *World Trade Information Service,* Part 1, No. 56–44, May, 1956.

—— "Economic Development in Kuwait 1956," *World Trade Information Service,* Part 1, No. 57–32, March, 1957.

—— "Laws on the Investment of Foreign Capital in Saudi Arabia," *World Trade Information Service,* Part 1, No. 57–75, September, 1957.

—— "Petroleum Law of Iran," *World Trade Information Service,* Part 1, No. 58–37, April, 1958.

—— Department of Commerce, "Saudi Arabia — Summary of Current Economic Information," *Business Information Service World Trade Series,* No. 345, 6–12, December, 1952.

—— Department of Commerce, Office of International Trade, "Arabian Peninsula Area — Summary of Economic Information," *International Reference Service,* V, No. 92, November, 1948.

—— "Economic Review of Iran, 1948," *International Reference Service,* VI, 1–8, July, 1949.

—— "Iran — Summary of Current Economic Information," *International Reference Service,* V, No. 85, November, 1948.

—— "Iraq — Summary of Basic Economic Information," *International Reference Service,* VII, 1–4, March, 1950.

—— "Iraq — Summary of Current Economic Information," *International Reference Service,* V, 1–5, December, 1948.

—— Department of State, *Mandate for Palestine,* Washington, 1927. State Department Publication Near Eastern Series, No. 1.

—— *Papers Relating to the Foreign Relations of the United States 1919, I & II* (beginning with 1932 titled *Foreign Relations of the United States Diplomatic Papers*), Washington, 1934.

—— *Papers . . .* 1920, I, II, III, Washington, 1935–36.

—— *Papers . . .* 1921, II, Washington, 1936.

—— *Papers . . .* 1922, II, Washington, 1938.

—— *Papers . . .* 1923, II, Washington, 1938.

—— *Papers . . .* 1924, II, Washington, 1939.

—— *Papers* . . . 1925, II, Washington, 1940.

—— *Papers* . . . 1926, II, Washington, 1941.

—— *Papers* . . . 1927, II, Washington, 1942.

—— *Papers* . . . 1929, III, Washington, 1944.

—— *Papers* . . . 1930, III, Washington, 1945.

—— *Papers* . . . 1931, II, Washington, 1946.

—— *Foreign Relations* . . . 1932, II, Washington, 1947.

—— *Foreign Relations* . . . 1934, I & II, Washington, 1951.

—— *Foreign Relations* . . . 1937, II, Washington, 1954.

—— *Rights of the United States of America and of Its Nationals in Iraq. Convention and Protocol between the United States of America and Great Britain and Iraq,* Washington, 1931. Treaty Series No. 835.

—— Department of State, *United States Policy in Middle East, September 1956– June 1957 Documents,* Washington. Near and Middle East Series 26, 1957.

—— Federal Trade Commission, *The International Petroleum Cartel, Staff Report submitted to the Subcommittee on Monopoly of the Select Committee on Small Business,* Washington, 1952. 82nd Congress, 2nd Session Committee Print No. 6.

—— *Report of the Federal Trade Commission on Foreign Ownership in the Petroleum Industry,* Washington, 1923.

—— Petroleum Administrator for War, *Petroleum in War and Peace,* Washington, 1945.

BOOKS AND ARTICLES

Abercrombie, Thomas J., "Saudi Arabia Beyond the Sands of Mecca," *National Geographic,* CXXIX, 1–53, January, 1966.

Abu al-Wafa, Ahmed, "Terms of Arbitration in the Oil Contracts," *Arab Oil Review,* 10–15, November–December, 1969.

"Abu Dhabi Areas for Mitsubishi," *Petroleum Press Service,* XXXV, 212–213, June, 1968.

'Abu Dahbi Fields and Finds," *Petroleum Press Service,* XXXVII, 245–247, July, 1970.

Abuhamdeh, Said, "Kuwait," *Middle East Journal,* XXXIII, 18–22, April, 1958.

"Action and Inaction in Kuwait," *Petroleum Press Service,* XXXIII, 136–138, April, 1966.

"The Active Middle East," *Petroleum Press Service,* XXXII, November, 1965, 406–409.

Acworth, Bernard, "Oil and Policy," *The English Review,* LVI, 119–128, February, 1933.

Adamiyat, Fereydoun, *Bahrein Islands: A Legal and Diplomatic Study of the British-Iranian Controversy,* New York, 1955.

Adams, T. W., "A Profile of the Turkish Oil Industry," *Lands East,* III, 7–11, October, 1958.

"Aden Refinery," *The Oil Forum,* VIII, 339–357, October, 1954.

Adler, Joseph L., "Difficulties Attending Exploration in Qatar," *The Oil Forum,* II, 178–179, May, 1948.

"After Fifty-Fifty," *Petroleum Press Service,* XXXI, 202–204, June, 1964.

"Agreement between the National Iran Oil Company and the Pan-American Petroleum Corporation," *Tehran Economist,* 2–15, May 5, 1958.

"Agreement in Sight," *Petroleum Press Service,* XXXI, 322–324, September, 1964.

Ajtony, M. A., *The Expanding Role of KNPC in the Oil Business,* Kuwait National Petroleum Co.

Alan, Ray, "Saudi Arabia: Oil, Sand, and Royalties," *The Reporter,* XIII, 18–20, December 1, 1955.

Al Abusi, Muhammad Jawad, *Albatrul fil balad al Arabiya* (Arabic: Petroleum in the Arab countries), Cairo, 1956.

Al Baki, Ahmad Abd, *Mizaniyat ad-dowla al-Iraqiya* (Arabic: The Budget of Iraq), Baghdad, 1947.

Ala'i, Heshmat, "How Not to Develop a Backward Country," *Fortune,* XXXVIII, 76–77, 145–147, August, 1948.

"Alaska – A Substitute for the Middle East?" (Hebrew) *Beolam Hadelek,* 92–94, July 1, 1969.

Al Husaini, Rajai, "Reconstruction and Development Projects in the Kingdom of Saudi Arabia," *Al Abhath,* VIII, 343–362, September, 1955.

Allen, George V., "American Advisers in Persia," *Department of State Bulletin,* XI, 88–93, July 23, 1944.

"All Maps Were Blank: Arabia's Great Desert Yields its Secrets to Oil Explorers," *The Lamp,* XLIII, 16–21, Fall, 1961.

Alnasrawi, Abbas, *Financing Economic Development in Iraq: The Role of Oil in a Middle Eastern Economy,* 1967, New York.

Al Witry, Hashim, *Health Services in Iraq,* Baghdad, 1944.

"American Oil Interests in Mesopotamia: The Turkish Petroleum Company," *Foreign Policy Association Information Service,* II, 68–78, May 22, 1926.

American Petroleum Institute, *Anglo-American Oil Agreement,* New York, 1944.

Amin, Ahmed, *Turkey in the World War,* New Haven, 1930.

Amin, Mahmoud S., "The Main Features of the Partnership Petroleum Agreements in the UAR and other Developing Countries," *L'Egypte Contemporaine,* LIX'eme Année, 165–186 (31–52), July, 1968.

Amiralaï, Chamseddine, *Le Pétrole et L'independence de L'Iran,* Aix-en-Provence, 1961.

"Ample Oil in Egypt," *Petroleum Press Service,* XXXV, 243–245, July, 1968.

Amps, L. W., "Kuwait Town Developments," *Journal of the Royal Central Asian Society,* XL, 234–240, July–October, 1953.

"Anglo-Iranian's Nationalization Losses Cut," *The Oil Forum,* VIII, 297–299, 307, September, 1954.

Apak, Kemal, "Turkey's Growing Role in Mid-East," *The Oil Forum,* VI, 380, November, 1952.

Apremont, B., "La politique Pétrolière de l'U.R.S.S.," *Politique Etrangere,* 25 Année, 572–583, November 6, 1960.

"The Arab Oil Conference," *AICC Economic Review,* XI, 37–38, May 1, 1959.

"Arab Oil. First Arab Petroleum Congress," *Arab World,* V, 1–15, May–June, 1959.

"Arab Petroleum Congress Espouses OPEC Program," *World Petroleum,* XXXIV, 44–45, December, 1963.

"Arab Wealth From Oil," *The Economist,* CLXX, 591–2, February 27, 1954.

"Arabia and the West," *Round Table,* 323–328, September, 1957.

"Arabian Middle Class Growing Rapidly," *The Oil Forum,* III, 207, May, 1949.

"The Arabs and the Companies," *Petroleum Press Service,* XXXII, 166–170, May, 1965.

"The Arabs and the Companies," *Petroleum Press Service,* XXXIX, 247–249, July, 1972.

"Aramco," *Life,* 62–79, March 28, 1949.

"Aramco's Success Story Continued," *Petroleum Press Service,* XXIV, 259–261, July, 1957.

"The Arrested Talks in Iraq," *Petroleum Press Service,* XXVIII, 164–166, May, 1961.

Aschner, Ernst, "Haifa Refinery Played Important Part in Supporting Allied Operations in Mediterranean Theatre," *World Petroleum,* XVI, 44–46, December, 1945.

"Asia–Middle East," *World Oil,* CLIX, 148–178, August 15, 1964.

"Asia–Middle East," *World Oil,* CLXI, 161–191, August 15, 1965.

"Asia–Middle East," *World Oil,* CLXIII, 166–187, August 15, 1966.

"Asia–Middle East," *World Oil,* CLXV, 132–160, August 15, 1967.

"Asia–Middle East," *World Oil,* CLXVII, 198–216, August 15, 1968.

Atiyeh, Munir, "Petrochemicals: World Trends and Arab Prospects," *Middle East Forum,* XXXVIII, 25–27, January, 1962.

"The Atlantic Report: Soviet Oil," *Atlantic Monthly,* CCVII, 12–22, February, 1961.

Atyeo, Henry C., "Political Developments in Iran 1951–1954," *Middle Eastern Affairs,* V, 249–259, August–September, 1954.

Audemar, Jean, *Les Maîtres de la Mer, de la Houille et du Pétrole. L'impéralisme Anglo-Saxon,* Paris, 1923.

Awsenew, M., *Der englisch-armikanische Kampf um das Erdöl nach dem zweiten Weltkrieg,* Berlin, 1956.

Azami, Zangneneh (Abdul Hamid), *Le Pétrole en Perse,* Paris, 1933.

"Back of the Explosion in Persia," *World Petroleum,* IV, 9–11, January, 1933.

Baer, Gabriel, "The Agrarian Problem in Iraq," *Middle Eastern Affairs,* III, 381–391, December, 1952.

Baker, Robert L., *Oil, Blood and Sand,* New York, 1942.

Baldwin, George B., *Planning and Development in Iran,* Baltimore, 1967.

Balfour, J. M., *Recent Happenings in Persia,* London, 1922.

Barber, C. T., "World Market Trends and the Role of Middle East Oil," *The Oil Forum*, X, 19–20, 22, January, 1956.

Barr, Robert J., "Iraq Today," *Foreign Commerce Weekly*, X, 6–9, 11, February 20, 1943.

Barrows, Gordon Hensley, *The International Oil Companies 1967*, New York, 1967.

—— *International Petroleum Industry*, Volume I, World/Europe/Middle East, New York, 1965.

Bauer, C. J., "Middle East Oil and World Markets," *Mining and Metallurgy*, XXIX, 436–42, August, 1948.

Bayne, Edward Ashley, "Crisis of Confidence in Iran," *Foreign Affairs*, XXIX, 578–590, July, 1951.

Belgrave, James H. D., "A Brief Survey of the History of the Bahrain Islands," *Journal of the Royal Central Asian Society*, XXXIX, 57–68, January, 1952.

—— *Welcome to Bahrain: A Complete Illustrated Guide for Tourists and Travellers*, London, 1953.

Bell, E. A., "Mineral Oil in Turkey," *World Petroleum*, XXIII, 94, 112, September, 1952.

Bell, James, "He Said Forward! To the Backward," *Life*, 157–8 etc., November 17, 1952.

Berent, Khsan Ruhi, "Door Open to Private Exploitation of Turkish Oil," *The Oil Forum*, II, 48–51, February, 1948.

Berreby, Jean-Jacques, *Histoire Mondiale du Pétrole*, Paris, 1961.

—— "Impératifs strategiques du Pétrole," *Politique Etrangère*, XXX, No. 6, 498–516, 1965.

—— *La Peninsula Arabique*, Paris, 1958.

—— *Le Gulfe Persique*, Paris, 1959.

—— "Les politiques pétrolières de l'Atlantique au golfe persique," *Orient*, XXIX, No. 1, 173–201, 1964.

—— "Pétrole et Panarabisme," *Politique Etrangère*, Année 23, No. 4, 402–413, July, 1958.

—— "Pétrole et Politique dans le Golfe Persique," *L'Afrique et L'Asie*, 15–24, 4e, Trimestre, 1958.

Berteloot, Joseph, "Aux Lueurs du Pétroles Gênes, La Haye, Lausanne," *Etudes*, 175, 307–331, April–May–June, 1923.

Better Oil Outlook in Egypt," *Petroleum Press Service*, XXXI, 463–465, December, 1964.

"Big International Consortium Will Build New Sumed Pipeline," *World Petroleum*, XLII, 13–14, November, 1971.

"The Billionaire Governments," *Petroleum Press Service*, XXXIX, 321–323, September, 1972.

Binder, Leonard, *Iran: Political Development in a Changing Society*, Berkeley, 1962.

Birdwood, C. B., "Oil in the Middle East," *World Affairs*, IV, 48–59, January, 1950.

—— "Reflections on the Persian Oil Dispute," *The Twentieth Century*, CLII, 125–132, August, 1952.

"The Birth of INOC," *Petroleum Press Service*, XXXI, 86–88, March, 1964.

Bleiber, Fritz, "Die Bahrain Inseln," *Zeitschrift für Geopolitik*, VI, 988–993, November, 1929.

Bochenski, Feliks, and William Diamond, "TVA's in the Middle East," *The Middle East Journal*, IV, 52–82, January, 1950.

Boesch, H., "Erdöl in Mittleren Osten," *Erdkunde*, III, 68–82, August, 1949.

—— *Wasser Oder Oel. Ein Buch über den nahen Osten*, Bern, 1944.

"Bold Expansion in Iraq," *Petroleum Press Service*, XXVI, 97–98, Mar., 1959.

Bomli, P. E. J., *L'Affaire de Mossoul*, Amsterdam, 1929.

"Boost for Iraq's Revenues," *Petroleum Press Service*, XXXVII, 442–444, December, 1970.

Boveri, Margret, *Minaret and Pipe-Line*, New York, 1939.

—— "Pétrole et Moyen-Orient," *En Terre d'Islam*, 3/40, 273–277, 4e Trimestre, 1947.

Brickhouse, A. A., "Tapline's Sidon Terminal," *World Petroleum*, XXVIII, 42–48, July, 1957.

"Britain, Persia and Oil," *The Economist*, CLX, 670, 671, March 24, 1951.

Brodie, Bernard, *Foreign Oil and American Security* (Mimeographed), New Haven, 1947.

Brodie, G. H., "Oil and Gas in Israel," *Economic Horizons*, XI, 8–11, February, 1959.

Brooks, Michael, *Oil and Foreign Policy*, London, 1949.

Brown, Edward Hoagland, *The Saudi Arabia Kuwait Neutral Zone*, Beirut, 1963.

Brown, Russel, M., "Schooling in Skills," *Oil Lifestream of Progress*, IV, 6–8, October, 1954.

Browne, Edward G., *The Persian Revolution of 1905–1909*, Cambridge (Eng.), 1919.

Brundage, H. T., "Saudi Arab Offshore Field to Go on Production Soon," *World Oil*, CXLIV, 238–239, 256–261, April, 1957.

B. S.- E., "The Middle East: Background to the Russian Intervention," *The World Today*, XI, 463–471, November, 1955.

Bullard, Reader, "Behind the Oil Dispute in Iran: A British View," *Foreign Affairs*, XXXI, 461–471, April, 1953.

—— *Britain and the Middle East, From the Earliest Times to 1950*, London, 1951.

—— "Persian Oil," *The Fortnightly*, CLXXI, N.S., 219–225, April, 1952.

Burck, Gilbert, "A Strange New Plan for World Oil," *Fortune*, 94–97, 146–148, Aug., 1959.

Burns, Norman, "The Dujaylah Land Settlement," *The Middle East Journal*, V, 362–366, Summer, 1951.

Bustani, Emile, "Sharing Oil Benefits," *Middle East Forum*, XXXIII, 9–13, 33–35, January, 1958.

Byrnes, James F., *Speaking Frankly*, New York, 1947.

Cadman, John, "Great Britain and Petroleum," *American Petroleum Institute Bulletin*, 20–23, December 21, 1921.

—— The World's Unexploited Resources," *The Petroleum Times*, VIII, 953–954, December 30, 1922.

"Cairo Forum," *Petroleum Press Service*, XXVI, 168–170, May, 1959.

"Cairo Oil Congress Worries Producers," *Petroleum Week*, V, 44–46, September 6, 1957.

"Calmer Arab Congress," *Petroleum Press Service*, XXVIII, 403–407, November, 1961.

Calvocoressi, Peter (Ed.), "The Middle East," *Survey of International Affairs 1951*, 255–337, London, 1954.

—— (Ed.), "The Middle East and the Arab West," *Survey of International Affairs 1952*, 191–298, London, 1955.

Campbell, Robert W., *The Economics of Soviet Oil and Gas*, Baltimore, 1968.

"Can Persia Plan," *The Economist*, CLVIII, 982–3, May 6, 1950.

"Can the West do without Middle Eastern Oil?", *Western World*, 16–23, September, 1958.

Carey, Jane Perry Clark and Andrew Galbraith, "Oil and Economic Development in Iran," *Political Science Quarterly*, LXXV, 66–86, Mar., 1960.

Carmical, J. H., *The Anglo-American Petroleum Pact*, New York.

Carmichael, Keith, "The Middle East Holds Its Own," *Aramco World*, XVII, 20–25, November–December, 1966.

Caroe, Olaf, *Wells of Power: The Oilfields of South-Western Asia*, London, 1951.

Carter, John, "The Bitter Conflict over Turkish Oilfields," *Current History*, XXIII, 492–495, January, 1926.

Case, Paul Edward, "Boom Time in Kuwait," *The National Geographic Magazine*, CII, 783–802, December, 1952.

Cattan, Henry, *The Evolution of Oil Concessions in the Middle East and North Africa*, Dobbs Ferry, 1967.

—— *The Law of Oil Concessions in the Middle East and North Africa*, 1967.

The Chase Manhattan Bank, *Annual Financial Analysis of a Group of Petroleum Companies*.

—— *Annual Financial Analysis of a Group of Petroleum Companies 1965*.

—— *1967 Annual Financial Analysis of a Group of Petroleum Companies*.

—— *Annual Financial Analysis of a Group of Petroleum Companies 1968*.

—— *Annual Financial Analysis of a Group of Petroleum Companies 1969*.

—— *Annual Financial Analysis of a Group of Petroleum Companies 1970*, n.p., n.d.

—— *Balance of Payments of the Petroleum Industry*.

—— *Capital Investments of the World Petroleum Industry 1965*.

—— *1968 Capital Investments of the World Petroleum Industry*, 1969.

—— *Capital Investments of the World Petroleum Industry 1971*, New York, 1972.

—— *1963 Financial Analysis of Thirty-Two Petroleum Companies*.

—— Petroleum Division, *1964 Financial Analysis of Thirty-One Petroleum Companies*, 1965.

—— *1966 Financial Analysis of a Group of Petroleum Companies*, 1967.
—— *1971 Financial Analysis of a Group of Petroleum Companies.*
—— *Future Growth of the World Petroleum Industry*, New York, 1957.
—— *Future Growth of the World Petroleum Industry*, New York, 1958.
—— *Future Growth of the World Petroleum Industry.*
—— Energy Division, *Outlook for Energy in the United States*, New York, 1958.
—— *Petroleum Industry 1958*, 1959.
—— Energy Economics Division, *Outlook for Energy in the United States to 1985*, 1972.
"Chemicals in the Persian Gulf," *Petroleum Press Service*, XXXVI, 56—57, February, 1969.
Cheney, Michael Sheldon, *Big Oil Man from Arabia*, New York, 1958.
Cheng, B., ' The Anglo-Iranian Dispute," *World Affairs*, V, 387—405, October, 1951.
Childs, James Rives, (Henry Filmer), *The Pageant of Persia*, New York, 1936.
Childs, Marquis, "All the King's Oil," *Collier's* 22, 47, August 18, 1945.
Chisholm, A. H., "Anglo-Iranian Answers Iran With Facts," *The Oil Forum*, VI, Special Insert i—xviii, April, 1952.
Churchill, Winston S., *The World Crisis*, I, New York, 1923.
Clair, Pierre, *L'Independance Pétrolière de La France: 1. Le Théâtre de Guerre*, Paris, 1968.
Clapp, Gordon R., "Iran: A TVA for the Khuzestan Region," *Middle East Journal*, XI, 1—11, Winter, 1957.
Clawson, Marian (Ed.), *National Resources and International Development*, Baltimore, 1964.
"Clouds along the Persian Gulf," *The Economist*, CLXVIII, 189—191, July 18, 1953.
Cluzeau, Marc, "Babagurgu la plus vieille Raffinerie du Monde?" *Pétrole Progrès*, 7—9, July, 1952.
Coffman, Paul B., *Iran's Seven-Year Development Plan*, Washington, 1949.
Cohen, Aaron, "Oil as an economic and political factor in the Middle East," (Hebrew), *Bashaar*, March—April, May—August, 1967, (Reprint).
"Congress in Kuwait," *Petroleum Press Service*, XXXVII, 177—179, May, 1970.
Constantino, D., "Petrolio Siriano," *Levante*, XIII, No. 1, 3—57, 1966.
"Construction du plus Haut Pipe-Line du Monde," *L'Asie Nouvelle*, 59—61, March, 1957.
"Contrasts in Oil Legislation," *Petroleum Press Service*, XXXI, 407—410, November, 1964.
Cooke, Hedley V., *Challenge and Response in the Middle East: The Quest for Prosperity, 1919—1951*, New York, 1952.
—— "Foreign Investment in the Middle Eastern Region, 1944—1953," *Middle Eastern Affairs*, V, 109—115, April, 1954.
Coon, Carleton S., *Caravan: The Story of the Middle East*, New York, 1951.

Coqueron, Frederic G. et al, *Annual Financial Analysis of the Petroleum Industry 1957*, New York, 1958.

—— *Annual Analysis of the Petroleum Industry 1959*, 1960.

—— *Annual Analysis of the Petroleum Industry 1960*, 1961.

—— *Annual Analysis of the Petroleum Industry 1961*, New York, 1962.

—— *Petroleum Industry 1960*, New York, 1961.

—— and J. E. Pogue, *Investment Patterns in the World Petroleum Industry*, New York, 1956.

"Countering the Suez Threat," *Petroleum Press Service*, XXIII, 318–320, September, 1956.

Crutians, Léon, *La Mésopotamie et la Lutte pour les Pétroles de Mossoul*, Paris, 1927.

Cullum, D. M., "The Kharg Story," *Asian Review*, LVIII, 84–92, July, 1962.

Cumming, Henry H., *Franco-British Rivalry in the Post-War Near East*, London, 1938.

Curzon, George N., *Persia and the Persian Question*, London, 1892.

Da Cruz, Daniel, "The Long Steel Shortcut," *Aramco World*, XV, 16–25, September–October, 1964.

Dalley, C. M., "The Middle East," *Petroleum Review*, XXV, 10–16, January, 1971.

Davenport, E. H. and S. R. Cooke, *The Oil Trusts and Anglo-American Relations*, London, 1923.

Davis, C. E., "Making up for Suez," *American Petroleum Institute Quarterly*, 10–12, Winter, 1956–1957.

Davis, Helen Miller, *Constitutions, Electoral Laws, Treaties of States in the Near and Middle East*, Durham, 1947.

Day, M. E., "Partnerships of British Companies," *World Petroleum*, I, 26–27, February, 1930.

De Candole, E. A. V., "Development in Kuwait," *Royal Central Asian Society Journal*, XLV, 27–38, January, 1959.

—— "Developments in Kuwait," *Royal Central Asian Society Journal*, XLII, 21–29, January, 1955.

DeGaury, Gerald, "The End of Arabian Isolation," *Foreign Affairs*, XXV, 82–89, October, 1946.

DeGolyer, E., "The Oil Fields of the Middle East," in *Problems of the Middle East – Proceedings of a Conference, School of Education, New York University*, 1947.

—— "Preliminary Report of the Technical Oil Mission," *Bulletin of the American Association of Petroleum Geologists*, XXVIII, 919–921, July, 1944.

—— and MacNaughton, *Twentieth Century Petroleum Statistics 1954*, Dallas, Texas, 1954.

Delaisi, Francis, *Oil, Its Influence on Politics*, London, 1922.

de Lavilleon, Patrick, "More Influence for France in Middle East," *World Petroleum*, XXXVII, 14, 82, November, 1966.

de Morgan, M.J., "Notes sur les gîtes des Naphte de Kend-é-Chirin – (governement de Ser-i-Roul)," *Annales des Mines,* 227–238, 1892.

Dennis, Alfred, "The United States and the New Turkey," *The North American Review,* CCXVII, 721–731, June, 1923.

Denny, Ludwell, *We Fight for Oil,* New York, 1928.

De Novo, John A., "Petroleum and American Diplomacy in the Near East, 1908–1928," Doctoral dissertation, Yale University, 1948.

Desprairies, P., "Europe's Oil – Today and Tomorrow," (Hebrew), *Beolam Hadelek,* 95–103, July, 1969.

"Development in Iraq," *The Economist,* CLXXXIII, 1–16, June 22, 1957.

Dickson, H. R. P., *Kuwait and her Neighbors,* London, 1956.

Dinstein, Zvi, "The Petroleum Industry after the Six-Day War" (Hebrew), *Beolam Hadelek,* 11–14, July, 1969.

Dixon, Homer, "Cool, Condense and Ship," *Aramco World,* XVI, 8–12, July–August, 1964.

"Djakarta's Green Light," *Petroleum Press Service,* XXXII, 2–4, January, 1965.

Djamalzadeh, M. A., "An Outline of the Social and Economic Structure of Iran," *International Labour Review,* LXIII, 24–39, January, 1951.

Drayton, Geoffrey, "The Travails of OPEC," *The World Today,* XIX, 485–491, November, 1963.

Duce, James Terry, "The Changing Oil Industry," *Foreign Affairs,* XL, 627–634, July, 1962.

–– *Middle East Oil Developments,* [New York], 1952.

E. M., "Cause and Effect in Persia," *The World Today,* VII, 329–338, August, 1951.

Earle, E. M., "The Secret Anglo-German Convention of 1914 regarding Asiatic Turkey," *Political Science Quarterly,* XXXVIII, 24–44, January, 1923.

Earle, Edward Mead, *Turkey, The Great Powers, and the Baghdad Railway. A Study in Imperialism,* New York, 1923.

–– "The Turkish Petroleum Company: A Study in Oleaginous Diplomacy," *Political Science Quarterly,* XXXIX, 265–279, June, 1924.

"East Europe's Five-Year Plans," *Petroleum Press Service,* XXXVIII, 287–289, August, 1971.

Ebel, Robert E., *Communist Trade in Oil and Gas – An Evaluation of the Future Export Capability of the Soviet Bloc,* New York, 1970.

–– *The Petroleum Industry of the Soviet Union,* n.p., n.d.

Eden, Anthony, *The Memoirs of Anthony Eden. Full Circle,* Boston, 1960.

Edgar, E. Mackay, "Britain's Hold on the World's Oil," *Sperling's Journal,* V, 38–49, September, 1919.

Edwards, A. C., "Persia Revisited," *International Affairs,* XXIII, 52–60, January, 1947.

–– "Persia Since Pahlevi," *The Geographical Magazine,* XXIV, 77–88, June, 1951.

"Egypt: The Production Picture," *Petroleum Press Service,* XXXVIII, 95–97, March, 1971.

"Egypt Wedded to Oil," *The Egyptian Economic & Political Review,* VII, 14–16, March, 1961.

"The Egyptian Oil Industry after the Israel Defense Forces Victory in Sinai," (Hebrew), *Beolam Hadelek,* XV, 51–54, September, 1967.

"Egypt," *World Petroleum Report,* III, 91–92, January 15, 1957.

Eigeland, Tor, "Iran Celebrates 2,500 years of History and the White Revolution," *The Lamp,* LIII. 4–12, Winter, 1972.

El Ayouty, Mustafa K., "Egypt's Oil Prospects Today," *World Petroleum,* XL, 36–44, May, 1969.

El-Bakkary, Farouk Muhamed, "The Legal, Economic and Political Effects Resulting From the Principle of the State Ownership of the Mineral Wealth in its Territory," *L'Egypte Industrielle,* 35e Année, 24–37, April, 1959.

El-Hashimi, Sayed, "Whither King Saud," *Contemporary Review,* CLXXXII, 14–17. July, 1957.

El-Husseiny, Farouk, "The Economics of the Petroleum in the Middle East," *L'Egypte Industrielle,* 35e Année, 6–23, April, 1959.

Eller, E. M., "Troubled Oil and Iran," *United States Naval Institute Proceedings,* LXXX, 1189–99, November, 1954.

Ellis, H. S., *United States Interests in the Middle East: Private Enterprise and Socialism in the Middle East,* n.p., 1970.

Elwell-Sutton, L. P., *Modern Iran,* London, 1941.

"The Emirs Unite," *Petroleum Press Service,* XXXIX, 7–9, January, 1972.

'The Empty Quarter," *Standard Oil Company of California Bulletin,* 1–7, April, 1961.

Evans. G. F., "Oil States of the Persian Gulf," *Contemporary Review,* 117–120, February, 1959.

Evans, William S., *Petroleum in the Eastern Hemisphere,* New York, 1959.

"Exit from Israel," *Petroleum Press Service,* XXIV, 326–328, September, 1957.

Fahmy, Hussein, "Egypt's Third Oil Boom Results from Legislative Incentives," *The Oil Forum,* VIII, 268–269, August, 1954.

Fanning, Leonard M. *American Oil Operations Abroad,* New York, 1947.

— — *Foreign Oil and the Free World,* New York, 1954.

— — "Growing Share of World's Oil is held by American Firms," *World Oil,* CXLI, 132–134, August 15, 1955.

Farmanfarmian, Khodadad, "The Oil Industry and Native Enterprise in Iran," *Middle Eastern Affairs,* VIII, 333–342, October, 1957.

Faroughy, Abbas, *The Bahrein Islands (1750–1951),* New York, 1951.

Fartache, Manoutchehr, "De la Compétence de la Cour Internationale de Justice dans l'Affaire de l'Anglo-Iranian Oil Co.," *Revue Générale de Droit International Public,* 57, 584–612, October–December, 1953.

— — "Le Développement économique et les Problèmes politiques du Moyen-Orient," *Politique Etrangère,* XVIII, 23–34, March–April, 1953.

Fatemi, Nasrollah Saifpour, *Diplomatic History of Persia 1917–1923,* New York, 1952.

—— *Oil Diplomacy: Powderkeg in Iran,* New York, 1954.

—— "Tensions in the Middle East," *The Annals of the American Academy of Political and Social Science,* CCLXXII, 53–59, July, 1952.

Faust, Jean-Jacques, "Koweit ou la sagacité," *Etude Méditerranéennes,* No. 6, 15–33, Winter, 1959.

Feis, Herbert, "Oil for Peace or War," *Foreign Affairs,* XXXII, 416–429, April, 1954.

—— *Petroleum and American Foreign Policy,* Stanford, 1944.

—— *Seen from E. A. Three International Episodes,* New York, 1947.

Fifth Arab Petroleum Congress. Information, Observation and Documentation, *Beirut, 1965.*

"Fifty Years of Iranian Oil," *Institute of Petroleum Review,* XII, 307–310, September, 1958.

Finnie, David, "Business in Basra and the Oil Industry," *Middle East Economic Papers, 1956,* 38–51, 1956.

—— *Desert Enterprise. The Middle East Oil Industry in Its Local Environment,* Cambridge, Mass., 1958.

—— "Recruitment and Training of Labor. The Middle East Oil Industry," *Middle East Journal,* XII, 127–143, Spring, 1958.

"The First Arab Petroleum Congress," *The Petroleum Times,* LXIII, 356–358, May 8, 1959.

"First Reckoning for Suez," *The Economist,* CLXXXI, 615–618, November 17, 1956.

Fischer, A. J., "Persia: A New Departure," *The Contemporary Review,* CLXXXIV, 261–264, November, 1953.

—— "Persia between Feudalism and Terror," *The Contemporary Review,* CLXXXII, 74–80, August, 1952.

Fischer, Louis, *Oil Imperialism. The International Struggle for Petroleum,* New York, 1926.

Fishman, Herzl, "A Chapter in the History of Oil Concessions in Palestine," (Hebrew), *Hamizrah Hehadash,* II, 16–20, October, 1950.

Fisk, Brad, "Dujaila: Iraq's Pilot Project for Land Settlement," *Economic Geography,* XXVIII, 343–355, October, 1952.

"Focus on Middle East Gas," *Petroleum Press Service,* XXXII, 46–50, February, 1965.

Fohs, F. Julius, "Resources of the Middle East," *Technion Yearbook 1948,* 49–59.

Fontaine, Pierre, *La Guerre occulte du Pétrole,* Paris, 1949.

Forbin, Victor, "Visite aux Puits de Pétroles de l'Irak," *Revue de Deux Mondes,* XX, 862–885, April 15, 1934.

Ford, Alan W., *The Anglo-Iranian Oil Dispute of 1951–1952,* Berkeley, 1954.

"Forty Years of Education in Bahrain," *The Islamic Review,* XLVII, 35–36, July–August, 1959.

Foster, Henry A., *The Making of Modern Iraq. A Product of World Forces,* Norman, Oklahoma, 1935.

Fourth Arab Petroleum Congress, Beirut, 1963.

Fox, A. F., "A Short History of Exploration in Kuwait," *World Petroleum,* XXVIII, 94–102, 107, September, 1957.

"France Still Odd Man Out," *Petroleum Press Service,* XXXIX, 327–329, September, 1972.

Frank, J. Helmut, *Crude Oil Prices in the Middle East. A Study in Oligopolistic Price Behavior,* New York, 1965.

Frankel, J., "The Anglo-Iranian Dispute," *The Year Book of World Affairs 1952,* 75 97.

Frantz, Joe B., "Stockpile Will Enable the West to Negotiate from Strength," *The Oil Forum,* XI, 224–225, July, 1957.

Freeth, Zahara, *Kuwait Was My Home,* London, 1956.

Frendzel, Donald J., *The Soviet Seven-Year Plan (1959–65) for Oil,* An Appendix to United States Congress, Senate, Committee on the Judiciary, 87th Congress, *Hearings before the Subcommittee to Investigate the Administration of the Internal Security Act and other Internal Security Laws. Soviet Oil in East-West Trade,* Washington, 1962.

"A Fresh Start for Aramco," *Petroleum Press Service,* XXX, 174–175, May, 1963.

Friedwald, E. M., *Oil and the War,* London-Toronto, 1941.

"From Caracas to Vienna," *Petroleum Press Service,* XXXIII, 42–44, February, 1966.

"From Riyadh to Geneva," *Petroleum Press Service,* XXX, 121–123, April, 1963.

Froozan, Mansur, "Can the Under-Exploitation of Iranian Oilfields be Justified on Economic Grounds," *News Letter,* 7–19, March, 1963.

Fry, H. I., "Egypt's new important Oil Field," *The Oil Forum,* III, 86, February, 1949.

Frye, Richard N., *Iran,* New York, 1953.

— (Ed.), *The Near East and the Great Powers,* Cambridge, Mass., 1951.

"Future of Middle East Development clarified by Red Line Settlement," *World Petroleum,* XIV, 46–48, December, 1948.

"Future of Russian Oil," *World Petroleum,* XXXII, 59–63, September, 1961.

Gardner, L. S., "Oil Possibilities in the Hashemite Kingdom of Jordan," *The Oil Forum,* VIII, 448–450, 454, March, 1954.

Garfias, Valentin R., *Petroleum Resources of the World,* New York, 1923.

Gass, Oscar, "Must the West Depend on Mideast Oil?" *Commentary,* XXVIII, 42–50, July, 1959.

Gerig, Benjamin, *The Open Door and the Mandate System,* London, 1930.

Ghosh, Sunil Kanti, *The Anglo-Iranian Oil Dispute,* Calcutta, 1960.

Gibbon, Anthony, "Western Desert, Egypt Find May Open New Oil Province," *World Oil,* CLXIV, 69–72, March, 1967.

Gibbs, Leo Vernon, *Oil and Peace,* Los Angeles, 1929.

Gibson, Ray, "Consumption Grows Normally Despite Suez Shutdown," *World Petroleum,* XXXIX, 20–22, January, 1968.

Gillman, William, "No Commercial Results from Million Dollar Campaign in

Turkey," *World Petroleum*, IX, 82–83, July, 1938.

"Good Days for Iran," *Petroleum Press Service*, XXXVIII, 249–251, July, 1971.

"Government Looks to International Oil," *World Petroleum*, XIV, 30–31, 62, November, 1943.

"The Government's Share," *Petroleum Press Service*, XXX, 322–324, September, 1963.

Grady, Henry F., "What Went Wrong in Iran?" *The Saturday Evening Post*, 30, etc., January 5, 1952.

Graf, George Engelbert, *Erdöl, Erdölkapitalismus und Erdölpolitik*, Jena.

"Grappling With Emergency," *Petroleum Press Service*, XXIII, 434–437, December, 1956.

Graves, Philip, *The Life of Sir Percy Cox*, London, 1941.

"The Great Oil Deals," *Fortune*, XXXV, 138–143, May, 1947.

Groseclose, Elgin, *Introduction to Iran*, New York, 1947.

Grover, David H., "Whither the Supertanker?", *The Oil Forum*, XI, 204–205, June, 1957.

Grunwald, K., "The Oilfields of the Middle East," *Economic News*, IV, 5–12, 67–83, October–December, 1951.

Gupta, Raj Narain, "Iranian Oil," *Journal Indian Institute of International Affairs*, III, 11–28, January, 1947.

–– *Oil in the Modern World*, Allahabad, 1950.

Habermann, Stanley John, "The Iraq Development Board: Administration and Program," *The Middle East Journal*, IX, 179–186, Spring, 1955.

Hakim, George, "Economic Development in the Middle East," *India Quarterly*, VIII, 207–217, April–June, 1952.

Hale, William Harlan, "Troubled Oil in the Middle East," *The Reporter*, XVIII, 8–12, January 23, 1958.

Hall, Harvey P., "The Arab States: Oil and Growing Nationalism," *Current History*, XXI, 19–23, July, 1951.

–– (Ed.), *Middle East Resources: Problems and Prospects*, Washington, 1954.

Hamilton, Charles W., *Americans and Oil in the Middle East*, Houston, 1962.

Hamzavi, A. H. K., *Persia and the Powers: An Account of Diplomatic Relations, 1941–46*, London, 1946.

Hanighen, Frank Cleary, *The Secret War*, New York, 1934.

Harding, C. L., "Major Shortage Without Middle East Oil," *The Oil Forum*, II, 182, 185, May, 1948.

Hardinge, Arthur H., *A Diplomatist in the East*, London, 1928.

Harrington, Charles W., "The Saudi Arabian Council of Ministers," *The Middle East Journal*, XII, 1–19, Winter, 1958.

Harrison, David Lakin, *Footsteps in the Sand*, London, 1959.

Hartshorn, J. E., *Oil Companies and Governments*, London, 1962.

Hassmann, Heinrich, *Erdöl aus dem Mittleren Osten*, Hamburg, 1950.

–– *Oil in the Soviet Union, History, Geography, Problems*, Princeton, 1953.

Hay, Rupert, "Great Britain's Relations with Yemen and Oman," *Middle Eastern*

Affairs, XI, 142–149, May, 1960.
—— "The Impact of the Oil Industry on the Persian Gulf Shaykdoms," *The Middle East Journal,* IX, 261–372, Autumn, 1955.
—— *The Persian Gulf States,* Washington, 1959.
—— "The Persian Gulf States and their Boundary Problems," *The Geographical Journal,* CXX, 431–445, December, 1954.
Hayden, Lyle S., "Living Standards in Rural Iran: A Case Study," *The Middle East Journal,* III, 140–150, April, 1949.
Hearn, Arthur, "Oil and the Middle East," *International Affairs* XXIV, 63–75, January, 1948.
Heller, Charles A., "The Role of Oil in the Middle East Economy 1961 to 1963," *World Petroleum,* XXXV, 82–84, October, 1964.
—— "The Strait of Hormuz – Critical in Oil's Future," *World Petroleum,* XL, 24–26, October, 1969.
—— "Ten Years OPEC," *World Petroleum,* XLI, 46–54, November, 1970.
Helmreich, P. C., "Oil and the negotiation of the Treaty of Sévres, December 1918–April 1920," *Middle East Forum,* XLII, No. 3, 67–76, 1966.
Hendryx, Frank, *"It is time for a New Approach to Arab Concession Negotiations,"* World Petroleum, XXXVII, 58–62, September, 1966.
Hesse, Fritz, *Die Mossulfrage,* Berlin, 1925.
Hewins, Ralph, *A Golden Dream. The Miracle of Kuwait,* London, 1963.
—— *Mr. Five Per Cent: The Story of Colouste Gulbenkian,* New York, 1958.
Heyworth-Dunne, J., "Report from Saudi Arabia," *Jewish Observer and Middle East Review,* IV, 13–14, January 14, 1955.
"High Payment to Kuwait Ruler by Aminoco," *The Oil Forum,* II, 314, August, 1948.
Himadeh, Said B., "Economic Factors Underlying Social Factors," *The Middle East Journal,* V, 269–283, Summer, 1951.
Hindus, Maurice, *In Search of a Future, Persia, Egypt, Iraq and Palestine,* New York, 1949.
Hirst, David, *Oil and Public Opinion in the Middle East,* London, 1966.
Hirszowicz, Lukaz, *Iran 1951–1953 Nafta, Imperializm, Nacjonalizm,* Warsaw, 1958.
Hoepli, Henry U., *England im Nahen Osten. Das Königreich Irak und die Mossulfrage,* Erlangen, 1931.
Hoffmann, Karl, *Olpolitik und Angelsächsischer Imperialismus,* Berlin, 1926.
Holden, David, "The Oil Genie and the Sheikh," *New York Times Magazine,* March 11, 1962.
—— "The Persian Gulf: After the British Raj," *Foreign Affairs,* XLIX, 721–735, July, 1971.
Holman, Eugene, *Oil. Product and Pillar of Freedom,* n.p., n.d.
Hoskins, Halford L., "Background of the British Position in Arabia," *The Middle East Journal,* I, 137–147, April, 1947.
—— *Middle East Oil and United States Foreign Policy,* (Mimeographed), Washing-

ton, 1950. Library of Congress, Legislative Reference Service.
—— *The Middle East: Problem Area in World Politics,* New York, 1954.
—— "Needed: A Strategy for Oil," *Foreign Affairs,* XXIX, 229–237, January, 1951.
Hotchkiss, Henry, "Petroleum Developments in Middle East and Adjacent Countries in 1956," *Bulletin of American Association of Petroleum Geologists,* XLI, 1591–1615, July, 1957.
Hourani, A. H., *Syria and Lebanon,* London, 1946.
"How BP Bypassed the Canal," *Petroleum Press Service,* XXXIV, 411–413, November, 1967.
"How Egypt is Faring," *Petroleum Press Service,* XXV, 11–13, January, 1958.
"How to Bring in an Oil Field," *Aramco World,* XIV, 3–6, August–September, 1963.
Howard, Bushrod, "Buraimi: A Study in Diplomacy by Default," *The Reporter,* XVIII, 13–16, January 23, 1958.
Hubbard, G. E., *From the Gulf to Ararat,* Edinburgh, 1916.
Hudson, G. F., "America, Britain and the Middle East," *Commentary,* XXI, 516–521, June, 1956.
Hull, Burton, E., "Tapline Presents Great Organization Problem," *The Oil Forum,* II, 449–454, November, 1948.
Hull, Cordell, *The Memoirs of Cordell Hull,* II, New York, 1948.
Hurewitz, J. C., *Middle East Dilemmas: The Background of United States Policy,* New York, 1953.
Ickes, Harold L., "Oil and Peace," *Collier's,* 21, etc., December 2, 1944.
—— "Persian Gulf Oil Furnishes Great Backlog for U.S. Reserves," *The Oil Weekly,* 13–15, etc., March 6, 1944.
—— "We're Running out of Oil," *The American,* 26–27, etc., January, 1944.
"Impact of Middle East Pipelines," *Petroleum Press Service,* XXXVIII, 134–136, April, 1971.
Independent Petroleum Association of America, *The Anglo-American Petroleum Agreement,* Washington, 1944.
—— *The Proposed Arabian Pipe Line – A Threat to our National Security,* 1944.
"The Independents in the Middle East," *Petroleum Press Service,* XXXIII, 99–103.
"In Egypt Now," *Petroleum Press Service,* XXVI, 217–20, June, 1959.
"Iran," *World Petroleum Report,* III, 114–118, January 15, 1957.
"Iran Plans Broad-Scale Progress. Program Means Rise in Imports," *Foreign Commerce Weekly,* XXXIV, 3–6, 39–41, February 21, 1949.
"Iran Presents Its Case for Nationalization," *The Oil Forum,* VI, 79–94, 107, March, 1952.
"Iranian Highlights," *Petroleum Press Service,* XXXIII, 175–177, May, 1966.
"Iran's Big Deal," *Petroleum Press Service,* XXXIX, 283–285, August, 1972.
"Iraq," *World Petroleum Report,* III, 119–122, January 15, 1957.
The Iraq Oil Negotiations, London, Iraq Petroleum Company, 1962.

"Iraq's Award to INOC," *Petroleum Press Service*, XXXIV, 324–325, September, 1967.

"Iraq's Development Plans," *Asian Review*, LVIII, 193–95, July, 1962.

"Iraq's Golden Opportunity," *Petroleum Press Service*, XX, 246–249, July, 1953.

"Iraq's Third Annual Development Week," *Iraq Petroleum*, VII, 5–11, July, 1958.

"Iraq: The Uses of Oil," *Fortune*, 120–127, September, 1958.

"The Incomparable Middle East," *Petroleum Press Service*, XXIII, 239–242, July, 1956.

Ireland, Philip Willard, *Iraq: A Study in Political Development*, New York, 1938.

Iskandar, Marwan, "OPEC in Less Than Unanimity?", *World Petroleum*, XLII, 30–31, 54, December, 1971.

Ismael, Tareq Y., (Ed.), *Governments and Politics of the Contemporary Middle East*, Homewood, Ill., 1970.

"Israel," *World Petroleum Report*, III, 124–128, January 15, 1957.

"Israel Looks to Oil Possibilities," *World Petroleum*, XXIII, 97–112, September, 1952.

"Israel, Middle Phase Operator," *Petroleum Press Service*, XXXVIII, 176–178, May, 1971.

"Israel: Projects and Supplies," *Petroleum Press Service*, XXXVI, 85–87, March, 1969.

"Israel Steps Up Drilling and Production Operations," *World Oil*, CXLIV, 166–169, February 1, 1957.

"Israel: Travail and Blessing," *Petroleum Press Service*, XXIII, 163–166, May, 1956.

"Israel's Oil Prospects Look Better Than Ever," *The Oil Forum*, IX, 92–94, March, 1955.

"Israel's Transit Trade Falters," *Petroleum Press Service*, XXXIX, 297–299, 1972.

Issawi, Charles, and Mohammed Yeganeh, *The Economics of Middle Eastern Oil*, New York, 1962.

"Issues at the Arab Congress," *Petroleum Press Service*, XXX, 443–446, December, 1963.

Itayim, Fuad, "The Organisation of Petroleum Exporting Countries," *Middle East Forum*, XXXVIII, 13–19, December, 1962.

Izzedin, Najla, "The Development of Kuwait," (Arabic), *Al Abhath*, VII, 39–53, March, 1954.

J. H. D. B., "Oil and Bahrain," *The World Today*, VII, 76–83, February, 1951.

Jalabert, Louis, "De la Conférence de Lausanne à la Paix Orientale," *Etudes*, 175, 21–41, April–May–June, 1923.

Jameel, Fuad, "Development Projects and Economic Progress," *International Review*, XLII, 14–16, March, 1954.

Jaraljmek, Edmund, *Das andere Iran*, Munich, 1951.

"The Japanese-Saudi Agreement," *Petroleum Press Service*, XXV, 46–48, February, 1958.

582 THE MIDDLE EAST, OIL AND THE GREAT POWERS

Jawdat, Nameer Ali, *Selected Documents of the Petroleum Industry 1967,* Vienna, 1968.

Jekiel, N. T., "L'Industrie Pétrolière Iranienne," *L'Asie Nouvelle,* 8e Année, 211–13, October–November, 1959.

Jessen, Sydney, *Die Weltinteressen der englischen Petroleumindustrie,* Berlin, 1925.

"Jogging Along in Iraq," *Petroleum Press Service,* XXXIII, 214–216, June, 1966.

Jones, J. H., "My Visit to the Persian Oilfields," *Journal of the Royal Central Asian Society,* XXXIV, 56–68, January, 1947.

Jones, K. Westcott, "The Miracle of Aramco," *The National and English Review,* CXXXVII, 344–349, December, 1951.

Jouan, R., *Le Pétrole Roi du Monde,* Paris, 1949.

Julien, Raymond C., "Le Pétrole dans l'Economie du Moyen-Orient," *Revue de Droit International pour le Moyen-Orient,* I, 51–65, May–June, 1951.

Kaissouni, A. M., "Oil in the Middle East," *L'Egypte Contemporaine,* XXXIX, 263–294, May, 1949.

Kapeliuk, Amnon, "The Tudeh Party in Iran" (Hebrew), *Hamizrah Hehadash,* IV, 244–254, Summer, 1953.

Kareem, Mohamed Zaki Abdul, "La Construction en Irak," *L'Asie Nouvelle,* VI, 123–129, June–July, 1956.

Katabi, Adnan S., "Iraq's Income Pyramids on Oil Revenues," *The Oil Forum,* IX, 392–394, November, 1955.

Keen, B. A., *The Agricultural Development of the Middle East,* London, 1946.

Kelly, J. B., "Buraimi Oasis Dispute," *International Affairs,* XXXII, 318–26, July, 1956.

—— "Sovereignty and Jurisdiction in Eastern Arabia," *International Affairs,* XXXIV, 16–24, January, 1958.

Kemp, Norman, *Abadan: A First-hand Account of the Persian Oil Crisis,* London, 1953.

Khadduri, Majid, "Iran's Claim to Sovereignty of Bahrein," *American Journal of International Law,* XLV, 631–647, October, 1951.

Kheirallah, George, *Arabia Reborn,* Albuquerque, 1952.

Kimche, Jon, "Oil and Arab Nationalism," *Journal of the Middle East Society,* I, 72–79, Spring, 1947.

Kinch, E. A., "Social Effects of the Oil Industry in Iraq," *International Labour Review,* LXXV, 193–206, March, 1957.

Kingsley, Robert, "Premier Amini and Iran's Problems," *Middle Eastern Affairs,* XIII, 194–198, August–September, 1962.

Kirk, George, *The Middle East in the War: Survey of International Affairs 1939–1946,* London, 1952.

—— *The Middle East 1945–1950: Survey of International Affairs,* London, 1954.

"Kirkuk to Paris: A Market for Mid-East Gas," *The Oil Forum,* VII, 22–26, January, 1953.

Kliewer, Don, "Will Western Europeans Reappraise Value of Oil?", *World Oil,* CXLIV, 191–192, January, 1957.

Knowles, Ruth Sheldon, "Arab Countries Continue to Drive for Bigger Role in Oil," *World Petroleum,* XXXVI, 52–54, 56, 94, June, 1965.

Koester, Edward E., "Turkey's Oil Prospects are Looking Better Now," *World Oil,* CLVI, 108–112, March, 1963.

Kogan, M., "The Struggle for Near Eastern Oil," *New Times,* XVIII, 8–11, September 15, 1946.

Kosloff, Israel R., "Twenty Years of Oil in Israel," (Hebrew), *Beolam Hadelek,* 27–35, July, 1969.

Kuschiar, Amir Hossein, *Der Einfluss der Erdölindustrie auf die politische Gestaltung des Irans,* Köln, 1959.

"Kuwait," *World Petroleum Report,* III, 133–134, January 15, 1957.

"Kuwait Branches Out," *Petroleum Press Service,* XXVIII, 246–248, July, 1961.

"Kuwait Looks Ahead," *Petroleum Press Service,* XXXIX, 131–133, April, 1972.

"Kuwait Takes Action," *Petroleum Press Service,* XXXIV, 204–206, June, 1967.

"Kuwait's Incredible Ten Years," *Petroleum Press Service,* XXIII, 288–291, August, 1956.

"Kuwait's New Facilities," *Petroleum Press Service,* XXXVI, 131–133, April, 1969.

Lacoste, Raymond, *La Russie Soviétique et la Question d'Orient. La Poussée Soviétique vers les Mers Chaudes Méditerranée et Golfe Persique,* Paris, 1946.

Lambton, A. K. S., "Some of the Problems facing Persia," *International Affairs,* XXII, 254–272, April, 1946.

Landis, Lincoln, "Soviet Interest in Middle East Oil," *The New Middle East,* 16–21, December, 1968.

Langley, Kathleen M., *The Industrialization of Iraq,* Cambridge, 1961.

Langley, S. J., "Oil Royalties and Economic Development in the Middle East," in *Middle East Economic Papers, 1954,* Beirut, 1954.

Laurens, Henri, *La Signification Economique du Moyen-Orient: La Phase du Pétrole,* Aalter, Belgium, 1956.

Law, J., "Reasons for Persian Gulf Oil Abundance," *Bulletin of American Association of Petroleum Geologists,* XLI, 51–69, January, 1957.

Leatherdale, D., "The Material Background of Life in Northern Iraq," *Journal of the Royal Central Asian Society,* XXXV, Part I, 63–73, January, 1948.

"Lebanon," *World Petroleum Report,* III, 135–138, January 15, 1957.

Lebkicher, Roy, "America's Greatest Middle East Oil Venture," *The Oil Forum,* VI, 389–396, November, 1952.

—— *Aramco and World Oil,* New York, 1952.

Lebkicher, and George Rentz and Max Steineke, *The Arabia of Ibn Saud,* New York, 1952.

Leduc, Gaston, "L'Evolution économique du Moyen-Orient," *Politique Etrangère,* XII, 285–300, June–July, 1947.

Leeman, Wayne A., *The Price of Middle East Oil: An Essay in Political Economy,* Ithica, 1962.

Lees, G. M., "The Middle East as a Whole," in *World Geography of Petroleum,* Princeton, 1950.

—— "Oil in the Middle East," *Journal of the Royal Central Asian Society,* XXXIII, 45–57, January, 1946.

Le Fèvre, Georges, *Sa Majesté le Pétrole,* 1950.

Lenczowski, George, "Iran's Deepening Crisis," *Current History,* XXIV, 23–37, April, 1953.

—— *Oil and State in the Middle East,* Ithaca, 1960.

—— "Oil in the Middle East," *Current History,* XXXVIII, 262–7, May, 1960.

—— *Russia and the West in Iran 1918–1948 – A Study in Big-Power Rivalry,* Ithaca, New York, 1949.

Lentz, Wolfgang, *Iran 1951/1952,* Heidelberg, 1952.

—— "Sondervollmachten für Mossadegh," *Zeitschrift für Geopolitik,* XXIII, 680–697, November, 1952.

—— "Die Verstaatlichung der Olindustrie in Iran," *Zeitschrift für Geopolitik,* XXIII, 608–631, October, 1952.

LePage, Louis, *L'Impérialisme du Pétrole,* Paris, 1921.

L'Espagnol de la Tramerye, Pierre, *The World Struggle for Oil,* New York, 1924.

"The Lesson of Persia: Social Aspiration and the Oil Dispute," *The Round Table,* XLI, 307–312, September, 1951.

Lesueur, Emile, *Les Anglais en Perse,* Paris, 1921.

"The Levant: Oil Corridor," *Petroleum Press Service,* XX, 51–54, February, 1954.

Levy, Walter J., "Economic Problems Facing a Settlement of the Iranian Oil Controversy," *The Middle East Journal,* VIII, 91–96, Winter, 1954.

—— "Oil Power," *Foreign Affairs,* XLIX, 652–668, July, 1971.

—— "Issues in International Oil Policy," *Foreign Affairs,* XXXV, 454–469, April, 1957.

Lichtblau, John H., "Oil: How Long Can Nasser Strangle the Flow?", *The Reporter,* XVI, 29–32, February 7, 1957.

Liebesny, Herbert J., "International Relations of Arabia: The Dependent Areas," *The Middle East Journal,* I, 148–168, April, 1947.

—— "Legislation on the Sea Bed and Territorial Waters of the Persian Gulf," *Middle East Journal,* IV, 94–98, January, 1950.

—— "U.S.S.R. and Iranian Oil," *American Perspective,* I, 407–423, December, 1947.

"The Limits of the Squeeze," *Petroleum Press Service,* XXXV, 362–364, October, 1968.

Lind, John H., "Global Economic Impact of the Suez Crisis," *The Magazine of Wall Street,* XCIX, 311–313, 344, 348, December 8, 1956.

Lisle, B. Orchard, "First Unwise Step of Sa'udi Arabia," *The Oil Forum,* VIII, 236–237, July, 1954.

—— "Turkish Oil Achieving accelerated Development," *The Oil Forum,* VII, 123–124, April, 1953.

Lloyd George, David, *The Truth about the Peace Treaties,* II, London, 1938.

Lockhart, Laurence, "The Causes of the Anglo-Persian Oil Dispute," *Journal of the Royal Central Asian Society,* XL, 134–151, April, 1953.

—— "Khuzistan Past and Present," *Asiatic Review*, XLIV, 410–416, October, 1948.
—— "Outline of the History of Kuwait," *Journal of the Royal Central Asian Society*, XXXIV, 262–274, July–October, 1947.
Loftus, John A., "Middle East Oil: The Pattern of Control," *The Middle East Journal*, II, 17–32, January, 1948.
Lomax, E. Lawson, "Kuwait Completes Model Oil Port," *World Petroleum*, XXI, 36–39, October, 1950.
—— "Kuwait Oil Company Provides Social Services and Training for its Workers," *World Petroleum*, XXIII, 36–39, June, 1952.
Longrigg, Stephen Hemsley, *Iraq 1900 to 1950*, London, 1953.
—— "The Liquid Gold of Arabia," *Journal of the Royal Central Asian Society*, XXXVI, 20–33, January, 1949.
—— "Middle Eastern Oil Blessing or Curse?", *Royal Central Asian Society Journal*, XLII, 150–164, April, 1955.
—— "Oil in the Middle East," *Current History*, XXX, 353–359, June, 1956.
—— *Oil in the Middle East: Its Discovery and Development*, London, 1968. (Third Edition).
—— "Prospects for Iraq," *The Geographical Magazine*, XXVI, 276–290, September, 1953.
Lotz, J. D., "The Iranian Seven-Year Development Plan: Problems and Proposals," *The Middle East Journal*, IV, 102–105, January, 1950.
Lubell, Harold, "Middle East Crises and World Petroleum Movements," *Middle Eastern Affairs*, IX, 338–348, November, 1958.
—— *Middle East Oil Crises and Western Europe's Energy Supplies*, Baltimore, 1963.
—— *The Soviet Oil Offensive and Inter-Bloc Economic Competition*, Prepared for U.S. Air Force. An appendix in: U.S. Congress, Senate Committee on the Judiciary, 87th Congress, *Hearings before the Subcommittee to Investigate the Administration of the Internal Security Act and other Internal Security Laws. Soviet Oil in East-West Trade*, Washington, 1962.
Luke, Harry Charles, *Mosul and Its Minorities*, London, 1925.
Lutfi, Ashraf T., *Arab Oil: A Plan for the Future*, Beirut, 1960.
—— *OPEC Oil*, Beirut, Lebanon, 1968.
Lyautey, Pierre, *Le Duel en Orient*, Paris, 1957.
"Making Headway in Turkey," *Petroleum Press Service*, 207–210, June, 1965.
Mallakh, Ragaei, "Economic Development Through Cooperation: The Kuwayt Fund," *Middle East Journal*, XVIII, 405–420, Autumn, 1964.
—— *Some Dimensions of Middle East: The Producing Countries and the United States*, New York, 1970.
Mann, Clarence, *Abu Dhabi: Birth of an Oil Sheikhdom*, Beirut, 1964.
Manning, Van H., "International Aspects of Petroleum Industry," *Mining and Metallurgy*, 1–10, February, 1920.
Marcus, Alfred, "Palästina und die internationale Erdölpolitik," *Preussische Jahrbücher*, CCXXVII, 69–73, January, 1932.

Margulies, Heinrich, "Oil Economics in the Near East," *Economic News,* II, 3—11, February, 1950.

Marlowe, John, *The Persian Gulf in the Twentieth Century,* London, 1962.

Martin, Lawrence, "The Chester Concession," *Annals of the American Academy of Political and Social Science,* CXII, 186—188, March, 1924.

Massudi-Toiserkan, Schapur, *Historische Entwicklung der persischen Erdöl-problems,* Bonn, 1960.

Mattei, Enrico, "Problèmes et Perspectives du Ravitaillement de L'Europe en Pétrole," *Politique Etrangère,* 22e Année, 505—522, November—December, 1957.

"A Matter of Foresight," *Aramco World,* XIII, 3—6, April, 1962.

Mayer, Ferdinand, *Erdöl Weltatlas,* Brunswick, Germany, 1966.

Mazour, Anatole G., "Russia, the Middle East and Oil," *World Affairs Interpreter,* XXII, 415—423, January, 1952.

McKitterick, T. E. M., "Politics and Economics in the Middle East," *The Political Quarterly,* XXVI, 65—69, January—March, 1955.

McNutt, Russell A., and Joseph A. Ferenz, "Kharg Island Terminal One of World's Largest," *World Petroleum,* XXXVIII, 48—57, July, 1967.

Melamid, Alexander, "Boundaries and Petroleum Developments in Saudi Arabia," *Geographical Review,* XLVII, 589—591, October, 1957.

—— "The Buraimi Oasis Dispute," *Middle Eastern Affairs,* VII, 56—63, February, 1956.

—— "Economic Changes in Yemen, Aden and Dhofar, *Middle Eastern Affairs,* V, 88—91, March, 1954.

—— "The Geographical Pattern of Iranian Oil Development," *Economic Geography,* XXXV, 199—218, July, 1959.

—— "Governments' Offshore Claims Seriously Affect Potential Oil Exploration," *The Oil Forum,* IX, 46—47, February, 1955.

—— "The Oil Fields of the Sinai Peninsula," *Middle Eastern Affairs,* X, 191—194, May, 1959.

—— "Political Geography of Trucial 'Oman and Qatar," *The Geographical Review,* XLIII, 194—206, April, 1953.

—— "Territorial Rights of Persian Gulf Waters," *The Oil Forum,* VIII, 90—91, March, 1954.

"Menhall Concession Viewed as Valid Despite Coup," *The Oil Forum,* III, 208—209, XI, 150—152, May, 1949.

"The Mesopotamian Oilfields," *Oil Engineering and Finance,* 197—205, February 17, 1923.

Meyer, Albert J., "The Iran Consortium: Solution or Stopgap?", *Middle East Economic Papers 1955,* 62—71, 1955.

—— "North African Oil and the World Economy," *Foreign Policy Bulletin,* XXXIX, 93—5, March 1, 1960.

Michaelis, Alfred, "Economic Recovery and Development of Iraq," *Middle Eastern Affairs,* III, 101—106, April, 1952.

—— "Iran: Economic Structure," *Middle Eastern Affairs,* V, 312–318, October, 1954.

Michel, Roland C., "Turkey Offers Good to Excellent Oil Prospects," *World Oil,* CLXVII, 92–95, December, 1968.

Michler, Gordon H., "Oil Transportation Must Face Security Issue," *The Oil Forum,* XI, 150–152, May, 1957.

"Middle East," *World Oil,* CLXIX, 185–207, August 15, 1969.

—— CLXXI, 166–175, August, 1970.

—— CLXXIII, 109–120, August, 1970.

—— CLXXV, 112–128, August 15, 1972.

"Middle East," *World Petroleum Report,* V, 29–62, February 15, 1959.

—— VI, 71–90. February 15, 1960.

—— VIII, 43–56, February 15, 1962.

—— X, 169–70, 74, 76, 78–82, 83, March 15, 1964.

—— XI, 68–86, 128, March 15, 1965.

—— XII, 74–84, March 15, 1966.

—— XIII, 66–72, 1967.

—— XIV, 78–87, 1968.

—— XV, 78–92, 1969.

—— XVI, 70–86, 1970.

—— XVII, 64–77, 1971.

—— XVIII, 52–65, 1972.

"The Middle East After Abadan," *The Round Table,* XLIV, 32–41, December, 1953.

"The Middle East: An Economic Appraisal," *United Nations Review,* 18–23, July, 1957.

"The Middle East: An Introductory Survey," *Petroleum Times,* LVII, 331–350, April 3, 1953.

Middle East Basic Oil Laws and Concession Contracts, Petroleum Legislation, New York, 1959, 2 vols.

"Middle East Offers No Threat to International Oil Trade," *World Petroleum,* XVIII, 64–65, September, 1947.

"Middle East Oil Discoveries add New Reserves," *World Petroleum,* XXXIII, 40–44, February, 1962.

"Middle East Oil Europe's Lifeblood?", *The Banker,* CVI, 680–706, November, 1956.

"The Middle East: Oil Politics and Intelligence," *Eastern World,* XVII, 14–17, July, 1963.

"Middle East Oil: Problems that Wealth Brings," *Newsweek,* 81–92, July 13, 1955.

"Middle East Production Will reach More than 7 million Barrels per Day next year," *World Petroleum,* XXXIV, 28–32, July, 1963.

"The Middle East Squeeze on the Oil Giants," *Newsweek,* 56–62, July 29, 1972.

"Middle East Terms Tabulated," *Petroleum Press Service,* XXXVI, 99–102, March, 1969.

"Mid-East Awakening Spurred by Oil Industry," *The Oil Forum,* II, 175—177, May, 1948.

"Mid-East Becoming Hub of International Oil," *World Petroleum,* XIX, 54—57, September, 1948.

"Mid-East Concession Grievances are Serious," *The Oil Forum,* III, 77—79, February, 1949.

"Middle Eastern Oil," *The National and English Review,* CXLVI, 212—215, April, 1956.

Midgley, John, "Two Oil Sheikdoms," *The Geographical Magazine,* XXIX, 143—48, July, 1956.

Mikdashi, Zuhayr, *A Financial Analysis of Middle Eastern Oil Concessions: 1901—65,* New York, 1966.

—— (Ed.), *Continuity and Change in the World Oil Industry,* Beirut, 1970.

Mikesell, Raymond, F., and Hollis B. Chenery, *Arabian Oil: America's Stake in the Middle East,* Chapel Hill, N. C., 1949.

Miller, E. Willard, "The Role of Petroleum in the Middle East," *The Scientific Monthly,* LVII, 240—248, September, 1943.

Millspaugh, A. C., *The American Task in Persia,* New York, 1925.

—— *Americans in Persia,* Washington, 1946.

—— "The Persian-British Oil Dispute," *Foreign Affairs,* XI, 521—525, April, 1933.

Mineau, Wayne, *The Go Devils,* London, 1958.

"The Mining, Quarrying and Petroleum Industries in 1957," *L'Egypte Industrielle,* XXXIV, 6—14, July, 1958.

"Mining, Quarrying and Petroleum Institute in Egypt," *L'Egypte Contemporaine,* XLVIII, 77—86, October, 1957.

"The Miriella Case," *Petroleum Press Service,* XX, 122—123, April, 1953.

"Mission to Jiddah," *Aramco World,* XIII, 19—21, December, 1962.

Mohr, Anton, *The Oil War,* New York, 1926.

Möhrmann, D., "Bohrkonzessionen im Persischen Golf in betriebs- und volkwirtschaftlicher Beurteilung," *Wirtschafdienst* (Hamburg), XXXVIII, 387—392, July, 1958.

Monroe, Elizabeth, "British Interests in the Middle East," *The Middle East Journal,* II, 129—146, April, 1948.

—— "Iran et Irak Comparaisons et Contrastes," *Politique Etrangère,* XVII, 5—12, April—May, 1952.

—— "The Shaikhdom of Kuwait," *International Affairs,* XXX, 271—284, July, 1954.

Mookerjee, E., "Schmieriges Gold-Das Nahes-Ol-Ausgangspunkt der Welkrise," *Zeitschrift für Geopolitik in Gemeinschaft und Politik,* XXIX, 1—80, July—August, 1958.

Morris, James, *Sultan in Oman,* London, 1957.

Mostofi, Baghir, "Petrochemical Resources and Potential of the Persian Gulf," *World Petroleum,* XL, 42—48, 56, August 15, 1969.

Mostofi, B., and August Gansser, "The Story Behind the 5 Albroz," *Oil and Gas*

Journal, LV, 78–84, January 21, 1957.

"Much Activity in Iran," *Petroleum Press Service,* XXXVI, December, 1969.

Mughraby, Muhammad A., *Permanent Sovereignty Over Oil Resources: A Study of Middle East Oil Concessions and Legal Change,* Beirut, 1966.

—— "Reflections on the Fourth Arab Petroleum Congress," *Arab Journal,* X, 40–4, Winter, 1964.

Murphy, Charles J. V., "The Great Tanker Dilemma," *Fortune,* 125–127, 266–272, November, 1956.

—— "Oil East of Suez," *Fortune,* 131–138, 252–266, October, 1956.

Murray, John, "Oil Revenues and Middle East Economics," *Petroleum Times,* LXV, 766–769, December 1, 1961, 794–797, December 15, 1961.

Najmabadi, F., "A History of Naft-i-Shah Oilfield and Kermanshah Refinery – Past, Present and Future," *News Letter,* (NIOC), 1–6, March, 1963.

Nakasian, Samuel, "The Security of Foreign Petroleum Resources," *Political Science Quarterly,* LXVIII, 181–202, June, 1953.

Nakhai, M., *Le Pétrole en Iran,* Bruxelles, 1938.

National Petroleum Council, *Impact of Oil Exports From the Soviet Bloc,* 2 Vols., Washington, D. C., 1962.

"La Nationalisation de l'Anglo-Iranian Oil Company," *Chronique de Politique Etrangère,* 601–71, September, 1951.

Nauwelaerts, Louis, *Petroleum: Macht der Erde,* Leipzig, 1937.

Nawratzki, Curt, "Oil Prospecting in the Middle East and in Israel," *Economic News,* V, 40–41, 1953.

"The Near East," *The Round Table,* XII, 319–337, March, 1922.

"Near East Becomes Strategic Oil Center," *World Petroleum,* VII, 399, August, 1936.

Netschert, Bruce, *The Future Supply of Oil and Gas: A Study of the Availability of Crude Oil, Natural Gas, and Natural Gas Liquids in the United States in the Period Through 1975,* Baltimore, 1958.

Neumann, Robert G., "L'Irak de Kassem," *Politique Etrangère,* 25e Année, 476–483, 1960.

"Neutral Zone," *World Petroleum Report,* III, 209, January 15, 1957.

Neville-Bagot, G. H., "Kuwait. Its Spectacular Economic Development," *The Islamic Review,* XL, 22–26, October, 1952.

"New Daura Plant Meets Iraq's National Lube Requirements," *World Petroleum,* XXVIII, 52, October, 1957.

"New Deal in the Middle East," *Petroleum Press Service,* XL, 44–46, February, 1973.

"New Mining Law in Egypt," *Petroleum Press Service,* XXII, 127–129, April, 1953.

"New Moves in the Middle East," *Petroleum Press Service,* XXX, 324–326, September, 1963.

"New Perspectives in Iraq," *Petroleum Press Service,* XXV, 288–291, August, 1958.

"New Plant for Daura," *Iraq Petroleum,* VII, 4–9, February, 1958.

"New Setting for Soviet Exports," *Petroleum Press Service,* XXXIII, 165–168, May, 1966.

The New York Times, "A Century of Oil 1859–1959," May 31, 1959. Special Section.

Nibley, Preston P., and Alan C. Nelson, "Economics of Oil Transportation Middle East to Western Europe," *World Petroleum,* XXXII, 56–65, November, 1961.

Nicolesco, C. P., *Gisements Pétrolifères de l'Irak,* Paris, 1933.

Nikpay, Gholam Reza, *The Role of the Oil Industry in Iran's Economy and Future Programmes,* Teheran, 1962.

"NIOC Tackles Paternalism," *Petroleum Press Service,* XXX, 103–105, March, 1963.

"No Agreement Reached in Iraq," *Petroleum Press Service,* XXVIII, 409–411, November, 1961.

"No Oil Bonanza for Turkey," *Petroleum Press Service,* XXIX, 55–58, February, 1962.

Normand, Suzanne and Jean Acker, *La Route du Pétrole au Moyen-Orient,* Paris, 1956.

"Novel Terms in new Iranian Oil Concession," *World Petroleum,* X, 44, August, 1939.

Novomeysky, M. A., "To Find Oil, You have to Drill," *Technion Yearbook,* XVI, 77–9, 1959.

Odell, Peter, *Oil and World Power,* London, 1969.

"Offshore Lull in Persian Gulf," *World Petroleum,* XXVIII, 48–49, February, 1957.

"Offshore Production to Reach One Million Barrels Daily in Persian Gulf This Year," *World Petroleum,* XXXV, 28–31, July, 1964.

"Oil Abroad," *Aramco World,* 6–9, January, 1958.

"Oil and Politics in the Trucial Coast," (Hebrew), *Beolam Hadelek,* 84–91, July 1, 1969.

"Oil and Social Change in the Middle East," *The Economist,* CLXXVI, 1–16, July 2, 1955.

Oil Industry in Iran, Teheran, Iranian Petroluem Institute, 1963.

"Oil in Israel," *Economic Horizons* (special issue), VIII, 3–14, January, 1956.

"Oil in Sinai," *The Bulletin* (Egyptian Education Bureau), 11–19, July, 1952.

"Oil Possibilities in Israel," *Davar,* June 14, 1957.

"Oil Production in Iraq," *Iraq Petroleum,* VIII, 2–9, March, 1959.

"Oil Section," *The Egyptian Economic & Political Review,* I, 21–36, 41–44, 50–51, November, 1954.

"Oil: The First Agreement," *Fortune,* XXX, 113–114, October, 1944.

Okyar, Osman, "The National Income of Turkey," *Middle Eastern Affairs,* II, 361–366, November, 1951.

"Oman Operations Analysed," *Petroleum Press Service,* XXXVIII, 215–217, June, 1971.

"OPEC Allowances to End," *Petroleum Press Service,* XXXV, 44–47, February, 1968.

"OPEC Goes to Work," *Petroleum Press Service,* XXVIII, 446–448, December, 1961.

"OPEC Recommends," *Petroleum Press Service,* XXXIII, 242–245, July, 1966.

"OPEC Seeks a Plan," *Petroleum Press Service,* XXXVII, 318–320, September, 1970.

"OPEC Seeks a Yardstick," *Petroleum Press Service,* XXXV, 164–167, May, 1968.

"OPEC's Next Objective," *Petroleum Press Service,* XXXV, 282–283, August, 1968.

"OPEC's Recipe," *Petroleum Press Service,* XXXII, 328–331, September, 1965.

"The Other Discovery," *Aramco World,* XIII, 3–7, 8–11, January, February, 1962.

"Outlook for Suez," *Petroleum Press Service,* XXX, 48–49, February, 1963.

Overseas Consultants, Inc., *Report on Seven-Year Development Plan for the Plan Organization of the Imperial Government of Iran,* New York, 5 Vols., 1949.

Padover, Saul K., "Oil Behind the Politics," *'48,* 57–67, May, 1948.

Page, Howard W., "The Iranian Oil Agreement," *The Oil Forum,* VIII, 470–471, 478, December, 1954.

—— "What the Middle East Means to Us," *The Lamp,* 2–5, November, 1956.

Parra, R., *The Development of Petroleum Resources under the Concession System in Non-Industrialized Countries,* Geneva, OPEC, 1964.

Parra, Francisco R., "A Role for OPEC," *World Petroleum,* XXXII, 29–31, August, 1961.

"Participation-Dollar Parity-BP Libyan Nationalisation OPEC v. Oil Companies," *Petroleum Times,* LXXV, 8–9, December 17, 1971.

Patrick, Martin, "Oil and the Middle East," *Political Quarterly,* XXVIII, 168–178, April–June, 1957.

"Peak Year Profile," *Aramco World,* XIII, 3–11, June–July, 1962.

Pegourier, Y., "Les Problèmes d'exploitation dans le Code pétrolier saharien," *Revue Juridique et Politique D'Outre-Mer,* 13e Année, 531–54, October–December, 1959.

Penrose, Edith, *The International Oil Industry in the Middle East,* Cairo, National Bank of Egypt, 1968.

—— *The Large International Firm in Developing Countries: The International Petroleum Industry,* London, 1968.

—— "Profit Sharing Between Producing Countries and Oil Companies in the Middle East," *Economic Journal,* LXIX, 238–54, June, 1959.

"Persia, the Keystone," *The Round Table,* XLIII, 28–41, December, 1952.

"Persia Presses On," *Petroleum Press Service,* XXV, 169–172, May, 1958.

"The Persian Crisis," *Petroleum Press Service,* XVIII, 101–104, April, 1951.

"The Persian Gulf," *Petroleum Press Service,* XX, 161–167, May, 1953.

"The Persian Gulf – a Romance," *The Round Table,* XXXIX, 131–137, March, 1949.

"Persian Gulf Land," *Aramco World,* XIV, 14–17, October, 1963.

"Persian Gulf Patchwork," *Petroleum Press Service,* XXXVII, 365–367, October, 1970.

"The Persian Settlements," *The Round Table,* XLIV, 326–335, September, 1954.

"Persia's Petroleum Policy," *Petroleum Press Service,* XXIV, 359–362, October, 1957.

Perspective: A Review of Iran's Oil Industry, National Iranian Oil Company, [1961].

"Le Pétrole du Moyen-Orient," *Le Commerce du Levant,* October 17, 21, 24, 28, 1953.

"Petroleum Act 31st July, 1957," *Bank Melli Iran Bulletin,* 302–315, August–September, 1957.

"Petroleum Industry in Israel," *Economic News,* V, 1953. Entire Issue.

Petroleum Industry War Council, *A National Oil Policy for the United States,* Washington, 1944.

—— *Action by Petroleum Industry War Council on the Anglo-American Oil Agreement – December 6, 1944,* [Washington].

—— *United States Foreign Oil Policy and Petroleum Reserves Corporation. An Analysis of the Effect of the Proposed Saudi Arabian Pipeline,* Washington, 1944.

Petroleum Investment Forum for Insurance Company Executives, New York, The Chase Manhattan Bank, [1959].

"Petroleum, Mines and Quarries Industries in 1964," *Industrial Egypt,* 27–53, July–September, 1965.

Philby, H. St. J. B., *Arabian Jubilee,* London, 1952.

—— *Arabian Oil Ventures,* Washington, 1964.

—— "The New Reign in Saudi Arabia," *Foreign Affairs,* XXXII, 446–458, April, 1954.

Platt, Stanley G., *Financial Analysis of 33 Petroleum Companies 1962,* [New York], 1963.

Pogue, Joseph E., "Must an Oil War follow this War?" *The Atlantic Monthly,* CLXXIII, 41–47, March, 1944.

—— *Oil and National Policy,* 1948.

Polk, William R., *The United States and the Arab World,* Cambridge, 1969. (Revised Edition.)

Pollack, Hubert, "Middle East Oil – Recent Developments and Outlook" (Hebrew), *Hamizrah Hehadash,* IX, 264–84, 1959.

Polyani, G., "The Taxation of Profits from Middle East Oil Production: Some Implications for Oil Prices and Taxation Policy," *Economic Journal,* LXXVI, 668–685, December, 1966.

Powers, W. L., "Soil and Land Use Capabilities in Iraq: A Preliminary Report," *The Geographical Review,* XLIV, 373–381, July, 1954.

The Present Situation in Iran: A Survey of Political and Economic Problems Confronting the Country (Mimeographed), Washington, 1953.

Price, M. Philips, "The Persian Situation," *The Contemporary Review,* 1–5, July, 1951.

"Process Goals High in Middle East," *World Petroleum,* XXXV, 35—37, July, 1964.

"Processing Grows in the Middle East," *World Petroleum,* XXXIV, 33—38, July, 1963.

"Producer and Transit Countries in Middle East Suffer," *The Oil Forum,* VII, 16—17, January, 1957.

"Qatar," *World Petroleum Report,* III, 227, January 15, 1957.

"Qatar: From Poverty to Opulance," *Petroleum Press Service,* XXV, 380—382, October, 1958.

"Qatar's New Fields Offshore," *Petroleum Press Service,* XXXII, 12—14, January, 1965.

Qubain, Fahim I., *The Reconstruction of Iraq: 1950—1957,* New York, 1958.

R. W. B., "Dr. Mussaddiq and After," *The World Today,* IX, 421—429, October, 1953.

Rabinowitz, Stanley, "Pipelines for the Middle East," *Jewish Frontier,* XXV, 21—5, March, 1959.

Ramchandran, R. V., "European Dependence on Middle East Oil," *Asian Studies,* III, 42—9, February—March, 1959.

"RAP Moves into Saudi Arabia," *Petroleum Press Service,* XXXII, 185—187, May, 1965.

Rawlinson, Henry, *England and Russia in the East,* London, 1875.

Rayner, Charles, "Anglo-American Oil Policy: Basis of Multilateral Trade," *Department of State Bulletin,* XV, 867—870, November 10, 1946.

Razwy, Akhtar Adil, "The Anglo-Iranian Oil Dispute," *Pakistan Horizon,* VI, 75—85. June, 1953.

Rea, Henry Carter, "Suez Crisis Reveals U.S. Oil Weaknesses," *The Oil Forum,* VII, 80—83, March, 1957.

"Realism at the Arab Congress," *Petroleum Press Service,* XXXIV, 124—126, April, 1967.

"Receipts of Host Governments," *Petroleum Press Service,* XXXVII, 323—325, September, 1970.

"Recent Development in the Field of Pipe-Lines and Pipe-Line Transportation," *Transportation and Communication Review,* VIII, 17—32, January, 1955.

Reed, Gordon W., "Russia's Oil Drive," *World Petroleum,* XXXII, 54—58, September, 1961.

"Refining in Turkey to Go Ahead," *Petroleum Press Service,* XXIII, 170—171, May, 1957.

"Refining Issues in the Lebanon," *Petroleum Press Service,* XXXVIII, 55—57, February, 1971.

"Reformer's Zeal in Egypt," *Petroleum Press Service,* XXIX, 11—14, January, 1962.

"Relieving the Suez Bottleneck," *Petroleum Press Service,* XXIII, 201—203, June, 1956.

Rentz, George, *A Sketch of the Geography, People and History of the Arabian Peninsula,* Dhahran.

"Repercussions of the Syrian Shutdown," *Petroleum Press Service,* XXXIV, 4–5, January, 1967.

"Resurgent Persia," *Petroleum Press Service,* XXIII, 171–173, May, 1957.

"Review of Middle East Oil," *Petroleum Times,* LII, 1–115, June, 1948.

"Review of Middle East Oil 1948–1959," *The Petroleum Times,* LXIII, 219–95, April 10, 1959.

"Reviving Interest in Egypt," *Petroleum Press Service,* XXXIX, 125–127, April, 1972.

Ricasoli, I., "Il Petrolio in Iraq," *Levante,* XIII, Nos. 3–4, 3–33, December, 1966.

Ricci, Carlo, "Note sulle communicazione stradoli e ferroviarie nella provincia di el Hasa," *Oriente Moderno,* XXXIV, 293–303, July, 1954.

"The Riches of Kuwait," *Petroleum Press Service,* XXIV, 323–325, September, 1957.

Richter, Kurt, "Erdöl und Weltpolitik," *Deutsche Aussenpolitik,* VI, Jahrgang . 1961, Heft 2, 151–161.

"The Right Home for Everyone," *Aramco World,* 2–5, November, 1955.

Rihani, Ameen, *Maker of Modern Arabia (Ibn Sa'oud of Arabia),* New York, 1928.

"The Road from Riyadh," *Petroleum Press Service,* XXXI, 42–44, February, 1964.

"The Role of Petromin and Its Future in Saudi Arabia," *World Petroleum,* XXXVII, 12, 68, March, 1966.

Rondot, Jean, *La Compagnie Française des Pétroles,* Paris, 1962.

—— "Les Intérêts pétroliers français dans le Proche Orient," *Politique Etrangère,* XVII, 267–291, October, 1952.

Rondot, Pierre, "Quelques Aspects de la crise iranienne des Pétroles," *L'Afrique et l'Asie,* 20–30, 3rd Quarter, 1951.

Roosevelt, Kermit, *Arabs, Oil and History: The Story of the Middle East,* New York, 1949.

Royal Institute of International Affairs, *The Middle East, A Political and Economic Survey,* London, 1954.

"Russia and Arab Oil," *Petroleum Press Service,* XXXV, 52–54, February, 1968.

"Russia Buys Gas – and Sells It," *Petroleum Press Service,* XXXVII, 207–209, June, 1970.

"Russian Exports Lose Momentum," *Petroleum Press Service,* XXX, 123–126, April, 1963.

"Russian Reserves are Inadequate," *Petroleum Press Service,* XXXVI, 122–123, April, 1969.

"Russia's New Export Drive," *Petroleum Press Service,* XXXII, 367–369, October, 1965.

"Russia's New Posture," *Petroleum Press Service,* XXXIX, 323–325, September, 1972.

Sablier, Edouard, "La Signification de l'Affaire du Pétrole iranienne," *Politique Etrangère,* XVIII, 17–22, March–April, 1953.

"The Safaniya Field," *Aramco World,* XIII, 3–7, August–September, 1962.

Sahwell, Aziz S., "The Buraimi Dispute," *Islamic Review,* XLIV, 13–17, April, 1956.

Salter, Lord, *The Development of Iraq. A Plan of Action,* 1955.
"Sanctity of Petroleum Concession Contracts," *Petroleum Times,* LXIII, 528–530, August 28, 702–703, November 6, 1959.
Sanger, Richard H., *The Arabian Peninsula,* Ithaca, New York, 1954.
—— "Ibn Saud's Program for Saudi Arabia," *The Middle East Journal,* I, 180–190, April, 1947.
Sarel, Benno, "L'Iran, clef du sud-ouest asiatique," *Politique,* XX, 44–53, January, 1946.
Sarkis, Nicolas, *Le Pétrole et Les Economies Arabes,* Paris, 1963.
"Saudi Arabia," *The Lamp,* XXXV, 4–5, March, 1953.
"Saudi Arabia," *World Petroleum Report,* III, 233–236, January 15, 1957.
"Saudi Arabia and Iraq Warn Middle East: Hands Off our Oil Flow!", *Petroleum Week,* 28–31, October 12, 1956.
"The Saudi Arabian Budget," *The Islamic Review,* XLIV, 12–15, November, 1955.
"Saudi Arabia's Rising Resources," *Petroleum Press Service,* XXV, 341–344, September, 1958.
Saunders, R. M., "The Middle East," *External Affairs,* IV, 362–370, November, 1952.
Sayegh, Kamal S., *Oil and Arab Regional Development,* New York, 1968.
Sayigh, Yusuf, "Arab Oil: A Second Look," *Middle East Forum,* XXXII, 10–14, January, 1957.
Schaffer, Robert, *Tents and Towers of Arabia,* New York, 1952.
Schmidt, Walther, "Erdölanzeichen in Anatolien, Syrien, Palästina und Arabien," *Petroleum Zeitschrift,* XXI, 1601–12, September 10, 1925.
Schurr, Sam H., and Paul T. Homan, *Middle Eastern Oil and the Western World,* New York, 1971.
Schweer, Walther, *Die Türkisch-persischen Erdölvorkommen,* Hamburg, 1919.
Scott-Reid, Don, "After Oil-Grain," *Iraq Petroleum,* VII, 4–9, April, 1958.
"The Search Began in 1933," *Aramco World,* XIV, 17–19, February, 1963.
"Second Seven-Year Development Plan of Iran," *Bank Melli Iran Bulletin,* 192–202, May–June, 1956.
Sercey, Laurent de Comte, "Les Anglais en Perse – L'Anglo-Persian Oil C'y," *Le Correspondant,* N. S. 294, 274–286, January 25, 1933.
Seton, Lloyd, *Twin Rivers: A Brief History of Iraq from the Earliest Times to the Present Day,* London, 1947.
Shafaq, S. Rezazadeh, "The Iranian Seven-Year Development Plan Background and Organization," *The Middle East Journal,* IV, 100–102, January, 1950.
Shaffer, Edward H., *The Oil Import Program of the United States: An Evaluation,* New York, 1968.
"Shaikhdoms of the Persian Gulf," *Petroleum Press Service,* XXVIII, 294–297, August, 1961.
"Shale Oil on the Threshold," *Petroleum Press Service,* XXXII, 446–449, December, 1965.
Shamma, Samir, (compiler), *The Law of Income Tax and Zakat in the Kingdom of*

Saudi Arabia, Jedda, 1951.

Shamma, Samir, *The Oil of Kuwait: Present and Future,* Beirut, 1959.

Sharif, Amer A., and Mohammed F. Dandashi, *Oil Economics of Syria,* Damascus, 1963, Second Revised Edition.

Shehab, Fakhri, "Kuwait: A Superaffluent Society," *Foreign Affairs,* XLIII, 461–74, April, 1964.

Sherwood, Sidney, "Economic Problems in the Middle East," *Middle Eastern Affairs,* II, 115–125, April, 1951.

Shor, Jean and Franc, "Iraq – Where Oil and Water Mix," *The National Geographic Magazine,* CXIV, 443–489, October, 1958.

Shuster, Morgan, *The Strangling of Persia,* New York, 1912.

Shwadran, Benjamin, "The Kuwait Incident," *Middle Eastern Affairs,* XIII, 2–13, January, 43–53, February, 1962.

—— "Middle East Oil in 1959," *Middle Eastern Affairs,* XI, 218–238, August–September, 1960.

—— "Middle East Oil in 1960," *Middle Eastern Affairs,* XII, 162–179, June–July, 1961.

—— "Middle East Oil in 1961," *Middle Eastern Affairs,* XIII, 226–235, October, 258–269, November, 1962.

—— "Middle East Oil in 1962," *Middle Eastern Affairs,* XIV, 194–200, August–September, 226–235, October, 1963.

—— "Oil in the Middle East Crisis," *Middle Eastern Affairs,* VIII, 126–133, April, 1957.

—— *The Power Struggle in Iraq,* New York, 1960.

Siksek, Simon G., *The Legal Framework of Oil Concessions in the Arab World,* Beirut, 1960.

Simpson, Dwight J., "Impact of the Oil Industry on the Middle East," *World Affairs Quarterly,* XXVIII, 36–61, April, 1957.

Sinclair, Angus, "Iranian Oil," *Middle Eastern Affairs,* II, 213–224, June–July, 1951.

—— "The Resources and Their Potentials III, Petroleum and Minerals," in *Middle East Resources,* Washington, 1954.

Singer, H. W., "Capital Requirements for the Economic Development of the Middle East," *Middle Eastern Affairs,* III, 35–40, February, 1952.

"La Situation économique dans les Emirats du Golfe Persique," *L'Asie Nouvelle,* V, 154–158, June, 1955.

Siv, Z., "Oil is Streaming to Our Refineries" (Hebrew), *Avodah,* 12th Year, 25–7, January, 1960.

Skrine, Clarmont, "New Trends in Iran," *Royal Central Asian Society Journal,* XLII, 100–115, April, 1955.

Smith, Richard Austin, "The Greatest Oil King of Them All," *Fortune,* 152–178, March, 1957.

Snodgrass, C. Stribling, and Arthur Kuhl, "U.S. Petroleum's Response to the Iranian Situation," *The Middle East Journal,* V, 501–504, Autumn, 1951.

Sousa, Ahmed, *Irrigation in Iraq Its History and Development*, 1945.

Southwell, C. A. P., "Kuwait," *Journal of the Royal Society of Arts*, CII, 24–41, December 11, 1953.

—— "Oil in Kuwait," *Journal of the Royal Central Asian Society*, XXXVI, 221–227, July–October, 1949.

"Soviet Bloc Turns to Middle East for Oil," *World Petroleum*, XXXIX, 46–48, 65, September, 1968.

"Soviet Mideast Dilemma: Oil from Persian Gulf," *World Oil*, CLXXII, 69–70, January, 1971.

Sparling, Richard C., and Norman J. Anderson, *Captial Investments of the World Petroleum Industry 1966*, [New York], 1967.

—— *1969 Capital Investments of the World Petroleum Industry*, [New York], 1970.

—— *Capital Investments of the World Petroleum Industry, 1970*, [New York], 1971.

Speiser, E. A., "Cultural Factors in Social Dynamics in the Near East," *The Middle East Journal*, VII, 133–152, Spring, 1953.

—— *The United States and the Near East*, Cambridge, 1950.

Spillmann, Georges, "Force et faiblesses du Pétrole Arab," *L'Afrique et L'Asie*, 3e Trimestre, 3–10, 1962.

—— "Le Pétrole en 1962," *L'Afrique et L'Asie*, Année 1963–3e Trimestre, No. 63, 27–32.

Spriggs, Dillard, "Pressures for Change in Middle East Oil Arrangements," *World Petroleum*, XXX, 37–8, April, 1959.

"Stability and Advance in Turkey," *Petroleum Press Service*, 377–379, October, 1968.

Stafford, Lawrence, "Washington very close-lipped over Tapline," *The Oil Forum*, III, 83–85, February, 1949.

Stahmer, Alfred M., *Erdöl Mächte und Probleme*, Kevelaer, Rheinland, 1950.

"Stalemate in the Lebanon," *Petroleum Press Service*, XXIII, 410–412, November, 1956.

"State Crude for Export," *Petroleum Press Service*, 456–458, December, 1966.

"The State Takes All in Syria," *Petroleum Press Service*, XXXII, 134–135, April, 1965.

Steineke, Max, and M. P. Yachel, "Saudi Arabia and Bahrein," in *World Geography*, Princeton, 1950.

Stevens, George P. Jr., "Saudi Arabia's Petroleum Resources," *Economic Geography*, XXV, 216–225, July, 1949.

Stevens, Harley C., "Side by Side – and Beyond," *World Petroleum*, XXXI, 39–43, February, 1960.

—— "Some Reflections on the First Arab Petreoleum Congress," *The Middle East Journal*, XIII, 273–280, Summer, 1959.

Steward, C. E., *Report on the Petreolum Districts Situated on the Red Sea Coast*, Cairo, 1888.

Stocking, George W., *Middle East Oil. A Study in Political and Economic Contro-versy*, Nashville, 1970.

"Storm over the Anglo-U.S. Oil Agreement," *Petroleum Press Service*, XIII, 224–226, December, 1946.

"Substantial Exports from Oman," *Petroleum Press Service*, XXXV, 407–409, November, 1968.

Sumner, H. B., "Daurah's Power Plant Makes Refinery Self-Sufficient," *The Oil Forum*, X, 122–123 etc., April, 1956.

Sykes, Edward, "Some Economic Problems of Persia," *Journal of the Royal Central Asian Society*, XXXVII, 262–272, July–October, 1950.

Sykes, Percy, *A History of Persia*, II, London, 1930.

Symonds, Edward, *Oil Prospects and Profits in the Eastern Hemisphere*, New York, 1961.

"Syria," *World Petroleum Report*, III, 246, January 15, 1957.

"Syria Pushes Ahead," *Petroleum Press Service*, XXVI, 459–61, December, 1959.

"Syria Still in Turmoil over Transits," *The Oil Forum*, III, 203–205, May, 1949.

Tadmor, D., and I. Barkay, "Oil Prospecting in Israel," (Hebrew), *Rivon Lekhal-kala*, VIII, 273–83, June, 1961.

Taher, Abdulhady Hassan, *Income Determination in the International Oil Industry*, Oxford, 1966.

Tanzer, Michael, *The Political Economy of International Oil and the Underdevel-oped Countries*, Boston, 1969.

"Tapline Operations," *World Petroleum*, XXIII, 84–89, September, 1952.

Tariqi, Abdullah, "Arab Petroleum as an Instrument of Policy in the Palestine Problem" (Arabic), *Arab Journal*, X, 48–54, Winter, 1964.

—— "Saudi Arabia Demands," *International Oilman*, XII, 371, Mid-November, 1958.

Teagle, Walter C., "America," *American Petroleum Institute Bulletin*, 3–6, Decem-ber 10, 1920.

"Teheran Agreement Details – A Look into the Future," *Petroleum Times*, LXXV, 33–36, February 12/26, 1971.

Teilhac, Ernest, "Le Pétrole dans l'économie du Moyen-Orient," *Le Commerce du Levant*, May 27, 30, 1953.

Temperley, H. W. V., *A History of the Peace Conference*, VI, London, 1920–24.

"Ten Fruitful Years for Iran," *Petroleum Press Service*, XXXI, 372–374, October, 1964.

"Tensions in the Gulf," *Petroleum Press Service*, XXXV, 83–87, March, 1968.

"Tensions in the Middle East with Particular Reference to Iran," *Proceedings of the Academy of Political Science*, XXIV, 558, January, 1952.

"Terminal," *Aramco World*, XVIII, 28–33, September–October, 1966.

"Testing-Time in Turkey," *Petroleum Press Service*, XXV, 213–215, June, 1958.

Thesiger, Wilfred, *Arabian Sands*, London, 1959.

"Thinking Aloud," *Petroleum Press Service*, XXXI, 402–404, November, 1964.

"The Third Arab Petroleum Congress," *Petroleum Times*, LXV, 738–741, November 17, 1961.

"Third River Almost Trebles 'Iraq Production," *The Oil Forum*, VII, 21, 31, January, 1953.

Thomas, Lewis V., and Richard N. Frye, *The United States and Turkey and Iran*, Cambridge, Mass., 1951.

Thompson, Carol L., "American Policy in the Middle East," *Current History*, XXXV, 234–239, October, 1958.

Thralls, W. H., and R. C. Hasson, "Basic Pattern of Exploration in Saudi Arabia Remains Unchanged," *The Oil and Gas Journal*, 84–87, 91–96, July 15, 22, 1957.

"The Threat of Soviet Oil," *Aramco World*, XIII, 3–7, March, 1962.

"Three R's in Arabic," *Aramco World*, 3–7, June, 1956.

Todd, William F., "The Impact of Oil on Middle East Economics," *World Petroleum*, XL, 38–40, January, 1969.

Toynbee, Arnold J., "The Islamic World," *Survey of International Affairs 1928*, 188–374, London, 1929.

—— "The Middle East," *Survey of International Affairs 1930*, 169–332, London, 1931.

—— (Ed.), "The Middle East," *Survey of International Affairs 1934*, 95–321, London, 1935.

—— (Ed.), "The Middle East," *Survey of International Affairs 1936*, 662–803, London, 1937.

—— *Survey of International Affairs 1925*, I. *The Islamic World*, London, 1927.

Tracy, William, "Island of Steel," *Aramco World*, XVII, 1–7, May–June, 1966.

"Transforming Kuwait," *Petroleum Press Service*, XXI, 10–12, January, 1954.

Tugendhat, Christopher, *Oil: The Biggest Business*, New York, 1968.

Tugendhat, G., "A World Market in Upheaval," *Fortune*, LXII, 159–61, October, 1960.

—— "The Politics and Prospects of OPEC," *Petroleum Times*, LXIV, 676–677, October 7, 1960, 709–710, October 21, 1960.

"Turkey," *World Petroleum Report*, III, 248–254, January 15, 1957.

Twitchell, K. S., *Saudi Arabia*, Princeton, 1953.

"Two Saudi Agreements," *Petroleum Press Service*, XXXV, 4–6, January, 1968.

"Two Viewpoints," *Petroleum Press Service*, XXX, 362–363, October, 1963.

Uren, C. Lester, "The Struggle for Oil in the Near East," *Africa, the Near East and the War*, Berkeley, 1925.

"U.S. Engineers in Iran," *Fortune*, XLI, 70–3, February, 1950.

Van Den Heuvel, J. A., "Oil Crises Solved by International Cooperation," *World Petroleum*, XLI, 94–106, June, 1970.

Van der Meulen, D., *The Wells of Ibn Saud*, London, 1957.

Van Pelt, Mary Cubberly, "The Sheikhdom of Kuwait," *The Middle East Journal*, IV, 12–26, January, 1950.

Van Wagenen, Richard W., *The Iranian Case 1946*, New York, 1952.

Veatch, Arthur C., "Oil, Great Britain and the United States," *Foreign Affairs*, IX, 665–673, July, 1931.

Verlaque, C., "Le transport et le raffinage du pétrole dans le bassin Mediterranéen," *Mediterranée*, 91–158, April–June, 1966.

Vidal, F. S., *The Oasis of Al-Hasa*, 1955.

"Vigorous Oil Search in Sinai," *The Oil Forum*, IX, 48–49, 59, February, 1955.

"Vital Issues Hang on Trans-Arabian Pipeline," *The Oil Forum*, II, 61–64, February, 1948.

Wade, Arthur, "Oil Prospects in Palestine," *Oil News*, IX, 594–597, June 11, 1921.

Walden, Jerrold L., "The International Petroleum Cartel in Iran – Private Power and the Public Interest," *Journal of Public Law*, XI, 3–60, Spring, 1962.

Walter, Leo, "Product Line Spans Iran," *International Oilman*, XII, 20–23, January, 1958.

Warburg, James P., "Steps Toward A Middle Eastern Peace," *The Reporter*, XVI, 17–20, February 7, 1957.

Ward, Thomas Edward, *Negotiations for Oil Concessions in Bahrain, El Hasa (Saudi Arabia), the Neutral Zone, Qatar and Kuwait*, New York, 1965.

Warriner, Doreen, *Land and Poverty in the Middle East*, New York, 1948.

"Water for Kuwait," *The Economist*, CLXVII, 28–29, April 4, 1953.

Watt, D. C., "Russians Need Middle East Oil," *The New Middle East*, 21–23, December, 1968.

W. D. P., "New Oil Agreements in the Middle East," *World Today*, XIV, 135–143, April, 1958.

Weinberger, Siegbert J., "The Suez Canal Issue," *Middle Eastern Affairs*, VIII, 46–56, February, 1957.

Wells, Michael J., "Abu Dhabi Capacity Expands Both on and Offshore," *World Petroleum*, XXXVIII, 40–44, July, 1967.

—— "Bu Hasa Opening Boosts Output from Abu Dhabi's Murban Field," *World Petroleum*, XXXVI, 34–36, July, 1965.

—— "Dubai – The Middle East's Newest Oil Producing State," *World Petroleum*, XL, 26–28, July, 1969.

—— "Egypt Makes Up for Sinai Loss But Seeks New Strikes," *World Petroleum*, XL, 16–19, March, 1969.

—— "Egypt's New Fields – Speculation and Reality," *World Petroleum*, XXXVIII, 36–38, 70, July, 1967.

—— "Iran Expands Fields, Terminals as Demands Rise for More Revenue," *World Petroleum*, XL, 32–35, July, 1969.

—— "Iran – Well Prepared for Combined Rising Demand," *World Petroleum*, XLI, 22–27, September, 1970.

—— "Kuwait Prepares for Mammoth Tankers," *World Petroleum*, XXXIX, 26–29, July, 1968.

—— "The Middle East – A Year After Suez," *World Petroleum*, XXXIX, 30–32, August 1, 1968.

—— "New Concessions and New Strikes Spotlight the Arabian Gulf," *World Petroleum*, XXXVI, 22—27, July, 1965.

—— "Offshore Finds Expand Middle East Oil Potential," *World Petroleum*, XXXVII, 20—21, 61, July, 1966.

—— "Oil Production Brings Oman to an Uncertain Future," *World Petroleum*, XXXVIII, 22—24, July, 1967.

—— "Saudi Arabia: Beginning of a New Era," *World Petroleum*, XXXVI, 28—30, July, 1965.

—— "Saudi Arabia Broadens Industrial Base," *World Petroleum*, XXXIX, 24—26, August, 1968.

—— "Saudi Arabia — Coming Power in the Middle East," *World Petroleum*, XXXVII, 41—49, February, 1966.

—— "Saudi Arabia Ready For Giant Tanker Era," *World Petroleum*, XL, 22—24, August, 1969.

"What the Middle East Means to Us," *The Lamp*, XXXVIII, 2—5, Winter, 1956.

White, David, "Our Future Oil Supply," *Engineering and Mining Journal*, 951—955, June 4, 1921.

White, Leigh, "Allah's Oil: World's Richest Prize," *Saturday Evening Post*, 18 etc., November 20, 1948; 30—31 etc. November 27, 1948.

"Who's Where in the Middle East," *Petroleum Press Service*, XXXVII, 447—449, December, 1970.

"Who Next for the Middle East," *Petroleum Press Service*, XXXI, 331—332, September, 1964.

Wilbur, Donald N., *Iran: Past and Present*, Princeton, 1948.

Williams, Kenneth, "Middle Eastern Oil," *The Fortnightly*, CLXV, N. S., 223—227, April, 1949.

—— "Oil and the Persian Gulf," *The Fortnightly*, CLXVII, N. S., 289—294, May, 1950.

Williams, Maynard Owen, "Bahrein: Port of Pearls and Petroleum," *The National Geographic Magazine*, LXXXIX, 195—210, February, 1946.

Williams, R., "The Trans-Iranian Oil Products Pipeline System," *Petroleum Times*, LXVI, 229—232, April 6, 1962.

Williams, Randall S., "Iran: Operations Commence on $650,000,000 Development Program," *Foreign Commerce Weekly*, XXXVII, 3, 39, November 14, 1949.

Williamson, J. W., *In a Persian Oil Field. A Study in Scientific and International Development*, London, 1930.

Wilson, Arnold T., *Loyalties, Mesopotamia 1914—1917*, London, 1930.

—— *Mesopotamia 1927—20: A Clash of Loyalties, a Personal and Historical Record*, London, 1931.

—— *Persia*, London, 1932.

—— *SW Persia: A Political Officer's Diary 1907—1914*, London, 1941.

Wilson, Charles Morrow, *Oil Across the World: The American Saga of Pipelines*, New York, 1946.

Winslow, Hall, "Backward and Forward in Iran," *Middle East Forum*, XXXI,

16–18, 36, July, 1956.

—— "Homes Within Reach," *Middle East Forum,* XXXI, 12–16, 36–40, April, 1956.

World Energy Demands and the Middle East. The 26th Annual Conference of the Middle East Institute, Washington, 1972.

"World's Richest Oil Region Continues to Grow Richer," *World Oil,* CLVII, 169–188, August 15, 1963.

Yaari, S., "The Markets for Middle East Oil," *Middle Eastern Affairs,* VII, 213–221, June–July, 1956.

Yeganeh, Mohammad, "Investment in the Petroleum Industry of the Middle East," *The Middle East Journal,* VI, 241–246, Spring, 1952.

Yinam, S., "Iraqi Politics – 1948–1952," *Middle Eastern Affairs,* III, 349–359, December, 1952.

Young, Arthur N., "Saudi Arabian Currency and Finance," *The Middle East Journal,* VII, 361–380, 539–556, Summer and Winter, 1953.

Young, T. Cuyler, "The Problem of Westernization in Modern Iran," *The Middle East Journal,* II, 47–59, January, 1948.

—— "The Race between Russia and Reform in Iran," *Foreign Affairs,* XXVIII, 282, January, 1950.

—— "The Social Support of Current Iranian Policy," *The Middle East Journal,* VI, 125–143, Spring, 1952.

Zaidi, Ali Nasir, "The Arab Oil Case and Pakistan," *The Pakistan Review,* IX, 16–18, 21, March, 1961.

Zeinaty, Afif, *L'Industrie du Pétrole au Liban,* Paris, 1970.

Zischka, Anton, *Die Auferstehung Arabiens: Ibn Sauds Weg und Ziel,* Leipzig, 1942.

—— *La Guerre Secrète pour le Pétrole,* Paris, 1933.

—— *Ölkrieg Wandlung Einer Weltmacht,* Leipzig, 1944.

PERIODICALS AND NEWSPAPERS

L'Afrique et L'Asie – Paris.

American Petroleum Institute Bulletin – New York.

The Annals of the American Academy of Political and Social Science – Philadelphia.

Arab News and Views – Beirut.

Arab Oil Review – Tripoli, Libya.

Aramco World – New York.

Beolam Hadelek – Tel Aviv.

Cahiers de L'Orient Contemporain – Paris.

The Christian Science Monitor – Boston.

CIPO (Centre d'Information du Proche-Orient et de l'Afrique) – Paris.

Le Commerce du Levant – Beirut.

Contemporary Review – London.

Current History — New York-Philadelphia.
Daily Radio Broadcasts.
Davar — Tel Aviv.
Department of State Bulletin — Washington.
Economic Bulletin (National Bank of Egypt) — Cairo.
Economic Geography — Worcester, Mass.
Economic News — Tel Aviv.
The Economist — London.
L'Egypte Contemporaine — Cairo.
The Egyptian Economic & Political Review — Cairo.
Foreign Affairs — New York.
Foreign Commerce Weekly — Washington.
The Fortnightly — London.
Fortune — New York.
The Geographical Journal — London.
The Geographical Magazine — London.
The Geographical Review — New York.
Haaretz — Tel Aviv.
Hadashoth Haneft — Tel Aviv.
Hamizrah Hehadash — Jerusalem.
Industrial Egypt — Cairo.
International Affairs — London.
International Labor Review — Geneva.
Journal of the Royal Central Asian Society — London.
The Manchester Guardian.
Middle East Economic Digest — London.
Middle East Economic Survey — Beirut.
Middle Eastern Affairs — New York.
The Middle East Journal — Washington.
The National Geographic Magazine — Washington.
The New York Herald Tribune.
The New York Times.
The Oil and Gas Journal — Tulsa.
The Oil Forum — New York-Fort Worth.
Oil Journal — Teheran.
Oil News — London.
Oriente Moderno — Rome.
Petroleum Press Service — London.
Petroleum Times — London.
Petroleum Zeitschrift — Vienna-Berlin.
Political Science Quarterly — New York.
Politique Etrangère — Paris.
The Round Table — London.
Statistical Summary — Jidda.

Teheran Economist.
The Times — London.
Weekly News Bulletin — Baghdad.
World Affairs — London.
World Oil (until 1947 *The Oil Weekly*) — Houston.
World Petroleum — New York.
World Petroleum Report — New York.
The World Today — London.
Zeitschrift für Geopolitik — Heidelberg.

BIBLIOGRAPHIES

American University of Beirut, Economic Research Institute, *A Selected and Annotated Bibliography of Economic Literature on the Arabic Speaking Countries of the Middle East, 1938–1952,* Beirut, 1954. 1953 Supplement to above, 1955.

The Arabian Peninsula, Washington, 1951, mimeographed, Library of Congress.

Cahiers de L'Orient Contemporain, bibliographical sections.

Ettinghausen, Richard, (Ed.), *A Selected and Annotated Bibliography of Books and Periodicals in Western Languages Dealing with the Near and Middle East, with Special Emphasis on Mediaeval and Modern Times,* Washington, 1952. Supplement to above, 1954.

Iran. A Selected and Annotated Bibliography, Washington, 1951, mimeographed, Library of Congress.

Middle Eastern Affairs, bibliographies in the January, April, June–July and October issues.

The Middle East Journal, bibliographical sections.

Stevens, Curtis (Ed.), *Petroleum Sourcebook 1958,* Amarillo, 1958.

—— *Petroleum Sourcebook 1959,* Amarillo, 1960.

INDEX

Abadan (island), 18, 96, 97, 103,
104, 106, 107, 173, 178, 179;
question of expulsion of British
technicians from, 105; refinery of,
18, 28, 92, 99, 170, 242, 243, 410,
506, 545; refinery facilities of expanded, 128; refinery growth of,
129; refinery of completed, 127
Abadan Institute of Technology, 174
Abadan Petrochemicals, 177
Abdullah, Emir, 215, 226, 459, 461;
became Emir of Transjordan, 227
Abdullah, Shaikh of Kuwait, 422
Abqaiq, 349, 376; LPG plant at, 378
Abu Dhabi, 162, 182, 435, 436,
438–442, 464, 510, 512, 514;
barter agreement with Rumania,
442; Five-Year Plan for social
services, 442; oil and development,
441–442; oil payments and
relinquishments, 439–441
Abu Dhabi Fund for Arab Economic
Development, 441
Abu Dhabi Marine Areas (ADMA),
439
Abu Dhabi National Petroleum
Company, 441, 442
Abu Dhabi Oil Company (Japanese
group), 440
Abu Dhabi Petroleum Company
(ADPC), 439
Abu Gharadig, 487
Abu Hadriya, 349
Abu Madi, 487
Abu Musa, 402
Abu Qir, 487
Abu Rudais, 482
Abu Safah, 371, 400
Acheson, Dean, 97, 107, 111, 112,
113
Actual Costs Agreement (Iraq-IPC),
278

Adelman, M.A., 180
Aden Petroleum Refinery, 151
Aden refinery, 520
ADMA, 440, 441
ADPC, 440, 441
Afghanistan, 79
Africa, 1971; share in world's oil
reserves, 525
Agha Jari, 129, 176
Agha Jari-Ganaveh pipelines, 173
AGIP, 181–440; terms of concession
in Saudi Arabia, 376–378; terms
of Iranian concession of, 164
Agrarian Reform Law (Iraq), 291
Ahmadi, 411, 415
Ahwaz, 17; properties of AIOC in,
96
Ahwaz-Ganaveh pipeline, 174
Ahwaz-Teheran pipeline, 175
Ain-Dar, 349
Ain Sukhna, 488
Ain Zalah, 242, 247, 282, 287
Ain Zalah-Station K2 pipeline, 242
Anglo-Iranian Oil Company (AIOC,
formerly Anglo-Persian Oil
Company and latterly British
Petroleum Company (BP)), 27, 92,
96, 97, 100, 102, 106, 107, 108,
157, 242, 244, 246, 248, 309, 321,
352, 353, 453; British personnel of,
97; commercial operations of, 99;
controlled by British Government,
245; different attitudes to by Iran
and Great Britain, 116ff.; granted
concession in Abu Dhabi, 439
Ajman, 443
Akins, James E., 384
Ala, Hussein, 42, 56, 57, 58, 59, 77,
91, 92, 103
Alai, Hussein, see Ala, Hussein
Alaska, petroleum reserves of, 548
Al-Bahra, 409

605